Undergraduate Texts in Mathematics

Editors
S. Axler
F.W. Gehring
K.A. Ribet

Springer

New York
Berlin
Heidelberg
Barcelona
Budapest
Hong Kong
London
Milan
Paris
Singapore
Tokyo

Undergraduate Texts in Mathematics

(continued after index)

Francis J. Flanigan   Jerry L. Kazdan

Calculus Two
Linear and Nonlinear Functions

Revised by David L. Frank, Bert E. Fristedt, and Lawrence F. Gray

With 207 Illustrations

 Springer

Francis J. Flanigan
Dept. of Mathematics
San Jose State University
San Jose, CA 95192
USA

Jerry L. Kazdan
Dept. of Mathematics
University of Pennsylvania
Philadelphia, PA 19104-6395
USA

Original edition published in 1971 by Prentice Hall.

Cover illustration is a paraboloid in $\mathbb{R}^3$, with a tangent plane. This plane is the "best affine approximation" to the nonlinear parabolic surface.

Mathematics Subject Classification: 26-01, 15-01, 26-B12, 26-B15, 15-A03, 15-A04, 26-B20, 53-A04, 53-A05, 26-B10, 15-A06

Library of Congress Cataloging-in-Publication Data
Flanigan, Francis J.
    Calculus two : linear and nonlinear functions / Francis J.
Flanigan, Jerry L. Kazdan. — 2nd ed.
        p. cm. — (Undergraduate texts in mathematics)
    ISBN 0-387-97388-5. — ISBN 3-540-97388-5(Berlin)
    1. Calculus. I. Title. II. Title: Calculus II. III. Series.
QA303.F583 1990
515—dc20   90-45205

Printed on acid-free paper.

Photocomposed copy prepared by authors using LaTeX.
Printed and bound by R.R. Donnelley & Sons, Inc., Harrisonburg, Virginia.
Printed in the United States of America.

9 8 7 6 5 4 3 2 (Corrected second printing, 1998)

ISBN 0-387-97388-5 Springer-Verlag New York Berlin Heidelberg
ISBN 3-540-97388-5 Springer-Verlag Berlin Heidelberg New York SPIN 10681280

To Our Parents

Preface to the Second Edition

This second edition remains faithful to the guiding idea of the first edition: use geometric and algebraic concepts to help illuminate calculus of several variables. The revisions were actually written by a few dedicated and generous mathematicians at the University of Minnesota, Professors Bert E. Fristedt and Lawrence F. Gray, with assistance from Professor David Frank. While retaining the strong geometric flavor of the first edition, they made many changes, both large and small, to clarify the concepts and make the text more usable for students. For instance, the study of curves in space and vector-valued functions of one variable (the old Chapter 7) has been revised and now appears as Chapter 4 as well as in the revised Section 5.3. Both Section 5.2, "Some Challenging Problems," and Section 5.3 are optional, although we hope that every class will at least look at some of this material. In the previous edition we had a number of "blurbs" which introduced supplementary topics in an informal way. Many of these are now incorporated into the text, while others now appear at the ends of the appropriate chapters.

There are five major additions in the present edition:

1. Gauss-Jordan elimination, including its use for finding inverse matrices, is treated explicitly — although the technique is discussed long before matrices are formally introduced.

2. Projections and isometries are introduced. In particular the matrices corresponding to rotations and reflections in two dimensions are identified.

3. The curvature of a curve is discussed following the computation of arc length.

4. Change of variables for multiple integrals is introduced, especially for use in polar, cylindrical, and spherical coordinates.

5. Surfaces are defined parametrically. The formulas for surface area and the integral of a vector field over an oriented surface are given.

We have continued to resist the temptation to treat eigenvalues and eigenvectors, as well as a discussion of ordinary differential equations and linear partial differential equations (particularly the technique of separation of variables using Fourier series). These topics are basic and powerful. Historically they served as important motivation for the development of linear analysis, and they are essential to many applications both within mathematics as well as to other areas of science. But including them in this book would have increased its length considerably and might have made it less useful. Nonetheless we hope that every student sees these topics in subsequent courses and learns to appreciate the abstract role of linearity in them.

We are grateful to Professors Fristedt, Gray, and Frank for their thoughtful and unselfish contributions. They have taken charge of this revision completely. It is they who are responsible for any improvements.

The prerequisites for this book consist of basic single-variable differential and integral calculus, including the calculus of trigonometric, exponential, and logarithmic functions, as well as an introduction to some of the central results of the subject, such as the Mean Value Theorem and the Intermediate Value Theorem. We do not assume familiarity with parametric equations, polar coordinates, or infinite series. We envision the book being used for 12 quarter credits (or 8 semester credits). At the University of Minnesota, the first nine chapters are used for two 5-credit quarter courses, one at the end of the freshman year and the second at the beginning of the sophomore year. The tenth chapter constitutes part of a later course in the sophomore year.

Francis J. Flanigan
Jerry L. Kazdan

Preface to the First Edition

The heart of this book is the study of nonlinear functions by means of the simpler linear and affine functions. Our guide in this is the elementary calculus procedure of constructing the tangent line (a linear object) to a curve (a nonlinear object). Thus, we begin with a study of linear objects.

Linear algebra is just this study; it should be viewed as analytic geometry in modern dress. Without it, geometric ideas are lost in a fog of equations. We have developed the linear algebra by first studying the algebra of $\mathbb{R}^n$ (ordinary n dimensional space), especially $\mathbb{R}^2$ and $\mathbb{R}^3$, in Chapter 1. Geometry of $\mathbb{R}^n$, the inner product, is the subject of Chapter 2. Then we introduce the concept of a linear map in Chapter 3. And only after some familiarity with linear maps is acquired do we represent them as matrices and interpret the algebraic operations on maps in terms of the corresponding operations on matrices. This concludes the "linear" part of the book. Chapter 4 serves as a bridge from linear to nonlinear.

The differential calculus of several variables is taken up in Chapters 5, 6, 7, and 8. One measure of the success of approaching calculus of several variables using linear algebra is that students *anticipate* ideas and theorems. The criterion $f'(X) = 0$ for a maximum or minimum is but one example. This notation vastly simplifies formulas, enabling the concepts to emerge clearly.

Chapter 9 discusses multiple integrals. Instead of pushing the difficult ideas of existence of the integral, we have chosen to show how to estimate numerically the value of the integral. In this way students grasp the concepts and understand that a proof of the existence of the integral consists precisely in showing that the elementary estimation procedure can be made to yield as close an approximation as they desire. Some teachers may wish to use computers at this point of the course.

Line integrals and Stokes' Theorem are the subject of Chapter 10. We keep to $\mathbb{R}^2$, since higher dimensions force well-known complications we wish to avoid. One temptation was to introduce differential forms and exhibit the elegant modern notation for this theorem—at least in $\mathbb{R}^2$. Unfortunately, the formalism seems to swamp all but the very best students. Thus, we have taken the classical approach. However, we have tried to give some nontrivial applications. All too many students know the modern version with differential forms yet they know no genuine use of the result and have no idea why it is important (except that their teachers say so).

Scattered throughout the book are a number of "blurbs." These are very short sections that either illuminate some idea in the text, present an application, or else introduce a more advanced topic. We encourage the student to read these on his own.

This book is intended for the—possibly mythical—"average" student. Two and three dimensional cases are emphasized. We have chosen to write in an informal lecture note style with questions directed to the reader. Hopefully, we have not insulted the reader's intelligence and "driven in the hammer along with the nail." But the reader should feel free to skip. This skipping is important, both over material that seems too difficult and that which seems dull and repetitive. Some teachers (including us) will often go much faster than the text, leaving the gap as reading material. Our goal throughout has been to give a feeling for the ideas involved, not just computational facility—although that is certainly not underplayed. We also hope this book is interesting. In order to keep the level of difficulty even, we have deleted most of the technical proofs, such as equality of mixed partial derivatives.

We place a great importance on the organization of this book. The logical development should be clear from the Table of Contents. This is actually an outline of the subject and should itself be used for review purposes.

Sins of omission: We do not discuss change of basis, eigenvalues, determinants (except the 2×2 case), function spaces (except for a blurb), technicalities about real numbers, inverse or implicit function theorems, change of variables in multiple integrals (except for polar coordinates), or give the proofs for the harder theorems. These topics are more difficult and are left for more advanced courses in algebra or analysis.

This book originated as a more difficult book which turned out to be unrealistic for many second year students, mainly due to the presence of function spaces. A thorough rethinking has led to the present volume. Many colleagues and students have generously assisted us. We give special thanks to Lawrence Corwin, Jack Gomberg, and David Ragozin. Robert Martin and James Walsh of Prentice-Hall have also been a pleasure to know and to work with.

Francis J. Flanigan
Jerry L. Kazdan

Contents

0

Remembrance of Things Past

0.1 Introduction

This chapter is included for reference. In it, we will discuss sets, functions, and some related concepts that you may have seen previously. You may discover it is unnecessary to read the chapter thoroughly and only refer to it occasionally, when you encounter an unfamiliar term or notation.

0.2 Sets

0.2A EXAMPLES AND NOTATION

A **set** is any collection of objects, together with a criterion for deciding whether a given object is in the set. The objects in the set are called the **elements** or **members** of the set.

Examples are (1) the set of all animals with brown fur and claws and (2) the (less picturesque) set of all positive even integers.

A set may sometimes be specified by actually listing all its elements. Thus, the set of all students in some class is specified by the list of names in the roll book. The **empty set** is the set $\emptyset$ that has no elements at all.

There is a standard notation we sometimes use to describe sets. Here are some examples:

1. $\mathcal{A} = \{x : x \text{ is an odd integer}\}$ is the set of all odd integers; the colon may be read "such that".

2. $\mathcal{S} = \{(x, y) : x^2 + y^2 = 1\}$ is the set of all points (x, y) on the unit circle $x^2 + y^2 = 1$.

3. $\mathcal{B} = \{-3, 1, 2, 7\}$ is the set whose elements are the integers -3, 1, 2, and 7.

Note that we often use capital script letters to denote sets.

Here is some further notation. If $\mathcal{S}$ is any set (not necessarily the unit circle above), we write

1. $x \in \mathcal{S}$ if x is an element of $\mathcal{S}$;

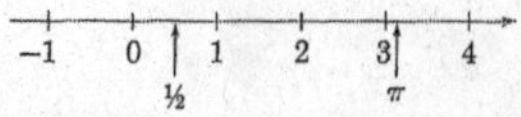

FIGURE 0.1. The real line $\mathbb{R}$

2. $x \notin S$ if x is not an element of S.

For example, if $\mathcal{A}$ is the set of odd integers mentioned above, then

$$3 \in \mathcal{A}, \qquad -11 \in \mathcal{A}, \qquad 4 \notin \mathcal{A}, \qquad \frac{2}{3} \notin \mathcal{A}.$$

0.2B SUBSETS

The term "subset" is a way of referring to a portion of a given set. Formally, $\mathcal{A}$ is a **subset** of S, or $\mathcal{A}$ is **contained** in S, written

$$\mathcal{A} \subseteq S,$$

if and only if every element of $\mathcal{A}$ is also an element of S. Note that, according to this definition, the set S is a subset of itself, that is, $S \subseteq S$. If $\mathcal{A}$ is a subset of S that is different from S, we say that $\mathcal{A}$ is a **proper subset** of S. The empty set is a subset of every set, and a proper subset of every set except itself.

For example, if $S = \{1, 2, 3, 4, 5\}$, then some proper subsets of S are

$$\{1, 2, 3\}, \qquad \{1, 4\}, \qquad \{3\}, \qquad \{3, 4\}, \qquad \emptyset.$$

There are others. Can you find others?

0.2C THE SET OF REAL NUMBERS; THE SETS $\mathbb{R}^n$

We assume that you have some familiarity with the very important set of real numbers. We designate this set throughout this book by $\mathbb{R}$. You may think of $\mathbb{R}$ as the set of all decimals (positive and negative) such as

$$-2 = -2.000\ldots, \qquad \frac{1}{3} = 0.333\ldots, \qquad \pi = 3.14159\ldots.$$

We will often be regarding $\mathbb{R}$, however, as the set of all points on a line or axis, the so-called "real line" or "x-axis" from one-variable calculus. See Figure 0.1. Every point on this line corresponds to a unique real number, and vice versa. This correspondence between numbers and points on a line is a source of geometric insight we intend to exploit.

We define the **cartesian product** (after Descartes) of $\mathbb{R}$ with itself to be the set, denoted $\mathbb{R} \times \mathbb{R}$, of all ordered pairs (x, y), where $x \in \mathbb{R}$ and $y \in \mathbb{R}$. We remark that the order in (x, y) is important. Thus $(1, 0) \neq (0, 1)$ and, in general, $(x, y) = (a, b)$ if and only if $x = a$ and $y = b$. The elements of $\mathbb{R} \times \mathbb{R}$ are often called **ordered pairs**.

Another notation for $\mathbb{R} \times \mathbb{R}$ is $\mathbb{R}^2$, a notation that we shall use repeatedly.

Note that we need not stop at the product of $\mathbb{R}$ with itself. We may form cartesian products with more copies of $\mathbb{R}$:

$$\begin{aligned} \mathbb{R}^3 &= \mathbb{R} \times \mathbb{R} \times \mathbb{R} = \{(x, y, z) : x, y, z \in \mathbb{R}\}, \\ \mathbb{R}^4 &= \mathbb{R} \times \mathbb{R} \times \mathbb{R} \times \mathbb{R} = \{(x_1, x_2, x_3, x_4) : x_1, x_2, x_3, x_4 \in \mathbb{R}\}, \end{aligned}$$

and so on. Thus, if n is any positive integer, we define

$$\mathbb{R}^n = \{(x_1, \ldots, x_n) : x_1, \ldots, x_n \in \mathbb{R}\}.$$

Some elements of $\mathbb{R}^3$ are $(1, 0, 0)$, $(0, 0, 0)$, and $(4, -\sqrt{2}, \frac{1}{2})$, for example, and $(1, -1, 2, 0, 5, \frac{1}{3}) \in \mathbb{R}^6$. We do speak of $\mathbb{R}^n$ when $n = 1$. Thus we have another name for $\mathbb{R}$, namely, $\mathbb{R} = \mathbb{R}^1$.

In Chapter 1, we will call the elements of $\mathbb{R}^n$ **points** or **vectors** and learn a way to add them.

0.2D SOME NOTATION FROM LOGIC

The symbol $\Rightarrow$ is to be read "implies". For example, x is an even integer between 1 and 3 $\Rightarrow x = 2$. As another example, y is an integer of the form $10x + 1$, where x is an integer $\Rightarrow y$ is odd.

We also use the symbol $\Leftrightarrow$, which is a quick way of writing "if and only if". Thus a real number y is nonnegative $\Leftrightarrow y = x^2$ for some real number x. You might regard the symbol $\Leftrightarrow$ as being composed of $\Rightarrow$ and $\Leftarrow$. Thus "$P \Leftrightarrow Q$" means "P implies Q" and also "Q implies P". In other words, the two statements P and Q are **equivalent**, that is, one of them is true if and only if the other is true.

0.3 Functions

0.3A CONCEPTS

Let $\mathcal{A}$ and $\mathcal{B}$ be sets. A **function** f from $\mathcal{A}$ to $\mathcal{B}$, written

$$f : \mathcal{A} \to \mathcal{B} \qquad \text{or} \qquad \mathcal{A} \xrightarrow{f} \mathcal{B}$$

is a rule that assigns to each $x \in \mathcal{A}$ one and only one element $y = f(x) \in \mathcal{B}$. Some synonyms for "function" are "map", "mapping", "transformation", and "operator". The set $\mathcal{A}$ is called the **domain** of the function f, and $\mathcal{B}$ the **target** of f. The subset of $\mathcal{B}$ given by

$$\{y \in \mathcal{B} : y = f(x) \text{ for some } x \in \mathcal{A}\}$$

is the **image** of f. A good notation for this set is $f(\mathcal{A})$. We will also use the notation $\mathcal{D}(f)$ for the domain of f and $\mathcal{I}(f)$ for the image of f.

For example, let $f : \mathbb{R} \to \mathbb{R}$ be given by the rule $f(x) = x^2$. Thus f assigns to each real number x its square. The domain of f is $\mathbb{R}$, the target is $\mathbb{R}$, and the image is the set of all nonnegative real numbers $y \geq 0$, since $x^2 \geq 0$. In symbols: $\mathcal{D}(f) = \mathbb{R}$ and $\mathcal{I}(f) = f(\mathbb{R}) = [0, \infty)$, where the interval $[0, \infty)$ is the set of all nonnegative real numbers.

Here is another example, this time of a function whose target is $\mathbb{R}^2$. Let $g : \mathbb{R} \to \mathbb{R}^2$ be given by $g(x) = (x + 1, x - 1)$. What are $\mathcal{D}(g)$ and $\mathcal{I}(g)$?

0.3B THE COMPOSITION OF FUNCTIONS

Suppose that $f : \mathcal{A} \to \mathcal{B}$ and also $g : \mathcal{B} \to \mathcal{C}$ are two functions. We write $\mathcal{A} \xrightarrow{f} \mathcal{B} \xrightarrow{g} \mathcal{C}$. Since, for each $x \in \mathcal{A}$, the value $f(x)$ is an element of $\mathcal{B} = \mathcal{D}(g)$, we may apply g to $f(x)$ to get $g(f(x)) \in \mathcal{C}$. The function that assigns to each $x \in \mathcal{A}$ the value $g(f(x)) \in \mathcal{C}$ is the **composition** of g **on** f. Note how important it is in the definition of the composition that the image of f be contained in the domain of g. We will use the notation $g \circ f$ for the composition of g on f.

For example, let $f(x) = x^2$ and $g(x) = (x + 1, x - 1)$ be the functions given above. Then we have $\mathbb{R} \xrightarrow{f} \mathbb{R} \xrightarrow{g} \mathbb{R}^2$ and $(g \circ f)(x) = (x^2 + 1, x^2 - 1)$ is a formula for the composition of g on f. Thus, setting $x = 2$,

$$(g \circ f)(2) = (2^2 + 1, 2^2 - 1) = (5, 3).$$

1

The Algebra of $\mathbb{R}^n$

1.0 Introduction

Let us begin with an overview. The study of differential calculus of a function of a single variable, which you have already completed, may be divided into four parts:

1. Basic arithmetic and algebra. You learned the usual operations with numbers and equations.

2. Geometry and distance. You studied geometry of the real number line and of the xy-plane by using coordinates. In particular, you worked with formulas, in terms of coordinates, for distance. This concept is essential to the notion of limit in calculus, although it is sometimes disguised with algebraic symbolism such as $|x - x_0|$.

3. Linear functions. You studied relatively simple functions of the form $y = mx + b$ and the corresponding geometrical objects, lines in the xy-plane.

4. Nonlinear functions. Finally, you studied more general functions, for example, $y = f(x) = x^2$. The graphs of such functions were typically not lines. Nevertheless, the tangent line to the graph of $y = f(x)$ above a point x_0 was found to be quite useful. See Figure 1.1.

The slope m of this tangent line is given by the first derivative $m = f'(x_0)$. Having approximated the curved graph $y = f(x)$ by the tangent line, you were able to deal successfully with several natural problems: maximum and minimum, rate of change, and the approximate value of $f(x)$ for x near some particular point x_0 at which f is easily evaluated.

In the chapters that follow, we repeat this four-stage process. Now, however, we wish to study functions of *several* variables. A useful example is

$$f(x_1, x_2) = x_1^2 + x_2^2 \,,$$

which is a real-valued function of the two real variables x_1 and x_2. It is customary to think of the pair (x_1, x_2) as a point, denoted by X, in a plane. Therefore we may write $f(x_1, x_2)$ as $f(X)$, which should remind us of one-variable calculus. Now, however, our variable is a point $X = (x_1, x_2)$. We

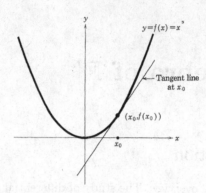

FIGURE 1.1.

will also want to consider functions that take *values* which are points in a plane; for instance,

$$(y_1, y_2) = f(x) = (x + 1, x^2).$$

Sometimes points are called **vectors**. We have, in the preceding paragraph, described a real-valued function of a vector variable and a vector-valued function of a real variable. Even more generally, we will study functions which are vector-valued functions of a vector variable.

The four-stage process in which we will engage is as follows:

1. Vector algebra. We will examine addition, subtraction, and the solution of equations involving vectors in Chapter 1.

2. Vector geometry. A natural concept of distance between vectors will be defined in Chapter 2. This will help us later to define the notion of limit, essential to calculus, and to give meaning to such phrases as "let the vector X approach the vector X_0".

3. Linear functions. We will examine these, and the slightly more general affine functions, in Chapter 3. Lines and planes are familiar objects that are graphs of certain affine functions.

4. Nonlinear functions. Finally, we will have developed enough machinery to study differential calculus of functions involving vectors, such as $f(X) = f(x_1, x_2) = x_1^2 + x_2^2$. Just as in one-variable calculus, in which we constructed tangent lines to graphs which were curves, we will learn how to construct tangents to the graphs of functions involving several variables. For example, we will find that the tangent to a vector-valued function of a real variable is a line (Chapter 4), while the tangent to the graph of a real-valued function of a vector variable is a plane or, in higher dimensions, a generalization of a plane (Chapter 6). See Figure 1.2. Once we can construct tangents to graphs of

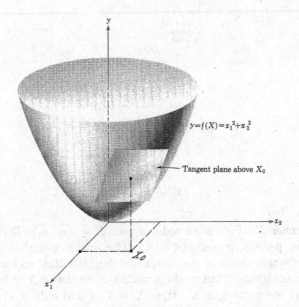

FIGURE 1.2.

functions, we will be equipped to deal with the standard issues such as maxima and minima, rates of change, approximation of functions, etc.

Some words of encouragement: The brief comments given here outline an ambitious and important program. The concepts of vector and linear and affine functions are of great importance in modern mathematics, physics, engineering, economics and many other fields. You should not be disheartened if certain statements in the outline seem bizarre or incomprehensible. As we pass through the four stages, the outline will be repeated and amplified.

We have oversimplified the statement of our goals somewhat. For instance, in the study of linear functions we will treat certain important topics not directly related to our desire to study nonlinear functions, as we also treat topics that are critical for understanding nonlinear functions.

1.1   The Space $\mathbb{R}^2$

1.1A   Points in $\mathbb{R}^2$

As discussed in Section 0.2c, we use $\mathbb{R}$ to denote the real number line and $\mathbb{R}^2$, or equivalently $\mathbb{R} \times \mathbb{R}$, to denote the cartesian product of $\mathbb{R}$ with itself:

$$\mathbb{R}^2 = \{(x_1, x_2) : x_1, x_2 \in \mathbb{R}\}$$

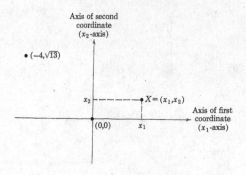

FIGURE 1.3.

Thus a member X of $\mathbb{R}^2$ is an ordered pair $X = (x_1, x_2)$. These ordered pairs are the **points** or **vectors** in $\mathbb{R}^2$. The words "point" and "vector" are logically synonymous in the current setting, but they do carry slightly different connotations in informal discussion. Note that if $X = (x_1, x_2)$ and $Y = (y_1, y_2)$ are vectors in $\mathbb{R}^2$, then $X = Y$ if and only if $x_1 = y_1$ and $x_2 = y_2$.

Some examples of vectors are $(3, 0)$, $(-4, \sqrt{13})$, $(\frac{1}{2}, -10)$, $(1, 1)$, $(1, 3)$, and $(3, 1)$. The last two vectors in the list are two different vectors.

If $X = (x_1, x_2)$, the real numbers x_1 and x_2 are the **coordinates** of the vector X. For example, if $X = (-4, \sqrt{13})$, then its first coordinate is -4 and its second coordinate is $\sqrt{13}$.

You are probably familiar with the pictorial representation of $\mathbb{R}^2$ as a plane, with axes intersecting at right angles at the point $(0, 0)$. It is customary to use the horizontal axis as the axis of first coordinates. See Figure 1.3.

1.1B ALGEBRA IN $\mathbb{R}^2$

So far $\mathbb{R}^2$ is merely a set. Let us propose a reasonable algebraic structure for $\mathbb{R}^2$. For $X = (x_1, x_2)$ and $Y = (y_1, y_2)$ any two vectors and α any real number—or as we will say from now on, a **scalar**–we define the addition of vectors by

$$X + Y = (x_1 + y_1, x_2 + y_2)$$

and the multiplication of a vector by a scalar by

$$\alpha X = (\alpha x_1, \alpha x_2) \, .$$

Note how these two operations involving vectors take advantage of the fact that we already know how to add and multiply real numbers.

EXAMPLE:

1. Let $X = (1, 5)$, $Y = (2, -1)$, and $\alpha = 7$. Then, for instance,

$$
\begin{aligned}
X + Y &= (1 + 2, 5 - 1) = (3, 4)\,, \\
\alpha X &= (7, 35)\,, \\
\alpha Y &= (14, -7)\,, \\
X + \alpha Y &= (15, -2)\,.
\end{aligned}
$$

Here are some consequences of these definitions which are valid for all vectors X, Y, and Z in $\mathbb{R}^2$:

1. Addition is **associative**: $(X + Y) + Z = X + (Y + Z)$.

2. Addition is **commutative**: $X + Y = Y + X$.

3. There is an **additive identity** or **zero element**, namely the vector $\mathbf{0} = (0, 0)$, satisfying $X + \mathbf{0} = X$ for all X.

4. Each vector X has an **additive inverse** $-X$ ("minus X") satisfying $X + (-X) = \mathbf{0}$. If $X = (x_1, x_2)$, then $-X = (-x_1, -x_2)$. (Thus the additive inverse of $(1, -3)$ is $(-1, 3)$.)

The following two properties govern multiplication by scalars, α and β being arbitrary scalars in the first property:

5. $\alpha(\beta X) = (\alpha\beta)X$.

6. $1X = X$.

Finally, we state, in terms of arbitrary X, Y, α, and β, the **distributive laws**, which describe the relationships between the two different operations:

7. $(\alpha + \beta)X = \alpha X + \beta X$.

8. $\alpha(X + Y) = \alpha X + \alpha Y$.

To convince you that these properties are obvious, let us prove one, for example, Property 7. We have

$$
\begin{aligned}
(\alpha + \beta)X &= (\alpha + \beta)(x_1, x_2) = ((\alpha + \beta)x_1, (\alpha + \beta)x_2) \\
&= (\alpha x_1 + \beta x_1, \alpha x_2 + \beta x_2) \\
&= (\alpha x_1, \alpha x_2) + (\beta x_1, \beta x_2) \\
&= \alpha(x_1, x_2) + \beta(x_1, x_2) \\
&= \alpha X + \beta X
\end{aligned}
$$

as desired. Note that the proof consisted of replacing X by its coordinate representation (x_1, x_2) and then using the definition of multiplication by

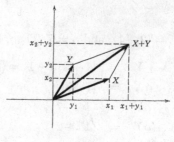

FIGURE 1.4.

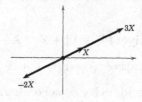

FIGURE 1.5.

scalars and the well-known distributive law for real numbers in each co-ordinate slot. You should try proving one or two of the other properties yourself.

Here we are viewing Properties 1 through 8 as properties that can be proved for $\mathbb{R}^2$. Later we will see that corresponding properties hold in other settings as well. Accordingly we will be motivated to give a name to any collection of objects satisfying these properties; the name will be "vector space" and Properties 1 through 8 will be called the "vector-space axioms". These properties are not spectacular, but are absolutely essential and we will meet them throughout this chapter.

1.1C   VECTORS AS ARROWS

Instead of thinking of the member (x_1, x_2) of $\mathbb{R}^2$ as a point, it is sometimes useful to think of it as a directed line segment or arrow with tail at the origin $(0,0)$ and head at the point (x_1, x_2). Verify that $X + Y$ is the vector that is the diagonal of the parallelogram determined by the arrows X and Y. See Figure 1.4.

Multiplication of a vector X by a scalar α may be visualized as a stretching or shrinking (a "scaling") of the arrow X. Thus $3X$ may be thought of as an arrow starting at $(0,0)$ and pointing in the same direction as X but three times as long (admittedly, we do not yet have a rigorous definition of the length of a vector). And if $\alpha < 0$, then αX points in the direction opposite to that of X. See Figure 1.5.

Exercises

Practice with arithmetic and algebra in $\mathbb{R}^2$.

1. Let $X = (1,2)$, $Y = (-1,3)$, and $Z = (0,4)$. Compute

   (a) $X + Y$

   (b) $X - Y$

   (c) $X + Y + Z$

   (d) $4X$

   (e) $-3Y$

   (f) $4X - 3Y$

   (g) $X + 2Y - Z$

   (h) $4(X + Y) - 5Z$.

2. Given $Y = (2,3)$ and $Z = (1,-4)$. For each equation below, compute a vector X that satisfies it:

   (a) $3X + 2Y - Z = 0$

   (b) $Z - X = Y$

   (c) $Z - 2X = 5Y$

   (d) $Z - 2X - 5Y = 0$.

3. Modify the preceding exercise by letting Y and Z be fixed but arbitrary vectors.

4. Let $X = (1,2)$ and $Y = (-1,3)$ be as in Exercise 1. Locate them as points in the plane (sketch). Then locate $X + Y$, $X - Y$, and $2X - 2Y$. Finally, verify that the points 0, X, Y, and $X + Y$ are the corners of a parallelogram in $\mathbb{R}^2$.

5. As a preview of distance, answer the following. How long is the vector $(1,2)$; that is, what is the distance from $(0,0)$ to $(1,2)$? Can you think of a formula for the length of a typical vector $X = (x_1, x_2)$?

6. As a preview of higher-dimensional spaces, answer the following. How should we define $\mathbb{R}^3$; $\mathbb{R}^4$? What should a typical X in $\mathbb{R}^3$ be? How should we define the sum $X + Y$ if X and Y are vectors in $\mathbb{R}^3$?

7. Let $e_1 = (1,0)$ and $e_2 = (0,1)$. Show that, for every $X \in \mathbb{R}^2$, there exist unique scalars α and β such that $X = \alpha e_1 + \beta e_2$. This property of e_1 and e_2 will be used frequently.

8. (a) Let $Y = (3,0)$ and $Z = (2,0)$. Is it possible, for each vector $X \in \mathbb{R}^2$, to find α and β such that $X = \alpha Y + \beta Z$ (compare Exercise 7)?

   (b) For $Y = (1,2)$ and $Z = (-4,-8)$ answer the same question as in the preceding part. A sketch might clarify the issue here.

9. Let $X = (3,2)$, $Y = (1,2)$, and $Z = (-2,-3)$. Find scalars α and β such that $X = \alpha Y + \beta Z$.

10. Find a scalar α and vectors X and Y such that $X \neq Y$, but $\alpha X = \alpha Y$.

11. Let α and β be scalars, and X and Y be vectors. Use only Properties 1 to 8 to prove the following (that way, the proofs will be valid for any vector space):

(a) $0X = \mathbf{0}$ for every vector X.

(b) $\alpha\mathbf{0} = \mathbf{0}$ for every scalar α.

(c) If $\alpha \neq 0$ and $\alpha X = \mathbf{0}$, then $X = \mathbf{0}$.

(d) If $\alpha \neq 0$ and $\alpha X = \alpha Y$, then $X = Y$.

(e) If $X \neq \mathbf{0}$ and $\alpha X = \mathbf{0}$, then $\alpha = 0$.

(f) If $X \neq \mathbf{0}$ and $\alpha X = \beta X$, then $\alpha = \beta$.

12. Let $X = (1,0)$ and $Y = (1,1)$. On graph paper plot the points X, $X + \frac{1}{2}Y$, $X + Y$, $X + \frac{3}{2}Y$, $X + 2Y$, and $X - 5Y$. On the basis of your picture, make a conjecture.

1.2   The Space $\mathbb{R}^n$

1.2A   Points in $\mathbb{R}^n$

We turn to the generalization of $\mathbb{R}^2$ described in Section 0.2c. A typical element X of the cartesian product $\mathbb{R}^n$ of $\mathbb{R}$ with itself n times is an ordered n-tuple $X = (x_1, x_2, \ldots, x_n)$ with $x_j \in \mathbb{R}$ for $j = 1, 2, \ldots, n$. As in the case of $\mathbb{R}^2$, X is called a **point** or **vector**. If $X = (x_1, x_2, \ldots, x_n)$ and $Y = (y_1, y_2, \ldots, y_n)$, then $X = Y$ if and only if the corresponding **coordinates** x_j and y_j are equal, that is, $x_j = y_j$ for each subscript $j = 1, 2, \ldots, n$.

1.2B   Algebraic operations in $\mathbb{R}^n$

The discussion here follows the pattern set with $\mathbb{R}^2$. For members $X = (x_1, x_2, \ldots, x_n)$ and $Y = (y_1, y_2, \ldots, y_n)$ of $\mathbb{R}^n$ and a scalar α, we define addition of vectors by

$$X + Y = (x_1 + y_1, x_2 + y_2, \ldots, x_n + y_n)$$

and multiplication of a vector by a scalar by

$$\alpha X = (\alpha x_1, \alpha x_2, \ldots, \alpha x_n).$$

EXAMPLE:

1. Let $X = (1, 0, 1/2)$, $Y = (-1, 4, 4)$, and $\alpha = 7$. Then, for instance,

$$X + Y = (0, 4, \frac{9}{2}),$$

$$\alpha X = (7, 0, \frac{7}{2}),$$

$$\alpha Y \;=\; (-7, 28, 28),$$
$$\alpha X + Y \;=\; (6, 4, \frac{15}{2}).$$

REMARK:

- We do not add vectors together from different spaces $\mathbb{R}^n$ and $\mathbb{R}^q$ with $n \neq q$. Thus, $(1, 0, 1/2) \in \mathbb{R}^3$ and $(0, 6, 1, -1) \in \mathbb{R}^4$ *cannot* be added together.

1.2C   THE VECTOR SPACE PROPERTIES

Note that Properties 1 to 8 listed for $\mathbb{R}^2$ are easily modified to fit $\mathbb{R}^n$, with the proofs essentially unchanged (just add dots ... inside the parentheses). This is because we have defined $X + Y$ and αX using coordinates. For instance, $(x_1, x_2, x_3) + (y_1, y_2, y_3) = (x_1 + y_1, x_2 + y_2, x_3 + y_3)$. Therefore all the proofs can be reduced to proofs within a single coordinate slot, that is, within the real numbers.

In the preceding section, we mentioned informally the term "vector space" and referred to Properties 1 to 8 as the "vector space axioms". Now we will be more precise. A set $\mathcal{S}$ is a **vector space** provided that

I Addition is defined: given X, $Y \in \mathcal{S}$, an element $X + Y$ is uniquely specified in $\mathcal{S}$.

II Multiplication by scalars is defined: given $X \in \mathcal{S}$ and $\alpha \in \mathbb{R}$, an element αX is uniquely specified in $\mathcal{S}$.

III The operations defined in (I) and (II) satisfy Properties 1 through 8, that is, for all X, Y, $Z \in \mathcal{S}$ and $\alpha, \beta \in \mathbb{R}$, we have

1. Addition is associative: $(X + Y) + Z = X + (Y + Z)$.

2. Addition is commutative: $X + Y = Y + X$.

3. Existence of zero: $\mathcal{S}$ contains an element $\mathbf{0}$ satisfying $X + \mathbf{0} = X$.

4. Existence of additive inverses: for each $X \in \mathcal{S}$ there is an element $-X$ satisfying $X + (-X) = \mathbf{0}$.

5. $\alpha(\beta X) = (\alpha\beta)X$.

6. $1X = X$.

7. $(\alpha + \beta)X = \alpha X + \beta X$.

8. $\alpha(X + Y) = \alpha X + \alpha Y$.

We already know that $\mathbb{R}^2$ is a vector space, and in fact, that $\mathbb{R}^n$ is a vector space for any positive integer n. In particular, $\mathbb{R}^1 = \mathbb{R}$, so the set of real numbers is itself a vector space. If these were the only examples we might encounter, we would not need the abstract definition of vector space.

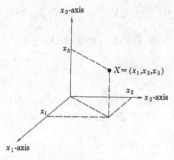

FIGURE 1.6.

But we will be dealing with many more vector spaces. A humble but very useful specimen is the space consisting of the single element **0**, the origin of $\mathbb{R}^2$, for example. You might check that this is a vector space. Moreover, we will see that such natural objects as (straight) lines through the origin in $\mathbb{R}^2$, lines and planes through the origin in $\mathbb{R}^3$, and so on, are all vector spaces. The notion of vector space serves to describe a variety of important mathematical objects.

1.2D PICTURES

In the preceding section we viewed $\mathbb{R}^2$ as a plane. Is there a similar representation for $\mathbb{R}^n$ with $n > 2$?

If $n = 3$, the answer is yes. The standard picture for $\mathbb{R}^3$, with $X = (x_1, x_2, x_3)$ a typical vector, is illustrated in Figure 1.6.

If $n > 3$, most of us (except for a few mathematicians) cannot visualize $\mathbb{R}^n$. Nonetheless, it is possible and sometimes necessary to deal with such higher-dimensional spaces. In fact, by scientific necessity, much of modern mathematics occurs in infinite-dimensional spaces in order to understand aspects of our three-dimensional world (or is it four?). In dealing with these higher-dimensional spaces, we must rely on two things: our spatial intuition in $\mathbb{R}^2$ and $\mathbb{R}^3$, used with care, and our algebraic precision. You should note that we have defined the spaces $\mathbb{R}^n$ as *algebraic* structures. We will reduce many geometric problems therein to algebra and, later, calculus. It is natural for us to picture three dimensions and impossible for us to picture five. But algebraically we can solve linear equations in x_1, x_2, x_3, x_4, x_5 almost as easily as in x_1, x_2, x_3; there is essentially no difference.

REMARKS:

- In the terminology of the Introduction, Section 1.0, we are still in stage one, acquiring the basic vector algebra we will need for our purposes. Recall that eventually we wish to study functions $f(X)$ of a *vector* variable $X = (x_1, \ldots, x_n)$.

- We have defined the sum, but not the product, of two vectors in $\mathbb{R}^n$; we have no rule for multiplying X by Y to get a third vector. In general, there is no such well-behaved product, although there is a natural product that we will define in $\mathbb{R}^3$. We will soon encounter, however, the "inner product" $\langle X, Y \rangle$ of vectors X and Y, but this product is a number, not a vector.

Exercises

Some basic calisthenics in vector arithmetic and algebra.

1. Let $X = (1, 2, -2)$, $Y = (0, 2, 1)$, and $Z = (1, -2, 2)$ in $\mathbb{R}^3$. Compute

    (a) $X + Y$

    (b) $X - Y$

    (c) $X + Z$

    (d) $X - Y - Z$

    (e) $2X + 4Y$

    (f) $3X - 2Y + Z$

    (g) $7(X - Y)$.

2. Let $Y = (1, 2, -1, 1)$ and $Z = (0, 2, -1, 0)$ in $\mathbb{R}^4$. Compute a vector X satisfying

    (a) $X = Y + Z$

    (b) $2X + Y - Z = 0$

    (c) $Y - X = 2Z$

    (d) $Z - 3X - Y = 0$

    (e) $7X + 7Y + 7Z = 0$.

3. Sketch (as in Figure 1.6) a portion of $\mathbb{R}^3$ as seen from the **first octant**

$$\{(x_1, x_2, x_3) : x_1 \geq 0, x_2 \geq 0, x_3 \geq 0\} \, .$$

    Locate the vectors $\mathbf{0} = (0, 0, 0)$, $(4, 1, 0)$, $(0, 1, 0)$, $(0, 0, -1)$, $(1, 1, 0)$, $(0, 1, 2)$, $(1, 1, 1)$, and $(1, -1, 1)$.

4. (a) Define the vectors

$$\mathbf{e}_1 = (1, 0, 0), \quad \mathbf{e}_2 = (0, 1, 0), \quad \mathbf{e}_3 = (0, 0, 1).$$

    Show that given any $W = (w_1, w_2, w_3)$, there exist scalars α, β, and γ such that $W = \alpha \mathbf{e}_1 + \beta \mathbf{e}_2 + \gamma \mathbf{e}_3$. Note that these scalars are uniquely determined by W.

    (b) Let $W = (1, -2, 3)$. Find α, β, and γ so that $W = \alpha \mathbf{e}_1 + \beta \mathbf{e}_2 + \gamma \mathbf{e}_3$.

5. Let $X = (1, 2, 0)$, $Y = (1, 1, 0)$, and $Z = (-1, 4, 0)$. Is it possible to find, for every $W \in \mathbb{R}^3$, scalars α, β, and γ so that $W = \alpha X + \beta Y + \gamma Z$? For instance, what if $W = (1, 3, 1)$?

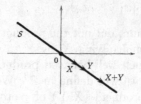

FIGURE 1.7.

6. (a) Let $S = \{X = (x_1, x_2, x_3) : x_1 - x_2 = 0\}$. Show that the sum of any two vectors in S is also in S.

    (b) Is this true if the equation determining the set is $x_1 + 2x_2 - x_3 = 0$? State your reasons.

    (c) What if the equation is $x_1 = 4$?

    (d) For what kinds of linear equations of the general form $ax_1 + bx_2 + cx_3 = d$ is it true that if the coordinates of two vectors X and Y satisfy this equation, then the coordinates of the sum $X + Y$ satisfy the same equation? Give conditions on a, b, c, and d.

7. Sketch in $\mathbb{R}^2$ the set of all points X such that $x_1 + x_2 = 0$. Likewise for $x_1 + x_2 = 1$.

8. Let e_1, e_2, and e_3 be as in Exercise 4. Let S be the set of all vectors X of the form $X = \alpha_1 e_1 + \alpha_2 e_2$.

    (a) Sketch $\mathbb{R}^3$ and locate the set S in your sketch.

    (b) If $X \in S$, what simple equation must its coordinates satisfy?

    (c) Locate in your sketch of $\mathbb{R}^3$ the set of all vectors satisfying $x_2 = 0$.

    (d) Also locate three more sets of vectors, those satisfying: $x_1 = 0$, $x_1 = 1$, and $x_3 = 5$.

9. Let $X = (2, 0, 0)$, $Y = (0, 1, 0)$, and $Z = (0, 0, 1)$. Plot the following points in a sketch of $\mathbb{R}^3$: X, $X + Y$, $X + Z$, $X + \frac{1}{2}Y$, $X + \frac{1}{2}Y + \frac{1}{4}Z$, $X + \frac{3}{2}Y + \frac{1}{2}Z$, and $X + 2Z$. On the basis of your picture make a conjecture.

1.3 Linear Subspaces

1.3A Definition

Now we delve into the internal anatomy of the vector spaces $\mathbb{R}^n$. We begin with a simple example.

Let $\mathbb{R}^2$ be the plane as usual, and consider a line (extended infinitely in both directions) through the origin $0 = (0, 0)$. We denote this line by the script letter S. See Figure 1.7. Now each point on S is a vector, as a member of $\mathbb{R}^2$.

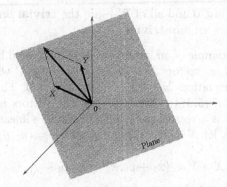

FIGURE 1.8.

It is crucial to note the following:

1. The sum of any two vectors on S is also on S. Thus $X, Y \in S$ implies $X + Y \in S$. See Figure 1.7.

2. If $X \in S$ and $\alpha \in \mathbb{R}$, then $\alpha X \in S$.

It follows that if $X \in S$, then $-X = -1X \in S$. Therefore, if $Y \in S$, then $Y + (-X)$ (which we will often write as $Y - X$) is in S. So, when we add, subtract, or multiply by scalars any vectors in S, the sum, difference, or product is again a vector in S. In fact, S itself is a vector space; not only are vector addition and multiplication by scalars possible in S, but also the vector-space axioms (Properties 1 through 8) are satisfied.

Definition 1.3.1 The subset S of the vector space $\mathbb{R}^n$ is a **linear subspace** of $\mathbb{R}^n$ if and only if S is itself a vector space (see Section 1.2c) under the addition and multiplication by scalars defined in $\mathbb{R}^n$.

EXAMPLES:

1. As we have just seen, any line through the origin in $\mathbb{R}^2$ is a linear subspace of $\mathbb{R}^2$.

2. Any line or plane through the origin in $\mathbb{R}^3$ is a linear subspace of $\mathbb{R}^3$. Note that if we add two vectors X and Y that are in a plane through the origin (an unbounded plane, of course, not just a piece of one), then the sum $X + Y$ is also in that plane. See Figure 1.8.

3. The subset of $\mathbb{R}^n$ consisting of $\mathbf{0} = (0, \dots, 0)$ alone is a linear subspace, the **zero linear subspace**, often denoted by $\{\mathbf{0}\}$.

4. We should note that since $\mathbb{R}^n$ is a subset of itself—not a **proper** subset, to be sure—then $\mathbb{R}^n$ is a linear subspace of $\mathbb{R}^n$. These linear

subspaces—just $\mathbf{0}$ and all of $\mathbb{R}^n$—are the **trivial** linear subspaces of $\mathbb{R}^n$. All others are **nontrivial**.

5. Here is an example of an important type motivated by algebra. Let $\mathcal{S}$ be the set of all vectors $X = (x_1, x_2, x_3, x_4)$ in $\mathbb{R}^4$ whose coordinates satisfy the equation $2x_1 - x_2 + x_3 + 5x_4 = 0$. Thus, for example, $(0, 1, 1, 0)$ and $(5, 1, 1, -2)$ are among the vectors in $\mathcal{S}$. You should check that $\mathcal{S}$ is a vector space and, therefore, a linear subspace of $\mathbb{R}^4$. For instance, let $X = (x_1, x_2, x_3, x_4)$ and $Y = (y_1, y_2, y_3, y_4)$ be in $\mathcal{S}$. Is

$$X + Y = (x_1 + y_1, x_2 + y_2, x_3 + y_3, x_4 + y_4)$$

also in $\mathcal{S}$? To see that it is, note that

$$2(x_1 + y_1) - (x_2 + y_2) + (x_3 + y_3) + 5(x_4 + y_4)$$
$$= (2x_1 - x_2 + x_3 + 5x_4) + (2y_1 - y_2 + y_3 + 5y_4) = 0 + 0 = 0$$

since both X and Y are in $\mathcal{S}$.

REMARK:

- Example 5 is the forerunner of important ideas, namely, the relations between linear subspaces of $\mathbb{R}^n$ and equations, or systems of equations, of the form $a_1 x_1 + \cdots + a_n x_n = 0$. Such equations are called **homogeneous linear equations**. The word "linear" refers to the form in which the left-hand side of the equation appears, and the word "homogeneous" refers to the fact that the right-hand side is zero. You should become aware that most of the basic computational problems in the vector algebra of $\mathbb{R}^n$ reduce to solving some system of linear equations. Actually, this connection between linear subspaces and equations should not be completely surprising. If the line in Example 1 has slope m in the sense of calculus, then it is the graph of $x_2 = mx_1$, that is, the set of solutions (x_1, x_2) of the homogeneous linear equation $mx_1 - x_2 = 0$. We will learn more about systems of linear equations and their solutions in Section 1.6.

1.3B A CRITERION FOR LINEAR SUBSPACES

Suppose that we are given a subset $\mathcal{S}$ of $\mathbb{R}^n$. How can we tell if it is actually a linear subspace? We can, of course, check to see if $\mathcal{S}$ satisfies the definition of vector space. This involves checking the associativity and commutativity of addition, the existence of a zero vector in $\mathcal{S}$, and so on, through the eight properties. But reflection should convince us that much of this is superfluous. Since addition in $\mathcal{S}$ is addition in $\mathbb{R}^n$, it is certainly associative and commutative and well behaved with respect to multiplication by scalars. The real issue is whether the vectors in $\mathcal{S}$ comprise a self-contained algebraic system themselves, that is, whether sums and scalar multiples of vectors in S are themselves in S. We will say that S is **closed under addition** if the sum of two vectors in S is itself a vector in S, and we will say that S is **closed under multiplication by scalars** if every scalar multiple of a vector in S is a vector in S.

Theorem 1.3.1 Let S be a nonempty subset of $\mathbb{R}^n$. Then S is a linear subspace $\Leftrightarrow$ S is closed under addition and closed under multiplication by scalars.

PROOF: We must verify that each statement implies the other. It is best to begin with the easier.

($\Rightarrow$) This is simply a matter of definitions. Given that S is a linear subspace, it is certainly a vector space and thereby closed under both operations. Done.

($\Leftarrow$) We are given that S is closed under both operations. We check the vector-space axioms 1 through 8 in turn.

1. Associativity is clear, because addition in S is addition in $\mathbb{R}^n$.

2. Commutativity is clear for the same reason.

3. Why is $\mathbf{0} \in S$? The reason is that S is nonempty, and so there is some X in S. Also S is closed under multiplication by scalars. Therefore $0 \cdot X$ is in S. But $0 \cdot X = \mathbf{0}$, the zero vector.

4. Additive inverses exist because if $X \in S$, then $-1X \in S$ and $-1X = -X$.

The remaining properties, 5 through 8, hold in S because they hold in $\mathbb{R}^n$. Done. $<<$

REMARKS:

- The preceding theorem implies that if S_1 and S_2 are linear subspaces of $\mathbb{R}^n$, then their intersection $S_1 \cap S_2$ is also a linear subspace of $\mathbb{R}^n$. Can you see why?

- Suppose that S_1 is the solution set of the equation

$$3x_1 - 2x_2 + 5x_3 = 0 \, ,$$

and S_2 is the solution set of

$$x_1 + x_2 - x_3 = 0 \, .$$

Both of the equations are homogeneous and linear, so S_1 and S_2 are linear subspaces of $\mathbb{R}^3$. Their intersection is the solution set of the system of homogeneous linear equations

$$
\begin{aligned}
3x_1 - 2x_2 + 5x_3 &= 0 \\
x_1 + x_2 - x_3 &= 0 \, .
\end{aligned}
$$

According to the previous remark, the solution set of the system is a linear subspace of $\mathbb{R}^3$. This example is easily generalized: the solution set of any system of homogeneous linear equations in n unknowns is a linear subspace of $\mathbb{R}^n$.

- If the goal is to check that S is *not* a linear subspace, one might be successful by merely showing that $\mathbf{0} \notin S$. For instance, the vertical line $x_1 = 2$ in $\mathbb{R}^2$ is not a linear subspace since it does not contain the point $(0,0)$. The quick reasoning used here will be useful in the exercises.

1.3C WHERE WE ARE

In the preceding sections we have defined the concepts of "vector space", "linear subspace" and presented several examples. The discussion tended towards abstract and axiomatic language building. The term "linear subspace", for example, is a wonderfully useful bag in which we will place the diverse notions of "line through $\mathbf{0}$", "plane containing $\mathbf{0}$", "set of solutions of a system of homogeneous linear equations", and so on. From now on we will be moving gradually toward problems arising in geometry or linear equations, rather than those prompted by our own definitions. Nonetheless, this careful process of definition, observation, and theorem will reveal the underlying similarity of the diverse problems to be encountered, and ultimately will help lead to their solutions.

Exercises

Practice in identifying and working with linear subspaces of $\mathbb{R}^n$.

1. Sketch the following subsets of $\mathbb{R}^2$. Which are linear subspaces? Here $X = (x_1, x_2)$, as usual, and the sets are defined by way of coordinates.

    (a) $\{X \in \mathbb{R}^2 : x_1 = 0\}$

    (b) $\{X \in \mathbb{R}^2 : x_1 \geq 0\}$

    (c) $\{X \in \mathbb{R}^2 : x_1 \neq 0\}$

    (d) $\{X \in \mathbb{R}^2 : x_1 = 1\}$

    (e) $\{X \in \mathbb{R}^2 : x_1 - x_2 = 0\}$

    (f) $\{X \in \mathbb{R}^2 : x_1 - x_2 = 2\}$

    (g) $\{X \in \mathbb{R}^2 : x_1^2 - x_2 = 0\}$.

2. Decide which of the following subsets of $\mathbb{R}^3$ are linear subspaces:

    (a) $\{X \in \mathbb{R}^3 : x_1^2 + x_2^2 + x_3^2 = 1\}$

    (b) $\{X \in \mathbb{R}^3 : x_1 - x_2 = 0, x_3 = 0\}$

    (c) $\{X \in \mathbb{R}^3 : x_1 + x_2 + x_3 = 1\}$

    (d) $\{X \in \mathbb{R}^3 : x_1 - x_2 = 0\}$

    (e) $\{X \in \mathbb{R}^3 : x_1 + x_2 + x_3 = 0\}$

(f) $\{X \in \mathbb{R}^3 : x_2 = 0\}$

(g) $\{X \in \mathbb{R}^3 : x_1 \leq x_2 \leq x_3\}$

(h) $\{X \in \mathbb{R}^3 : x_1 = x_2 = x_3 = 0\}$

(i) $\{X \in \mathbb{R}^3 : x_1 - x_2 = 0, \, 2x_1 + x_2 + x_3 = 0\}$

(j) $\{X \in \mathbb{R}^3 : x_1 = x_2 + 3x_3, \, x_2 = 2x_1 + x_3\}$

(k) $\{X \in \mathbb{R}^3 : x_1^2 = x_2, \, x_1 = x_3\}$.

3. Let Y and Z be given fixed vectors in $\mathbb{R}^3$. Is the set

$$S = \{X \in \mathbb{R}^3 : X = \alpha Y + \beta Z, \, \alpha, \beta \text{ arbitrary scalars}\}$$

a linear subspace of $\mathbb{R}^3$? Say why or why not. Does your reasoning depend essentially on $\mathbb{R}^3$?

4. Select true or false for the following:

   (a) Every line in $\mathbb{R}^2$ is a linear subspace of $\mathbb{R}^2$.

   (b) Every linear subspace of $\mathbb{R}^n$ contains the origin of $\mathbb{R}^n$.

   (c) Every linear subspace of $\mathbb{R}^3$ that contains a nonzero vector X also contains the line through $\mathbf{0}$ and X.

   (d) The union of two linear subspaces of $\mathbb{R}^n$ is also a linear subspace.

   (e) $\mathbb{R}^1$ has a nontrivial linear subspace.

   (f) All nontrivial linear subspaces of $\mathbb{R}^3$ are planes through the origin.

   (g) Every nontrivial linear subspace of $\mathbb{R}^2$ is the set of all solutions $X = (x_1, x_2)$ to some homogeneous linear equation $\alpha_1 x_1 + \alpha_2 x_2 = 0$.

   (h) The set of all $X = (x_1, x_2, x_3)$ satisfying $x_3 = 0$ is a line through the origin in $\mathbb{R}^3$.

   (i) If a linear subspace S of $\mathbb{R}^n$ contains two different lines through the origin, then S contains the plane containing both lines.

   (j) $\mathbb{R}^2$ has a nontrivial linear subspace S that consists of a line through the origin as well as certain points not on this line.

   (k) A subset S of $\mathbb{R}^2$ that can be enclosed within the unit circle is not a nontrivial linear subspace of $\mathbb{R}^2$.

   (l) There exist nontrivial linear subspaces of $\mathbb{R}^n$ with finitely many members.

5. Suppose that $X \in \mathbb{R}^n$ may be written $X = \alpha_1 Y + \beta_1 Z = \alpha_2 Y + \beta_2 Z$ with $\alpha_1 \neq \alpha_2$ and $\beta_1 \neq \beta_2$. Find γ and δ, both different from 0, such that $\mathbf{0} = \gamma Y + \delta Z$ (note also that $\mathbf{0} = 0Y + 0Z$).

6. One way to specify a particular linear subspace of $\mathbb{R}^n$ is to give a homogeneous linear equation (or equations) of which the linear subspace is the set of solutions. Can you describe another way of specifying or building linear subspaces? See Exercise 3.

7. (a) Describe geometrically all the nontrivial linear subspaces of $\mathbb{R}^2$.

(b) Describe geometrically all the nontrivial linear subspaces of $\mathbb{R}^3$. Can you prove your assertions? See the following section.

8. Describe the linear subspaces of $\mathbb{R}^2$ that contain the point $(2, -1)$.

9. The set $\{X \in \mathbb{R}^4 : x_4 = 0\}$ is a linear subspace of $\mathbb{R}^4$. Describe it visually as best as you can. In what ways (if any) do you think it looks different from the linear subspace $\{X \in \mathbb{R}^4 : x_2 = 0\}$?

1.4   The Linear Subspaces of $\mathbb{R}^3$

1.4A   THE PROBLEM

In this section we grapple with the following problem: classify all linear subspaces of the vector space $\mathbb{R}^3$. The study of this problem will give us some useful practice in working with the concepts previously introduced. In addition, it will enable us to introduce some new ideas and, finally, will yield some information about $\mathbb{R}^3$ we will use later.

1.4B   THE STANDARD BASIS

First let us look at $\mathbb{R}^3$ itself. Consider the vectors

$$\mathbf{e}_1 = (1, 0, 0), \quad \mathbf{e}_2 = (0, 1, 0), \quad \mathbf{e}_3 = (0, 0, 1).$$

If $X = (x_1, x_2, x_3)$ is any vector in $\mathbb{R}^3$, then note that we may write $X = x_1\mathbf{e}_1 + x_2\mathbf{e}_2 + x_3\mathbf{e}_3$. See Figure 1.9. For instance, if $X = (2, -1, -7)$, then $X = 2\mathbf{e}_1 - \mathbf{e}_2 - 7\mathbf{e}_3$; if $Y = (1, 0, 4)$, then $Y = \mathbf{e}_1 + 4\mathbf{e}_3$. The important thing here is that *every* vector X in $\mathbb{R}^3$ may be written *uniquely* as a sum involving $\mathbf{e}_1$, $\mathbf{e}_2$, and $\mathbf{e}_3$, with scalars as coefficients. These three vectors comprise the **standard basis** of $\mathbb{R}^3$.

REMARK:

- Roughly speaking, $\mathbb{R}^3$ is three-dimensional, because its standard basis consists of three vectors, no more, no less.

Throughout this discussion, we will be using expressions like $x_1\mathbf{e}_1 + x_2\mathbf{e}_2 + x_3\mathbf{e}_3$. Hence, we can make the following general definition. Let $X_1, \ldots, X_k$ be vectors in $\mathbb{R}^n$. Then $Y \in \mathbb{R}^n$ is a **linear combination** of $X_1, \ldots, X_k$ if and only if $Y = \alpha_1 X_1 + \cdots + \alpha_k X_k$ for some scalars $\alpha_1, \ldots, \alpha_k$. The set

$$\mathcal{S} = \{Y : Y \text{ is a linear combination of } X_1, X_2, \ldots, X_k\}$$

is called the **linear span** of the vectors $X_1, X_2, \ldots, X_k$. We also say that the vectors $X_1, X_2, \ldots, X_k$ **span** the set $\mathcal{S}$. The following fact is so easy to

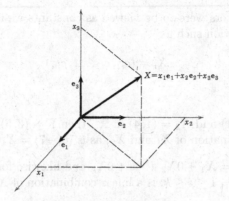

FIGURE 1.9.

prove that it hardly deserves to be called a theorem, but we state it as a theorem in order to emphasize its usefulness.

Theorem 1.4.1 The linear span of a finite set of vectors in $\mathbb{R}^n$ is a linear subspace of $\mathbb{R}^n$.

PROOF: Let $\mathcal{S}$ denote the linear span of vectors $X_1, X_2, \ldots, X_k$. The following calculation shows that $\mathcal{S}$ is closed under vector addition:

$$
\begin{aligned}
(\alpha_1 X_1 + &\cdots + \alpha_k X_k) + (\beta_1 X_1 + \cdots + \beta_k X_k) \\
&= \alpha_1 X_1 + \beta_1 X_1 + \cdots + \alpha_k X_k + \beta_k X_k \\
&= (\alpha_1 + \beta_1) X_1 + \cdots + (\alpha_k + \beta_k) X_k .
\end{aligned}
$$

Closure under scalar multiplication follows from the calculation

$$
\gamma(\alpha_1 X_1 + \cdots + \alpha_k X_k) = \gamma \alpha_1 X_1 + \cdots + \gamma \alpha_k X_k .
$$

Done. $<<$

REMARK:

- In the preceding discussion we treated an arbitrary set of vectors. It was, therefore, not convenient for us to use different letters for the different vectors. Instead, we have used the same letter with different subscripts. If we talk about the coordinates of these vectors, we use double subscripts. For example, the vector X_3 is written as

$$
X_3 = (x_{31}, x_{32}, \ldots, x_{3n}) ,
$$

and, more generally, we have

$$
X_i = (x_{i1}, x_{i2}, \ldots, x_{in}) .
$$

If these vectors were to be viewed as constants we might use a suggestive notation such as

$$X_i = (c_{i1}, c_{i2}, \ldots, c_{in}) \, .$$

EXAMPLES:

1. If $X_1 = (2,1)$ and $X_2 = (1,4)$ in $\mathbb{R}^2$, then $Y = (3,5) = X_1 + X_2$ is a linear combination of X_1 and X_2, as is $(0,-7) = X_1 - 2X_2$.

2. Clearly $X_1 = X_1 + 0X_2 + \cdots + 0X_k$ and likewise for $X_2, \ldots, X_k$, so that each X_i, $1 \leq i \leq k$, is a linear combination of $X_1, \ldots, X_k$.

3. We indicated above that $\mathbb{R}^3$ is the linear span of the standard basis vectors e_1, e_2, and e_3. In addition, we stated that there is only one way to write each vector in $\mathbb{R}^3$ as a linear combination of the standard basis vectors. You should convince yourself of these facts.

4. For $\mathbb{R}^n$, we define $e_j, j = 1, 2, \ldots, n$, to be the vector in $\mathbb{R}^n$ whose j^{th} coordinate is 1 and whose remaining coordinates are 0. Then the set $\{e_1, e_2, \ldots, e_n\}$ is the **standard basis** of $\mathbb{R}^n$. Thus, the standard basis of $\mathbb{R}^2$ is $\{(1,0), (0,1)\}$, while the standard basis of $\mathbb{R}^4$ is

$$\{(1,0,0,0), (0,1,0,0), (0,0,1,0), (0,0,0,1)\} \, .$$

Note that with this definition, the symbols $e_1, e_2, \ldots$ all have multiple meanings: $e_1 = (1,0)$ in $\mathbb{R}^2$, but $e_1 = (1,0,0,0)$ in $\mathbb{R}^4$. In most cases, the correct meaning will be clear from the context. Again, you should convince yourself that every vector X in $\mathbb{R}^n$ is a unique linear combination of the standard basis vectors $e_1, e_2, \ldots, e_n$.

1.4C LINES AND PLANES REVISITED

When asked to list *all* the linear subspaces of $\mathbb{R}^3$, we are likely to mention lines and planes through the origin. We might then recall the trivial linear subspaces $\{0\}$ and $\mathbb{R}^3$. Are there any others?

As we ponder this question, we are brought to the realization that its answer requires that we be more precise about the notions we have been using. What is a line? How can we tell if a subset given by such-and-such a condition or property is a plane? In the preceding section, we used lines and planes pictorially, in $\mathbb{R}^2$ and $\mathbb{R}^3$. But what is a plane in $\mathbb{R}^n$ when n is arbitrary?

Lines through the origin. In the special case of $\mathbb{R}^2$, a line through the origin may be described as the set of all (x_1, x_2) such that $a_1 x_1 + a_2 x_2 = 0$, where a_1 and a_2 are scalars not both zero. This definition involving one equation, however, does not work in $\mathbb{R}^n$, for example, you can see that the

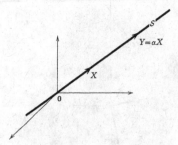

FIGURE 1.10. A one-dimensional linear subspace of $\mathbb{R}^3$

set of solutions (x_1, x_2, x_3) to the single equation $x_3 = 0$ in $\mathbb{R}^3$ is the *plane* consisting of all points $(x_1, x_2, 0)$ and *not* a line.

Here is a natural definition of a line through the origin whose charm is that it works in $\mathbb{R}^n$ for arbitrary n. A subset S of $\mathbb{R}^n$ is a **line through the origin** (also called a **one-dimensional linear subspace**) if and only if there is a nonzero vector X in S such that

$$S = \{Y \in \mathbb{R}^n : Y = \alpha X, \ \alpha \in \mathbb{R}\}.$$

In other words, a line through the origin is a linear subspace S which is spanned by a single nonzero vector X. See Figure 1.10.

REMARKS:

- A line S is "one-dimensional" precisely because it can be built up by taking *one* vector X and forming the set of all linear combinations (that is, scalar multiples). Every nonzero vector X determines a line through the origin in this way.

- The line spanned by X is the same as the line spanned by βX for any scalar $\beta \neq 0$, since if $Y = \alpha X$, then $Y = (\alpha/\beta)\beta X$. Thus, a line through the origin is spanned by *any* nonzero vector lying on that line.

Planes through the origin. A subset S of $\mathbb{R}^n$ is a **plane through the origin** (also called a **two-dimensional linear subspace**) if and only if it is the linear span of two vectors X, Y that do not lie on the same line through the origin; that is, if and only if

$$S = \{Z \in \mathbb{R}^n : Z = \alpha X + \beta Y \text{ with } \alpha, \beta \in \mathbb{R}\}.$$

Vectors that lie on a common line through the origin are said to be **collinear** with the origin. Thus the requirement on X and Y in the preceding definition is that X and Y be non-collinear with the origin. Two vectors in $\mathbb{R}^n$

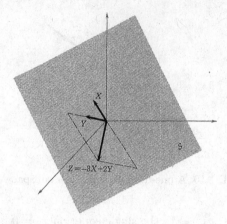

FIGURE 1.11. Every $Z \in S$ is a linear combination of X and Y

which are non-collinear with the origin span a plane in $\mathbb{R}^n$. It is also true, as we will see in the next section, that a plane S is spanned by *any* two vectors in S which are non-collinear with the origin.

REMARKS:

- A plane through the origin in $\mathbb{R}^n$ is a linear subspace of $\mathbb{R}^n$ since it is the linear span of a pair of vectors.

- We have called the plane S a "two-dimensional linear subspace" because every element of S is a linear combination of *two* vectors (for examples, $X + Y$, $X - 2Y$, X, and $6X$), and, crucially, *both* vectors are required. See Figure 1.11. If only one vector were required, the linear subspace would be a line through the origin.

1.4D   CLASSIFYING THE LINEAR SUBSPACES OF $\mathbb{R}^3$

Let S be a linear subspace of $\mathbb{R}^3$. What must S look like? Must S be one of the linear subspaces that we already know about, namely, $\{\mathbf{0}\}$, a line through the origin, a plane through the origin, or $\mathbb{R}^3$ itself?

We deal with these possibilities as follows:

1. Since S is a linear subspace, it contains the zero vector. If that is the only vector that S contains, then $S = \{\mathbf{0}\}$.

2. If $S \neq \{\mathbf{0}\}$, then it contains a nonzero vector X. Being a linear subspace, S therefore contains all vectors of the form αX, $\alpha \in S$. These form a line through the origin. If S contains no other vectors, then S is itself this line through the origin.

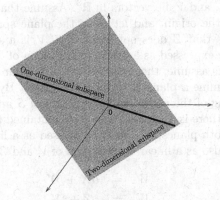

FIGURE 1.12. Which linear subspaces of $\mathbb{R}^3$ contain a plane?

3. On the other hand, suppose S contains a vector Y not of the form
   αX. Then S also contains all vectors of the form $\alpha X + \beta Y$, since S is
   a linear subspace and therefore closed under the formation of linear
   combinations of its elements. If S contains no other vectors, then S
   is a plane through the origin.

4. But S might contain a vector Z *not* of the form $\alpha X + \beta Y$. Then S
   is not the plane spanned by X and Y. See Figure 1.12. How may we
   describe S? Does $S = \mathbb{R}^3$, or might it be possible that S is bigger
   that the plane but smaller that all of $\mathbb{R}^3$?

This question is answered by the following unglamorous but useful lemma.

Lemma 1.4.1 Let X and Y be vectors in $\mathbb{R}^3$ which are non-collinear with
the origin, and let Z be a vector in $\mathbb{R}^3$ that does not lie on the plane spanned
by X and Y. Then it is possible to express any other vector $V \in \mathbb{R}^3$ as a linear
combination of X, Y, and Z.

REMARK:

- Logically speaking a "lemma" is a "theorem". The two terms reflect
  different viewpoints. The term "theorem" is used when the statement
  is viewed as a central feature of the subject. The term "lemma" is
  used when the major interest in the statement is in its usefulness for
  proving theorems and other lemmas.

PROOF: The proof will be based on the following fact: if two planes in $\mathbb{R}^3$
both contain the origin, then their intersection contains a line through the
origin. In the next section, we will give a proof of this fact based entirely
upon the algebra of $\mathbb{R}^3$. For now, you can accept it as something that could
be proved using the axioms of Euclidean geometry.

Let X, Y, Z, and V be vectors in $\mathbb{R}^3$. Assume that X and Y are non-collinear with the origin, and let S be the plane spanned by X and Y. Further assume that Z does not lie on S. If V is a scalar multiple of Z, then V may be expressed as a linear combination of X, Y, and Z and we are done. Let us assume, then, that V is not a scalar multiple of Z. Then V and Z determine a plane T through the origin. By the fact mentioned at the beginning of the proof, the intersection of S and T contains a line. It follows that there is a nonzero vector W contained in both S and T. As a member of both planes, W can be expressed as a linear combination of X and Y and also as a linear combination of V and Z as follows:

$$W = \alpha X + \beta Y$$
$$W = \gamma Z + \delta V .$$

If, in the second equation, δ equals 0, then γ would not be 0, and we could write Z as a multiple of W: $Z = \frac{1}{\gamma} W$. It would then follow from the first equation that $Z = \frac{\alpha}{\gamma} X + \frac{\beta}{\gamma} Y$, which contradicts the hypothesis that Z does not lie on the plane spanned by X and Y. Therefore, δ cannot be 0, and we can solve for V in the second equation:

$$V = \frac{1}{\delta} W - \frac{\gamma}{\delta} Z .$$

After substituting into this equation our expression for W from the first equation, we obtain

$$V = \frac{\alpha}{\delta} X + \frac{\beta}{\delta} Y - \frac{\gamma}{\delta} Z .$$

Done. $<<$

The lemma may be phrased as follows: the only linear subspace of $\mathbb{R}^3$ larger than a plane is the entire space $\mathbb{R}^3$. The complete story concerning the linear subspaces of $\mathbb{R}^3$ is summarized in the following theorem.

Theorem 1.4.2 Let S be a linear subspace of $\mathbb{R}^3$. Then S must be one of the following: the zero linear subspace $\{0\}$, a line through the origin (one-dimensional linear subspace), a plane through the origin (two-dimensional linear subspace), or the entire space $\mathbb{R}^3$ itself.

1.4E LINEAR SUBSPACES AND HOMOGENEOUS LINEAR EQUATIONS

The discussion leading up to the theorem in the preceding section indicates that every linear subspace, at least in $\mathbb{R}^3$, can be built by carefully choosing some vectors from the linear subspace and then forming all linear combinations of them. The resulting set of all linear combinations is the linear subspace itself. However, a question arises. It was easy to see (Example 5 in Section 1.3a) that the set of all solutions to a homogeneous linear equation in the variables x_1, x_2, x_3 form a linear subspace of $\mathbb{R}^3$. Now that we claim to know all the linear subspaces of $\mathbb{R}^3$, we ought relate an equation like $x_1 - x_2 + 5x_3 = 0$ to the discussion of Section 1.4d and especially Theorem 1.4.2.

Let us restate this problem. We are given the homogeneous linear equation $x_1 - x_2 + 5x_3 = 0$. What can we say about the set S of solutions (x_1, x_2, x_3), considered as a subset of $\mathbb{R}^3$?

1. S is a linear subspace of $\mathbb{R}^3$. You should review for yourself the justification of this fact.

2. It is easily checked that $(5, 0, -1)$ is in S and that $(0, 0, 1)$ is not in S. Thus, S does not equal $\{0\}$ or $\mathbb{R}^3$, so S is either a line or a plane.

3. Is S a line or is it a plane? We have already seen that $X = (5, 0, -1)$ is in S. Are all other vectors in S scalar multiples of this vector X? In other words, is S a line? The answer is no. For we readily find another vector $Y = (0, 5, 1)$ in S such that $Y \neq \alpha X$ for any α. To check this inequality, note that $(0, 5, 1) = Y = \alpha X = (5\alpha, 0, -\alpha)$ would imply $5\alpha = 0$. But $Z = 0Y$ would imply $Z = 0$, which is false.

And this does it. The solution set S must be the linear span of $X = (5, 0, -1)$ and $Y = (0, 5, 1)$.

REMARK:

- We now have two ways of describing the solution set of the equation $x_1 - x_2 + 5x_3 = 0$. It is the linear span of the two vectors $(5, 0, -1)$ and $(0, 5, 1)$. Alternatively, we can also say that the solution set is the plane containing the origin and the two vectors $(5, 0, -1)$ and $(0, 5, 1)$ (although it is only in the next section that we will be able to conclude that there is only one such plane). For someone who knows very little about linear algebra a description of the solutions may require the use of more symbols. We can say that the solutions are those vectors that can be written in the form

$$\alpha(5, 0, -1) + \beta(0, 5, 1) = (5\alpha, 5\beta, -\alpha + \beta)$$

for some real numbers α and β. For someone who does not even know vector notation we can write the coordinates explicitly (and not use the term "coordinates"):

$$\begin{aligned} x_1 &= 5\alpha \\ x_2 &= 5\beta \\ x_3 &= -\alpha + \beta \,. \end{aligned}$$

For each choice of α and β we get a unique solution of the homogeneous equation and, as we will see later, different choices give different solutions. The fact that there are two numbers to choose underscores the two-dimensionality of the set S of solutions. The numbers α and β are sometimes called "parameters" and the set of solutions is a "two-parameter family".

Although we considered the special case $x_1 - x_2 + 5x_3 = 0$, you should be convinced that the following theorem is true:

Theorem 1.4.3 Let $a_1x_1 + a_2x_2 + a_3x_3 = 0$ be a homogeneous linear equation with at least one of the coefficients a_1, a_2, and a_3 different from zero. Then the set of solutions $X = (x_1, x_2, x_3)$ is a plane through the origin (two-dimensional linear subspace) in $\mathbb{R}^3$.

PROOF: The proof is left to you. Note that the key in the special case above was our ability to find at least one non-solution, together with the two solutions X and Y that do not lie on the same line through the origin. See the examples below. <<

EXAMPLES:

5. Consider $3x_1 + 4x_2 + 5x_3 = 0$. Then one solution is $Y = (5, 0, -3)$. This was easy. To get a second solution Z not of the form αY, we begin by putting the zero in the first slot this time: $Z = (0, ?, ?)$. Except for the zero vector $(0, 0, 0)$, all vectors of the form $(0, ?, ?)$ that satisfy the homogeneous equation will *not* be of the form αY. Think about this. Hence $Z = (0, 5, -4)$ is a solution that does not lie on the line through the origin that contains Y. The solution set of the equation is the linear span of Y and Z.

6. Consider $x_1 + x_3 = 0$. Then $Y = (1, 0, -1)$ and $Z = (0, 1, 0)$ are solutions which are non-collinear with the origin. Any other solution is of the form $\alpha Y + \beta Z$.

Exercises

Practice with linear combinations in and linear subspaces of $\mathbb{R}^3$.

1. Which pairs $X, Y \in \mathbb{R}^3$ lie on the same line through the origin $\mathbf{0} = (0, 0, 0)$? For those pairs that do, write one of the two vectors as a scalar multiple of the other.

   (a) $X = (1, 2, 4)$, $Y = (-1, -4, -2)$

   (b) $X = (1, 2, 4)$, $Y = (3, 6, 12)$

   (c) $X = (1, 5, 2)$, $Y = (0, 0, 0)$

   (d) $X = (1, 0, 1)$, $Y = (2, 0, -2)$

   (e) $X = (1, 0, 2)$, $Y = (2, 1, 4)$

   (f) $X = (0,0,0)$, $Y = (3,-1,2)$

   (g) $X = (\frac{1}{3}, \frac{2}{7}, \frac{9}{5})$, $Y = (\frac{1}{4}, \frac{3}{14}, \frac{27}{20})$.

2. Observe that $X_1 = (1,1,0)$ and $X_2 = (0,1,1)$ span a two-dimensional linear subspace S of $\mathbb{R}^3$. Which of the following vectors are in S? Write those that are as linear combinations of X_1 and X_2. In this exercise the arithmetic is simple enough to obtain the solutions by inspection.

   (a) $(2,2,0)$

   (b) $(1,2,1)$

   (c) $(1,0,0)$

   (d) $(0,0,1)$

   (e) $(1,1,1)$

   (f) $(0,0,0)$

   (g) $(0,-1,-1)$

   (h) $(0,1,1)$

   (i) $(0,1,-1)$

   (j) $(1,-1,-2)$.

3. Verify that the three given vectors all lie on the same line through the origin in $\mathbb{R}^3$; that is, show that two of them are each scalar multiples of the remaining one.

   (a) $(2,0,8)$, $(1,0,4)$, $(-4,0,-16)$

   (b) $(0,0,0)$, $(1,-1,2)$, $(-2,2,-4)$

   (c) $(4,-4,4)$, $(-3,3,-3)$, $(5,-5,5)$.

4. Verify that the three given vectors all lie in a two-dimensional linear subspace (a plane through the origin) in $\mathbb{R}^3$, that is, write one of them as a linear combination of the other two. Again, it is preferable to do these by inspection. In Section 1.5, we will learn a more mechanical method.

   (a) $(1,1,0)$, $(1,0,1)$, $(3,1,2)$

   (b) $(0,0,0)$, $(1,1,1)$, $(1,1,-1)$

   (c) $(3,3,1)$, $(1,1,1)$, $(1,1,-1)$

   (d) $(3,-1,5)$, $(6,4,-17)$, $(12,2,-7)$.

5. In each case below let S be the linear subspace of $\mathbb{R}^3$ spanned by the given X_1, X_2, and X_3. Decide what type of linear subspace S is (e.g. a line, a plane). In the case of a line identify it as the linear span of one of the vectors X_1, X_2, and X_3. In the case of a plane identify it as the linear span of two of these three vectors.

   (a) $X_1 = (1,0,0)$, $X_2 = (1,1,0)$, $X_3 = (1,1,1)$

   (b) $X_1 = (0,0,0)$, $X_2 = (0,0,0)$, $X_3 = (0,0,0)$

   (c) $X_1 = (1,2,0)$, $X_2 = (2,1,0)$, $X_3 = (1,-1,0)$

   (d) $X_1 = (1,2,3)$, $X_2 = (-3,-6,-9)$, $X_3 = (2,4,6)$

    (e) $X_1 = (0, 1, 0)$, $X_2 = (0, 2, 0)$, $X_3 = (0, 3, 0)$

    (f) $X_1 = (1, 2, 3)$, $X_2 = (1, 2, -3)$, $X_3 = (1, 4, 0)$.

    (g) $X_1 = (1, 2, 3)$, $X_2 = (-1, -2, -3)$, $X_3 = (1, 4, 0)$.

6. For the homogeneous linear equations below (for all of which the variables are x_1, x_2, and x_3) compute two vectors X_1 and X_2 such that every solution $X = (x_1, x_2, x_3)$ may be written uniquely as $X = \alpha_1 X_1 + \alpha_2 X_2$ for suitable α_1 and α_2.

    (a) $x_3 = 0$

    (b) $x_1 + x_2 = 0$

    (c) $x_1 - x_2 + x_3 = 0$

    (d) $x_1 + x_2 + x_3 = 0$

    (e) $2x_1 - x_2 - x_3 = 0$

    (f) $4x_1 + 3x_2 - 2x_3 = 0$.

1.5 Systems of Equations

1.5A THE PROBLEM

Before we can learn more about $\mathbb{R}^3$ and its linear subspaces, we need to develop an efficient algorithm (procedure) for finding the set of solutions of a system of linear equations. This algorithm is called **Gauss-Jordan elimination**. It is simply an organized way of solving systems of equations by eliminating variables. A geometric interpretation of the results of this section will be given in Section 1.6.

1.5B ONE EQUATION, SEVERAL UNKNOWNS

Consider the equation

$$a_1 x_1 + a_2 x_2 + a_3 x_3 = b .$$

The constants a_1, a_2, and a_3 are real numbers, called the coefficients of the equation. The equation shown has three variables, also called unknowns, denoted by x_1, x_2, and x_3. The right-hand side of the equation, is a real constant called b. If $b = 0$, the equation is homogeneous. Even though there is only one equation, we may still speak of a "system" of one linear equation in three unknowns.

Solving a single linear equation in several unknowns is simple, and does not require any elimination of variables. In the equation given above, let us assume that the first coefficient a_1 is not zero. Then we can solve for x_1 in terms of the other two unknowns:

$$x_1 = \frac{b}{a_1} - \frac{a_2}{a_1} x_2 - \frac{a_3}{a_1} x_3 .$$

When the equation is put into this form, it is clear that the value of one of the unknowns (in this case x_1) is completely determined by the values of the other two unknowns. We may freely choose the values of x_2 and x_3, but once we do so, we have no choice as to the value of x_1.

EXAMPLES:

1. Let us determine the set of real numbers x_1, x_2, and x_3 such that $3x_1 - x_2 + 2x_3 = 5$. We solve the equation for x_1, and then express the solution set in the following form:

$$x_1 = \frac{5}{3} + \frac{1}{3}x_2 - \frac{2}{3}x_3 , \quad x_2, x_3 \text{ arbitrary} .$$

Note that we can equally well solve this equation for x_2, in which case the solution set takes the form

$$x_2 = -5 + 3x_1 + 2x_3 , \quad x_1, x_3 \text{ arbitrary} .$$

Despite the different formulas, both solution sets are identical. They both give the same set of values for x_1, x_2, and x_3 (you should check this). Yet another path leading to the same set of solutions would be to solve for x_3. In order to be systematic in our examples, we will usually solve our equations for the **leading variable**, that is, for the first unknown that appears in the equation with a nonzero coefficient.

2. Find the set of vectors (x_1, x_2, x_3) in $\mathbb{R}^3$ whose coordinates satisfy $x_2 + x_3 = 1$. We may think of this equation as a linear equation with three unknowns in which one of the unknowns has a zero coefficient:

$$0x_1 + x_2 + x_3 = 1 .$$

We solve the equation for the leading variable. Since the coefficient of x_1 is zero, the leading variable is x_2:

$$x_2 = 1 - x_3 , \quad x_1, x_3 \text{ arbitrary} .$$

Can you see why we are allowed to choose x_1 arbitrarily?

3. Let us express the solutions of the preceding two examples in vector form. For the first example, the vectors (x_1, x_2, x_3) whose coordinates satisfy the equation are of the form

$$(\frac{5}{3} + \frac{1}{3}x_2 - \frac{2}{3}x_3, x_2, x_3) ,$$

where x_2 and x_3 are arbitrary real numbers. This can be rewritten as

$$(\frac{5}{3}, 0, 0) + x_2(\frac{1}{3}, 1, 0) + x_3(-\frac{2}{3}, 0, 1) .$$

Alternatively, we can write an expression for the solution set:

$$\{X : X = (\frac{5}{3}, 0, 0) + \alpha(\frac{1}{3}, 1, 0) + \beta(-\frac{2}{3}, 0, 1) \text{ for some } \alpha \text{ and } \beta\}.$$

In the second example, the solutions are of the form

$$(x_1, x_2, x_3) = (x_1, 1 - x_3, x_3) = (0, 1, 0) + x_1(1, 0, 0) + x_3(0, -1, 1).$$

4. We will find the set of vectors (x_1, x_2, x_3) whose coordinates satisfy the homogeneous equation

$$2x_1 + x_2 - 4x_3 = 0.$$

Solving for the leading variable x_1, we obtain

$$x_1 = -\frac{1}{2}x_2 + 2x_3,$$

so

$$(x_1, x_2, x_3) = (-\frac{1}{2}x_2 + 2x_3, \ x_2, \ x_3) = x_2(-\frac{1}{2}, 1, 0) + x_3(2, 0, 1),$$

where x_2 and x_3 are arbitrary real numbers. The solution set is the set of all linear combinations of $(-1/2, 1, 0)$ and $(2, 0, 1)$, or in other words, it is the plane spanned by these two vectors (cf. Theorem 1.4.3).

5. Here is a trivial example that is nevertheless worth thinking about:

$$0x_1 + 0x_2 + 0x_3 = b.$$

If $b = 0$, the solution set is all of $\mathbb{R}^3$; otherwise, there are no solutions.

Note that in each of the previous examples except for the last, we have been able to write the solution set in vector form as the sum of a vector Z in $\mathbb{R}^3$ plus an arbitrary linear combination of two other vectors in $\mathbb{R}^3$. You should be able to convince yourself that this will always be the case when there is one equation in three unknowns, provided at least one of the coefficients in the equation is nonzero. In case of homogeneous equations, Z can be chosen equal to 0.

So far, our examples have involved $n = 3$ unknowns. There should be no difficulty in extending the method to other values of $n \geq 1$, provided that there is exactly one linear equation and that equation has at least one nonzero coefficient. In general, the solution set will be expressed in vector form as the sum of a vector Z and an arbitrary linear combination of $n - 1$ vectors $X_1, X_2, \ldots, X_{n-1}$ in $\mathbb{R}^n$. Try examples of your own with $n = 1, 2$, and 4.

1.5C TWO EQUATIONS, SEVERAL UNKNOWNS

Consider the system

$$a_{11}x_1 + a_{12}x_2 + a_{13}x_3 = b_1$$
$$a_{21}x_1 + a_{22}x_2 + a_{23}x_3 = b_2 .$$

This is a system of two linear equations in the three unknowns x_1, x_2, and x_3. The coefficients a_{ij} are real constants, as are the numbers b_1 and b_2. If $b_1 = b_2 = 0$, the system is homogeneous. The method of solution is to eliminate one of the variables from one of the equations, then eliminate a different variable from the other equation. We illustrate with an example.

EXAMPLE:

6. Let us solve the system

$$3x_1 + x_2 - x_3 = -1$$
$$-2x_1 - x_2 + 2x_3 = 0 .$$

We add the 2/3 times the first equation to the second equation to eliminate the variable x_1:

$$3x_1 + x_2 - x_3 = -1$$
$$-\frac{1}{3}x_2 + \frac{4}{3}x_3 = -\frac{2}{3} .$$

Next, add 3 times the second equation to the first to eliminate x_2:

$$3x_1 + 3x_3 = -3$$
$$-\frac{1}{3}x_2 + \frac{4}{3}x_3 = -\frac{2}{3} .$$

Then we solve each equation for its leading variable:

$$x_1 = -1 - x_3$$
$$x_2 = 2 + 4x_3 .$$

The solutions of this new system of equations can be expressed as

$$x_1 = -1 - x_3 , \quad x_2 = 2 + 4x_3 , \quad x_3 \text{ arbitrary} ,$$

or in vector form,

$$(x_1, x_2, x_3) = (-1 - x_3 , 2 + 4x_3 , x_3) = (-1, 2, 0) + x_3(-1, 4, 1) .$$

These are also the solutions of the original system of equations.

Let us review what we have just done. Starting with a system of linear equations, we carried out a series of steps, each of which produced a new system of linear equations. Since each step involves adding a multiple of one equation to another, any solution of the original system of equations is also a solution of each of the new systems of equations. The reverse is also true: a series of similar steps could be used to transform any of the new systems of equations back into the original system, so any solution of one of the new systems is also a solution of the original system. This is an important feature of the Gauss-Jordan elimination method: each step produces a system whose solutions are identical to the solutions of the original system. Furthermore, each time we complete a step of the method, the new system obtained is simpler (because of the elimination of variables) than the previous system, until finally we obtain a system whose solution may be written down immediately.

In our example, the values of the variables x_1 and x_2 are determined in the end by the value of the variable x_3, which may be chosen arbitrarily. We will find ourselves repeatedly solving for some of the variables in terms of the other variables, so it will be convenient to introduce some terminology. The variables that may be chosen arbitrarily are called **free variables**, and the remaining variables are called **basic variables**. Once the values of the free variables are chosen, the values of the basic variables are determined. In Example 6, x_3 is a free variable and x_1 and x_2 are basic variables. Of course, we can change which of the variables are free and which are basic by the manner in which we choose to carry out the Gauss-Jordan method. In Example 6, we could just as easily have made x_1 (or x_2) the free variable by carrying out the elimination procedure with the remaining two variables. In Example 1, we showed how to make either x_1 or x_2 the basic variable, with the remaining two variables being free. We could also have chosen x_3 for the basic variable. In Example 2, either x_2 or x_3 may be chosen as the basic variable, but not x_1. In Example 5, all three variables are free. As we will soon see, there are also cases in which there are no free variables.

EXAMPLES:

7. Let us solve a system of two equations in two unknowns.

$$2x_1 - x_2 = 0$$
$$x_1 + x_2 = 3.$$

Subtract 1/2 of the first equation from the second to obtain

$$2x_1 - x_2 = 0$$
$$\frac{3}{2}x_2 = 3.$$

Now add 2/3 of the second equation to the first:

$$2x_1 = 2$$
$$\frac{3}{2}x_2 = 3 .$$

Finally, solve each equation for its leading (and only) variable:

$$x_1 = 1 \quad \text{and} \quad x_2 = 2 .$$

This is the solution. It is unique — there are no free variables. Both x_1 and x_2 are basic variables. In vector form, the solution is simply

$$(x_1, x_2) = (1, 2) .$$

8. Here is a system of two equations in four unknowns:

$$x_1 - x_2 - x_4 = 3$$
$$x_1 + x_2 + 2x_3 = -1 .$$

Subtract the first equation from the second:

$$x_1 - x_2 - x_4 = 3$$
$$2x_2 + 2x_3 + x_4 = -4 ,$$

and then add 1/2 of the second equation to the first:

$$x_1 + x_3 - \frac{1}{2}x_4 = 1$$
$$2x_2 + 2x_3 + x_4 = -4 .$$

Finally, solve each equation for its leading variable:

$$x_1 = -x_3 + \frac{1}{2}x_4 + 1$$
$$x_2 = -x_3 - \frac{1}{2}x_4 - 2 .$$

The variables x_3 and x_4 are free, and x_1 and x_2 are basic. In vector form, the solutions are

$$(x_1, x_2, x_3, x_4) = (1, -2, 0, 0) + x_3(-1, -1, 1, 0) + x_4(\frac{1}{2}, -\frac{1}{2}, 0, 1)$$

with x_3 and x_4 arbitrary. The reader should choose various values for the free variables, calculate the corresponding values of the basic variables, and then plug those values into the original system to verify that they work. Even a better check is to plug in the general expressions for x_1 and x_2 and check that the equations are satisfied for all x_3 and x_4. Also, note that we chose to ignore a shortcut for this system: we could have solved the two equations immediately for x_3 and x_4, making them the basic variables. The same solution set is obtained, although its description is quite different.

After studying the examples, the reader should become convinced that the procedure given can be carried out in exactly the same way for almost any other system of two equations in several unknowns. Occasionally, however, the procedure must be modified to accommodate the unexpected appearance of zero as a coefficient. The following examples in three variables illustrate these exceptional cases.

EXAMPLES:

9. We will try to solve the system

$$-x_1 + x_2 + x_3 = 5$$
$$3x_1 - 3x_2 - 3x_3 = 2 .$$

We add three times the first equation to the second, obtaining

$$-x_1 + x_2 + x_3 = 5$$
$$0 = 17 .$$

When we eliminated the first variable from the second equation, it just happened that the other two variables were eliminated as well. Obviously, this new system has no solution. Therefore, the original system has no solution.

10. Consider a system with the same coefficients as in the preceding example, but with different values on the right-hand side:

$$-x_1 + x_2 + x_3 = 4$$
$$3x_1 - 3x_2 - 3x_3 = -12 .$$

Now when we add three times the first equation to the second, we obtain

$$-x_1 + x_2 + x_3 = 4$$
$$0 = 0 .$$

The second equation is obviously satisfied no matter what values the unknowns take, and so we may ignore it. We are left with a single equation in three unknowns, which can be solved as in Section 1.5b:

$$x_1 = -4 + x_2 + x_3 , \quad x_2, x_3 \text{ arbitrary} .$$

11. If the coefficient of x_1 is zero in one equation but not the other, then we only need to carry out one elimination step:

$$x_2 - 3x_3 = 5$$
$$-2x_1 + x_2 - 4x_3 = 0 .$$

If we subtract the first equation from the second and then solve for the leading variable in each, we obtain (after switching the order of the equations)

$$x_1 = \frac{5}{2} - \frac{1}{2}x_3$$
$$x_2 = 5 + 3x_3 \, .$$

The free variable is x_3. Its value determines the values of x_1 and x_2.

12. If both of the coefficients of x_1 are zero, then we must let x_1 be a free variable. For example, if we are interested in the set of vectors (x_1, x_2, x_3) such that

$$x_2 - x_3 = 0$$
$$2x_2 + x_3 = 1 \, ,$$

we first eliminate x_2 from the second equation by subtracting twice the first equation from the second:

$$x_2 - x_3 = 0$$
$$3x_3 = 1 \, .$$

Next we add 1/3 of the second equation to the first equation to eliminate x_3, leading to the following solution set:

$$x_1 \text{ arbitrary} \quad x_2 = \frac{1}{3} \quad x_3 = \frac{1}{3} \, .$$

13. The reader should also consider the case in which all the coefficients of x_1 and x_2 are zero, as well as the case in which all the coefficients of x_1, x_2, and x_3 are zero!

It is important to realize that Examples 6-8 are the most typical; the principles they illustrate should be mastered first. Examples 9-13 represent those situations in which the basic pattern must be altered somewhat because of special circumstances.

1.5D THREE OR MORE EQUATIONS IN THREE OR MORE UNKNOWNS

It is quite simple to extend the methods of the preceding sections to systems of three or more equations with three or more unknowns. Here is a description of the complete algorithm in the case of three unknowns, which we will denote by x_1, x_2, and x_3.

STEP 1: Determine whether x_1 appears as a leading variable (that is, with a nonzero coefficient) in any of the equations. If not, skip to Step 2.

Otherwise, choose one of the equations in which x_1 is the leading variable, and add appropriate multiples of that equation to the other equations to eliminate x_1 from them.

STEP 2: After Step 1, at most one of the equations contains the variable x_1. Determine whether x_2 appears as the leading variable in any of the remaining equations. If not, skip to Step 3. Otherwise choose one of the equations in which x_2 is the leading variable, and add appropriate multiples of that equation to the other equations to eliminate x_2 from them.

STEP 3: After Step 2, there is at most one equation containing x_1 and at most one equation containing x_2. Determine whether x_3 appears as the leading variable in any of the remaining equations. If not, skip to the final step. Otherwise, choose one of the equations in which x_3 is the leading variable, and add appropriate multiples of that equation to the other equations to eliminate x_3 from them.

FINAL STEP: If any of the equations produced by the preceding steps is of the form $0 = c$ for some real number $c \neq 0$, then there is *no solution*. Otherwise, discard all equations of the form $0 = 0$. Solve each of the remaining equations for its leading variable. These leading variables are the basic variables. The other variables are the free variables. The values of the free variables may be chosen arbitrarily. The values of the basic variables are determined by the values of the free variables.

Once this procedure is understood, it should be easy to see how to handle systems of equations with more than three unknowns. In general, if there are n unknowns, the procedure will include $n + 1$ steps. Steps 1 through n will involve elimination, with one step for each unknown. The Final Step will be exactly as above.

EXAMPLES:

14. We will solve the system

$$2x_1 - 6x_2 = 3$$
$$-x_1 + x_2 - 2x_3 = -1$$
$$3x_1 - 5x_2 + 2x_3 = 9 .$$

Step 1 asks us to determine if x_1 is the leading variable in any of the equations. It appears in all three. We choose the first one, and eliminate x_1 from the other two:

$$2x_1 - 6x_2 = 3$$
$$-2x_2 - 2x_3 = \frac{1}{2}$$
$$4x_2 + 2x_3 = \frac{9}{2} .$$

Following the instructions of Step 2, we note that x_2 is the leading variable in the second and third equations; we choose the second and eliminate x_2 from the first and third:

$$2x_1 + 6x_3 = \frac{3}{2}$$
$$-2x_2 - 2x_3 = \frac{1}{2}$$
$$-2x_3 = \frac{11}{2}.$$

In Step 3, we eliminate x_3 from the first two equations:

$$2x_1 = 18$$
$$-2x_2 = -5$$
$$-2x_3 = \frac{11}{2}.$$

Now we solve each equation for its leading variable. The unique solution of the system is

$$(x_1, x_2, x_3) = (9, \frac{5}{2}, -\frac{11}{4}).$$

There are no free variables.

15. We solve the system

$$-x_2 + 2x_3 = 1$$
$$x_1 + x_3 = 5$$
$$2x_1 + x_2 = 0.$$

Step 1 asks us to find an equation in which x_1 appears. We notice that x_1 appears in the second equation. To keep things organized, we switch the first two equations. We then eliminate x_1 from the third equation to obtain

$$x_1 + x_3 = 5$$
$$-x_2 + 2x_3 = 1$$
$$x_2 - 2x_3 = -10.$$

We now move on to Step 2. Upon examining the remaining two equations, we find that the variable x_2 appears in both of them. Add the second equation to the third to eliminate x_2 from the third. We obtain the system

$$x_1 + x_3 = 5$$
$$-x_2 + 2x_3 = -1$$
$$0 = -9.$$

In Step 3 we find that x_3 does not appear as the leading variable of any of the equations. Thus we move on to the Final Step. In this step we note that our system contains the equation $0 = -9$, so we conclude that there is no solution.

16. Let us solve a homogeneous system of three equations in five unknowns:

$$-x_2 + 2x_3 + x_4 + x_5 = 0$$
$$2x_1 + x_3 - x_4 - 2x_5 = 0$$
$$x_2 - 2x_3 + 3x_4 = 0.$$

Since x_1 only appears in one equation, Step 1 is already completed, although it is perhaps useful to switch the first two equations to keep things orderly:

$$2x_1 + x_3 - x_4 - 2x_5 = 0$$
$$-x_2 + 2x_3 + x_4 + x_5 = 0$$
$$x_2 - 2x_3 + 3x_4 = 0.$$

To complete Step 2, add the second equation to the third to eliminate x_2:

$$2x_1 + x_3 - x_4 - 2x_5 = 0$$
$$-x_2 + 2x_3 + x_4 + x_5 = 0$$
$$4x_4 + x_5 = 0.$$

The variable x_3 is not a leading variable in any of the equations, so we can move onto Step 4 (not the Final Step, since we have five unknowns), in which we eliminate x_4 from the first two equations:

$$2x_1 + x_3 - \frac{7}{4}x_5 = 0$$
$$-x_2 + 2x_3 + \frac{3}{4}x_5 = 0$$
$$4x_4 + x_5 = 0.$$

Since x_5 is not a leading variable in any of the equations, we have now completed the five elimination steps. In the Final Step, we solve each equation for its leading variable:

$$x_1 = -\frac{1}{2}x_3 + \frac{7}{8}x_5$$
$$x_2 = 2x_3 + \frac{3}{4}x_5$$
$$x_4 = -\frac{1}{4}x_5.$$

The basic variables are x_1, x_2, and x_4, and the free variables are x_3 and x_5. In vector form, the solutions can be written as

$$(x_1, x_2, x_3, x_4, x_5) = x_3(-\frac{1}{2}, 2, 1, 0, 0) + x_5(\frac{7}{8}, \frac{3}{4}, 0, -\frac{1}{4}, 1) ,$$

where x_3 and x_5 are arbitrary. It is common to rewrite this as

$$(x_1, x_2, x_3, x_4, x_5) = \alpha(-\frac{1}{2}, 2, 1, 0, 0) + \beta(\frac{7}{8}, \frac{3}{4}, 0, -\frac{1}{4}, 1) ,$$

where α and β are arbitrary scalars. Some people would replace the arbitrary numbers x_3 and x_5 by symbols such as α and β as we have just done, so as to place the original five unknowns on "equal footing"; others see no reason for such a proliferation of symbols. Note that the solution set is a plane in $\mathbb{R}^5$. We can describe the solution set accurately and concisely as the span of $(-\frac{1}{2}, 2, 1, 0, 0)$ and $(\frac{7}{8}, \frac{3}{4}, 0, -\frac{1}{4}, 1)$. In describing it in this manner one might want to add the assertion that the solution set is a plane, thereby saving the reader the trouble of checking whether one of the spanning vectors is a multiple of the other.

17. The system

$$\begin{aligned} -x_2 + 2x_3 + x_4 + x_5 &= 1 \\ 2x_1 + x_3 - x_4 - 2x_5 &= -2 \\ x_2 - 2x_3 + 3x_4 &= 7 \end{aligned}$$

is identical to the one in the preceding example, except that the quantities on the right-hand side of each equation are not zero. If we carry out the identical Gauss-Jordan steps as in that example, we obtain the following solutions:

$$\begin{aligned} &(x_1, x_2, x_3, x_4, x_5) \\ &= (0, 1, 0, 2, 0) + \alpha(-\frac{1}{2}, 2, 1, 0, 0) + \beta(\frac{7}{8}, \frac{3}{4}, 0, -\frac{1}{4}, 1) , \end{aligned}$$

where α and β are arbitrary scalars. The only difference between the solutions of the two examples is the vector $(0, 1, 0, 2, 0)$. We will have more to say about this observation in Section 1.6.

REMARKS:

- You may have noticed that in the preceding examples there was a considerable amount writing of the variables $x_1, x_2, \ldots$ that only served to "hold" the places of the various coefficients. In Chapter 3 we will introduce a notation that will cut down on the amount of repetitive writing in Gauss-Jordan elimination. Your teacher may choose to introduce that notation earlier. We are concerned that notational shortcuts too quickly used in one's studies may hide major ideas before those ideas are fully understood.

- It should be clear from the examples that in every case in which there is a solution, there exist vectors Z and $X_1, \ldots, X_k$ such that the solution may be expressed in vector form as

$$Z + \alpha_1 X_1 + \ldots + \alpha_k X_k,$$

where $\alpha_1, \ldots, \alpha_k$ are arbitrary scalars. There is one arbitrary scalar for each free variable, that is, k equals the number of free variables.

- In Example 15, as well as in some of the examples given earlier, there is no solution. This situation cannot occur in a system of homogeneous equations in n unknowns, because such a system always has at least the **trivial solution** $x_1 = 0, x_2 = 0, \ldots, x_n = 0$.

- It is apparent from the method that whenever a solution exists, there cannot be more basic variables than there are equations. Therefore, a system of m equations in more than m unknowns either has no solution, or else there must be at least one free variable, and thus infinitely many solutions. In particular, a system of m homogeneous linear equations in more than m unknowns always has infinitely many solutions, since it has the trivial solution. This fact is quite useful, and deserves to be stated as a theorem.

Theorem 1.5.1 A system of m homogeneous linear equations in more than m unknowns has infinitely many solutions.

1.5E   SOME APPLICATIONS

Let us use what we have learned to tie up a few loose ends that were left in preceding sections. First, we will fill in a gap that was left in the proof of Lemma 1.4.1, namely, we will prove that the intersection of two planes through the origin in $\mathbb{R}^3$ must contain a line.

PROOF: Let S_1 be spanned by vectors X and Y, and let S_2 be spanned by vectors U and V. We wish to find a nonzero vector W that can be written as a linear combination of X and Y and also as a linear combination of U and V. In other words, we are looking for scalars α, β, γ, and δ, not all zero, such that $\alpha X + \beta Y = \gamma U + \delta V$. If we express this equality in terms of the coordinates of X, Y, U, and V, we obtain the following three equations:

$$\begin{aligned}
\alpha x_1 + \beta y_1 &= \gamma u_1 + \delta v_1 \\
\alpha x_2 + \beta y_2 &= \gamma u_2 + \delta v_2 \\
\alpha x_3 + \beta y_3 &= \gamma u_3 + \delta v_3 \,.
\end{aligned}$$

The coordinates of X, Y, U, and V are "knowns", since these vectors are given to us. The unknowns in these equations are the scalars α, β, γ, and δ. Let us rewrite the equations to emphasize this point:

$$x_1\alpha + y_1\beta - u_1\gamma - v_1\delta = 0$$
$$x_2\alpha + y_2\beta - u_2\gamma - v_2\delta = 0$$
$$x_3\alpha + y_3\beta - u_3\gamma - v_3\delta = 0$$

We have a system of three homogeneous linear equations in four unknowns. Since there are more unknowns than equations, the system has infinitely many solutions (Theorem 1.5.1). In particular, we may choose α, β, γ, and δ, not all zero, to satisfy the equations. The corresponding vector $W = \alpha X + \beta Y = \gamma U + \delta V$ lies in both planes. The intersection of two linear subspaces is a linear subspace, so any scalar multiple of W also lies in both planes: the intersection of the two planes S_1 and S_2 contains the line spanned by W. Done. $<<$

REMARKS:

- We usually think of the unknowns in a system of equations as the coordinates of some vector. But in the proof just given, the unknowns were originally the scalars in certain linear combinations involving the four vectors X, Y, U, and V. The coordinates of these four vectors ended up as the coefficients of the system of equations. Thus we have quantities which play the roles of coefficients in one expression and the roles of unknowns or vector coordinates in another expression. This duality, which may at first be confusing to the reader, occurs frequently in applications of linear equations to vector geometry. See also the proof of Theorem 1.5.2 below.

- Our argument involved a system of three equations in the four unknowns α, β, γ, and δ. We have taken the point of view that the solution set of such a system is a subset of $\mathbb{R}^4$, that is, we think of the unknowns as the four coordinates of a point in $\mathbb{R}^4$. Thus, we have used some algebraic properties of $\mathbb{R}^4$ to prove a geometric fact about $\mathbb{R}^3$.

For our next application, we state and prove the converse to Theorem 1.4.3.

Theorem 1.5.2 Let S be a plane through the origin in $\mathbb{R}^3$. Then S is the solution set of a homogeneous linear equation in three variables with at least one nonzero coefficient.

PROOF: By definition, S is the set of all linear combinations of two vectors Y and Z in $\mathbb{R}^3$ which do not lie on the same line through the origin. Write $Y = (y_1, y_2, y_3)$ and $Z = (z_1, z_2, z_3)$. We are looking for an equation of the form

$$a_1 x_1 + a_2 x_2 + a_3 x_3 = 0$$

with at least one nonzero coefficient such that both Y and Z are solutions. In other words, we want to find a_1, a_2, and a_3, not all zero, such that

$$\begin{aligned} a_1 y_1 + a_2 y_2 + a_3 y_3 &= 0 \\ a_1 z_1 + a_2 z_2 + a_3 z_3 &= 0 \, . \end{aligned}$$

This is a system of two equations in the three unknowns a_1, a_2, and a_3. By Theorem 1.5.1 we know that such a system has infinitely many solutions. In particular, there is a solution for which a_1, a_2, and a_3 are not all zero.

We have shown that there exists a homogeneous equation with at least one nonzero coefficient whose solution set includes Y and Z, and our proof indicates that such an equation can be found by using Gauss-Jordan elimination on an appropriate system of two equations in three unknowns. It remains be seen that every solution to the equation is in the plane spanned by Y and Z. Let V be a vector that is not in the plane spanned by Y and Z. By Lemma 1.4.1, the linear span of V, Y, and Z is all of $\mathbb{R}^3$. Since the solution set of a homogeneous equation in three unknowns is a linear subspace of $\mathbb{R}^3$, any equation satisfied by V, Y, and Z must be satisfied by every vector in $\mathbb{R}^3$. But it is easily seen that a homogeneous equation in three unknowns with at least one nonzero coefficient cannot be satisfied by every vector in $\mathbb{R}^3$; if $a_j \neq 0$, then the equation $a_1 x_1 + a_2 x_2 + a_3 x_3 = 0$ is not satisfied by $\mathbf{e}_j$. Done. $<<$

REMARKS:

- At the end of the proof, we showed that $\mathbb{R}^3$ is not the solution set of a homogeneous linear equation in three unknowns with at least one nonzero coefficient. It follows from Theorem 1.5.2 that $\mathbb{R}^3$ is not a plane. This fact may seem obvious, but remember that we have defined a plane in a certain precise way, and until now, we had not proved that $\mathbb{R}^3$ fails to satisfy that definition.

- Consider a plane S in $\mathbb{R}^3$ spanned by two vectors W and X and let Y and Z be two members of S. Suppose that neither Y nor Z is a multiple of each other. Let T denote the span of Y and Z. By now the reader will certainly suspect that $T = S$. Let us now prove this. What is certainly true is that $T \subseteq S$. If W and X are both members of T, then certainly T, containing all linear combinations of W and X, equals S. So, suppose that W or X —say W —does not belong to T. By Lemma 1.4.1, every member of $\mathbb{R}^3$ is a linear combination of W, Y, and Z. Since Y and Z are themselves linear combinations of W and X, we obtain the consequence that W and X span $\mathbb{R}^3$. We have the desired contradiction since the preceding remark shows that $\mathbb{R}^3$ is not a plane in disguise.

- Let us summarize some important ideas resulting from the discussion beginning in Section 1.4. There are four types of linear subspaces of $\mathbb{R}^3$: $\{\mathbf{0}\}$, lines through $\mathbf{0}$, planes through $\mathbf{0}$, and $\mathbb{R}^3$. A line through $\mathbf{0}$ is spanned by any one nonzero vector belonging to it. A plane through $\mathbf{0}$ is spanned by any two vectors that belong to it and are non-collinear with the origin. And $\mathbb{R}^3$ itself is spanned by any three vectors that do not lie on a common plane through the origin. There are also important facts relating solution sets of systems of equations to linear subspaces. The theorem preceding this sequence of remarks says that any plane through the origin in $\mathbb{R}^3$ is the solution set of an appropriate single homogeneous linear equation. It will be seen in the exercises that any line through the origin in $\mathbb{R}^3$ is the solution set of an appropriate system of two linear homogenous equations, and thus, that it is the intersection of two planes. But the line cannot be the solution set of just one such equation. The trivial linear subspace $\{\mathbf{0}\}$ of $\mathbb{R}^3$ is the solution set of an appropriate system of three homogeneous linear equations, but not the solution set of any system of just two such equations. (The homogeneous linear equations described in the last several sentences are not unique.)

EXAMPLE:

18. Let $\mathcal{S}$ be the plane spanned by $(1, 1, 3)$ and $(0, -1, 2)$. Then $\mathcal{S}$ is the solution set of the homogeneous linear equation whose coefficients satisfy

$$\begin{aligned} a_1 + a_2 + 3a_3 &= 0 \\ -a_2 + 2a_3 &= 0. \end{aligned}$$

Applying the Gauss-Jordan methods learned so far, we find that

$$a_1 = -5a_3 \quad \text{and} \quad a_2 = 2a_3.$$

If we let the free variable a_3 equal 1, then $a_1 = -5$ and $a_2 = 2$. Thus, $\mathcal{S}$ is the solution set of

$$-5x_1 + 2x_2 + x_3 = 0.$$

(Of course, there are many choices for a_3, and hence for the homogeneous linear equation whose solution set is $\mathcal{S}$. Different equations may have identical solution sets.)

Exercises

Systems of equations and their relationship to other concepts.

1. Decide whether the linear subspaces of solutions of each of the homogeneous systems below is a line or a plane in $\mathbb{R}^3$. Accordingly, in each case compute a vector X_1 or vectors X_1 and X_2 that span the linear subspace. Also, describe the solutions as vectors with one or two arbitrary constants:

(a)
$$\begin{aligned} x_1 &= 0 \\ x_3 &= 0 \, ; \end{aligned}$$

(b)
$$\begin{aligned} x_1 - x_3 &= 0 \\ x_1 + x_3 &= 0 \, ; \end{aligned}$$

(c)
$$\begin{aligned} x_1 - x_2 &= 0 \\ x_1 + x_2 + x_3 &= 0 \, ; \end{aligned}$$

(d)
$$\begin{aligned} x_1 - x_2 &= 0 \\ -x_1 + x_2 &= 0 \, ; \end{aligned}$$

(e)
$$\begin{aligned} 2x_1 - 4x_2 - 6x_3 &= 0 \\ -x_1 + 2x_2 - 3x_3 &= 0 \, ; \end{aligned}$$

(f)
$$\begin{aligned} 2x_1 - 4x_2 - 6x_3 &= 0 \\ -x_1 + 2x_2 + 3x_3 &= 0 \, ; \end{aligned}$$

(g)
$$\begin{aligned} (3\sqrt{2}+2)x_1 - (\sqrt{2}+1)x_2 + x_3 &= 0 \\ (4 - \sqrt{2})x_1 - x_2 + (\sqrt{2}-1)x_3 &= 0 \, . \end{aligned}$$

2. Solve these systems for the a's (see Exercise 4 also):

(a)
$$\begin{aligned} a_1 - a_2 + a_3 &= 0 \\ 2a_1 + a_2 + a_3 &= 0 \, ; \end{aligned}$$

(b)
$$\begin{aligned} 3a_1 + a_2 &= 0 \\ a_1 + a_3 &= 0 \, . \end{aligned}$$

3. Solve the following systems, expressing the solutions in vector form:

(a)
$$\begin{aligned} -x_1 - x_2 &= 3 \\ x_1 + 3x_2 &= -1 \, ; \end{aligned}$$

(b)
$$\begin{aligned} x_2 - x_3 &= 0 \\ x_1 + x_2 + x_3 &= 2 \, ; \end{aligned}$$

(c)
$$\begin{aligned} x_1 - 2x_2 + x_4 &= 5 \\ 2x_1 - x_4 &= 0 \, ; \end{aligned}$$

(d)
$$\begin{aligned} x_1 - x_3 &= 3 \\ x_2 + 3x_3 &= -2 \\ 2x_1 - x_2 - 5x_3 &= 1 \, ; \end{aligned}$$

(e)
$$\begin{aligned} x_1 - x_3 &= 3 \\ x_2 + 3x_3 &= 5 \\ 2x_1 - x_2 - 5x_3 &= 1 \, ; \end{aligned}$$

(f)
$$\begin{aligned} x_1 - x_3 &= 3 \\ x_2 + 3x_3 &= 5 \\ 2x_1 - x_2 - 4x_3 &= 1 \, ; \end{aligned}$$

(g) $3x_1 - 2x_2 + 4x_3 - x_4 + 2x_5 = 1$.

4. Find a linear equation $a_1x_1 + a_2x_2 + a_3x_3 = 0$ (compute the a's) that determines the plane spanned by the given X_1 and X_2:

   (a) $X_1 = (1, -1, 1)$, $X_2 = (2, 1, 1)$
   (b) $X_1 = (3, 1, 0)$, $X_2 = (1, 0, 1)$.

5. Let S_1 be the plane spanned by $(0, 1, -1)$ and $(3, 0, 1)$ and let S_2 be the plane spanned by $(-2, 2, 1)$ and $(0, 0, 1)$. Find a vector X such that the line spanned by X is contained in the intersection of S_1 and S_2.

6. For each equation, compute two solutions which do not line on the same line through the origin; that is, find two solutions, neither of which is a scalar multiple of the other (see Exercise 7 also):

   (a) $a_1 + 2a_2 + a_3 = 0$
   (b) $a_1 - a_2 - 3a_3 = 0$.

7. From lines to equations. For each given vector X_0, find a system of two homogeneous linear equations in x_1, x_2, x_3 whose space of solutions is the one-dimensional linear subspace spanned by X_0 (use Exercise 6):

   (a) $X_0 = (1, 2, 1)$
   (b) $X_0 = (1, -1, -3)$,

8. Invent systems of linear equations with the following properties:

   (a) two inhomogeneous equations, three unknowns, no solution;
   (b) two inhomogeneous equations, three unknowns, with at least one solution;
   (c) three inhomogeneous equations, two unknowns, with at least one solution;
   (d) one inhomogeneous equation in x_1, x_2, and x_3 with no solution;
   (e) three homogeneous equations, two unknowns, with an infinite number of distinct solutions.

9. Show that any line through the origin in $\mathbb{R}^3$ is the solution space of a system of two homogeneous linear equations.

1.6 Affine Subspaces

1.6A SOME DEFINITIONS

So far the lines, planes, and so on that we have discussed have all contained the origin of $\mathbb{R}^n$; they have all been linear subspaces. We know from single-variable calculus and other sources, however, that we often must consider lines that do not necessarily contain the origin. The same is true of planes, and so on in higher dimensions, as we will see when we study calculus of functions of more than one variable.

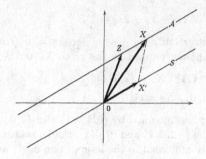

FIGURE 1.13. $X = X' + Z \quad A = S + Z$

Let us take a look at what we know about lines in $\mathbb{R}^2$. The equation of a line in $\mathbb{R}^2$ is

$$a_1 x_1 + a_2 x_2 = b \,,$$

where a_1, a_2, and b are constants and a_1 and a_2 are not both zero. You have learned methods in previous courses for determining exactly the line to which the equation corresponds, for example, by finding the slope and x- or y-intercept. Furthermore, given any line in $\mathbb{R}^2$, you learned how to find a corresponding equation. The slope is determined by the ratio between a_1 and a_2, so lines corresponding to the same values of a_1 and a_2 but different values of b have the same slope, that is, they are parallel.

Now let us take the linear algebra point of view. If we solve the equation $a_1 x_1 + a_2 x_2 = b$, assuming that $a_1 \neq 0$, we obtain the solution set $x_1 = b/a_1 - (a_2/a_1)x_2$, with x_1 as the basic variable and x_2 as the free variable. In vector form, the solution set is

$$\{\alpha(-a_2/a_1, 1) + (b/a_1, 0) : \alpha \in \mathbb{R}\} \,.$$

That is, any vector which satisfies the equation can be obtained by adding the vector $Z = (b/a_1, 0)$ to the vectors on the line spanned by $X' = (-a_2/a_1, 1)$. If we write S for the one-dimensional linear subspace spanned by X' and A for the solution set, it is natural to write $A = S + Z$. The set A is a line which is parallel to S, with the vector Z indicating the shift between the two lines. See Figure 1.13. Note that if $b = 0$, then $Z = \mathbf{0}$, in which case the two lines are the same.

These observations in $\mathbb{R}^2$ lead to the following notation and definition. If S is a subset of $\mathbb{R}^n$ and Z is a vector in $\mathbb{R}^n$, then we write

$$S + Z = \{X' + Z : X' \in S\} \,.$$

A subset A of $\mathbb{R}^n$ is called an **affine subspace** if and only if A is of the form $S + Z$ for some linear subspace S and some vector Z in $\mathbb{R}^n$. In this case, A and S are said to be **parallel**. We call A a **line** if S is a line through the origin, or a **plane** if S is a plane through the origin. Figure 1.14 shows

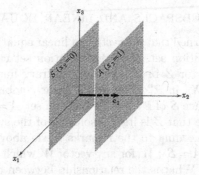

FIGURE 1.14. $\mathcal{A} = \mathcal{S} + \mathbf{e}_2$

the affine subspace $\mathcal{A}$ in $\mathbb{R}^3$ corresponding to $Z = \mathbf{e}_2$ and $\mathcal{S}$ equal to the solution set of $x_2 = 0$. Note that $\mathcal{A}$ is the solution set of $x_2 = 1$.

REMARKS:

- Since $\mathbf{0}$ is a member of $\mathcal{S}$, it follows that Z is a member of $\mathcal{S} + Z$.

- If Z happens to be a member of $\mathcal{S}$ then $\mathcal{S} + Z = \mathcal{S}$, since $\mathcal{S}$ is closed under vector addition.

- If W is a member of $\mathcal{S} + Z$, then $W = X + Z$ for some $X \in \mathcal{S}$, and so $W - Z = X$ is a member of $\mathcal{S}$.

- Consider affine subspaces $\mathcal{S} + Y$ and $\mathcal{T} + Z$ of $\mathbb{R}^n$, where $\mathcal{S}$ and $\mathcal{T}$ are linear subspaces. We ask: What are necessary and sufficient conditions for the two affine subspaces to be the same? An answer is this: $\mathcal{S}$ must equal $\mathcal{T}$ and Z must be a member of $\mathcal{S} + Y$. The proof that these two conditions imply that the affine spaces are identical is left for Exercise 8. Here we prove the other direction; so, we assume that $\mathcal{S} + Y = \mathcal{T} + Z$. Then, since $\mathbf{0} \in \mathcal{T}$, we have $Z \in \mathcal{T} + Z = \mathcal{S} + Y$, and therefore, $Z - Y \in \mathcal{S}$. It remains to prove that $\mathcal{S} = \mathcal{T}$. Let $W \in \mathcal{T}$. Then $W + Z \in \mathcal{S} + Y$. Therefore, there is a member V of $\mathcal{S}$ such that $W + Z = V + Y$ or, equivalently, $W = V - (Z - Y)$, the difference of two members of $\mathcal{S}$. Therefore, $\mathcal{T} \subseteq \mathcal{S}$. By symmetry (or by a similar argument), $\mathcal{S} \subseteq \mathcal{T}$. Hence, $\mathcal{S} = \mathcal{T}$, as desired.

- If Z is not a member of $\mathcal{S}$, then $\mathcal{S}$ and $\mathcal{S} + Z$ have no elements in common; for, if W were a member of both $\mathcal{S}$ and $\mathcal{S} + Z$, the preceding two remarks would imply that $\mathcal{S} = \mathcal{S} + W = \mathcal{S} + Z$, contradicting the assumption that Z is not a member of $\mathcal{S}$. In other words, if Z is not in $\mathcal{S}$ the two spaces $\mathcal{S}$ and $\mathcal{S} + Z$ do not meet.

- If Z is not a member of $\mathcal{S}$, then since $\mathcal{S}$ and $\mathcal{S} + Z$ have no elements in common, $\mathbf{0}$ is not a member of $\mathcal{S} + Z$. Thus in this case the affine subspace $\mathcal{S} + Z$ is not a linear subspace.

1.6B AFFINE SUBSPACES AND LINEAR EQUATIONS

In Section 1.5, we learned that if a system of linear equations in n unknowns has a nonempty solution set, then that solution set can be expressed in vector form as a vector Z in $\mathbb{R}^n$ plus an arbitrary linear combination of vectors $X_1, X_2, \ldots, X_k$ in $\mathbb{R}^n$. The set of linear combinations of the X's forms a linear subspace S of $\mathbb{R}^n$, so the solution set $A = S + Z$ is an affine subspace of $\mathbb{R}^n$. Note that Z is itself a solution of the system of equations, and furthermore, according to the remarks made above, the solution set could be written as $A = S + W$ for any vector W which is a solution of the system of equations. What is the relationship between the linear subspace S and the system of equations? The following example will help us answer this question.

EXAMPLE:

1. Let us use Gauss-Jordan elimination on the pair of equations

$$
\begin{aligned}
x_1 - 2x_2 + x_3 + x_4 &= -2 \\
-x_1 + x_2 - x_3 + x_4 &= 1 .
\end{aligned}
$$

Add the first equation to the second, then subtract twice the resulting equation from the first. Solve for the basic variables x_1 and x_2 in terms of the free variables x_3 and x_4. In vector form, the solutions are

$$
(x_1, x_2, x_3, x_4) = (0, 1, 0, 0) + \alpha(-1, 0, 1, 0) + \beta(3, 2, 0, 1) ,
$$

where α and β are arbitrary scalars. The vectors $(-1, 0, 1, 0)$ and $(3, 2, 0, 1)$ are non-collinear with the origin, so they determine a plane S in $\mathbb{R}^4$. The solution set is the plane $A = S + (0, 1, 0, 0)$. Note that $(0, 1, 0, 0)$ is a solution of the system. Now let us carry out the *same* sequence of Gauss-Jordan steps on the corresponding homogeneous system of equations:

$$
\begin{aligned}
x_1 - 2x_2 + x_3 + x_4 &= 0 \\
-x_1 + x_2 - x_3 + x_4 &= 0 .
\end{aligned}
$$

It should come as no surprise that the vector form of the solution set is

$$
(x_1, x_2, x_3, x_4) = \alpha(-1, 0, 1, 0) + \beta(3, 2, 0, 1) ,
$$

just the same as with the inhomogeneous system, except that the vector $(0, 1, 0, 0)$ does not appear. The linear subspace S is the solution set of the homogeneous system.

The preceding example (see also Example 17 in Section 1.5d) illustrates the fact that if a sequence of Gauss-Jordan elimination steps leads to a solution set of the form $Z + \alpha_1 X_1 + \ldots + \alpha_k X_k$ for some inhomogeneous system of linear equations, then the same sequence of steps will lead to $\alpha_1 X_1 + \ldots + \alpha_k X_k$ as the solution set of the corresponding homogeneous system. We summarize our discussion in the following theorem.

Theorem 1.6.1 The solution set of a system of linear equations may be written as the affine subspace $S + Z$, where Z is any particular solution of the system, and S, a linear subspace, is the solution set of the corresponding homogeneous system of linear equations.

In view of the preceding theorem, the solution set of a system of linear equations is also called its **solution space**. The term "space" is used in place of "set" in mathematics in certain situations in which the set of interest has some special structure.

1.6C PLANES IN $\mathbb{R}^3$

The preceding theorem relates the solution sets of inhomogeneous systems to the solution spaces of homogeneous systems. We found earlier, in Theorems 1.4.3 and 1.5.2, that there is a close relationship between the solution spaces of homogeneous equations in three unknowns and planes through the origin in $\mathbb{R}^3$. If we combine the three theorems, we obtain the following theorem.

Theorem 1.6.2 A subset $\mathcal{A}$ of $\mathbb{R}^3$ is a plane if and only if $\mathcal{A}$ is the affine subspace consisting of the solutions of a linear equation $a_1 x_1 + a_2 x_2 + a_3 x_3 = b$ with at least one of the coefficients a_1, a_2, and a_3 different from zero. Moreover, in this case, $\mathcal{A}$ is parallel to the two-dimensional linear subspace S of solutions of the homogeneous equation $a_1 x_1 + a_2 x_2 + a_3 x_3 = 0$, that is, $\mathcal{A} = S + Z$, where Z is any solution of $a_1 x_1 + a_2 x_2 + a_3 x_3 = b$.

EXAMPLE:

2. Let us find an equation representing the plane $S + Z$, where $Z = (3, 2, 1)$ and S is the plane through the origin that contains $(0, 1, -3)$ and $(1, 4, -1)$. We first find an equation representing S, as in Section 1.5, by solving the system

$$
\begin{aligned}
a_2 - 3a_3 &= 0 \\
a_1 + 4a_2 - a_3 &= 0
\end{aligned}
$$

for the unknowns a_1, a_2, a_3. Eliminate a_2 from the second equation to obtain the following solution set:

$$a_1 = -11a_3 , \quad a_2 = 3a_3 , \quad a_3 \text{ arbitrary} .$$

We may choose $a_3 = 1$, which gives $a_1 = -11$ and $a_2 = 3$. Thus, S is the set of vectors (x_1, x_2, x_3) such that $-11x_1 + 3x_2 + x_3 = 0$. Now we want to find a inhomogeneous equation with the same coefficients such that Z is a solution. Plugging $Z = (z_1, z_2, z_3) = (3, 2, 1)$ into the left-hand side of the homogeneous equation, we get

$$(-11 \times 3) + (3 \times 2) + (1 \times 1) = -26 \,,$$

so $\mathcal{S} + Z$ is the set of solutions of

$$-11x_1 + 3x_2 + x_3 = -26 \,.$$

Since the solution set of a single linear equation in three unknowns is a plane in $\mathbb{R}^3$, the solution set of a system of k linear equations in three unknowns must be the intersection of k planes in $\mathbb{R}^3$. A review of the examples in Section 1.5 now leads to the following observations.

REMARKS:

- A typical system of two equations in three unknowns has a solution set with one free variable, and so the vector form of the solution set is $Z + \alpha V$ for some nonzero vector V. In other words, two typical planes in $\mathbb{R}^3$ intersect in a line.

- A typical system of three equations in three unknowns has a solution set with no free variables, that is, the solution consists of a single vector Z: three planes in $\mathbb{R}^3$ typically intersect in a point. Note that a point in $\mathbb{R}^3$ is an affine subspace: it is a shifted copy of the trivial linear subspace $\{\mathbf{0}\}$.

- Systems of four or more equations in three unknowns typically have no solution. Four or more planes in $\mathbb{R}^3$ typically have an empty intersection.

- Atypical situations arise when two or more of the planes corresponding to a system of equations are parallel. If two of the planes are parallel and distinct, they obviously have an empty intersection, so it is possible for two linear equations in three unknowns to have no solution, rather than the usual one-dimensional solution set. On the other hand, if two of the planes are identical, then in effect, the number of equations in the system is reduced, leading to a higher dimensional solution set than is typically to be expected.

- It is also possible for a system of three equations in three unknowns to have no solution, even though no two of the corresponding planes are parallel. The system

$$
\begin{aligned}
x_1 &= 0 \\
x_2 &= 0 \\
x_1 + x_2 &= 1
\end{aligned}
$$

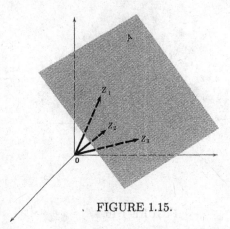

. FIGURE 1.15.

is an example. Draw a picture of the three planes corresponding to this system in order to see what happens. (It is easiest to draw this picture if you view it from above, that is, looking down along the x_3-axis onto the plane spanned by e_1 and e_2. The three planes corresponding to the system of equations will then be viewed edge-on.)

- If we replace the plane $x_1 = 0$ by the plane $x_1 = 1$ in the previous remark (which corresponds to shifting the plane $x_1 = 0$ one unit), we obtain three distinct planes in $\mathbb{R}^3$ whose intersection is a line.

1.6D THREE POINTS DETERMINE A PLANE

Consider three points Z_1, Z_2, and Z_3 in $\mathbb{R}^3$, and suppose that they do not lie on a common line. Then, as can be proved, there is exactly one plane that passes through these three points. See Figure 1.15. Although we will not give the formal proof we will show how to find the plane in a particular case, and it will be rather clear that the method could be turned into a general proof. Any plane $\mathcal{A}$ that contains the point Z_3 must be of the form $\mathcal{S} + Z_3$ for some plane $\mathcal{S}$ through the origin. See Figure 1.16. The problem is to show that the plane $\mathcal{S}$ is determined by the condition that $\mathcal{A}$ also contain Z_1 and Z_2. Also, we will want to find an equation for $\mathcal{A}$. Actually, we have accomplished all the preliminaries needed to handle this problem. Let us solve it in the special case where $Z_1 = (1, 2, 4)$, $Z_2 = (3, 1, -2)$, and $Z_3 = (0, 2, 2)$.

The first step is the determination of a plane $\mathcal{S}$ containing the origin, so that $\mathcal{A} = \mathcal{S} + Z_3$ contains Z_1, Z_2, and Z_3. Since Z_1 and Z_2 are to be in $\mathcal{A}$ the vectors $Z_1 - Z_3 = (1, 0, 2)$ and $Z_2 - Z_3 = (3, -1, -4)$ must be in $\mathcal{S}$. These two vectors do not lie on the same line through the origin, so they span a plane through the origin. No other plane through the origin contains

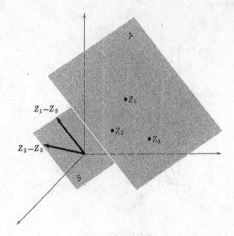

FIGURE 1.16.

both $(1, 0, 2)$ and $(3, -1, -4)$ (recall that we proved in Section 1.5e that two vectors which are non-collinear with the origin and which are contained in a plane through the origin determine that plane). Thus, the plane through the origin spanned by $(1, 0, 2)$ and $(3, -1, -4)$ is the plane $\mathcal{S}$ that we are seeking, and no other plane through the origin will work. In other words, $\mathcal{A}$ is uniquely determined by the three points.

We learned how to find the equation corresponding to $\mathcal{A}$ in the example in Section 1.6c. We first find a homogeneous linear equation satisfied by the two vectors $(1, 0, 2)$ and $(3, -1, -4)$ by solving the system

$$a_1 + 2a_3 = 0$$
$$3a_1 - a_2 - 4a_3 = 0 .$$

The solutions may be written as $a_1 = -2a_3$, $a_2 = -10a_3$, with a_3 arbitrary.

We choose $a_3 = -1$, so $\mathcal{S}$ is the set of solutions of the equation

$$2x_1 + 10x_2 - x_3 = 0 .$$

Now plug $Z_3 = (0, 2, 2)$ into the left-hand side:

$$(2 \times 0) + (10 \times 2) - 2 = 18 .$$

The plane $\mathcal{A}$ is the solution set of

$$2x_1 + 10x_2 - x_3 = 18 .$$

You may check that the vectors Z_1, Z_2, and Z_3 all satisfy this equation.

Exercises

Some elementary computations with affine subspaces in the same spirit as the preceding sections.

1. **Lines in $\mathbb{R}^2$.** Let $\mathcal{S}$ be the linear subspace of $\mathbb{R}^2$ spanned by the vector $X = (2, -1)$. Let $\mathcal{A}$ be the affine subspace parallel to $\mathcal{S}$ through the point $Z = (3, 0)$.

   (a) Compute a homogeneous linear equation in x_1, x_2 determining $\mathcal{S}$.

   (b) Compute a linear equation determining $\mathcal{A}$.

   (c) Does $\mathcal{A} = \mathcal{S} + Z$?

   (d) Find two other points on $\mathcal{A}$.

   (e) Sketch $\mathbb{R}^2$, $\mathcal{S}$, $\mathcal{A}$, X, Z, and the points found in Exercise 1(d).

2. **Planes in $\mathbb{R}^3$.** Let $\mathcal{S}$ be the linear subspace of $\mathbb{R}^3$ spanned by $X_1 = (1, 1, 0)$ and $X_2 = (0, 1, 1)$. Let $\mathcal{A}$ be the affine subspace parallel to $\mathcal{S}$ through the point $Z = (2, 2, 2)$.

   (a) Compute a homogeneous linear equation in x_1, x_2, and x_3 determining $\mathcal{S}$.

   (b) Compute a linear equation determining $\mathcal{A}$.

   (c) Does $\mathcal{A} = \mathcal{S} + Z$?

   (d) Find some other points on $\mathcal{A}$.

   (e) Sketch.

3. Let $\mathcal{S}$ be the two-dimensional linear subspace of $\mathbb{R}^3$ spanned by the standard basis vectors $\mathbf{e}_1$ and $\mathbf{e}_3$.

   (a) Compute an equation determining $\mathcal{S}$.

   (b) Compute an equation determining the affine subspace $\mathcal{A}$ parallel to $\mathcal{S}$ through the point $Z = (-3, 7, 4)$; through the point $Z_1 = (4, 7, -6)$; through the point $Z_2 = (4, 8, -5)$.

4. Let $\mathcal{A}$ be the plane in $\mathbb{R}^3$ containing the three points $\mathbf{e}_1$, $\mathbf{e}_2$, and $2\mathbf{e}_3$. Let $\mathcal{S}$ be the plane in $\mathbb{R}^3$ parallel to $\mathcal{A}$ and containing the origin.

   (a) Compute a homogeneous equation in x_1, x_2, amd x_3 determining $\mathcal{S}$.

   (b) Compute an equation determining $\mathcal{A}$.

   (c) Do the same for the affine plane $\mathcal{A}'$ determined by $2\mathbf{e}_1$, $2\mathbf{e}_2$, and $2\mathbf{e}_3$.

5. (a) Let $\mathcal{A}$ be the plane which is the solution set of $x_2 = 5$ in $\mathbb{R}^3$. What is an equation determining the linear subspace $\mathcal{S}$ parallel to $\mathcal{A}$?

   (b) Same question where the equation for $\mathcal{A}$ is $x_1 + x_2 + x_3 = 5$.

6. **Lines in $\mathbb{R}^3$.** Let $Z_1 = (1, -1, 2)$ and $Z_2 = (3, 0, 1)$.

   (a) Find a line $\mathcal{S}$ through the origin and a vector Z such that the line $\mathcal{S} + Z$ contains the two points Z_1 and Z_2. *Hint:* Let $\mathcal{S}$ be the line spanned by $Z_1 - Z_2$.

   (b) Show that the points Z_1 and Z_2 determine a unique line in $\mathbb{R}^3$.

   (c) Find two different planes that contain the points Z_1 and Z_2.

   (d) Find a pair of linear equations whose solution set is the line determined by Z_1 and Z_2.

(e) Sketch.

7. Show that any line in $\mathbb{R}^3$ is the solution space of a system of two linear equations.

8. Suppose that $\mathcal{S}$ is a linear subspace and Y and Z are two vectors such that $Z \in \mathcal{S} + Y$. Prove that $\mathcal{S} + Y = \mathcal{S} + Z$.

9. Let $X_1 = (1,0)$ and $X_2 = (-2,1)$.

   (a) Show that the set

   $$\mathcal{A} = \{X : X = \alpha_1 X_1 + \alpha_2 X_2, \text{ where } \alpha_1 + \alpha_2 = 1\}$$

   is the line containing X_1 and X_2.

   (b) Draw a sketch of the line defined in part (a), and indicate on your sketch the parts of the line corresponding to (i) $\alpha_1 < 0$; (ii) $\alpha_1 \geq 0$; (iii) $\alpha_2 < 0$; (iv) $\alpha_2 \geq 0$; and (v) $\alpha_1, \alpha_2 \geq 0$. Which of these five pieces of the line might be called the "line segment" connecting X_1 and X_2? Which might be called the "ray" emanating from X_1 through X_2?

   (c) Based on part (b), how might you define the line segment connecting two points X_1 and X_2 in $\mathbb{R}^n$?

   (d) Based on part (b), how might you define the ray emanating from X_1 through X_2 in $\mathbb{R}^n$?

10. (a) Let $\mathcal{A} \subseteq \mathbb{R}^3$ be an affine subspace containing vectors X, Y, and Z. If $\alpha + \beta + \gamma = 1$ and $W = \alpha X + \beta Y + \gamma Z$, prove that W is in $\mathcal{A}$.

    (b) Let X, Y, and Z be vectors in $\mathbb{R}^3$. Show that the set

    $$\mathcal{A} = \{W = \alpha X + \beta Y + \gamma Z, \text{ where } \alpha + \beta + \gamma = 1\}$$

    is an affine subspace.

    (c) Show that, if the three vectors X, Y, and Z do not lie on a common line, then the set $\mathcal{A}$ defined in part (b) is a plane.

    (d) Draw a sketch for the case $X = (1,1,0)$, $Y = (0,1,0)$, and $Z = (0,0,2)$. Indicate on your sketch the portion of $\mathcal{A}$ corresponding to $\alpha, \beta, \gamma \geq 0$.

1.7 The Dimension of a Vector Space

1.7A Linear dependence

In Section 1.4, we classified the linear subspaces of $\mathbb{R}^3$ in terms of spanning sets of vectors. One-dimensional linear subspaces are spanned by one vector and two-dimensional linear subspaces are spanned by two vectors. Of course, not every pair of vectors X and Y in $\mathbb{R}^3$ spans a plane. If X and Y lie on the same line through the origin, they only span that line; there is a type of dependence between X and Y that prevents them from spanning a plane. In general, we will classify a linear subspace of $\mathbb{R}^n$ according to the number of vectors it takes to span that linear subspace, but in order to do so, we must take a closer look at the concept of "dependence" between vectors.

Let us find another way to express the condition that two vectors X and Y do not lie on the same line through the origin. If $X \neq \mathbf{0}$, then the line through the origin containing X is the set of all scalar multiples αX of the vector X. Therefore, saying that X and Y lie on the same line through the origin is the same as saying that there is a scalar α such that $\alpha X = Y$, or equivalently, such that $-\alpha X + Y = \mathbf{0}$. Similarly, if Y is not the zero vector, X and Y lie on the same line through the origin if and only if there is a scalar β such that $X - \beta Y = \mathbf{0}$. Finally, if X and Y are both the zero vector, then they automatically lie on the same line through the origin, and obviously $X + Y = \mathbf{0}$. In all three cases, we find that the zero vector can be written as a linear combination of X and Y, with at least one of the scalar coefficients not equal to zero. We have proved the "only if half" of the following statement:

X and Y lie on the same line through the origin if and only if there exist scalars α and β, not both zero, such that $\alpha X + \beta Y = \mathbf{0}$.

To prove the "if half", assume that $\alpha X + \beta Y = \mathbf{0}$ with, say, $\alpha \neq 0$. Then $Y = (-\alpha/\beta)X$, so Y lies on the line spanned by X. A similar argument works if $\beta \neq 0$. Done.

The preceding discussion leads to an important definition. A set of r vectors $X_1, \ldots, X_r$ in $\mathbb{R}^n$ is **linearly dependent** (or simply **dependent**) if and only if there exist scalars $\alpha_1, \alpha_2, \ldots, \alpha_r$, not all zero, such that

$$\alpha_1 X_1 + \alpha_2 X_2 + \cdots + \alpha_r X_r = \mathbf{0} \, .$$

If $X_1, X_2, \ldots, X_n$ are not dependent, then we say they are **linearly independent** or simply **independent**. It is not hard to see that the standard basis vectors $\mathbf{e}_1 = (1, 0, \ldots, 0), \ldots, \mathbf{e}_n = (0, \ldots, 0, 1)$ are independent in $\mathbb{R}^n$.

EXAMPLES:

1. Let us show that $X_1 = (1, 1, 0)$, $X_2 = (2, 0, 3)$, and $X_3 = (1, -1, 3)$ are linearly dependent. We seek α_1, α_2, and α_3, not all zero, such that

$$\alpha_1(1, 1, 0) + \alpha_2(2, 0, 3) + \alpha_3(1, -1, 3) = (0, 0, 0) \, .$$

If we separate this expression into coordinates, we obtain

$$\begin{aligned} \alpha_1 + 2\alpha_2 + \alpha_3 &= 0 \\ \alpha_1 - \alpha_3 &= 0 \\ 3\alpha_2 + 3\alpha_3 &= 0 \, . \end{aligned}$$

This system of three linear equations in the three unknowns α_1, α_2, and α_3 is easily solved using the Gauss-Jordan method. The solutions are given by

$$\alpha_1 = \alpha_3 , \quad \alpha_2 = -\alpha_3 , \quad \alpha_3 \text{ arbitrary.}$$

Any choice of α_3 different from 0 will do, since we are interested in any solution other than $(0,0,0)$. For example, choose $\alpha_3 = 1$. The corresponding linear combination of the three vectors equals the zero vector:

$$\begin{aligned}
&\alpha_1(1,1,0) + \alpha_2(2,0,3) + \alpha_3(1,-1,3) \\
&= \ (1,1,0) - (2,0,3) + (1,-1,3) \ = \ (0,0,0) .
\end{aligned}$$

2. Let us apply the method of the preceding example to the vectors $(1,1,0)$, $(2,0,3)$, and $(1,-1,2)$. The system of equations is

$$\begin{aligned}
\alpha_1 + 2\alpha_2 + \alpha_3 &= 0 \\
\alpha_1 - \alpha_3 &= 0 \\
3\alpha_2 + 2\alpha_3 &= 0 ,
\end{aligned}$$

which has the unique solution $\alpha_1 = \alpha_2 = \alpha_3 = 0$. There is no non-trivial linear combination of the three vectors that equals the zero vector, so the three vectors are linearly independent.

Suppose that

$$\alpha_1 X_1 + \alpha_2 X_2 + \ldots + \alpha_r X_r = 0$$

with, say, $\alpha_1 \neq 0$. Then we may write

$$X_1 = (-1/\alpha_1)(\alpha_2 X_2 + \cdots + \alpha_r X_r) .$$

Similarly, if any of the other coefficients is nonzero, then the corresponding vector may be expressed as a linear combination of the remaining vectors. It follows that if a set of vectors is linearly dependent, then at least one of the vectors may be written as a linear combination of the remaining vectors. On the other hand, if one of the vectors, say X_1, can be written as a linear combination of the remaining vectors,

$$X_1 = \beta_2 X_2 + \ldots + \beta_r X_r ,$$

then $X_1, X_2, \ldots, X_r$ are linearly dependent, since

$$X_1 - \beta_2 X_2 - \ldots - \beta_r X_r = \mathbf{0} .$$

We have proved the following:

A set of vectors $X_1, X_2, \ldots, X_r$ in $\mathbb{R}^n$ is linearly dependent if and only if at least one of the vectors may be written as a linear combination of the remaining vectors.

EXAMPLE:

3. Consider the vectors

$$(1, 3, -2, 0, 0), \quad (0, -1, 5, 1, 0), \quad (0, 2, 3, 0, 1).$$

Note that the second vector, which has a 1 in the fourth coordinate position, cannot be written as a linear combination of the first and third vectors, because any such linear combination will have a 0 in the fourth coordinate position. Similarly, since each of the vectors has a 1 in a coordinate position in which the remaining two vectors have 0's, no one of the three vectors may be written as a linear combination of the remaining two vectors. According to the statement proved above, the three vectors are linearly independent. Of course, one may also prove this fact directly using the definition of linear independence.

1.7B DIMENSION

For a pair of vectors X and Y, linear independence means that the two vectors do not lie on the same line through the origin. Thus, two linearly independent vectors span a two-dimensional linear subspace. In general, we say that a linear subspace S of $\mathbb{R}^n$ is k-**dimensional** if S is the linear span of k linearly independent vectors $X_1, X_2, \ldots, X_k$. In this case, the set of vectors $\{X_1, X_2, \ldots, X_k\}$ is called a **basis** of S. Note that the standard basis vectors $\mathbf{e}_1, \mathbf{e}_2, \ldots, \mathbf{e}_n$ form a basis for $\mathbb{R}^n$, so, according to our definition, $\mathbb{R}^n$ is n-dimensional. We call an affine subspace $\mathcal{A}$ k-**dimensional** if $\mathcal{A} = S + Z$ and S is k-dimensional. In particular, lines are one-dimensional and planes are two-dimensional, consistent with our earlier definitions. The zero linear subspace is said to be **0-dimensional**, and any single point X in $\mathbb{R}^n$ forms by itself a 0-dimensional affine subspace. (We either do not speak of a basis for the zero linear subspace, or else we say that the empty set of vectors is its basis.)

A linear subspace of positive dimension has infinitely many different bases. For example, any nonzero vector X contained in a line through the origin is a basis for that line, and any two linearly independent vectors X and Y contained in a plane through the origin form a basis for that plane (why?). A problem arises: is it possible for a linear subspace S to have two different bases, one containing r vectors and the other containing s vectors, with $s \neq r$? If this were possible, then according to our definitions, S would be both r-dimensional and s-dimensional, an unsatisfactory situation. Let us state the problem another way. Suppose $X_1, X_2, \ldots, X_r$ is a linearly independent set of vectors that spans a linear subspace S, and suppose that $Y_1, Y_2, \ldots, Y_s$ is another set of vectors in S, with $s > r$. Is it possible for $Y_1, Y_2, \ldots, Y_s$ to be a basis for S? In particular, is it possible for $Y_1, Y_2, \ldots, Y_s$ to be linearly independent? The following lemma says that the answer is "no". Thus, all bases of a linear subspace S have the same size.

Lemma 1.7.1 Suppose that $X_1, \ldots, X_r$ span a vector space S and $Y_1, \ldots, Y_s$ are in S. If $s > r$, then $Y_1, \ldots, Y_s$ are linearly dependent.

PROOF: Given $s > r$. We want to find scalars $\gamma_1, \ldots, \gamma_s$, not all zero, such that $\gamma_1 Y_1 + \cdots + \gamma_s Y_s = \mathbf{0}$. Since each Y_j is in the linear subspace spanned by the X's, we can write Y_j as a linear combination of the X's: $Y_j = \alpha_{1j} X_1 + \cdots + \alpha_{rj} X_r$, for some scalars $\alpha_{1j}, \ldots \alpha_{rj}$. If we replace each Y_j by the corresponding linear combination of the X's and collect like terms, we obtain the following condition on $\gamma_1, \ldots \gamma_s$:

$$(\alpha_{11}\gamma_1 + \alpha_{12}\gamma_2 + \cdots + \alpha_{1s}\gamma_s)X_1 + \cdots + (\alpha_{r1}\gamma_1 + \cdots + \alpha_{rs}\gamma_s)X_r = \mathbf{0}.$$

This is the same as the original equation in the Y's. We now ask: Can we solve this long equation (recall that we want a nontrivial solution, i.e. at least one $\gamma_j \neq 0$)? It is certainly enough to find γ_j's, not all zero, such that the coefficient of each X_i is zero:

$$
\begin{array}{ccccc}
\alpha_{11}\gamma_1 & + & \cdots & + & \alpha_{1s}\gamma_s & = & 0 \\
\vdots & & \vdots & & \vdots \\
\alpha_{r1}\gamma_1 & + & \cdots & + & \alpha_{rs}\gamma_s & = & 0.
\end{array}
$$

This is a system of r equations in the s unknowns $\gamma_1, \ldots, \gamma_s$. It has a nontrivial solution by Theorem 1.5.1, since $s > r$. The corresponding linear combination $\gamma_1 Y_1 + \cdots + \gamma_s Y_s$ equals the zero vector. Therefore, the Y's are dependent. Done. $<<$

REMARKS:

- The assertion "if $s > r$, then $Y_1, \ldots, Y_s$ are linearly dependent" is equivalent to the statement "if $Y_1, \ldots, Y_s$ are independent, we must have $s \leq r$".

- The lemma implies that any two vectors contained in a line through the origin are dependent, since a line is spanned by a single vector. Similarly, any three vectors contained in a plane through the origin, any seven vectors in $\mathbb{R}^6$, etc., are linearly dependent.

- Suppose $X_1, X_2, \ldots, X_r$ span a linear subspace S of $\mathbb{R}^n$. If the X's are independent, then according to our definition, S is r-dimensional.

What can we say about the dimension of S if the X's are not independent? Dependence of the X's means that at least one of them, say X_r, is a linear combination of the remaining ones. In any expression involving X_r, we may replace X_r by its corresponding linear combination involving $X_1, \ldots, X_{r-1}$. It follows that any linear combination of $X_1, \ldots, X_r$ can be replaced by a linear combination of $X_1, \ldots, X_{r-1}$. Thus, S is the linear span of the $r - 1$ vectors $X_1, \ldots, X_{r-1}$. If this smaller set of $r - 1$ vectors is still linearly dependent, we may remove another vector, and S will be the linear span of $r - 2$ vectors. In this way we can eventually produce an independent set that spans S. Conclusion: if S can be spanned by r vectors, then the dimension of S is $\leq r$, and if the r vectors are dependent, then the dimension of S is $< r$.

- If a vector is adjoined to a linearly dependent set of vectors, the larger set of vectors is also linearly dependent (can you see why?). On the other hand, suppose $X_1, \ldots, X_r$ are linearly independent, and that Y is not in the linear span of $X_1, \ldots, X_r$. Do the vectors $X_1, \ldots, X_r, Y$ form a linearly independent set? The answer is "yes". Proof: suppose

$$\alpha_1 X_1 + \ldots + \alpha_r X_r + \beta Y = 0 .$$

If $\beta \neq 0$, then we could solve for Y as a linear combination of the X's, contradicting our assumption about Y. Therefore $\beta = 0$, and then it follows that all the α's equal zero because of the linear independence of the X's. Thus, all the scalar coefficients must be zero, so the vectors $X_1, \ldots, X_r, Y$ are linearly independent.

- The previous remark indicates how a basis for a linear subspace may be constructed. Let S be a linear subspace of $\mathbb{R}^n$. If S contains a nonzero vector, call that vector X_1. If there is a vector in S that is not a scalar multiple of X_1, call that vector X_2. If there is a vector in S which is not in the linear span of X_1 and X_2, call that vector X_3, and so on. Since S is a linear subspace of $\mathbb{R}^n$, this procedure must eventually stop after producing $r \leq n$ vectors $X_1, X_2, \ldots, X_r$ whose linear span is S; for if it did not stop, we would eventually have found more than n linearly independent vectors in $\mathbb{R}^n$. This discussion shows that linear subspaces of $\mathbb{R}^n$ always have dimension $\leq n$. Similarly, a linear subspace of a vector space of dimension k always has dimension $\leq k$.

- If $S_1 \subseteq S_2$ are two linear subspaces of $\mathbb{R}^n$, then the dimension of S_1 is less than or equal to the dimension of S_2 (see the previous remark). We can build a basis for S_2 by following the procedure just outlined, first building a basis for S_1, and then continuing until we have a basis for S_2. If S_2 is a proper subset of S_1, the basis that we construct for

S_2 will be strictly larger than the basis for S_1: a nontrivial linear subspace of a vector space S cannot have the same dimension as S.

These remarks indicate that the dimension of a linear subspace is a legitimate measure of the size of that linear subspace. We will use the notation

$$\dim(S)$$

for the dimension of a linear subspace S. The concepts of basis and dimension provide us with a convenient language for classifying and comparing linear subspaces of $\mathbb{R}^n$.

EXAMPLE:

4. We may now quickly classify the linear subspaces of $\mathbb{R}^4$. They are: the zero linear subspace; lines through the origin; planes through the origin; three-dimensional linear subspaces; and $\mathbb{R}^4$ itself.

In $\mathbb{R}^n$, a **hyperplane** is an affine subspace of dimension $n - 1$. In spaces with dimension less than 4, we have other, more familiar names for hyperplanes. Thus, in $\mathbb{R}^3$, hyperplanes are just ordinary planes, while in $\mathbb{R}^2$, hyperplanes are lines, and in $\mathbb{R}$, hyperplanes are single points.

1.7C LINEAR EQUATIONS AND DIMENSION

The solution set of a single linear equation in three unknowns with at least one nonzero coefficient is a plane in $\mathbb{R}^3$ (Theorem 1.6.2). If there are only two unknowns, the solution set is a line in $\mathbb{R}^2$, and if there is only one unknown, the solution set is a point in $\mathbb{R}$. In each case, the solution set is an affine subspace whose dimension is one lower than the number of unknowns, in other words, a hyperplane. The following result generalizes Theorem 1.6.2.

Theorem 1.7.1 (The Hyperplane Theorem) An affine subspace $\mathcal{A}$ in $\mathbb{R}^n$ is a hyperplane $\Leftrightarrow$ $\mathcal{A}$ is the set of solutions $X = (x_1, \ldots, x_n)$ in $\mathbb{R}^n$ of a linear equation $a_1 x_1 + \cdots + a_n x_n = b$ with at least one nonzero coefficient.

PROOF: ($\Rightarrow$) Let $\mathcal{A} = S + Z$, where S is an $(n - 1)$-dimensional linear subspace of $\mathbb{R}^n$ and Z is a vector in $\mathbb{R}^n$. Since $\dim(S) = n - 1$, there is a basis $X_1, \ldots, X_{n-1}$ for S. Let us write each X_i in coordinates: $X_i = (c_{i1}, c_{i2}, \ldots, c_{in})$, $i = 1, \ldots, n - 1$. We seek a homogeneous linear equation in n unknowns whose solution set is S, and whose coefficients $a_1, \ldots, a_n$ are not all zero. Form the system

$$
\begin{array}{cccccc}
c_{11}a_1 & + & \cdots & + & c_{1n}a_n & = & 0 \\
\vdots & & & & \vdots & & \vdots \\
c_{n-1,1}a_1 & + & \cdots & + & c_{n-1,n}a_n & = & 0.
\end{array}
$$

This is just what we did with the two vectors X and Y in the proof of Theorem 1.5.2. Again we have more unknowns than equations, and so we know that there exist $a_1, \ldots, a_n$, not all zero, satisfying the system. It follows that each X_i is a solution of the linear equation whose coefficients are the a's:

$$a_1 x_1 + \cdots + a_n x_n = 0 .$$

Since $X_1, \ldots, X_{n-1}$ is a basis for S, every $X = (x_1, \ldots, x_n)$ in S satisfies this equation. According to the remarks in Section 1.7b, any linear subspace of $\mathbb{R}^n$ which is strictly larger than S would have dimension n, so the solution space of the equation is either S or all of $\mathbb{R}^n$. But because not all of the a's are zero, the solution space cannot be all of $\mathbb{R}^n$ (compare the proof of Theorem 1.5.2), so S is precisely the solution space of the equation. Now, as in the proof of Theorem 1.6.2, write $Z = (z_1, \ldots, z_n)$ and let $b = a_1 z_1 + \ldots + a_n z_n$, and it will follow that $\mathcal{A}$ is the solution space of the equation $a_1 x_1 + \ldots + a_n x_n = b$.

($\Leftarrow$) We already know that the solution set of a linear equation in n unknowns is an affine subspace $\mathcal{A}$ of $\mathbb{R}^n$. In order to show that $\dim(\mathcal{A}) = n - 1$, it is enough to show that the corresponding homogeneous equation

$$a_1 x_1 + \ldots + a_n x_n = 0$$

has an $(n-1)$-dimensional solution set S. Let us apply the Gauss-Jordan method to solve the equation. By relabeling the variables if necessary, we may assume that $a_1 \neq 0$. Then we may choose x_1 as the basic variable, and $x_2, \ldots, x_n$ as the free variables:

$$x_1 = (-a_2/a_1)x_2 + \ldots + (-a_n/a_1)x_n .$$

In vector form, the solution set may be expressed as

$$(x_1, x_2, \ldots, x_n) = x_2 X_2 + \ldots + x_n X_n ,$$

where for each $i = 2, \ldots, n$,

$$X_i = (-a_i/a_1, 0, \ldots, 1, \ldots, 0) ,$$

with the 1 in the ith coordinate position. The vectors $X_2, \ldots, X_n$ are linearly independent, since each of them has a 1 in a coordinate position in which the remaining vectors all have a 0. The solution set S is the linear span of this set of $n - 1$ vectors, so S is $(n-1)$-dimensional. Done. $<<$

REMARKS:

- The proofs of the Hyperplane Theorem and Lemma 1.7.1 both use the algebraic fact that a system of homogeneous linear equations with sufficiently many unknowns always has a nontrivial solution. Again and again in linear algebra, questions about subspaces (linear and affine), planes, dimension, and so forth, are settled by translating the questions into ones about the solvability of such a system of equations.

- Note how the Gauss-Jordan method automatically produced an independent set of vectors in the proof of the second half of the Hyperplane Theorem. A review of the examples and exercises of Section 1.5 shows the following. For any system of linear equations that has a solution, the Gauss-Jordan method leads to a vector solution of the form $\alpha_1 X_1 + \ldots + \alpha_k X_k + Z$, with k being the number of free variables. Each vector X_i matches up with one of the free variables and has a 1 in the coordinate position corresponding to that free variable, with 0's in the coordinate positions corresponding to the remaining free variables. Thus, the vectors $X_1, \ldots, X_k$ are linearly independent. They form a basis for the linear subspace $\mathcal{S}$ of solutions to the homogeneous system corresponding to the original system. The dimension of $\mathcal{S}$ is therefore equal to the number of free variables, as is the dimension of the solution set $\mathcal{A} = \mathcal{S} + Z$ of the original system. In the case that there are no free variables, the solution set consists of a single point, and so has dimension 0.

For future reference, we summarize the last remark as a theorem.

Theorem 1.7.2 Suppose that a system of linear equations has a nonempty solution set $\mathcal{A}$. Further suppose that when the Gauss-Jordan method is applied to the system, the number of free variables is k. Then $\dim(\mathcal{A}) = k$.

EXAMPLE:

5. Using the Gauss-Jordan method, we find that the solution set of the system

$$x_1 - x_2 + x_3 - 2x_4 = -1$$
$$x_2 + 3x_3 = 0$$

is the set of vectors of the form

$$(x_1, x_2, x_3, x_4) = (-1, 0, 0, 0) + x_3(-4, -3, 1, 0) + x_4(2, 0, 0, 1) ,$$

with x_3 and x_4 arbitrary. (We have chosen x_1 and x_2 as the basic variables, and x_3 and x_4 as the free variables.) Thus, the solution set is the two-dimensional affine subspace $\mathcal{A}$ of $\mathbb{R}^4$ which contains the vector $(-1, 0, 0, 0)$ and is parallel to the plane $\mathcal{S}$ through the origin spanned by the two linearly independent vectors $(-4, -3, 1, 0)$ and $(2, 0, 0, 1)$.

Exercises

1. Determine the dimensions of the solution sets of the following systems. In addition, in each case, find a basis for the linear subspace which is parallel to the solution set. Assume that the number of variables equals the largest subscript that appears.

(a)
$$x_1 + x_2 + x_3 - x_4 - x_5 = 1$$
$$5x_1 + x_3 - 2x_4 + 2x_5 = 2 \, ;$$

(b)
$$-x_1 - x_2 + x_3 = 2$$
$$2x_1 + x_2 - 2x_3 = -5$$
$$x_1 - 5x_2 = -4 \, ;$$

(c) $x_1 + x_3 + x_5 - x_{11} = 0 \, ;$

(d)
$$2x_1 + 2x_2 - x_3 + x_4 = 0$$
$$x_1 + x_4 = 0 \, .$$

2. Find examples of (that is, find bases for) the following:

   (a) two planes through the origin in $\mathbb{R}^4$ whose intersection is a single point (the origin);

   (b) two planes through the origin in $\mathbb{R}^4$ whose intersection is a line;

   (c) a hyperplane through the origin in $\mathbb{R}^4$ that does not contain any of the coordinate axes;

   (d) four distinct linearly dependent vectors in $\mathbb{R}^4$ which are noncoplanar with the origin.

   Which of the above are possible in $\mathbb{R}^3$? in $\mathbb{R}^2$? in $\mathbb{R}^5$?

3. Determine whether each of the following sets of vectors is linearly independent:

   (a) $(1, 1, 1)$, $(2, 0, 3)$, $(-1, 2, -5)$

   (b) $(5, 1, 2)$, $(5, 1, 1)$

   (c) $(1, 1, 1)$, $(8, 9, 10)$, $(\pi, \sqrt{2}, 0)$, $(3, -99, 4)$

   (d) $(0, 0, 0)$, $(1, 1, 1)$, $(0, 1, 0)$

   (e) $(1, 1, 1)$, $(2, 2, 2)$, $(3, 2, 1)$

   (f) $(3, 0, -1)$, $(4, -4, 2)$, $(3, 12, -11)$.

4. Find the dimension of the linear span of the vectors

$$(0, 1, 0), \, (1, 1, 1), \, (2, 1, 0), \, (-5, -1, -1) \, .$$

5. Let $\mathcal{A}_1$ and $\mathcal{A}_2$ be two hyperplanes in $\mathbb{R}^n$. Use Theorem 1.7.2 to show that one of the following statements must be true: (i) $\mathcal{A}_1 = \mathcal{A}_2$; (ii) $\mathcal{A}_1 \cap \mathcal{A}_2 = \emptyset$; (iii) $\mathcal{A}_1 \cap \mathcal{A}_2$ is an affine subspace with dimension $n - 2$.

6. Show that if a system of m equations in n unknowns has a nonempty solution set $\mathcal{A}$, then $\dim(\mathcal{A}) \geq n - m$.

7. (a) Let $\{X_1, \ldots, X_r\}$ be a basis of a linear subspace S and $\{Y_1, \ldots, Y_s\}$ a basis of a linear subspace T, all in $\mathbb{R}^n$. Show that if $S \cap T = \{0\}$, then the set $\{X_1, \ldots, X_r, Y_1, \ldots, Y_s\}$ is linearly independent.

   (b) Use part (a) to give another proof of the fact that the intersection of two distinct planes through the origin in $\mathbb{R}^3$ is a line. Also note that either of the two preceding exercises may be used to prove this fact.

Extra: Function Spaces

Throughout this book, the formal discussion of vector spaces is limited to $\mathbb{R}^n$ and its linear subspaces. The power and significance of the vector space idea, however, is due to the fact that many different sets in mathematics satisfy Properties 1 through 8 and hence are vector spaces.

EXAMPLES:

1. $\mathcal{P}_n$, the space of polynomials

$$p(x) = a_0 + a_1 x + \cdots + a_n x^n$$

of degree at most n. Observe that if p and q are in $\mathcal{P}_n$, then so is $p + q$ and cp, where c is any scalar. Properties 1 to 8 are evidently satisfied, and so $\mathcal{P}_n$ is a vector space. Can we find a basis for $\mathcal{P}_n$? Absolutely. Notice that a polynomial $p \in \mathcal{P}_n$ is uniquely determined by its coefficients $a_0, a_1, \ldots, a_n$. If we let $\mathbf{e}_0 = 1$, $\mathbf{e}_1 = x$, $\mathbf{e}_2 = x^2, \ldots, \mathbf{e}_n = x^n$, then any $p \in \mathcal{P}_n$ can be written uniquely as

$$a_0 \mathbf{e}_0 + a_1 \mathbf{e}_1 + \cdots + a_n \mathbf{e}_n .$$

Thus the vectors $\mathbf{e}_0, \ldots, \mathbf{e}_n$ form a basis for $\mathcal{P}_n$ (can you see why they are linearly independent?). Since there are $n + 1$ vectors in the basis, the dimension of $\mathcal{P}_n$ is $n + 1$. Note: In considering $\mathcal{P}_n$ as a vector space, we *ignore* the fact that we can multiply polynomials as well as add them.

2. The set $\mathcal{C}$ of functions f continuous on the interval $-\infty < x < \infty$. Now if f and g are continuous, so are $f + g$ and cf, where c is any scalar. Properties 1 to 8 of a vector space are the standard algebraic rules for functions. Thus $\mathcal{C}$ is a vector space whose elements (the "vectors") are continuous functions.

Since every polynomial is a continuous function, it is clear that $\mathcal{P}_n \subseteq \mathcal{C}$. But $\mathcal{P}_n$ is also a vector space. Therefore, $\mathcal{P}_n$ is a linear subspace of $\mathcal{C}$. Now the dimension of $\mathcal{P}_n$ is $n + 1$. Since n can be any, possibly huge, integer, we are led to the fact that $\mathcal{C}$ is *infinite-dimensional*. Although this is a bit terrifying, one must not get upset. It just means that the space has lots of room. The virtue of considering $\mathcal{C}$ as a vector space is that many features of $\mathbb{R}^n$ depend only on the fact that it is a vector space and *not* on its dimension. Therefore, these vector-space results immediately carry over to the more complicated space $\mathcal{C}$. Both $\mathcal{C}$ and $\mathcal{P}_n$ are examples of vector spaces of functions, that is, **function spaces**.

Only in this century have mathematicians realized that many seemingly unrelated sets of objects are all vector spaces, and they have grown to know and cherish them. Physicists and other applied scientists often use infinite-dimensional vector spaces (they are fundamental to quantum mechanics). Anyone going on to use higher mathematics will soon find that he or she must make peace with many kinds of vector spaces. $\mathbb{R}^n$ is fine, but it is just a beginning.

2

The Geometry of $\mathbb{R}^n$

2.0 Introduction

With our treatment of affine subspaces of $\mathbb{R}^n$ we have completed the first stage of our four-stage approach to differential vector calculus. In this chapter we traverse the second stage, which emphasizes distance and geometry. The last two stages of the journey, linear and nonlinear functions, are confronted in subsequent chapters.

We now introduce geometry by developing the notion of the length of a vector. Using this idea, we may define the distance between points. The concept of distance will be useful later in obtaining a theory of limits in $\mathbb{R}^n$; the phrase "X approaches X_0" will mean that as X varies, the distance between X and X_0 approaches zero in the sense of real numbers. Just as in calculus, we need the notion of limit to deal with the continuity and differentiability of functions. But we are getting ahead of ourselves.

In this chapter we also study angles, not just in $\mathbb{R}^2$, but also in $\mathbb{R}^3$ and, more generally, in $\mathbb{R}^n$.

2.1 The Norm of a Vector

2.1A $\|X\|$ IN THE PLANE

First we work in the familiar $\mathbb{R}^2$. Let $X = (x_1, x_2)$. If we think of X as a directed line segment or arrow in the plane, then the **length** of X, denoted $\|X\|$, is

$$\|X\| = \sqrt{x_1^2 + x_2^2} \ .$$

We use **norm** as a synonym for length, favoring it when we are interpreting X as a point rather than an arrow. Figure 2.1 should convince you that this is in agreement with Pythagoras' famous theorem on right triangles: $x_1^2 + x_2^2 = \|X\|^2$.

EXAMPLES:

1. Let us calculate the length of $X = (1, -1)$. We have

$$\|X\| = \sqrt{1^2 + (-1)^2} = \sqrt{2} \ .$$

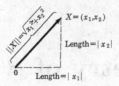

FIGURE 2.1. Pythagoras' Theorem and the length of a vector

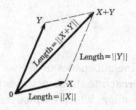

FIGURE 2.2. $\|X + Y\| \leq \|X\| + \|Y\|$

2. The norm of $Y = (5, 2)$ is given by $\|Y\| = \sqrt{29}$.

Three basic properties of $\|X\|$ are as follows:

1. Positivity: $\|X\| \geq 0$; and $\|X\| = 0$ (scalar) if and only if $X = \mathbf{0}$ (vector).

2. Homogeneity: $\|\alpha X\| = |\alpha| \, \|X\|$, where $|\alpha|$ denotes the absolute value of the scalar α.

3. Triangle inequality: $\|X + Y\| \leq \|X\| + \|Y\|$.

The first property is immediate. The second says that "stretching" the vector X by the factor α changes its length as expected. You will be asked to prove this property in the exercises. The third property compares the distance traversed in going between two points, namely $\mathbf{0}$ and $X + Y$, along two different routes. It says that the direct route along the line segment connecting $\mathbf{0}$ and $X + Y$ is shorter than the indirect route which runs first from $\mathbf{0}$ to X and then to $X + Y$. See Figure 2.2. Although the triangle inequality seems self-evident, a picture is not a proof. We will give a proof after introducing the inner product in Section 2.2.

2.1B $\|X\|$ IN $\mathbb{R}^n$

We generalize the notion of length to vectors in $\mathbb{R}^n$. If $X = (x_1, \ldots, x_n)$, we define the **length** or **norm** of X as

$$\|X\| = \sqrt{x_1^2 + \cdots + x_n^2} .$$

EXAMPLES:

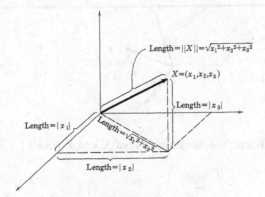

FIGURE 2.3. Length in $\mathbb{R}^3$

3. If $X = (1, -1, 1, -1, 1)$ in $\mathbb{R}^5$, then $\|X\| = \sqrt{5}$.

4. If $X = (2, 3, 4)$ in $\mathbb{R}^3$, then $\|X\| = \sqrt{29}$.

5. $\|\mathbf{e}_j\| = 1$ in $\mathbb{R}^n$ for $j = 1, 2, \ldots, n$. Vectors with length 1 are called **unit vectors**. Other examples of unit vectors are $(\frac{1}{2}, -\frac{\sqrt{3}}{2})$ in $\mathbb{R}^2$ and $(\frac{1}{2}, -\frac{1}{2}, \frac{1}{2}, \frac{1}{2})$ in $\mathbb{R}^4$.

This definition of $\|X\|$ clearly generalizes that given for $\mathbb{R}^2$ in the paragraph above. It is sometimes called the **Pythagorean** norm, since $\|X\|^2 = x_1^2 + \cdots + x_n^2$, in analogy with Pythagoras' sums of squares. See Figure 2.3.

The space $\mathbb{R}^n$, equipped with the Pythagorean norm, is usually called **Euclidean n-space**, in honor of the famous geometer. (Euclidean n-space is sometimes denoted $\mathbf{E}$ or $\mathbf{E}^n$, but we will continue to use $\mathbb{R}^n$.)

What is the "distance" between two points X and Y in $\mathbb{R}^n$? We define it to be the length of their difference, that is, the **distance** from Y to X is $\|X - Y\|$. This definition is reasonable in light of Figure 2.4 where it may be seen that the length of the vector $X - Y$ is the same as the length of the line segment from Y to X. In fact, the three quantities $\|X\|$, $\|Y\|$, and $\|X - Y\|$ are the lengths of the three sides of the triangle with vertices $\mathbf{0}$, X, and Y. We will make use of this fact later when we define the angle between two vectors in $\mathbb{R}^n$ (Section 2.2d).

An immediate consequence of the definition is the "distance formula". If $X = (x_1, \ldots, x_n)$ and $Y = (y_1, \ldots, y_n)$, then the distance between X and Y is given by

$$\|X - Y\| = \sqrt{(x_1 - y_1)^2 + \cdots + (x_n - y_n)^2}\,.$$

EXAMPLE:

6. The distance from $(1, 2, 3, 4)$ to $(2, 3, 1, 4)$ in $\mathbb{R}^4$ is $\sqrt{6}$.

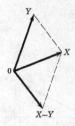

FIGURE 2.4. The distance between X and Y equals $\|X - Y\|$

REMARK:

- It is worth noting again that we have transformed a geometric notion (length or distance) into algebra (the distance formula), just as we did with the geometric notions of space and dimension.

Exercises
Practice with norms and distances.

1. Verify that $\|X\| = \sqrt{5}$ for $X = (1, -1, 1, -1, 1)$ and $\|X\| = \sqrt{29}$ if $X = (2, 3, 4)$ (see the examples in Section 2.1b).

2. Compute the norms of the following vectors in $\mathbb{R}^2$:

   (a) $(0, -4)$

   (b) $(1, -1)$

   (c) $(1, 1)$

   (d) $(3, 4)$.

3. Compute the norms of the following vectors in $\mathbb{R}^3$:

   (a) $(0, 0, -4)$

   (b) $(1, 1, 1)$

   (c) $(1, 2, -4)$

   (d) $\left(\frac{1}{\sqrt{3}}, \frac{1}{\sqrt{3}}, \frac{1}{\sqrt{3}} \right)$.

4. Compute the norms of the following vectors in $\mathbb{R}^4$:

   (a) $(0, 0, 0, 4)$

   (b) $(1, 1, 1, 1)$

   (c) $\left(\frac{1}{2}, \frac{1}{2}, \frac{1}{2}, \frac{1}{2} \right)$.

5. Compute the norms of the following vectors in $\mathbb{R}^n$:

   (a) $(1, 1, 1, \ldots, 1, 1)$

   (b) $\left(\frac{1}{\sqrt{n}}, \frac{1}{\sqrt{n}}, \frac{1}{\sqrt{n}}, \ldots, \frac{1}{\sqrt{n}}, \frac{1}{\sqrt{n}} \right)$

   (c) $\left(\frac{1}{2}, \frac{1}{4}, \frac{1}{8}, \ldots, \frac{1}{2^{n-1}}, \frac{1}{2^n} \right)$.

6. Derive the answer to part (d) of Exercise 3 directly from part (b) of that exercise and one of the properties of $\|X\|$. Do likewise for parts (c) and (b) of Exercise 4 and parts (b) and (a) of Exercise 5.

7. Compute the distance between X and Y, where

   (a) $X = (1,5)$, $Y = (1,1)$ in $\mathbb{R}^2$

   (b) $X = (3,4,5)$, $Y = (2,3,4)$ in $\mathbb{R}^3$

   (c) $X = (1,0,0,0)$, $Y = (0,0,1,0)$ in $\mathbb{R}^4$

   (d) $X = (1,0,0,\ldots,0,0)$, $Y = (0,1,0,\ldots,0,0)$ in $\mathbb{R}^n$.

8. Prove the homogeneity property of $\|X\|$ (Property 2).

9. (a) Given nonzero $X \in \mathbb{R}^n$, we wish to find a positive scalar α so that the vector αX, which points in the same direction as X, is a unit vector, that is, $\|\alpha X\| = 1$. Show that $\alpha = 1/\|X\|$ works.

   (b) Find a second value of α such that αX is a unit vector. In which direction does αX point for this value of α?

10. Using the formula in the previous exercise, find unit vectors that have the same direction as the vectors in part (a) of Exercise 3 and part (b) of Exercise 4.

11. Find unit vectors that have the same directions as the vectors in part (b) and (c) of Exercise 3 and part (a) of Exercise 5.

12. Give an algebraic proof of the triangle inequality for $\mathbb{R}^2$. Some algebraic fortitude is needed.

2.2 The Inner Product

2.2A DEFINITION

Let $X = (x_1, \ldots, x_n)$ and $Y = (y_1, \ldots, y_n)$ in $\mathbb{R}^n$. We define their **inner product**, denoted by $\langle X, Y \rangle$, to be the number

$$\langle X, Y \rangle = x_1 y_1 + x_2 y_2 + \cdots + x_n y_n.$$

EXAMPLE:

1. If $X = (1, -2, 1)$ and $Y = (-3, 1, 4)$ in $\mathbb{R}^3$, we immediately compute that

$$\langle X, Y \rangle = (1 \times (-3)) + ((-2) \times 1) + (1 \times 4) = -1.$$

Also, $\langle X, X \rangle = 6$ and $\langle Y, Y \rangle = 26$.

REMARKS:

- Since $\langle X, X \rangle = x_1^2 + \cdots + x_n^2 = \|X\|^2$, we conclude that $\|X\| = \sqrt{\langle X, X \rangle}$, a frequently used formula.

- Another notation and other names are sometimes used for the inner product. It is often called the "scalar product", and when the notation $X \cdot Y$ is used it is called the "dot product".

- It is a fact of mathematical life that a new mathematical notion can be presented in various ways, such as the following:

  ⋆ It can be given by a formula that is particularly useful for calculation, such as that for $\langle X, Y \rangle$ above.

  ⋆ It can be required to satisfy certain axioms. This was the approach we used in defining a vector space.

  ⋆ It can be given as a concept described in terms of previously understood concepts in a manner useful for interpretation and visualization. For example, our definitions of lines and planes were motivated by geometric notions.

We have defined the inner product via a formula designed for easy calculation. Now the burden is on us to discover its properties and give it meaning. If we had started with the definition $\langle X, Y \rangle = \|X\| \|Y\| \cos \theta$, where θ is the angle between X and Y, we would have had a nice geometric interpretation, but we would have had to labor in order to find a formula useful for calculation. For any interesting mathematical object the following *three* kinds of information are necessary: the computational, the axiomatic, and the interpretive, will be necessary. Given one, we must labor to obtain the other two. (These statements constitute an oversimplification, since in many situations the distinctions between the three kinds of information are fuzzy.)

2.2B ALGEBRAIC PROPERTIES OF $\langle X, Y \rangle$

We obtain some properties of the inner product, using the definition

$$\langle X, Y \rangle = x_1 y_1 + \cdots + x_n y_n$$

and basic properties of the real numbers. We have, for all X, Y, $Z \in \mathbb{R}^n$ and $\alpha \in \mathbb{R}$, the following four properties:

1. Positivity: $\langle X, X \rangle \geq 0$. Also $\langle X, X \rangle = 0$ if and only if $X = \mathbf{0}$.

2. Symmetry: $\langle X, Y \rangle = \langle Y, X \rangle$.

3. Homogeneity: $\langle \alpha X, Y \rangle = \alpha \langle X, Y \rangle$.

4. Distributivity: $\langle X, Y + Z \rangle = \langle X, Y \rangle + \langle X, Z \rangle$.

The first statement follows from the fact that $x_1^2 + \cdots + x_n^2$ is zero if and only if each of the x_i is zero. The rest are just as immediate.

Is it true, as we would hope, that

$$\langle X + Y, Z \rangle = \langle X, Z \rangle + \langle Y, Z \rangle?$$ (Compare Property 4 above.)

This is true, because $\langle X + Y, Z \rangle = \langle Z, X + Y \rangle$ (by symmetry), which equals $\langle Z, X \rangle + \langle Z, Y \rangle$ (by Property 4), which equals $\langle X, Z \rangle + \langle Y, Z \rangle$ (by symmetry), as claimed. You can prove similarly that $\langle X, \alpha Y \rangle = \alpha \langle X, Y \rangle$. Many theorems are proved by appealing to these properties, rather than to the definition itself.

EXAMPLES:

2. Prove that if $\langle Z, X \rangle = 0$ for *all* $X \in \mathbb{R}^n$, then $Z = \mathbf{0}$. Proof: This is easier than it looks. Let $X = Z$. Then $\langle Z, Z \rangle = 0$, whence $Z = \mathbf{0}$ by Property 1. Done.

3. Prove that if $\langle Z_1, X \rangle = \langle Z_2, X \rangle$ for *all* $X \in \mathbb{R}^n$, then $Z_1 = Z_2$. Proof: We have $\langle Z_1, X \rangle - \langle Z_2, X \rangle = 0$. We compute, using homogeneity and the distributive law:

$$\begin{aligned} 0 &= \langle Z_1, X \rangle - \langle Z_2, X \rangle \\ &= \langle Z_1, X \rangle + \langle -Z_2, X \rangle \\ &= \langle Z_1 - Z_2, X \rangle \end{aligned}$$

for all X. By Example 1, $Z_1 - Z_2 = \mathbf{0}$; thus $Z_1 = Z_2$. Done.

REMARK:

- As in the preceding example, often the neatest way to prove that two objects are equal is to prove that their *difference* is zero.

2.2C   AN INTERPRETATION IN $\mathbb{R}^2$

Now we relate $\langle X, Y \rangle$ to the geometry (and trigonometry) of the plane $\mathbb{R}^2$. The following theorem is the key. It gives an expression for $\langle X, Y \rangle$ involving geometric concepts, *not* coordinates (a "coordinate-free" expression).

Theorem 2.2.1 Let X, Y be nonzero vectors in $\mathbb{R}^2$. Then

$$\langle X, Y \rangle = \|X\| \, \|Y\| \cos \theta \, ,$$

where θ is the angle between X and Y. (Since $\cos \theta = \cos(-\theta)$, the sense in which we take the angle does not matter.)

PROOF: We know $\langle X, Y \rangle = x_1 y_1 + x_2 y_2$, where $X = (x_1, x_2)$ and $Y = (y_1, y_2)$ as usual. We will translate $x_1 y_1 + x_2 y_2$ into trigonometry. Let ω

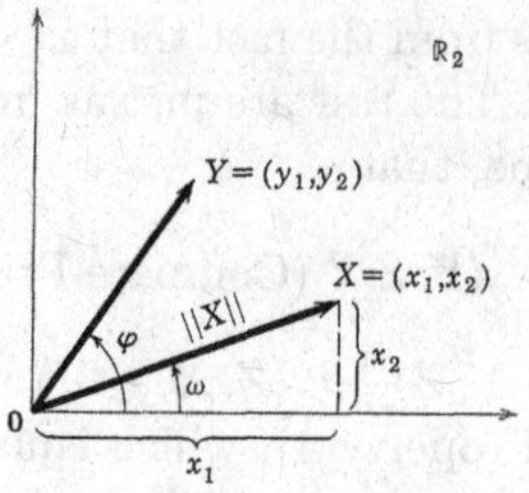

FIGURE 2.5.

and φ be the angles from the horizontal axis to X and Y, respectively. Now we note that

$$\cos \omega = \frac{x_1}{\|X\|} \quad \text{and} \quad \sin \omega = \frac{x_2}{\|X\|},$$

whence $x_1 = \|X\| \cos \omega$ and $x_2 = \|X\| \sin \omega$. Likewise $y_1 = \|Y\| \cos \varphi$ and $y_2 = \|Y\| \sin \varphi$. See Figure 2.5.

Thus

$$\langle X, Y \rangle = x_1 y_1 + x_2 y_2 = \|X\| \, \|Y\| \, (\cos \omega \cos \varphi + \sin \omega \sin \varphi) .$$

We recall

$$\cos \theta = \cos(\varphi - \omega) = \cos \omega \cos \varphi + \sin \omega \sin \varphi .$$

Thus $\langle X, Y \rangle = \|X\| \, \|Y\| \cos \theta$, as claimed. $<<$

EXAMPLE:

4. Let us calculate $\cos(\pi/4)$ (recall that $\pi/4$ radians equals $45°$). Let $X = \mathbf{e}_1 = (1, 0)$ and $Y = (1, 1)$. Then the angle θ between X and Y is $\pi/4$, and hence

$$\cos \frac{\pi}{4} = \frac{1 \cdot 1 + 0 \cdot 1}{1 \cdot \sqrt{2}} = \frac{1}{\sqrt{2}} = \frac{\sqrt{2}}{2},$$

as is well known.

REMARKS:

- Orthogonality in $\mathbb{R}^2$. The nonzero vectors (arrows) X and Y are at right angles (i.e. perpendicular), or as we will say, **orthogonal** if and only if the cosine of the angle between them is 0. We also agree to say that X and Y are **orthogonal** if either X or Y is $\mathbf{0}$. Hence X is orthogonal to Y, written $X \perp Y$, if and only if $\langle X, Y \rangle = 0$. (You should work out the connection here with the "slopes are negative reciprocals" requirement for the perpendicularity of lines in plane analytic geometry.)

- The Cauchy-Schwarz inequality in $\mathbb{R}^2$. This inequality, which is very useful and far-reaching, says that, for all X and Y in $\mathbb{R}^2$,

$$|\langle X, Y \rangle| \leq \|X\| \, \|Y\| \, .$$

If X or Y is $\mathbf{0}$, this is obvious. For nonzero X and Y, the inequality follows immediately from Theorem 2.2.1 and the fact that $|\cos\theta| \leq 1$; just take the absolute value of both sides in the expression $\langle X, Y \rangle = \|X\| \, \|Y\| \cos\theta$. We will see that "Cauchy-Schwarz" is true in $\mathbb{R}^n$ also.

- The triangle inequality in $\mathbb{R}^2$. We wish to prove that, for all X and Y in $\mathbb{R}^2$,

$$\|X + Y\| \leq \|X\| + \|Y\| \, .$$

Squaring would get rid of troublesome square roots, and we do get an equivalent inequality by squaring both sides since both sides are nonnegative. Thus we only need prove

$$\|X + Y\|^2 \leq (\|X\| + \|Y\|)^2 \, .$$

For the left-hand side we have

$$\begin{aligned}
\|X + Y\|^2 &= \langle X + Y, X + Y \rangle \\
&= \langle X, X \rangle + \langle X, Y \rangle + \langle Y, X \rangle + \langle Y, Y \rangle \\
&= \langle X, X \rangle + 2\langle X, Y \rangle + \langle Y, Y \rangle \, .
\end{aligned}$$

Here we have used the symmetry and distributivity of the inner product. For the right-hand side we have

$$\begin{aligned}
(\|X\| + \|Y\|)^2 &= \|X\|^2 + 2\|X\| \, \|Y\| + \|Y\|^2 \\
&= \langle X, X \rangle + 2\|X\| \, \|Y\| + \langle Y, Y \rangle \, .
\end{aligned}$$

Hence $\|X + Y\| \leq \|X\| + \|Y\|$ is true provided that $\langle X, Y \rangle \leq \|X\| \, \|Y\|$. But this is immediate from the Cauchy-Schwarz inequality described in the preceding remark.

2.2D   PROJECTION AND ANGLES IN $\mathbb{R}^n$

In Section 2.2c, we obtained the following results for all X and Y in $\mathbb{R}^2$:

1. Theorem 2.2.1: $\langle X, Y \rangle = \|X\| \, \|Y\| \cos\theta$

2. Cauchy-Schwarz inequality: $|\langle X, Y \rangle| \leq \|X\| \, \|Y\|$

3. Triangle inequality: $\|X + Y\| \leq \|X\| + \|Y\| \, .$

We found that "Cauchy-Schwarz" follows from Theorem 2.2.1 by taking absolute values on both sides and that the triangle inequality follows readily from Cauchy-Schwarz. Moreover, we observe that if Theorem 2.2.1 were true for X and Y in $\mathbb{R}^n$, $n \geq 2$, then the two inequalities would follow in $\mathbb{R}^n$ word for word as in $\mathbb{R}^2$.

We now claim that Theorem 2.2.1 is true in $\mathbb{R}^n$ for all $n \geq 1$ as formally stated in the next theorem.

Theorem 2.2.2 Let X and Y be nonzero vectors in $\mathbb{R}^n$. Then

$$\langle X, Y \rangle = \|X\| \, \|Y\| \cos \theta,$$

where θ is the angle between X and Y.

Strictly speaking, we have stated this theorem too early, since we have not yet made a careful definition of the angle between two vectors X and Y in $\mathbb{R}^n$. We will work our way up to such a definition in three steps. First we will give a definition of orthogonality (perpendicularity) in $\mathbb{R}^n$, based on a familiar theorem from geometry. Second we will find a way to drop a perpendicular from one vector to another, which will allow us to form a right triangle with two of its sides parallel to X and Y. Third, we will use this triangle and the usual definition of cosine to define the cosine of the angle between X and Y. Theorem 2.2.2 will be a trivial consequence of this definition.

Orthogonality in $\mathbb{R}^n$. Let X and Y be two vectors in $\mathbb{R}^n$. Recall that in the remarks that we made concerning Figure 2.4 in $\mathbb{R}^2$, we noted that the three sides of the triangle with vertices $\mathbf{0}$, X, and Y have lengths equal to the quantities $\|X\|$, $\|Y\|$, and $\|X - Y\|$. If X and Y are perpendicular, this triangle is a right triangle, and the Pythagorean theorem tells us that the **Pythagorean relationship** between X and Y holds:

$$\|X\|^2 + \|Y\|^2 = \|X - Y\|^2 \, .$$

On the other hand, it is a geometric fact that if the Pythagorean relationship between X and Y holds, then the corresponding triangle is a right triangle, with the right angle between X and Y. (Note that the Pythagorean relationship holds if either X or Y is the zero vector. We will agree to call a triangle a right triangle if one of its sides has zero length.)

We could use the Pythagorean relationship directly to define orthogonality in $\mathbb{R}^n$. However, it will be more convenient to have an equivalent definition involving the inner product. Using the properties of inner products, we compute as follows:

$$\begin{aligned}
\|X - Y\|^2 &= \langle X - Y, X - Y \rangle \\
&= \langle X, X \rangle - \langle X, Y \rangle - \langle Y, X \rangle + \langle Y, Y \rangle \\
&= \|X\|^2 - 2\langle X, Y \rangle + \|Y\|^2 \, .
\end{aligned}$$

Thus, the Pythagorean relationship between X and Y holds if and only if $\langle X, Y \rangle = 0$. This fact motivates the following definition:

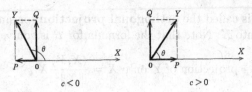

$c < 0$ $\qquad\qquad\qquad$ $c > 0$

FIGURE 2.6. $Y = P + Q$, with $P = cX$ and $P \perp Q$

Definition 2.2.1 The vectors X and Y are **perpendicular**, or **orthogonal**, if and only if $\langle X, Y \rangle = 0$.

We have already found that vectors X and Y in $\mathbb{R}^2$ are orthogonal if and only if $\langle X, Y \rangle = 0$, so this definition in $\mathbb{R}^n$ certainly agrees with our earlier notion of orthogonality in $\mathbb{R}^2$. As in $\mathbb{R}^2$, our definition means, in particular, that the zero vector in $\mathbb{R}^n$ is orthogonal to every vector in $\mathbb{R}^n$, including itself.

Projection of Y onto X. Let us assume that X is not the zero vector. We wish to form a right triangle with sides parallel to X and Y. One of the sides of this triangle will be a vector P contained in the line spanned by the vector X. The hypotenuse of the triangle will be the vector Y. See Figure 2.6. We think of P as being formed by dropping a perpendicular from Y to the line spanned by X. Assuming that we can find such a vector P, the triangle with vertices $\mathbf{0}$, Y, and P will be a right triangle, and the vector $Q = Y - P$ will be perpendicular to the vector X, as shown in Figure 2.6.

How do we find the vector P? We want P to be on the line spanned by X, so P must be a scalar multiple of X. We also want the vectors $Y - P$ and X to be orthogonal. Thus, we are looking for a scalar c such that

$$\langle Y - cX, X \rangle = 0 \, .$$

We may easily solve this equation for c:

$$c = \frac{\langle X, Y \rangle}{\langle X, X \rangle} \, .$$

This value of c is the unique value that satisfies our requirements. The vector $P = cX$ is the only vector on the line spanned by X such that $Y - P$ is orthogonal to X. Thus we have proved the following:

Theorem 2.2.3 Let X and Y be vectors in $\mathbb{R}^n$, with X nonzero. Then there is a unique vector P on the line spanned by X and a unique vector Q orthogonal to X such that

$$Y = P + Q \, .$$

The vector P is given by the formula $P = cX$, where $c = \langle X, Y \rangle / \langle X, X \rangle$.

The vector P is called the **orthogonal projection** (or simply the **projection**) of Y onto X. Note that the formula for P is equivalent to

$$\text{projection of } Y \text{ onto } X = \frac{\langle X, Y \rangle}{\|X\|^2} X \ .$$

REMARKS:

- We find Q by using the formula $Q = Y - P$. The length of Q is the shortest distance between the point Y and the line spanned by X. We will discuss this further in Section 2.2e.

- Even though the formula for P seems to depend on X, it actually only depends on the line spanned by X. Proof: If $\alpha \neq 0$ is a scalar, then the projection of Y onto αX is

$$\frac{\langle \alpha X, Y \rangle}{\langle \alpha X, \alpha X \rangle} (\alpha X) = \alpha \frac{\langle X, Y \rangle}{\alpha^2 \langle X, X \rangle} X = \frac{\langle X, Y \rangle}{\langle X, X \rangle} X \ .$$

- If $c = 0$, which is the case if and only if X and Y are orthogonal, then $P = \mathbf{0}$ and $Q = Y$.

EXAMPLES:

5. Let $X = (1, 2, -1)$ and $Y = (2, 4, 1)$. The projection of Y onto X is the vector $P = cX$, where

$$c = \frac{\langle (1, 2, -1), (2, 4, 1) \rangle}{\langle (1, 2, -1), (1, 2, -1) \rangle} = \frac{3}{2} \ .$$

Thus $P = (\frac{3}{2}, 3, -\frac{3}{2})$. We may check our answer by verifying that the vector $Q = Y - P = (\frac{1}{2}, 1, \frac{5}{2})$ is orthogonal to X:

$$\left\langle \left(\frac{1}{2}, 1, \frac{5}{2} \right), (1, 2, -1) \right\rangle = \frac{1}{2} + 2 - \frac{5}{2} = 0 \ .$$

6. In the previous example, let us find the projection of X onto Y. In the expression for c, the numerator is the same, but the denominator is $\langle Y, Y \rangle = 21$, so $c = 3/7$, and $cY = (\frac{6}{7}, \frac{12}{7}, \frac{3}{7})$. You should check that the vector $Q = X - cY$ is orthogonal to Y.

The angle between X and Y. We are now in a position to obtain a formula for the cosine of θ. Assume that Y is nonzero. Referring to Figure 2.6, we see that if $c > 0$, it is reasonable to define

$$\cos \theta = \|P\| / \|Y\| \ ,$$

while if $c < 0$,

$$\cos \theta = -\|P\| / \|Y\| \ .$$

From the formula for P, we find that

$$\|P\| = (|\langle X, Y \rangle| / \|X\|^2) \|X\| = |\langle X, Y \rangle| / \|X\| \ .$$

Since the sign of $c = \langle X, Y \rangle / \langle X, X \rangle$ is always the same as the sign of $\langle X, Y \rangle$, we may combine the above equations to obtain the following definition for the cosine of the angle between X and Y:

$$\cos \theta = \frac{\langle X, Y \rangle}{\|X\| \|Y\|} \ .$$

REMARKS:

- It is important to realize that we have not *derived* a formula for $\cos \theta$, but have instead made a *definition* that is motivated by some pictures and our knowledge of trigonometry. The definition is consistent with the formula we derived in $\mathbb{R}^2$ in Section 2.2c. Since the cosine is a function which is bounded between -1 and 1, it might be comforting to know that we can prove directly from our definition that $|\cos \theta| \leq 1$, without reference to any pictures. Recall that we chose c so that $P = cX$ and Q are orthogonal, or in other words so that the Pythagorean relationship holds for cX and Q: $\|cX\|^2 + \|Q\|^2 = \|cX + Q\|^2 = \|Y\|^2$. It follows that $|c|^2 \|X\|^2 \leq \|Y\|^2$. Since $c = \langle X, Y \rangle / \|X\|^2$, we have

$$\langle X, Y \rangle^2 / \|X\|^2 \leq \|Y\|^2 \ ,$$

which is essentially the Cauchy-Schwarz inequality. (By the way, there is a proof of the Cauchy-Schwarz inequality that uses no trigonometry whatever but rather the famous quadratic formula of high school algebra. You are led through this proof in the exercises.) Now divide both sides by $\|Y\|^2$, and look at the definition of

$\cos \theta$ to conclude that $(\cos \theta)^2 \leq 1$, or equivalently, that $|\cos \theta| \leq 1$. Done.

- Theorem 2.2.2 is now a trivial consequence of our definition of the cosine of the angle between X and Y. From Theorem 2.2.2 it follows that the triangle inequality is true in all the vector spaces $\mathbb{R}^n$. As mentioned above, the proof can be based on Theorem 2.2.2, exactly as in $\mathbb{R}^2$.

2.2E PROJECTION AND DISTANCE IN $\mathbb{R}^n$

In the previous section, we found it useful to project a vector Y onto another vector X. We remarked that this projection is the same if we project Y onto any nonzero scalar multiple of X. Thus, if S is the line spanned by X, we can speak of the projection of Y onto S. In this section, we will extend this idea to include projections of vectors onto linear subspaces other than lines.

Let Y be a vector and S a linear subspace of $\mathbb{R}^n$. Projecting Y onto S means writing Y as $P+Q$, where P is a vector in S and Q is perpendicular to every vector in S. We will show in the next theorem that this decomposition of Y into the sum $P+Q$ is unique. The vector P is called the **orthogonal projection** or simply the **projection** of Y onto S.

Theorem 2.2.4 Let Y be a vector and S a linear subspace of $\mathbb{R}^n$. Suppose that $Y = P+Q$, where P is a vector in S and Q is orthogonal to every vector in S. Then for any vector Z in S other than P, $\|Y - P\| < \|Y - Z\|$. The vector P is the unique vector in S such that $Y - P$ is orthogonal to every vector in S.

PROOF: Let Z be some vector in S other than P. Since S is a linear subspace and P is in S, the vector $P - Z$ is also in S. Since the vector $Q = Y - P$ is orthogonal to every vector in S, $Y - P$ and $P - Z$ are orthogonal. Thus, the Pythagorean relationship holds for $Y - P$ and $P - Z$:

$$\|Y - P\|^2 + \|P - Z\|^2 = \|(Y - P) + (P - Z)\|^2 = \|Y - Z\|^2 .$$

Since P is not equal to Z, $\|P - Z\|^2 > 0$. It follows that $\|Y - P\|^2 < \|Y - Z\|^2$, or equivalently, $\|Y - P\| < \|Y - Z\|$ for every vector Z in S other than P. We have shown that if P satisfies the hypothesis of the theorem, then P is closer to Y than any other vector in S. Therefore, P is the only vector that satisfies that hypothesis. Done. $<<$

Since the projection P of Y onto S is closer to Y than any other vector in S, we call the quantity $\|Y - P\| = \|Q\|$ the **distance** between the point Y and linear subspace S.

REMARKS:

- The preceding theorem does not assert that there exists a $P \in S$ such that $Y - P$ is orthogonal to every vector in S. It only asserts uniqueness, that is, that there can be at most one such P. Existence will have to wait until the next chapter, at which point it will be proved that there always exists such a P.

- If $Y \in S$, then $P = Y$ and $Q = \mathbf{0}$. The distance between Y and S is 0.

- If S is a line through the origin, we already know how to find P: simply project Y onto any vector X that spans S.

EXAMPLES:

7. Let us find the projection of $Y = (3, 4, 4, -1)$ onto the line S spanned by the vector $X = (1, 2, 0, -1)$. The projection of Y onto S is the same as the projection of Y onto X:

$$P = \frac{\langle (3, 4, 4, -1), (1, 2, 0, -1) \rangle}{\langle (1, 2, 0, -1), (1, 2, 0, -1) \rangle}(1, 2, 0, -1)$$
$$= 2(1, 2, 0, -1) = (2, 4, 0, -2) .$$

The distance between Y and S is

$$\|(3, 4, 4, -1) - (2, 4, 0, -2)\| = \|(1, 0, 4, 1)\| = 3\sqrt{2} .$$

You should verify that the vectors $(2, 4, 0, -2)$ and $(1, 0, 4, 1)$ are orthogonal and that their sum is Y.

8. Let us find the projection of the vector $Y = (-2, -6, -17)$ onto the plane S spanned by $X_1 = (1, 1, -2)$ and $X_2 = (1, -5, -4)$. We are looking for a vector P of the form $\alpha_1(1, 1, -2) + \alpha_2(1, -5, -4)$ such that the vector $Q = Y - P$ is orthogonal to every vector in S. Note that if Q is orthogonal to both X_1 and X_2, then Q is orthogonal to every linear combination of X_1 and X_2 (can you see why?), so Q is orthogonal to every vector in S. Thus, we only need P to satisfy

$$\langle Y - P, X_1 \rangle = 0 \quad \text{and} \quad \langle Y - P, X_2 \rangle = 0 .$$

Substituting in the values for Y, X_1, and X_2 and the expression for P, we obtain the following two equations in the unknowns α_1 and α_2:

$$\langle (-2, -6, -17) - \alpha_1(1, 1, -2) - \alpha_2(1, -5, -4), (1, 1, -2) \rangle = 0$$
$$\langle (-2, -6, -17) - \alpha_1(1, 1, -2) - \alpha_2(1, -5, -4), (1, -5, -4) \rangle = 0 ,$$

which, after the various inner products are computed, become

$$6\alpha_1 + 4\alpha_2 = 26$$
$$4\alpha_1 + 42\alpha_2 = 96 .$$

This system of two equations is easily solved using the Gauss-Jordan method. The solution is

$$\alpha_1 = 3 , \quad \alpha_2 = 2 .$$

The projection of Y onto S is

$$P = 3(1, 1, -2) + 2(1, -5, -4) = (5, -7, -14) .$$

The distance between Y and S is the norm of the vector

$$Q = Y - P = (-2, -6, -17) - (5, -7, -14) = (-7, 1, -3) ,$$

which is $\sqrt{59}$. You should verify that Q is orthogonal to the vectors P, X_1, and X_2.

The second example illustrates the general method for projecting a vector Y onto a plane S in $\mathbb{R}^n$. If S is spanned by X_1 and X_2, we look for a vector P of the form $\alpha_1 X_1 + \alpha_2 X_2$ such that $Y - P$ is orthogonal to both X_1 and X_2. In terms of inner products,

$$\langle Y - \alpha_1 X_1 - \alpha_2 X_2, X_1 \rangle = 0 \quad \text{and} \quad \langle Y - \alpha_1 X_1 - \alpha_2 X_2, X_2 \rangle = 0.$$

Using the properties of inner products, these two equations may be rewritten as

$$\alpha_1 \langle X_1, X_1 \rangle + \alpha_2 \langle X_1, X_2 \rangle = \langle Y, X_1 \rangle$$
$$\alpha_1 \langle X_1, X_2 \rangle + \alpha_2 \langle X_2, X_2 \rangle = \langle Y, X_2 \rangle.$$

Solve these two equations for the unknowns α_1 and α_2. Take the corresponding linear combination of X_1 and X_2 to obtain the projection P. The distance between Y and S is $\|Y - P\|$.

It is interesting to note that even though the vectors Y, X_1, and X_2 are in $\mathbb{R}^n$, we always obtain *two* equations in *two* unknowns when finding the projection of a vector onto a plane, whatever the value of n.

REMARKS:

- The method we have just outlined extends naturally to projections onto k-dimensional linear subspaces. If S is spanned by the vectors $X_1, \ldots, X_k$, then we want the vector P to satisfy $\langle Y - P, X_i \rangle = 0$ for $i = 1, \ldots, k$. Substitute $\alpha_1 X_1 + \cdots + \alpha_k X_k$ for P to obtain k equations in the k unknowns $\alpha_1, \ldots, \alpha_k$. Solve the equations and then compute P. The projection of Y onto S is P, and the distance between Y and S is $\|Y - P\|$.

- Do the systems of equations that arise in this method always have a solution? The answer is 'yes', but the proof is best delayed until after we have developed some more theory about systems of equations in Chapter 3. Another approach is to use calculus, as in Chapter 7, to show that there is a point in the linear subspace S whose distance from Y is smaller than the distance between Y and any other point in S.

Suppose one is interested in finding the "projection" of a vector Y onto an affine subspace $\mathcal{A}$ and the "distance" from Y to $\mathcal{A}$. Here is the issue described more precisely. We want to find P and Q such that $Y = P + Q$, $P \in \mathcal{A}$, and Q is perpendicular to the *difference* of any two members of $\mathcal{A}$. It develops that this can always be done, and in only one way. The vector P, thus uniquely determined, is the **projection** of Y onto $\mathcal{A}$, and $\|Q\|$ is the **distance** from Y to $\mathcal{A}$.

Write $\mathcal{A} = S + Z$ for some Z. Write $Y - Z = R + Q$, where R is the projection of $Y - Z$ onto the *linear* subspace S. Then $Y = P + Q$, where

$P = R + Z$. Now let us check that P is the projection of Y onto the *affine* subspace $\mathcal{A}$ and that, therefore, $\|Q\|$ is the distance from Y to $\mathcal{A}$. It is clear that $P \in \mathcal{A}$ since it is the sum of Z and a member of $\mathcal{S}$. That Q is perpendicular to the difference of any two members of $\mathcal{A}$ follows from the facts that Q is perpendicular to every member of $\mathcal{S}$ and that the difference of any two members of $\mathcal{A}$ is a member of $\mathcal{S}$. The drawing of an accurate picture for the following example can be of help in grasping these ideas.

EXAMPLE:

9. Let us calculate the projection of $(3, 5)$ onto the line $\mathcal{S} + (-1, 3)$ where $\mathcal{S}$ is the one-dimensional linear subspace spanned by $(3, 4)$. We subtract $(-1, 3)$ from $(3, 5)$ and find the projection of the difference $(4, 2)$ onto $(3, 4)$:

$$\frac{\langle (3, 4), (4, 2) \rangle}{\|(3, 4)\|^2} (3, 4) = \left(\frac{12}{5}, \frac{16}{5} \right).$$

Thus, the projection of $(3, 5)$ onto $\mathcal{S} + (-1, 3)$ is $(\frac{12}{5}, \frac{16}{5}) + (-1, 3) = (\frac{7}{5}, \frac{31}{5})$.

To get the distance from $(3, 5)$ to $\mathcal{S} + (-1, 3)$ we subtract its projection from it and take the norm:

$$\left\|\left(3 - \frac{7}{5}, 5 - \frac{31}{5}\right)\right\| = \sqrt{4} = 2.$$

Exercises

Practice with projections, angles, and distances.

1. Compute $\langle X, Y \rangle$ for the given vectors:

   (a) $X = (1, 3)$, $Y = (6, -2)$ in $\mathbb{R}^2$

   (b) $X = (1, 3, 1)$, $Y = (2, 0, 5)$ in $\mathbb{R}^3$

   (c) $X = (1, 4, 0, 1)$, $Y = (0, 1, 1, -1)$ in $\mathbb{R}^4$

   (d) $X = (1, 1, -1, 2, 0, 1)$, $Y = (1, -1, 1, 0, 4, 1)$ in $\mathbb{R}^6$.

2. (a) Compute $\langle X, X \rangle$, where $X = (1, -3, 1)$ in $\mathbb{R}^3$.

   (b) Now compute $\|X\|$, using part (a).

3. Compute $\cos \theta$, where θ is the angle between $X = (1, 1)$ and $Y = (0, 2)$ in $\mathbb{R}^2$. Note that $\cos \theta$ is unchanged, as we would expect, if we replace X by $2X$.

4. Do the following for each of the pairs X, Y in Exercise 1:

   (a) Find the cosine of the angle between X and Y.

   (b) Find the projection of Y onto X.

   (c) Find a vector P on the line spanned by X and a vector Q perpendicular to that line such that $Y = P + Q$.

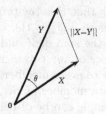

FIGURE 2.7. The law of cosines

(d) Find the distance between the point Y and the line spanned by X.

5. (a) Compute a nonzero vector $Y = (y_1, y_2)$ that is orthogonal to $X = (1, 5)$. Is Y unique?

   (b) How many unit vectors are orthogonal to X? Draw them.

6. For $X = (x_1, x_2) \in \mathbb{R}^2$, let $f(X) = 3x_1 + 4x_2$. Using the Cauchy-Schwarz inequality and the fact that $f(X) = \langle X, Y \rangle$ with $Y = (3, 4)$, prove that if $\|X\| \le 1$, then $|f(X)| \le 5$. Also find a V with $\|V\| \le 1$ for which $f(V) = -5$.

7. Let $Y = (2, 4, -1)$ in $\mathbb{R}^3$. True or false: X (in $\mathbb{R}^3$) and Y are orthogonal if and only if $2x_1 + 4x_2 - x_3 = 0$. Describe the set of all vectors X in $\mathbb{R}^3$ such that X is orthogonal to the fixed vector Y.

8. For each part find the projection of Y onto the plane S. Also find the distance between the point Y and the linear subspace S.

   (a) $Y = (-7, -17, -33)$; S is spanned by $X_1 = (6, -3, -2)$ and $X_2 = (3, 2, 6)$

   (b) $Y = (1, 1, 3)$; S is spanned by $X_1 = (1, 1, 1)$ and $X_2 = (2, 0, -1)$

   (c) $Y = (1, -1, -1)$; S is spanned by $X_1 = (1, 0, 2)$ and $X_2 = (3, -1, 3)$

   (d) $Y = (1, 1, 1, 1)$; S is spanned by $X_1 = (1, 1, 3, 3)$ and $X_2 = (1, 3, 1, 3)$

   (e) $Y = (2, 3, 0)$; S is the solution set of the equation $x_1 - x_2 + 2x_3 = 0$

   (f) $Y = (y_1, y_2, y_3)$, an arbitrary vector in $\mathbb{R}^3$, and S the span of $(1, 0, 0)$ and $(0, 1, 0)$. Express your answer in terms of the coordinates of Y.

9. High school algebra and Cauchy-Schwarz:

   (a) Recall that the quadratic polynomial $f(t) = at^2 + bt + c$, with $a, b, c \in \mathbb{R}$, is ≥ 0 for all t or ≤ 0 for all t if and only if it has either no real roots or one (double) real root. From the quadratic formula, this happens if and only if $b^2 - 4ac \le 0$.

   (b) Define $f(t) = \langle X + tY, X + tY \rangle$, where t is a real variable. Explain why $f(t) \ge 0$ for all t.

   (c) Using the algebraic properties of the inner product, write $f(t)$ as a quadratic polynomial in t with coefficients $a = \langle Y, Y \rangle$, $b = 2\langle X, Y \rangle$, and $c = \langle X, X \rangle$.

   (d) Using the last sentence of part (a), prove $|\langle X, Y \rangle| \le \|X\| \, \|Y\|$.

10. *The law of cosines*. Prove this generalization of Pythagoras' theorem (see Figure 2.7):

$$\|X - Y\|^2 = \|X\|^2 + \|Y\|^2 - 2\|X\|\,\|Y\|\cos\theta.$$

11. Use the law of cosines to calculate the angles of the triangles with vertices

   (a) $\mathbf{0}$, $(0, 0, 1)$, and $(1, 0, 0)$

   (b) $\mathbf{0}$, $(1, 0, 0)$, and $(1, 1, 0)$

   (c) $\mathbf{0}$, $(1, 1, 0)$, and $(0, 1, 1)$

   (d) $(1, 2, 3)$, $(4, 1, 4)$, and $(2, -1, 0)$

   (e) $\mathbf{0}$, $(1, 1, 1, 0)$, and $(0, 1, 1, 1)$

   (f) $(1, 2, 1, 2)$, $(2, 2, 3, 3)$, and $(3, 1, 1, 3)$.

12. Consider the triangle in $\mathbb{R}^n$ whose vertices are the origin, the point all of whose coordinates equal 1 except for the last which equals 0, and the point all of whose coordinates equal 1 except for the first which equals 0. Find formulas for its three angles. Then decide what happens as $n \to \infty$.

13. Let X and Y be two vectors in $\mathbb{R}^2$. Prove that X is perpendicular to $Y \Leftrightarrow \|X - Y\|^2 = \|X + Y\|^2 = \|X\|^2 + \|Y\|^2$.

14. Let X and $Y \in \mathbb{R}^n$. Using only the four basic properties of inner products, prove:

   (a) $4\langle X, Y \rangle = \|X + Y\|^2 - \|X - Y\|^2$

   (b) $\|X + Y\|^2 + \|X - Y\|^2 = 2\|X\|^2 + 2\|Y\|^2$. Show with a picture how this formula states that the sum of the squares of the diagonals of a parallelogram equals the sum of the squares of the sides.

15. For each part find the projection of Y onto $\mathcal{A} = \mathcal{S} + Z$ and the distance from Y to $\mathcal{A}$.

   (a) $Y = (1, 0)$; $Z = (0, 1)$; $\mathcal{S}$ is spanned by $(1, -1)$

   (b) $Y = (0, 0)$; $Z = (0, 1)$; $\mathcal{S}$ is spanned by $(1, -1)$

   (c) $Y = (1, 0, 2)$; $Z = (0, 1, 1)$; $\mathcal{S}$ is spanned by $(1, -1, -2)$

   (d) $Y = (1, 0, 2)$; $Z = (0, 1, 1)$; $\mathcal{S}$ is spanned by $(1, -1, -2)$ and $(1, 0, 1)$

   (e) $Y = (1, 0, 2)$; $Z = (0, 1, 1)$; $\mathcal{S}$ is spanned by $(1, -1, -2)$ and $(1, 0, 0)$

   (f) $Y = (1, 0, 0, 1)$; $Z = (0, 1, 1, 0)$; $\mathcal{S}$ is spanned by $(1, 1, 1, 1)$.

2.3   Hyperplanes and Orthogonality in $\mathbb{R}^n$

Let us remind you of the definition of orthogonality: two vectors X and Y in $\mathbb{R}^n$ are perpendicular, or orthogonal, denoted $X \perp Y$, if and only if $\langle X, Y \rangle = 0$. Recall that we agree to say that the zero vector is orthogonal to any vector.

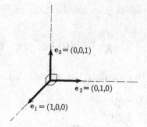

FIGURE 2.8. The standard basis vectors are pairwise orthogonal

Note that the standard basis vectors $\mathbf{e}_1$, $\mathbf{e}_2$, and $\mathbf{e}_3$ in $\mathbb{R}^3$ are pairwise orthogonal, $\mathbf{e}_1 \perp \mathbf{e}_2$, $\mathbf{e}_1 \perp \mathbf{e}_3$, and $\mathbf{e}_2 \perp \mathbf{e}_3$. See Figure 2.8. This agrees with our geometric intuition. Also, in $\mathbb{R}^4$, we readily compute, for instance, that $X = (1, -1, 2, 0)$ and $Y = (0, 6, 3, -2)$ are orthogonal.

Here is a problem. Given a fixed vector Z in $\mathbb{R}^n$, describe the set of all vectors X such that $X \perp Z$. For instance, is this set a linear subspace of $\mathbb{R}^n$? One description is

$$\{X : \langle X, Z \rangle = 0\} \, ;$$

that is, the set S of those $(x_1, \ldots, x_n)$ for which $x_1 z_1 + \cdots + x_n z_n = 0$. Theorem 1.7.1 tells us that S is a hyperplane provided that $Z \neq 0$. In particular, the set of vectors orthogonal to a given nonzero vector in $\mathbb{R}^3$ is a plane.

EXAMPLES:

1. Let $Z = (1, 1, 1)$. Then the set S of vectors X such that $X \perp Z$ is the set of solutions to $\langle X, Z \rangle = x_1 + x_2 + x_3 = 0$, a plane through the origin in $\mathbb{R}^3$. A basis for S is given by $X_1 = (1, 0, -1)$ and $X_2 = (0, 1, -1)$. Thus $X \perp Z$ if and only if there exist scalars α_1 and α_2 such that

$$X = \alpha_1(1, 0, -1) + \alpha_2(0, 1, -1) = (\alpha_1, \alpha_2, -\alpha_1 - \alpha_2) \, .$$

2. Let $Y = (0, 2, -1, 2)$ and let S be the hyperplane determined by the equation $x_1 + 2x_2 - x_3 - x_4 = 0$. According to the above discussion, this hyperplane is the set of all vectors orthogonal to $(1, 2, -1, -1)$. We know how to find the projection of Y onto $(1, 2, -1, -1)$. We will do so and then use the result to find the projection of Y onto S.

   Let P be the projection of Y onto $(1, 2, -1, -1)$:

$$\begin{aligned} P &= \frac{\langle (1, 2, -1, -1), (0, 2, -1, 2) \rangle}{\langle (1, 2, -1, -1), (1, 2, -1, -1) \rangle} (1, 2, -1, -1) \\ &= \frac{3}{7}(1, 2, -1, -1) = \left(\frac{3}{7}, \frac{6}{7}, -\frac{3}{7}, -\frac{3}{7} \right) \, . \end{aligned}$$

Let $Q = Y - P = (-\frac{3}{7}, \frac{8}{7}, -\frac{4}{7}, \frac{17}{7})$. Since $Q \perp P$ and P is a scalar multiple of $(1, 2, -1, -1)$, it follows that $Q \perp (1, 2, -1, -1)$, and, hence, that $Q \in S$, a fact that could also be checked by direct calculation. Thus, we have written Y as the sum of two vectors P and Q, with Q in S and P orthogonal to every vector in S. It follows that Q (not P!) is the projection of Y onto S, and that the norm of P (not the norm of Q) is the distance between Y and S. This distance is $\frac{3}{7}\sqrt{7}$.

The preceding example illustrates a useful method for projecting vectors onto hyperplanes. Given a vector Y and a hyperplane S in $\mathbb{R}^n$, we first find a vector Z which is orthogonal to every vector in S, and we project Y onto Z. Call this projection P, and let $Q = Y - P$. The vector Q is perpendicular to P, hence perpendicular to Z. The hyperplane is the set of all vectors orthogonal to Z, so Q is contained in S; Q is the projection of Y onto S. The distance between Y and S is $\|P\|$. Note how this method switches the roles of P and Q.

Consider the problem of finding the projection of a vector Y onto a linear subspace S of $\mathbb{R}^3$ and finding the distance from Y to S. The method described in Section 2.2d works if S is a line. The method described above works if S is a plane (since planes in $\mathbb{R}^3$ are hyperplanes). If $S = \{0\}$, then the projection of Y onto S is 0 and the distance from Y to S equals $\|Y\|$. If $S = \mathbb{R}^3$, then the projection of Y onto S is Y itself and the distance from Y to S is 0. Thus, when working in $\mathbb{R}^3$, we are able to very quickly calculate the projection of a given vector onto any linear subspace and its distance from that linear subspace.

Exercises

1. (a) Let $Z = (2, -1, 3)$ in $\mathbb{R}^3$. Compute the equation that determines the plane S of vectors $X = (x_1, x_2, x_3)$ such that $X \perp Z$.

    (b) Find vectors X_1 and X_2 in S that span S. Each of them will, of course, be perpendicular to Z.

2. (a) Let X_1 and X_2 be as in Exercise 1(b). Describe geometrically the set S' of all Y such that $Y \perp X_1$ and $Y \perp X_2$. Is this set a plane through the origin; a linear subspace? Is Z in this set (cf. Exercise 1(a))? What relation has S' to S?

    (b) Find a system of *two* linear equations whose set of solutions is precisely the set S' of those $Y = (y_1, y_2, y_3)$ which satisfy $Y \perp X_1$ and $Y \perp X_2$.

3. Without calculation, find a vector perpendicular to the plane through the origin in $\mathbb{R}^3$ determined by each of the following equations:

    (a) $2x_1 - 12x_2 + 11x_3 = 0$

    (b) $x_2 = 0$

    (c) $x_1 - x_2 + x_3 = 0$.

4. Given a vector Z in $\mathbb{R}^3$, outline a procedure for obtaining a system of two homogeneous equations in y_1, y_2, and y_3 whose set of solutions is the line (one-dimensional linear subspace) determined by Z. Use the methods in Exercises 1 and 2. Note that this procedure amasses information about the plane S orthogonal to Z; each linear equation is the statement that some inner product is zero.

5. (a) Given $X_1 = (1,0,2)$ and $X_2 = (2,2,3)$, compute a nonzero vector Z orthogonal to both. How unique is Z?

   (b) Use Z to find a homogeneous equation determining the plane (two-dimensional linear subspace) spanned by X_1 and X_2. How unique is this equation?

6. For each of the following give a proof if true and give a counterexample if false.

   (a) $Z = (1,4,-1)$ is orthogonal to the linear subspace determined by $x_1 + 4x_2 + x_3 = 0$.

   (b) Z is not orthogonal to any vectors in the linear subspace in part (a).

   (c) If X_1 and X_2 are nonzero vectors and if the set of all vectors in $\mathbb{R}^3$ orthogonal to both of them is a plane, then $X_1 = \alpha X_2$ for some scalar α.

   (d) $(1,0,1)$ and $(0,-1,0)$ are orthogonal to each other.

   (e) If $Z \perp X$ and $Z \perp Y$, then Z is perpendicular to any vector $\alpha X + \beta Y$.

   (f) If $X \perp Z$ and both are nonzero, then $X = \alpha Z$.

   (g) If $X \neq \alpha Z$ for every scalar α, then $X \perp Z$.

7. Let X and Y be nonzero vectors. If $X \perp Y$ and $aX + bY = \mathbf{0}$, prove that $a = b = 0$. In other words, if X and Y are orthogonal and nonzero, then they are linearly independent.

8. Let $Y = (1,1,1,1)$ and let S be the hyperplane in $\mathbb{R}^4$ determined by the equation $x_1 + x_2 - x_3 + x_4 = 0$. Find the projection of Y onto S and the distance between Y and S. Do the same for S equal to the hyperplane determined by $x_2 = 0$.

9. In $\mathbb{R}^n$, find the distance between the vector $Y = (1,1,\ldots,1)$ and the hyperplane determined by the equation $x_n = 0$.

10. Let $X, Y \in \mathbb{R}^n$.

    (a) If $X \perp Y$, prove that $\|X - aY\| \geq \|X\|$ for all $a \in \mathbb{R}$.

    (b) If $\|X - aY\| \geq \|X\|$ for all $a \in \mathbb{R}$, prove that $X \perp Y$.

2.4   The Cross Product in $\mathbb{R}^3$

As you grasped years ago, the world we inhabit appears to have three space dimensions. For this reason the material in this section is important in many applications. We intend to define a way to multiply two vectors X and Y in $\mathbb{R}^3$. Although the inner product of two vectors $\langle X, Y \rangle$ is a *scalar*, this product $X \times Y$, called the "cross product" (or "vector product"), is a *vector*. In physics, the ideas of angular momentum, torque, and many others are expressed using the cross product. Incidentally, there is no way to define an analogous product of two vectors in $\mathbb{R}^n$ if $n > 3$ (although one can define a "product" of $n - 1$ vectors in $\mathbb{R}^n$).

For the remainder of this one section, $\mathbf{i}$, $\mathbf{j}$, and $\mathbf{k}$ (and not $\mathbf{e}_1$, $\mathbf{e}_2$, and $\mathbf{e}_3$) will always denote the standard basis, so that $(x_1, x_2, x_3) = x_1\mathbf{i} + x_2\mathbf{j} + x_3\mathbf{k}$. Here is the definition of the cross product. It should be memorized (but see Exercise 5).

Definition 2.4.1 The **cross product** of vectors $X = x_1\mathbf{i} + x_2\mathbf{j} + x_3\mathbf{k}$ and $Y = y_1\mathbf{i} + y_2\mathbf{j} + y_3\mathbf{k}$ is defined as the vector

$$X \times Y = (x_2 y_3 - x_3 y_2)\mathbf{i} + (x_3 y_1 - x_1 y_3)\mathbf{j} + (x_1 y_2 - x_2 y_1)\mathbf{k}.$$

REMARK:

- Be careful with the sign of the $\mathbf{j}$ term in the preceding definition. It is the negative of what you might expect.

EXAMPLES:

1. $(1,0,0) \times (0,1,0) = (0,0,1)$, $(0,0,1) \times (0,1,0) = -(1,0,0)$, $(0,1,0) \times (0,1,0) = \mathbf{0}$; that is, $\mathbf{i} \times \mathbf{j} = \mathbf{k}$, $\mathbf{k} \times \mathbf{j} = -\mathbf{i}$, and $\mathbf{j} \times \mathbf{j} = \mathbf{0}$.

2. If $X = (1, 2, -1) = \mathbf{i} + 2\mathbf{j} - \mathbf{k}$ and $Y = (-3, 1, 1) = -3\mathbf{i} + \mathbf{j} + \mathbf{k}$, then

$$X \times Y = (2 + 1)\mathbf{i} + (3 - 1)\mathbf{j} + (1 + 6)\mathbf{k} = 3\mathbf{i} + 2\mathbf{j} + 7\mathbf{k},$$

but

$$Y \times X = (-1 - 2)\mathbf{i} + (1 - 3)\mathbf{j} + (-6 - 1)\mathbf{k} = -3\mathbf{j} - 2\mathbf{j} - 7\mathbf{k}$$

and so we immediately see that the cross product is not commutative. We will see that, in fact, it is always true that $X \times Y = -Y \times X$.

In order to understand the cross product, we will investigate its algebraic and geometric properties.

Theorem 2.4.1 (Algebra and the cross product) Let X, Y, and Z be vectors in $\mathbb{R}^3$. Then the following properties hold:

1. $X \times Y = -(Y \times X)$ (**anti-commutativity**).

2. $X \times (Y + Z) = (X \times Y) + (X \times Z)$ (**distributivity**).

3. $\alpha(X \times Y) = (\alpha X) \times Y$ for all scalars α.

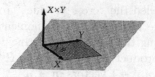

FIGURE 2.9. Geometric interpretation of $X \times Y$

PROOF: These all follow by writing $X = x_1\mathbf{i} + x_2\mathbf{j} + x_3\mathbf{k}$ and similarly for Y, and then using the definition of the cross product. Thus, Property 1 follows from

$$\begin{aligned} Y \times X &= (y_2x_3 - y_3x_2)\mathbf{i} + (y_3x_1 - y_1x_3)\mathbf{j} + (y_1x_2 - y_2x_1)\mathbf{k} \\ &= -(-y_2x_3 + y_3x_2)\mathbf{i} - (-y_3x_1 + y_1x_3)\mathbf{j} - (-y_1x_2 + y_2x_1)\mathbf{k} \\ &= -X \times Y. \end{aligned}$$

The proofs of Properties 2 and 3 are just as straightforward and just as tedious. $<<$

REMARK:

- Warning! There is (at least) one algebraic rule which you might have anticipated but which was not given above. We are thinking of the associative rule $(X \times Y) \times Z \overset{?}{=} X \times (Y \times Z)$. It is *false* for the cross product. For example,

$$(\mathbf{i} \times \mathbf{j}) \times \mathbf{j} = \mathbf{k} \times \mathbf{j} = -\mathbf{i}, \text{ but } \mathbf{i} \times (\mathbf{j} \times \mathbf{j}) = \mathbf{i} \times \mathbf{0} = \mathbf{0}.$$

In order to interpret the cross product geometrically, we have to visualize both its direction and its magnitude.

Theorem 2.4.2 (Geometry and the cross product) Let X and Y be nonzero vectors in $\mathbb{R}^3$.

1. $X \times Y$ is perpendicular to both X and Y, and in the case that X and Y are linearly independent, to the plane determined by X and Y.

2. $\|X \times Y\| = \|X\|\,\|Y\|\,|\sin\theta|$, where θ is the angle between X and Y (see Figure 2.9).

3. $X \times Y = \mathbf{0} \Leftrightarrow X = cY$ for some scalar c, that is, X and Y lie on the same line through the origin.

PROOF: 1. To prove that $X \perp (X \times Y)$, we show that $\langle X, X \times Y \rangle = 0$. Thus

$$\begin{aligned} \langle X, X \times Y \rangle &= x_1(x_2y_3 - x_3y_2) + x_2(x_3y_1 - x_1y_3) + x_3(x_1y_2 - x_2y_1) \\ &= 0. \end{aligned}$$

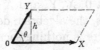

FIGURE 2.10. The parallelogram generated by X and Y

Similarly, one shows that $\langle Y, X \times Y \rangle = 0$, which implies that $Y \perp (X \times Y)$.

2. We compute $\|X \times Y\|^2$:

$$
\begin{aligned}
\|X \times Y\|^2 &= (x_2 y_3 - x_3 y_2)^2 + (x_3 y_1 - x_1 y_3)^2 + (x_1 y_2 - x_2 y_1)^2 \\
&= (x_1^2 + x_2^2 + x_3^2)(y_1^2 + y_2^2 + y_3^2) - (x_1 y_1 + x_2 y_2 + x_3 y_3)^2 \\
&= \|X\|^2 \|Y\|^2 - \langle X, Y \rangle^2 \\
&= \|X\|^2 \|Y\|^2 - \|X\|^2 \|Y\|^2 \cos^2 \theta \\
&= \|X\|^2 \|Y\|^2 \sin^2 \theta \ .
\end{aligned}
$$

Now take square roots.

3. From the formula just given it follows that $X \times Y = 0$ if and only if $\sin \theta = 0$ which, in turn, is true if an only if θ equals 0 or π, which itself is equivalent to X and Y being collinear with the origin. $<<$

Corollary 2.4.1 The area of the parallelogram whose adjacent sides are X and Y is $A = \|X \times Y\|$.

PROOF: By high school geometry, $A = \|X\| h$ (see Figure 2.10). But $h = \|Y\| |\sin \theta|$, and so $A = \|X\| \|Y\| |\sin \theta| = \|X \times Y\|$, where the last equality is part 2 of the theorem above. $<<$

To recapitulate the geometric interpretation, $X \times Y$ is a vector perpendicular to the plane determined by X and Y and its magnitude equals the area of the parallelogram whose sides are X and Y. See Figure 2.9.

Now that we have a formula for the area of parallelograms in $\mathbb{R}^3$ it would be nice to have a formula for the area of parallelograms in $\mathbb{R}^2$. Let (x_1, x_2) and (y_1, y_2) be two vectors in $\mathbb{R}^2$. We would like a formula for the area of the parallelogram with vertices at $(0, 0)$, (x_1, x_2), (y_1, y_2), and $(x_1 + y_1, x_2 + y_2)$. To obtain an answer we view $\mathbb{R}^2$ as the linear subspace of $\mathbb{R}^3$ spanned by $(1, 0, 0)$ and $(0, 1, 0)$. With this point of view the vertices of the parallelogram become $(0, 0, 0)$, $(x_1, x_2, 0)$, $(y_1, y_2, 0)$, and $(x_1+y_1, x_2+y_2, 0)$. So, the area equals

$$\|(x_1, x_2, 0) \times (y_1, y_2, 0)\| = |x_1 y_2 - x_2 y_1|.$$

Returning to our general discussion in $\mathbb{R}^3$ we note that there are two vectors which satisfy the geometric description of $X \times Y$, namely, of being perpendicular to both X and Y and having norm equal to the area of the parallelogram with vertices at 0, X, Y, and $X + Y$: if Z is one such vector, $-Z$ is the other. Nevertheless, the algebraic definition of $X \times Y$ picks out one of these two vectors. There is nothing sacred about this choice; in the definition of $X \times Y$ we could have replaced each component by its negative. However, the choice we have made is the standard one, and most people have a way of picturing this using the "right-hand rule". Hold your right hand with the fingers curled and the thumb extended. Now position your right hand so that your fingers curl from the vector X to the vector Y. Then your thumb will point in the direction of $X \times Y$, provided that the positive directions on the axes have been chosen according to the "right-hand rule", the thumb on the right hand pointing in the direction of $\mathbf{k}$ when the fingers are curled from $\mathbf{i}$ to $\mathbf{j}$. Figure 2.9 has an arrow showing the direction in which your fingers curl. Also see Exercise 10.

Undoubtedly the most useful property of the cross product is that, given two vectors X and Y, we can find a vector perpendicular to both of them without having to solve equations. That vector is $X \times Y$.

EXAMPLES:

3. If $X = (1, 1, -1)$ and $Y = (2, 1, 3)$, find all the vectors perpendicular to both X and Y in $\mathbb{R}^3$. Since X and Y span a two-dimensional linear subspace in $\mathbb{R}^3$, the set of vectors perpendicular to X and Y is a one-dimensional linear subspace in $\mathbb{R}^3$. Since $X \times Y = 4\mathbf{i} - 5\mathbf{j} - \mathbf{k} = (4, -5, -1)$ is perpendicular to both X and Y, it generates the line perpendicular to X and Y. Thus every vector Z perpendicular to both X and Y is of the form $Z = c(4, 5, -1)$, where c is a scalar.

4. Let us find the area of the parallelogram two of whose adjacent sides are the arrows $(2, -3)$ and $(-4, 1)$. The area is the norm of an appropriate vector in $\mathbb{R}^3$:

$$\|(2, -3, 0) \times (-4, 1, 0)\| = \|(0, 0, -10)\| = 10 \,.$$

As a final application of the cross product, we define the so-called "triple product". Suppose X, Y, and Z are three vectors in $\mathbb{R}^3$. Then we define the **triple product** $[X, Y, Z]$ to be the inner product $\langle X \times Y, Z \rangle$. We will give a geometric interpretation of this number.

Think of the vectors X, Y, and Z as directed line segments starting at the origin. These three vectors generate a parallelepiped with one corner at the origin, as follows: The four points 0, X, Y, and Z are the vertices of the parallelepiped, and the directed line segments corresponding to X, Y, and Z form three of the edges. The remaining vertices are at $X + Y$, $X + Z$, $Y + Z$, and $X + Y + Z$. See Figure 2.11.

Theorem 2.4.3 The volume of the parallelepiped generated by X, Y, and Z equals $|\langle X \times Y, Z \rangle|$.

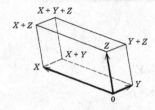

FIGURE 2.11. The parallelepiped generated by X, Y, and Z

PROOF: The volume of the parallelepiped equals the area of the base of the parallelepiped times its height. The base is the parallelogram generated by X and Y, so the area of the base is $\|X \times Y\|$. How do we find the height of the parallelepiped? Note that the height is equal to the norm of the projection of Z onto a vector perpendicular to the base. But $X \times Y$ is perpendicular to the base, so

$$\text{height} = \| \text{ projection of } Z \text{ onto } X \times Y\| .$$

The projection of Z onto $X \times Y$ equals

$$(\langle X \times Y, Z\rangle)/\|X \times Y\|^2)(X \times Y) ,$$

so height $= |\langle X \times Y, Z\rangle|/\|X \times Y\|$.

Thus, the volume of the parallelepiped = (base)(height) = $|\langle X \times Y, Z\rangle|$. Done. $<<$

REMARK:

- The significance of the sign of the triple product is given in Exercise 10.

Exercises

Cross products, triple products, areas, and volumes.

1. Let $X = \mathbf{i} + 2\mathbf{j} - 3\mathbf{k}$, $Y = \mathbf{i} - \mathbf{j}$, $Z = -\mathbf{i} - 2\mathbf{j} + \mathbf{k}$. Compute the following:

   (a) $X \times Y$

   (b) $Y \times X$

   (c) $X \times Z$

   (d) $X \times X$

   (e) $Y \times Y$

   (f) $Z \times Y$

   (g) $X \times (Y \times Z)$

   (h) $(X \times Y) \times Z$

   (i) $X \times (Z \times Y)$

(j) $X \times (X \times Y)$

(k) $X \times (2Y + Z)$

(l) $(X - 2Z) \times Y$

(m) $(X + Y) \times (X - Y)$

(n) $(X + Y) \times (Y + Z)$.

2. Complete the following multiplication table:

$X \times Y$

$X \downarrow$ $Y \rightarrow$	i	j	k
i			
j			
k			

3. Let X, Y, and Z be any vectors in $\mathbb{R}^3$. Prove the following identities:

(a) $(X + Y) \times Z = X \times Z + Y \times Z$

(b) $X \times X = 0$

(c) $X \times (X + Y) = X \times Y$

(d) $X \times (Y \times Z) = \langle X, Z \rangle Y - \langle X, Y \rangle Z$

(e) $(X \times Y) \times Z = \langle Z, X \rangle Y - \langle Z, Y \rangle X$

(f) $(X \times Y) \times (X \times Z) = \langle X, Y \times Z \rangle X$.

4. Prove the **triple product identity**: $\langle X \times Y, Z \rangle = \langle Y \times Z, X \rangle$. An equivalent form of this identity is $\langle X \times Y, Z \rangle = \langle X, Y \times Z \rangle$.

5. By writing $X \times Y = (x_1 \mathbf{i} + x_2 \mathbf{j} + x_3 \mathbf{k}) \times (y_1 \mathbf{i} + y_2 \mathbf{j} + y_3 \mathbf{k})$ and using the multiplication table in Exercise 2 along with Theorem 2.4.1, reconstruct the formula for $X \times Y$. (This shows that if you forget the formula for $X \times Y$, you can easily derive it.)

6. Find all unit vectors perpendicular to the plane spanned by $X = (0, 3, 2)$ and $Y = (1, 1, -1)$.

7. Let S be spanned by $(1, 2, 3,)$ and $(-1, 0, 4)$. Find a nonzero vector X that is perpendicular to every vector in S.

Explain why a vector that is perpendicular to every vector in S is necessarily a scalar multiple of X.

8. Find all unit vectors in the **ij-plane** that are perpendicular to the vector $X = -3\mathbf{i} - 4\mathbf{j} + \mathbf{k}$.

9. (a) Is there a vector X such that $X \times \mathbf{i} = \mathbf{i}$? Justify your statement.

(b) Given the vectors Y and Z, show that there is a vector X such that $X \times Y = Z$ if and only if Z is perpendicular to Y. Is the solution X unique? If not, what is the most general solution?

10. In $\mathbb{R}^3$, we define the vectors X, Y, and Z (the order is important) to be a **right-handed** coordinate system if $\langle X \times Y, Z \rangle > 0$. Similarly, we define the vectors X, Y, and Z be a **left-handed** coordinate system if $\langle X \times Y, Z \rangle < 0$. Prove the following assertions:

(a) $\mathbf{i}$, $\mathbf{j}$, and $\mathbf{k}$ is a right-handed coordinate system.

(b) $\mathbf{j}$, $\mathbf{i}$, and $\mathbf{k}$ is a left-handed coordinate system.

(c) If X, Y, and Z is a right-handed coordinate system, then Y, X, and Z is a left-handed coordinate system.

(d) If X and Y are linearly independent, then X, Y, and $X \times Y$ is a right-handed coordinate system. Also X, Y, and $Y \times X$ is a left-handed coordinate system.

(e) Suppose that X, Y, and Z are linearly independent. Then X, Y, and Z is a right-handed coordinate system if and only if Z and $X \times Y$ lie on the same side of the plane spanned by X and Y. *Hint:* this last part is difficult because it requires a formulation of a definition of "same side", and then an appropriate use of it.

11. Prove that the vectors X, Y, and $Z \in \mathbb{R}^3$ span $\mathbb{R}^3 \Leftrightarrow \langle X \times Y, Z \rangle \neq 0$.

12. Prove that $\|X \times Y\| = \|X\| \|Y\| \Leftrightarrow X \perp Y$.

13. Find the area of a parallelogram whose adjacent sides are the arrows $X = \mathbf{i} + \mathbf{j} - 2\mathbf{k}$ and $Y = 2\mathbf{i} - 3\mathbf{j}$.

14. Find the area of the parallelogram in $\mathbb{R}^2$ whose adjacent sides are the arrows $(8, -11)$ and $(-7, 3)$.

15. (a) If $X \times Z = \mathbf{0}$ for all vectors $Z \in \mathbb{R}^3$, what can you conclude?

    (b) If $X \times Z = Y \times Z$ for all vectors $Z \in \mathbb{R}^3$, what can you conclude?

16. Find all vectors in the plane spanned by $X = \mathbf{i} + \mathbf{j} - 2\mathbf{k}$ and $Y = -\mathbf{i} + \mathbf{j} + \mathbf{k}$ that are perpendicular to the vector $Z = 2\mathbf{i} + \mathbf{j} + 2\mathbf{k}$.

17. If the vertices of a parallelogram in the xy-plane all have integer coordinates, prove that the area is an integer.

18. Use the triple product to find the volume of the parallelepiped generated by the following sets of vectors:

    (a) $3\mathbf{i}, -\mathbf{j}, 4\mathbf{k}$

    (b) $(1, -1, 0), (2, 4, -1), (3, 3, 5)$

    (c) $(1, -1, 0), (2, 4, -1), (3, 3, -1)$.

Does your answer to part (c) seem reasonable? Can you give any additional geometric insight into the answer to part (a)?

Extra: Euclid using Vectors

After learning a new topic, it is fun to see how it applies to an old subject. We will prove some classical results from Euclidean geometry using vector methods, with emphasis on the norm and inner product.

A comment is in order about the pictures that will be used in this section. Until now we have drawn vectors as directed line segments (arrows) emanating from the origin. However, in many cases, the position of a vector is not so important as its length and direction. In this section, we will often draw vectors in positions that make the relationships between the

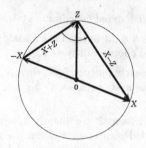

FIGURE 2.12.

various lengths and directions of these vectors easier to see. For example, the vector $X - Y$ is parallel to and the same length as the line segment from the point Y to the point X (refer back to Figure 2.4). In this section we will draw $X - Y$ as if it were identical to that line segment.

Proposition 2.5.1 An angle inscribed in a semicircle is a right angle.

PROOF: Say the circle has radius a and is centered at the origin. If one end of a diameter is X, then the other end is $-X$. Let Z be any other point on the circle. We must show (see Figure 2.12) that $X - Z \perp X + Z$. Use the inner product and the fact that $\|X\| = a = \|Z\|$ to compute the following:

$$\langle X - Z, X + Z \rangle = \|Z\|^2 + \langle Z, X \rangle - \langle X, Z \rangle - \|X\|^2$$
$$= a^2 - a^2 = 0.$$

Done. <<

Proposition 2.5.2 If the medians to two sides of a triangle have equal length, then the triangle is isosceles.

PROOF: Place the origin at the vertex that does not have one of the median lines. Then (see Figure 2.13) we are told that

$$\|X - \frac{1}{2}Y\| = \|Y - \frac{1}{2}X\|.$$

Now

$$\|X - \frac{1}{2}Y\|^2 = \langle X - \frac{1}{2}Y, X - \frac{1}{2}Y \rangle = \|X\|^2 - \langle X, Y \rangle + \frac{1}{4}\|Y\|^2$$

and

$$\|Y - \frac{1}{2}X\|^2 = \langle Y - \frac{1}{2}X, Y - \frac{1}{2}X \rangle = \|Y\|^2 - \langle Y, X \rangle + \frac{1}{4}\|X\|^2$$

Equating these two expressions we find that $\|X\| = \|Y\|$. Done. <<

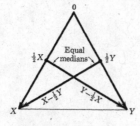

FIGURE 2.13.

Since mathematics is not a spectator sport, we offer some assertions for you to prove:

1. A parallelogram is a rectangle ⇔ the diagonals have equal length.

2. If a triangle is isosceles, then the medians to the two sides of equal length also have equal length.

3. The sum of the squares of the sides of a parallelogram equals the sum of the squares of the diagonals. (cf. Exercise 14 in Section 2.2).

4. The diagonals of a rhombus are perpendicular to each other.

5. Given any quadrilateral Q, construct a new one P by connecting the midpoints of successive sides of Q. Prove that P is a parallelogram.

6. The diagonals of a parallelogram bisect each other.

7. The medians of a triangle with vertices X, Y, and Z intersect at the point $\frac{1}{3}(X + Y + Z)$.

3

Linear Functions

3.0 Introduction

Having made our acquaintance with Euclidean spaces, we are now ready to discuss functions defined on these spaces. In this chapter we reexamine the familiar functions $y = mx + b$ and generalize them to higher-dimensional spaces. These functions, we recall, are essential to single-variable differential calculus because their graphs are the tangent lines to the graphs of other, more complicated functions. We are able to learn about these more complicated nonlinear functions by a study of their tangent lines. Recall, for example, that maxima and minima occur where the tangent line is horizontal. Our generalizations of $y = mx + b$ will be useful in an analogous study of nonlinear functions in higher dimensions.

This chapter should be viewed as much more than a preparation for the topics later in the book. There are many topics in this chapter that are important for reasons other than as preparation for the study of nonlinear functions. For instance, the last section of the chapter is concerned with motions that leave distances and angles fixed.

Let us write $T(x) = mx + b$. See Figure 3.1. Here m and b are constants and x is a real variable. Recall that m is the slope of the line $y = mx + b$ and b is the y-intercept, the height at which the line crosses the y-axis. The function $T(x)$ is our one-dimensional model for an "affine" function. A function of the form $L(x) = mx$ is our model of a "linear" function. Examples of linear functions are $L(x) = 6x$, $L(x) = -x$, and $L(x) = \frac{1}{2}x$. Examples of affine functions are $T(x) = 6x + 1$, $T(x) = -x - 5$, and $T(x) = \frac{1}{2}x + 1$. We note that affine functions are readily built from linear functions, simply by adding a constant b. Since $b = 0$ is a possibility, a linear function is also an affine function.

A linear function $L(x) = mx$ has two essential properties:

$$
\begin{array}{lrcl}
\text{Additivity:} & L(x + x') & = & L(x) + L(x')\,, \\
\text{Homogeneity:} & L(\alpha x) & = & \alpha L(x)\,.
\end{array}
$$

These properties, as a moment's reflection shows, are immediate consequences of the commutative law of multiplication and the distributive law for real numbers. Furthermore, it can be shown that any function $y = f(x)$ that is additive and homogeneous must be of the form $f(x) = mx$ for some constant m.

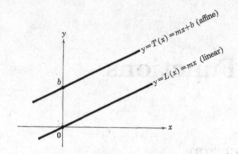

FIGURE 3.1. The case of a single variable x

We need functions in higher dimensions that act like the functions $y = L(x) = mx$. Such functions are of the form $Y = L(X)$, where X and Y are vectors. How should we define them? Now we attend to this issue.

3.1   Definition and Basic Properties

3.1A   A DEFINITION AND MANY EXAMPLES

Let $\mathbb{R}^n$ and $\mathbb{R}^q$ be Euclidean spaces. Perhaps $n = q$, perhaps not. A function F from $\mathbb{R}^n$ to $\mathbb{R}^q$ is a rule that assigns to each vector X in $\mathbb{R}^n$ some vector $F(X)$ in $\mathbb{R}^q$. Using notation described in Chapter 0, we write

$$F : \mathbb{R}^n \to \mathbb{R}^q .$$

Adapting the general terminology described in Chapter 0 to the present setting, we see that $\mathbb{R}^n$ is the domain of F, and $\mathbb{R}^q$ is the target of F. Now we single out a special kind of function, which is the subject of this chapter.

Definition 3.1.1 A function L from $\mathbb{R}^n$ into $\mathbb{R}^q$, that is, $L : \mathbb{R}^n \to \mathbb{R}^q$, is **linear** if and only if, for all $X, X' \in \mathbb{R}^n$ and $\alpha \in \mathbb{R}$, it satisfies

Additivity: $L(X + X') \;=\; L(X) + L(X')$;
Homogeneity: $L(\alpha X) \;=\; \alpha L(X)$.

It is important to keep in mind that X, X', $L(X)$, and $L(X')$ in this definition are all vectors. Thus, the sums on both sides of the first equation in the definition are vector sums, and the products with α in the second equation involve the multiplication of a vector by a scalar.

Linear functions are also called **linear mappings, linear maps, linear transformations**, and **linear operators**. The different names for "linear functions" are standard among researchers in different areas of science. Thus a physicist might speak of a "linear operator", but a geometer is more likely to say "linear mapping" or "linear transformation". We are

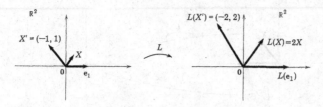

FIGURE 3.2. $L(X) = 2X$.

partial to "linear map" and "linear transformation", and will use the two phrases interchangeably. Remember: "linear map" = "linear transformation" = "linear function".

Here are some important linear maps $L : \mathbb{R}^n \to \mathbb{R}^q$. Note that in each case we first specify the domain $\mathbb{R}^n$ and target $\mathbb{R}^q$ and then give the rule for $L(X)$. Finally we verify that the rule determines a function that is indeed linear. We often illustrate with two coordinate systems, one for the domain and the other for the target, together with some feature that indicates the mapping accomplished by L.

EXAMPLES:

1. Let $L : \mathbb{R}^n \to \mathbb{R}^n$ (note $q = n$ in this example) be given by $L(X) = 2X$ for all $X \in \mathbb{R}^n$. Thus L doubles the length of each X in $\mathbb{R}^n$, and, in particular, $\|L(\mathbf{0})\| = 2\|\mathbf{0}\| = 0$. See Figure 3.2.

   To see that L is linear, we must verify the criteria of the definition. To verify additivity, note that $L(X + X') = 2(X + X') = 2X + 2X' = L(X) + L(X')$, as required.

   To verify the homogeneity of L, let α be any scalar. Then $L(\alpha X) = 2(\alpha X) = (\alpha 2)X = \alpha(2X) = \alpha L(X)$, as required. Thus $L(X) = 2X$ defines a linear map.

   In general, for γ any fixed scalar, $L(X) = \gamma X$ defines a linear map $L : \mathbb{R}^n \to \mathbb{R}^n$. You should check that this function is linear and describe geometrically what L does to each X.

2. Let $O : \mathbb{R}^n \to \mathbb{R}^q$ (any n, q) be given by $O(X) = \mathbf{0}$, the zero vector in $\mathbb{R}^q$, for *every* X in $\mathbb{R}^n$. Figure 3.3 is a picture in the case of $n = q = 2$. This map O is called the **zero map**.

3. Consider the map of $\mathbb{R}^n$ into $\mathbb{R}^n$ that assigns to each vector X itself. This is the **identity map** $I : \mathbb{R}^n \to \mathbb{R}^n$, $I(X) = X$. It is clearly linear. Figure 3.4 is a picture for the case $n = 2$. As you can see, the identity map leaves things exactly as they are. Contrast this with the zero map, which sends every vector X in $\mathbb{R}^n$ to the single point $\mathbf{0}$.

4. Now let $n = 2$ and $q = 1$ and define $L : \mathbb{R}^2 \to \mathbb{R}$ as follows. For each $X = (x_1, x_2)$ in $\mathbb{R}^2$, let $L(X) = x_1$. Thus L maps each vector

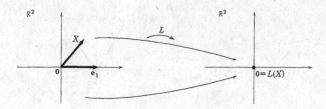

FIGURE 3.3. The zero map

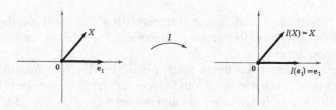

FIGURE 3.4. The identity map

X in $\mathbb{R}^2$ to its first coordinate x_1, which is a real number, and hence a vector in $\mathbb{R}$. To see that L is additive, write $X' = (x_1', x_2')$ and note that $L(X + X') = L((x_1 + x_1', x_2 + x_2')) = x_1 + x_1' = L(X) + L(X')$. Homogeneity is verified just as easily, and so L is linear. See Figure 3.5.

5. This time let $L : \mathbb{R}^2 \to \mathbb{R}$ be given by $L(X) = x_1 + 2x_2$, where $X = (x_1, x_2)$. Thus, if $X = (1, -3)$ then $L(X) = -5$. There is an interesting picture associated with this example. See Figure 3.6. Note that all points on the line $x_1 + 2x_2 = 0$ are mapped to 0 in the target $\mathbb{R}$ (drawn tilted) and that each line $x_1 + 2x_2 = \beta$ parallel to $x_1 + 2x_2 = 0$ is compressed and projected by L onto the single point β in $\mathbb{R}$.

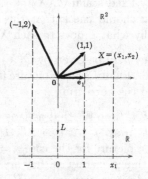

FIGURE 3.5. Projection onto the first coordinate

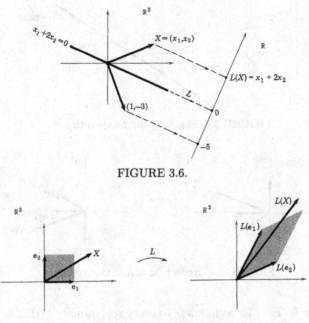

FIGURE 3.6.

FIGURE 3.7.

To prove L linear, note that $L(\alpha X) = L((\alpha x_1, \alpha x_2)) = \alpha x_1 + 2\alpha x_2 = \alpha(x_1 + 2x_2) = \alpha L(X)$, whence L is homogeneous. You may check additivity yourself.

6. Let $L : \mathbb{R}^2 \to \mathbb{R}^2$ be defined by $L(X) = (x_1 + 2x_2, 3x_1 + x_2)$, where, as usual, $X = (x_1, x_2)$. You should be able to verify that L is additive and homogeneous and thus linear. See Figure 3.7. Note that $L(\mathbf{e}_1) = (1, 3)$ and $L(\mathbf{e}_2) = (2, 1)$. It can be shown that L of any point in the shaded region in the left-hand picture in Figure 3.7 belongs to the shaded region in the right-hand picture.

Suppose we want to find those X for which $L(X) = (10/3, 5)$. Then we are looking for the set of those $X = (x_1, x_2)$ for which

$$x_1 + 2x_2 = 10/3$$
$$3x_1 + x_2 = 5 \, .$$

This system is easily solved: $x_1 = 4/3$, $x_2 = 1$. Thus, $L((4/3, 1)) = (10/3, 5)$, and there is no other vector X for which $L(X) = (10/3, 5)$.

We encourage you to ponder this example. It is often the case that linear equations are needed to answer questions about linear maps L. In this case, the question was of the following sort: given Y, for which X, if any, does $L(X) = Y$?

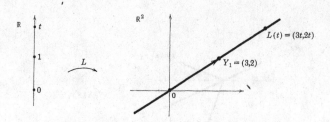

FIGURE 3.8. Mapping the t-axis into $\mathbb{R}^2$

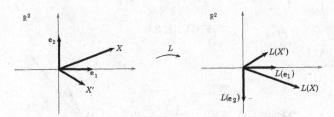

FIGURE 3.9. Reflection across the x_1-axis

7. Define $L : \mathbb{R} \to \mathbb{R}^2$ as follows. To every real number t (that is, to every vector in $\mathbb{R}$) we associate the point $L(t) = (3t, 2t)$ in the plane $\mathbb{R}^2$. See Figure 3.8. It is easily verified that L is linear. What does the set of points $\{(3t, 2t) : t \in \mathbb{R}\}$ look like? That is, what does the image $\mathcal{I}(L)$ of L look like? Note that $L(1) = (3, 2) \in \mathbb{R}^2$. If we write $Y_1 = (3, 2)$, then $L(t) = tY_1$. As t varies from $-\infty$ to ∞ in $\mathbb{R}$, $L(t)$ varies along the line (one-dimensional linear subspace) in $\mathbb{R}^2$ determined by the vector Y_1. Therefore, $\mathcal{I}(L)$ equals the one-dimensional linear subspace spanned by $(3, 2)$. This example indicates how a line through the origin in $\mathbb{R}^2$ may be represented as the image of a linear map.

8. Let $L : \mathbb{R}^2 \to \mathbb{R}^2$ be given by $L(X) = (x_1, -x_2)$. Figure 3.9 indicates that L reflects every vector in the horizontal axis, that is, L flips each X across the horizontal axis to its image $L(X)$, or in other words, transforms each vector into its mirror image, the horizontal axis being the mirror. We leave it to you to verify that L is linear.

9. The preceding two examples have important geometrical interpretations even though the functions were defined algebraically. Here, in contrast, is a transformation defined geometrically. Let $L : \mathbb{R}^2 \to \mathbb{R}^2$ rotate each vector X through $90°$ counterclockwise. See Figure 3.10. Note that lengths are not changed, $\|L(X)\| = \|X\|$, and the angle between $L(X)$ and $L(X')$ equals that between X and X'.

Now let us verify that the function L is indeed linear. To prove additivity, recall that addition of vectors in $\mathbb{R}^2$ obeys the parallelogram rule that is, $X + X'$ is a diagonal of the parallelogram determined by X

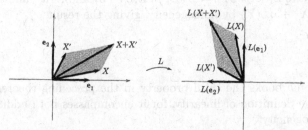

FIGURE 3.10. Rotation 90 degrees counteclockwise. Note: $L(X + X') = L(X) + L(X')$

and X'. Using Figure 3.10, you can convince yourself that $L(X + X')$ is the diagonal of the parallelogram determined by $L(X)$ and $L(X')$ and therefore that $L(X + X') = L(X) + L(X')$.

From the picture, the homogeneity of L is also readily apparent, and thus the rotation L is a linear map.

Linear functions are very special. Grab a function at random and it is not likely to be linear. Some nonlinear functions are not very interesting, but many are very important. For instance $f(X) = \|X\|$ and $g(X) = \|X\|^2$ are two very important nonlinear functions from $\mathbb{R}^n$ to $\mathbb{R}$. The reader can check that neither f nor g is additive or homogeneous.

3.1B SOME CONSEQUENCES OF LINEARITY

Going back to the definition of linear map, we derive some immediate consequences that will be used repeatedly.

Theorem 3.1.1 Let $L : \mathbb{R}^n \to \mathbb{R}^q$ be a linear map. Then L satisfies the following properties:

1. $L(0) = 0$.

2. $L(-X) = -L(X)$ for all $X \in \mathbb{R}^n$.

3. $L(\alpha X + \beta Y) = \alpha L(X) + \beta L(Y)$ for all $X, Y \in \mathbb{R}^n$ and $\alpha, \beta \in \mathbb{R}$.

PROOF:

1. First note that the **0** in the expression $L(0)$ is the zero vector in $\mathbb{R}^n$, and the **0** on the right-hand side of $L(0) = 0$ is the zero vector of $\mathbb{R}^q$ (q may or may not equal n). For the proof, just let X be arbitrary and $\alpha = 0$ in $L(\alpha X) = \alpha L(X)$.

2. We have $L(-X) = L(-1X) = -1L(X) = -L(X)$.

3. We have $L(\alpha X + \beta Y) = L(\alpha X) + L(\beta Y)$ by additivity, and this equals $\alpha L(X) + \beta L(Y)$ by homogeneity, giving the result.

Done. $<<$

REMARK:

- In some books the third property in the preceding theorem is used as the definition of linearity, for it encompasses both additivity and homogeneity.

3.1c LINEARITY AND THE STANDARD BASIS

Now we discuss a remarkable property of linear maps $L : \mathbb{R}^n \to \mathbb{R}^q$. As with many discoveries in mathematics, it is a consequence of relating two concepts. The two relevant ideas have already been introduced: standard bases and linear maps.

Suppose we know that $L : \mathbb{R}^n \to \mathbb{R}^q$ is linear and, moreover, that $L(\mathbf{e}_1) = Y_1, \ldots, L(\mathbf{e}_n) = Y_n$, where $Y_1, \ldots, Y_n$ are vectors in $\mathbb{R}^q$, that is, we know the value of L at the n points $\mathbf{e}_1, \mathbf{e}_2, \ldots, \mathbf{e}_n$. Then we assert the striking fact that we now know $L(X)$ for *all* vectors X.

Why is this? It is a simple consequence of the linearity of L. Let $X = (x_1, \ldots, x_n)$ be any vector in $\mathbb{R}^n$. Then, using the standard basis, we have $X = x_1\mathbf{e}_1 + \ldots + x_n\mathbf{e}_n$. By the linearity of L, we have

$$
\begin{aligned}
L(X) &= L(x_1\mathbf{e}_1 + \ldots + x_n\mathbf{e}_n) \\
&= x_1 L(\mathbf{e}_1) + \cdots + x_n L(\mathbf{e}_n) \\
&= x_1 Y_1 + \cdots + x_n Y_n \,.
\end{aligned}
$$

Thus $L(X)$ is a linear combination of $Y_1, \ldots, Y_n$ in $\mathbb{R}^q$, with the coefficients being the coordinates of X. If we know the image vectors $L(\mathbf{e}_1), \ldots, L(\mathbf{e}_n)$ in $\mathbb{R}^q$ and are given an explicit vector X in $\mathbb{R}^n$, then we can readily compute $L(X)$.

EXAMPLE:

10. Let $L : \mathbb{R}^3 \to \mathbb{R}^2$ be linear, and denote by $\mathbf{e}_1$, $\mathbf{e}_2$, and $\mathbf{e}_3$ the standard basis for $\mathbb{R}^3$. Suppose that

$$
L(\mathbf{e}_1) = (3, 1) \,, \quad L(\mathbf{e}_2) = (2, 4) \,, \quad L(\mathbf{e}_3) = (-1, 2) \,.
$$

These vectors in $\mathbb{R}^2$ are the values $L(X)$ for X equal to $\mathbf{e}_1$, $\mathbf{e}_2$, and $\mathbf{e}_3$, respectively. Now we ask: What is $L(X)$ in general, when $X = (x_1, x_2, x_3)$? To answer, we apply the general comments made in the preceding paragraph:

$$
\begin{aligned}
L(X) &= x_1 L(\mathbf{e}_1) + x_2 L(\mathbf{e}_2) + x_3 L(\mathbf{e}_3) \\
&= x_1(3, 1) + x_2(2, 4) + x_3(-1, 2) \\
&= (3x_1 + 2x_2 - x_3 \,, \, x_1 + 4x_2 + 2x_3) \,.
\end{aligned}
$$

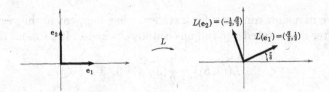

FIGURE 3.11. Rotation through $\frac{\pi}{6}$ (30 degrees) counterclockwise

This is the **coordinate form** of L. If $X = (1, 3, -2)$, then $L(X) = (11, 9)$; and if $X = (8, -7, 10)$, then $L(X) = (0, 0) = \mathbf{0}$.

Suppose that $L : \mathbb{R}^n \to \mathbb{R}^q$ and $M : \mathbb{R}^n \to \mathbb{R}^q$ are linear maps, and they have equal values at $\mathbf{e}_1, \ldots, \mathbf{e}_n$, that is, $L(\mathbf{e}_1) = M(\mathbf{e}_1), \ldots, L(\mathbf{e}_n) = M(\mathbf{e}_n)$. On the basis of the preceding discussion we see that $L(X) = M(X)$ for every $X \in \mathbb{R}^n$. Hence, L and M are the same function, that is, $L = M$. For instance, consider the following situation. Suppose that we have a function $M : \mathbb{R}^2 \to \mathbb{R}^2$ and we know that $M((1, 0)) = (0, 1)$, $M((0, 1)) = (-1, 0)$, and M is linear. Note that M does the same thing to each of the standard basis vectors as does the linear map L of Example 9 in Section 3.1a, namely, M rotates both standard basis vectors counterclockwise 90 degrees about the origin. We may conclude that M is the same as L, and thus that it rotates *every* vector X in $\mathbb{R}^2$ counterclockwise through a right angle.

EXAMPLE:

11. Let us calculate the rule for a linear map $L : \mathbb{R}^2 \to \mathbb{R}^2$ that rotates each vector $\pi/6$ radians counterclockwise. According to the preceding discussion we only need calculate $L(\mathbf{e}_1)$ and $L(\mathbf{e}_2)$. Looking at Figure 3.11, we see that we are able to use elementary trigonometry:

$$
\begin{aligned}
L(\mathbf{e}_1) &= (\cos(\pi/6), \sin(\pi/6)) = (\sqrt{3}/2, 1/2), \\
L(\mathbf{e}_2) &= (-\sin(\pi/6), \cos(\pi/6)) = (-1/2, \sqrt{3}/2).
\end{aligned}
$$

We then obtain the coordinate form of L:

$$
\begin{aligned}
L(X) &= x_1 L(\mathbf{e}_1) + x_2 L(\mathbf{e}_2) \\
&= x_1 \left(\frac{\sqrt{3}}{2}, \frac{1}{2} \right) + x_2 \left(-\frac{1}{2}, \frac{\sqrt{3}}{2} \right)
\end{aligned}
$$

$$= \frac{1}{2}(\sqrt{3}x_1 - x_2, \, x_1 + \sqrt{3}x_2) .$$

We may now compute, for instance, what happens to the vector $(3, 5)$ after it is rotated $\pi/6$ radians counterclockwise. It becomes the vector

$$L((3,5)) = \frac{1}{2}(3\sqrt{3} - 5, \, 3 + 5\sqrt{3}) .$$

3.1D MATRIX NOTATION

A linear map L is defined by formal properties, namely, $L(X + X') = L(X) + L(X')$ and $L(\alpha X) = \alpha L(X)$. Examples involve linear equations, rotations of the plane, projections from one space to another, and so on— a wide variety of phenomena. Faced with this formal definition on the one hand and the proliferation of diverse examples on the other, we feel the need for a uniform, concrete, almost mechanical way of representing and computing with linear maps, without regard to special geometric or algebraic interpretations.

Let us work with specific n and q. Let $L : \mathbb{R}^2 \to \mathbb{R}^3$ be a linear map. Denote the standard bases in $\mathbb{R}^2$ and $\mathbb{R}^3$ by $\mathbf{e}_1, \mathbf{e}_2$ and $\bar{\mathbf{e}}_1, \bar{\mathbf{e}}_2, \bar{\mathbf{e}}_3$, respectively. Remember that $\mathbf{e}_1 = (1, 0)$ and $\bar{\mathbf{e}}_1 = (1, 0, 0)$, and so on.

We know that L is determined by the vectors $L(\mathbf{e}_1)$ and $L(\mathbf{e}_2)$. These, being elements of $\mathbb{R}^3$, are linear combinations of $\bar{\mathbf{e}}_1$, $\bar{\mathbf{e}}_2$, and $\bar{\mathbf{e}}_3$, that is,

$$
\begin{aligned}
L(\mathbf{e}_1) &= a_1\bar{\mathbf{e}}_1 + a_2\bar{\mathbf{e}}_2 + a_3\bar{\mathbf{e}}_3 , \\
L(\mathbf{e}_2) &= b_1\bar{\mathbf{e}}_1 + b_2\bar{\mathbf{e}}_2 + b_3\bar{\mathbf{e}}_3 ,
\end{aligned}
$$

where the a's and b's are scalars. These six scalar coefficients determine L. Different coefficients, different maps. It is customary to arrange these coefficients in a rectangular array with three horizontal rows and two vertical columns:

$$
\begin{bmatrix}
a_1 & b_1 \\
a_2 & b_2 \\
a_3 & b_3
\end{bmatrix} .
$$

This array is termed the **matrix** of L and denoted by $[L]$. Since the a's and b's determine the map, the matrix provides an uncluttered and concise summary of L. Observe, for instance, that its first column lists the coefficients in the expression of $L(\mathbf{e}_1)$. The reason for writing these coefficients vertically, rather than horizontally, will become clear later, when we define matrix multiplication.

The matrix of a linear map from $\mathbb{R}^n$ to $\mathbb{R}^q$ has q rows and n columns and thus is a q-**by**-n **matrix**, the first integer giving the number of rows and the second the number of columns.

EXAMPLES:

12. Let $L : \mathbb{R}^2 \to \mathbb{R}^3$ be determined by

$$
\begin{aligned}
L(\mathbf{e}_1) &= \bar{\mathbf{e}}_1 + 8\bar{\mathbf{e}}_3 , \\
L(\mathbf{e}_2) &= 3\bar{\mathbf{e}}_1 - 8\bar{\mathbf{e}}_2 + \bar{\mathbf{e}}_3 .
\end{aligned}
$$

Note that these equations are equivalent to the statements that

$$
L((1,0)) = (1,0,8) \quad \text{and} \quad L((0,1)) = (3,-8,1) .
$$

The matrix of L is

$$
[L] = \begin{bmatrix} 1 & 3 \\ 0 & -8 \\ 8 & 1 \end{bmatrix} .
$$

13. Let $L : \mathbb{R}^2 \to \mathbb{R}^2$, be the map (introduced earlier) given by $L(X) = 2X$. In particular, $L(\mathbf{e}_1) = 2\mathbf{e}_1$ and $L(\mathbf{e}_2) = 2\mathbf{e}_2$. So,

$$
[L] = \begin{bmatrix} 2 & 0 \\ 0 & 2 \end{bmatrix} .
$$

14. Let $L : \mathbb{R}^2 \to \mathbb{R}^2$ be the map that rotates the plane counterclockwise through one right angle. Then $L(\mathbf{e}_1) = \mathbf{e}_2$ and $L(\mathbf{e}_2) = -\mathbf{e}_1$ (check this!), and so

$$
[L] = \begin{bmatrix} 0 & -1 \\ 1 & 0 \end{bmatrix} .
$$

15. Suppose that we are told that the matrix of a linear map L is

$$
[L] = \begin{bmatrix} 1 & 3 & 0 & -1 \\ 2 & 1 & 4 & 4 \end{bmatrix} .
$$

This means that $L : \mathbb{R}^4 \to \mathbb{R}^2$ is the linear map given by

$$
\begin{aligned}
L(\mathbf{e}_1) &= \bar{\mathbf{e}}_1 + 2\bar{\mathbf{e}}_2 , \\
L(\mathbf{e}_2) &= 3\bar{\mathbf{e}}_1 + \bar{\mathbf{e}}_2 , \\
L(\mathbf{e}_3) &= 4\bar{\mathbf{e}}_2 , \\
L(\mathbf{e}_4) &= -\bar{\mathbf{e}}_1 + 4\bar{\mathbf{e}}_2 .
\end{aligned}
$$

For instance,

$$
L((1,2,3,4)) = (1,2) + 2(3,1) + 3(0,4) + 4(-1,4) = (3,32) .
$$

We will soon see that matrices provide not only a concise way of *describing* a linear map but also a useful means of *computing* with these maps.

Exercises

Practice in deciding whether a given map is actually linear, writing down a linear map with prescribed properties, and using the matrix notation.

1. Which of the following $L : \mathbb{R}^2 \to \mathbb{R}$ are linear?

   (a) $L(X) = 2x_1 - x_2$

   (b) $L(X) = x_1 + x_2 + 2$

   (c) $L(X) = x_1 x_2$

   (d) $L(X) = \|X\|$.

2. Which of the following $L : \mathbb{R} \to \mathbb{R}^2$ are linear?

   (a) $L(t) = (\cos t, \sin t)$

   (b) $L(t) = (0, t)$

   (c) $L(t) = (2t, -t)$.

3. Which of the following $L : \mathbb{R}^2 \to \mathbb{R}^2$ are linear?

   (a) $L(X) = (x_1^2 - x_2^2,\ 2x_1 x_2)$

   (b) $L(X) = (x_1, -x_1)$

   (c) $L(X) = \|X\|^{-1} X,\quad X \neq 0$

   (d) $L(X) = (x_1 + x_2,\ x_1 - x_2)$

   (e) $L(X) = (x_1, e^{x_2})$.

4. Which of the following $L : \mathbb{R}^3 \to \mathbb{R}^3$ are linear?

   (a) $L(X) = (x_3, x_1, x_2)$

   (b) $L(X) = (x_1 + x_2, x_2 + x_3, 0)$

   (c) $L(X) = \|X\|(1, 2, 0)$.

5. Let $L : \mathbb{R}^2 \to \mathbb{R}^2$ be the linear map given by $L(X) = (x_1, x_1 + x_2)$. Draw a picture of the target and locate $L(\mathbf{0})$, $L(\mathbf{e_1})$, $L(\mathbf{e_2})$, and $L(\mathbf{e_1} + \mathbf{e_2})$.

6. Given: $L : \mathbb{R}^2 \to \mathbb{R}^2$ is linear, $L(\mathbf{e_1}) = (3, 4)$, and $L(\mathbf{e_2}) = (1, -1)$. Write a formula for $L((x_1, x_2))$. In particular, evaluate $L((13, -2))$.

7. Complete Example 4 in Section 3.1a by showing that L is homogeneous.

8. Complete Example 5 in Section 3.1a by showing that L is additive.

9. Let γ be a fixed scalar and define L by $L(X) = \gamma X$. Prove that L is linea·

10. Complete Example 6 in Section 3.1a by showing that L is linear.

11. Complete Example 8 in Section 3.1a by showing that L is linear.

12. Let $L : \mathbb{R}^2 \to \mathbb{R}^2$ be the linear map with matrix

$$[L] = \begin{bmatrix} 1 & -1 \\ 2 & 0 \end{bmatrix}.$$

   What are the following:

   (a) $L(\mathbf{e_1})$

   (b) $L(\mathbf{e_2})$

   (c) $L(X)$, where $X = (10, 7) = 10\mathbf{e_1} + 7\mathbf{e_2}$?

13. Given: $L : \mathbb{R}^2 \to \mathbb{R}^2$ is linear, $L(\mathbf{e}_1) = \mathbf{e}_1 + 2\mathbf{e}_2$, and $L(\mathbf{e}_2) = -\mathbf{e}_1$. Write the 2-by-2 matrix $[L]$.

14. Let $L : \mathbb{R}^2 \to \mathbb{R}$ be linear, $L(\mathbf{e}_1) = 4$, and $L(\mathbf{e}_2) = 3$. Write the matrix $[L]$.

15. Let $M : \mathbb{R}^2 \to \mathbb{R}$ be given by $M(X) = 2x_1 - 3x_2$. Write the matrix $[M]$. (First compute $M(\mathbf{e}_1)$ and $M(\mathbf{e}_2)$.)

16. Let L be a linear transformation for which $[L] = \begin{bmatrix} 2 & -3 \end{bmatrix}$. What are the following:

   (a) $L(\mathbf{e}_1)$

   (b) $L(\mathbf{e}_2)$

   (c) $L(3\mathbf{e}_1 + 2\mathbf{e}_2)$

   (d) $L(X)$, where $X = (x_1, x_2)$?

17. For L as defined in the preceding exercise, write a formula for $L(X)$ in terms of an inner product of X with something.

18. Let $L : \mathbb{R}^3 \to \mathbb{R}^2$ be determined by

$$[L] = \begin{bmatrix} 0 & 1 & 1 \\ 1 & 0 & 1 \end{bmatrix}.$$

   Find a formula for each of the following in terms of the standard basis $\bar{\mathbf{e}}_1, \bar{\mathbf{e}}_2$ of $\mathbb{R}^2$:

   (a) $L(\mathbf{e}_1)$

   (b) $L(\mathbf{e}_2)$

   (c) $L(\mathbf{e}_3)$

   (d) $L(\mathbf{e}_1 + \mathbf{e}_2 - \mathbf{e}_3)$.

19. Given: $L : \mathbb{R}^3 \to \mathbb{R}^2$ is linear, $L(\mathbf{e}_1) = \bar{\mathbf{e}}_2$, $L(\mathbf{e}_2) = \bar{\mathbf{e}}_1$, and $L(\mathbf{e}_3) = \bar{\mathbf{e}}_1 + \bar{\mathbf{e}}_2$. Let $X = (x_1, x_2, x_3)$ in $\mathbb{R}^3$ and $Y = (y_1, y_2)$ in $\mathbb{R}^2$.

   (a) If $L(X) = Y$, what is y_1 in terms of x_1, x_2, and x_3?

   (b) Same for y_2.

   (c) Find $[L]$.

20. Respond true or false to each statement.

   (a) If $L(\mathbf{0}) = \mathbf{0}$, then L is linear.

   (b) L is linear $\Leftrightarrow L(\alpha_1 X_1 + \alpha_2 X_2) = \alpha_1 L(X_1) + \alpha_2 L(X_2)$ for all $\alpha_1, \alpha_2 \in \mathbb{R}$ and $X_1, X_2 \in \mathbb{R}^n$.

   (c) If L is a rotation of $\mathbb{R}^2$ about the origin $\mathbf{0}$, then L is linear.

   (d) If $L : \mathbb{R}^2 \to \mathbb{R}$ is constant, say $L(X) = 1$ for all X, then L is linear.

   (e) If L is a rotation of $\mathbb{R}^2$ about the point $X_0 = (1, 1)$, then L is linear.

   (f) If L is linear and the values of $L(\mathbf{e}_1), \ldots, L(\mathbf{e}_n)$ are specified, then the value of $L(X)$ is determined for every $X \in \mathcal{D}(L)$.

   (g) If $L : \mathbb{R}^2 \to \mathbb{R}^2$ is linear, $L(\mathbf{e}_1) = \mathbf{e}_1$, and $L(\mathbf{e}_2) = \mathbf{e}_2$, then $L(X) = X$ for all $X \in \mathbb{R}^2$.

(h) If $L : \mathbb{R}^2 \to \mathbb{R}^2$ is linear, $L(\mathbf{e}_1) = \mathbf{e}_2$, and $L(\mathbf{e}_2) = -\mathbf{e}_1$, then L rotates the plane $90°$ counterclockwise about the origin.

(i) If $L : \mathbb{R}^2 \to \mathbb{R}^2$ is linear and the first column of the 2-by-2 matrix $[L]$ is

$$\begin{bmatrix} 1 \\ 0 \end{bmatrix},$$

then $L(\mathbf{e}_1) = \mathbf{e}_1$.

(j) If $L : \mathbb{R}^2 \to \mathbb{R}^2$ is linear and the first row of the 2-by-2 matrix $[L]$ is

$$\begin{bmatrix} 1 & 0 \end{bmatrix},$$

then $L(\mathbf{e}_1) = \mathbf{e}_1$.

(k) The map $L : \mathbb{R}^n \to \mathbb{R}$ given by $L(X) = \|X\|$ is linear.

(l) The matrix of a linear map L is uniquely determined by the vectors $L(\mathbf{e}_1), \ldots, L(\mathbf{e}_n)$.

(m) The matrix of a linear transformation $L : \mathbb{R}^n \to \mathbb{R}^q$ is square $\Leftrightarrow q = n$.

(n) Let $L : \mathbb{R}^2 \to \mathbb{R}^3$ be linear and satisfy $L((1,1)) = (1,2,3)$ and $L((0,1)) = (1,-2,-1)$. Then $L((2,3)) = (1,1,1)$.

(o) Let $L : \mathbb{R}^n \to \mathbb{R}^q$ be linear. If $L(X_1) = \mathbf{0}$ and $L(X_2) = \mathbf{0}$, then $L(3X_1 - 7X_2) = \mathbf{0}$.

3.2   Linear Maps and Linear Subspaces

3.2A   THE IMAGE OF A LINEAR MAP

Let $L : \mathbb{R}^n \to \mathbb{R}^q$ be linear and denote, as usual, the standard basis for $\mathbb{R}^n$ by $\mathbf{e}_1, \mathbf{e}_2, \ldots, \mathbf{e}_n$. We have previously seen that

$$L(X) = x_1 L(\mathbf{e}_1) + x_2 L(\mathbf{e}_2) + \cdots + x_n L(\mathbf{e}_n)$$

for every $X = (x_1, x_2, \ldots, x_n) \in \mathbb{R}^n$. Thus, when one operates with L one obtains a linear combination of the vectors $L(\mathbf{e}_1), L(\mathbf{e}_2), \ldots, L(\mathbf{e}_n)$ belonging to $\mathbb{R}^q$. As we vary X we obtain all linear combinations of $L(\mathbf{e}_1), \ldots, L(\mathbf{e}_n)$. Thus, the image $\mathcal{I}(L)$ is the linear span of the vectors $L(\mathbf{e}_1), \ldots, L(\mathbf{e}_n)$, and hence a linear subspace of $\mathbb{R}^q$. If one is trying to identify the image of a linear map L that is given by its matrix, it is useful to remember that the coordinates of each $L(\mathbf{e}_j)$ appear as the entries in the jth column of the matrix $[L]$.

In order to generalize this conclusion we introduce further notation. For a subset $\mathcal{Q}$ of $\mathbb{R}^n$ let

$$L(\mathcal{Q}) = \{Y : Y = L(X) \text{ for some } X \in \mathcal{Q}\} .$$

Thus, $L(\mathcal{Q})$ is the image of L when $\mathcal{Q}$, rather than $\mathbb{R}^n$, is regarded as its domain; in particular we have a new notation for $\mathcal{I}(L)$, namely $L(\mathbb{R}^n)$. We call $L(Q)$ the **image** of Q **under** L or the **image** of L **restricted to** Q. Thus, either a function or a set may appear following the phrase "image of", depending on emphasis; moreover, the phrase "under L" is sometimes dropped when only one linear map is being discussed and there is no danger of confusion. Our current interest is in the case where Q is a linear subspace of $\mathbb{R}^n$. The following theorem asserts that the image of a linear subspace under a linear map is a linear subspace.

Theorem 3.2.1 For any linear map L and any linear subspace S of its domain, $L(S)$ is a linear subspace.

PROOF: Let Y_1 and Y_2 be members of $L(S)$ and let α be a scalar. We want to prove that $Y_1 + Y_2$ and αY_1 are members of $L(S)$.

Since $Y_1, Y_2 \in L(S)$, there exist members X_1 and X_2 of S such that $Y_1 = L(X_1)$ and $Y_2 = L(X_2)$. Since S is a linear subspace, $X_1 + X_2$ and αX_1 are members of S. Thus, L of each of these is a member $L(S)$. We use the linearity of L to evaluate L of each of these:

$$L(X_1 + X_2) = L(X_1) + L(X_2) = Y_1 + Y_2 \,,$$
$$L(\alpha X_1) = \alpha L(X_1) = \alpha Y_1 \,.$$

We are done since the right-hand sides are the two things we wanted to show to be members of $L(S)$. $<<$

REMARK:

- The adjective "linear" is used for certain equations, systems of equations, subsets (which are called "linear subspaces"), and functions. It is important, on the one hand, to avoid confusing the various types of nouns that are used with "linear"; and, on the other hand, to understand the close relations among linear systems of equations, linear subspaces, and linear maps.

If $L : \mathbb{R}^n \to \mathbb{R}^q$ and $\mathcal{I}(L) = \mathbb{R}^q$, then we say that L is **onto** $\mathbb{R}^q$, or **surjective**.

REMARKS:

- There are two pairs of synonomous adjectives for functions: "one-to-one" and "injective"; and "onto" and "surjective". In this book we use the terms "one-to-one" and "onto".

- The term "onto" is used both as an adjective and as a preposition. We can speak of the function L as being an "onto function" if its image equals its target; but we can also use it as a preposition by saying that L is a map onto $\mathbb{R}^q$.

EXAMPLES:

1. Let $L : \mathbb{R} \to \mathbb{R}^2$ be defined by $L(t) = (3t, 2t)$. In Example 7 of Section 3.1a it has been shown that $\mathcal{I}(L)$ is a one-dimensional linear subspace of $\mathbb{R}^2$. If Y is *not* on this line, no value of t satisfies $L(t) = Y$. Thus, L is not onto. The line $\mathcal{I}(L)$ is said to be represented "parametrically" by L; t is the parameter, and $L(t)$ might be regarded as one's location on the line at time t. This point of view is fundamental to Chapter 4, where lines and curves will be explored in depth.

2. If $L : \mathbb{R}^n \to \mathbb{R}^q$ is the zero map, then clearly $L(X) = Y$ has no solution X unless $Y = \mathbf{0}$. Thus, $\mathcal{I}(L) = \{\mathbf{0}\}$ and L is not onto if $q > 0$. (Sometimes $\mathbb{R}^0$ is used to denote the zero-dimensional vector space consisting of $\mathbf{0}$ only, so if $q = 0$ then the zero map is an onto map.)

3. Suppose $L : \mathbb{R}^2 \to \mathbb{R}^2$ is a rotation through $\pi/2$ radians in the counterclockwise direction. Then every Y equals $L(X)$ for some X — namely, that X which is obtained from Y by a *clockwise* rotation through $\pi/2$ radians. Therefore, $\mathcal{I}(L) = \mathbb{R}^2$, that is, L is onto.

4. Let $L : \mathbb{R}^2 \to \mathbb{R}^2$ be the function which assigns to each $X \in \mathbb{R}^2$ its projection onto the linear subspace spanned by the vector $(1, 1)$. Let us find a formula for L. From the formula for the projection (Section 2.2d), we see that

$$L(X) = \frac{x_1 + x_2}{2} (1, 1) = \left(\frac{x_1 + x_2}{2}, \frac{x_1 + x_2}{2} \right)$$

in terms of the coordinates x_1 and x_2 of X. Naturally, the two coordinates of $L(X)$ are equal for every X since $L(X)$ lies on the line spanned by $(1, 1)$. Moreover, every point (u, u) on this line belongs to $\mathcal{I}(L)$ since $L((u, u)) = (u, u)$. We conclude that $\mathcal{I}(L)$ equals the one-dimensional linear subspace spanned by $(1, 1)$. Of course, we already knew from Theorem 3.2.1 that $\mathcal{I}(L)$ is a linear subspace of $\mathbb{R}^2$, so in order to conclude that $\mathcal{I}(L)$ is the line through the origin spanned by $(1, 1)$, it would be enough to note that $\mathcal{I}(L)$ contains the vector $(1, 1)$, and that it does not contain all of $\mathbb{R}^2$ (for example, it does not contain $(1, 0)$).

Let us find the image under L of various one-dimensional linear subspaces of $\mathbb{R}^2$. We label these linear subspaces according to the angle they make with the x-axis. For $\theta \in [0, \pi)$, let S_θ denote the line through $\mathbf{0} \in \mathbb{R}^2$ that makes an angle θ with the x-axis, measured counterclockwise from the x-axis. For each θ let us calculate the image of S_θ under L. If you draw some pictures, you will see that $\theta = 3\pi/4$ deserves special attention. When $\theta \neq 3\pi/4$, the nonzero vectors in S_θ

are not orthogonal to $(1, 1)$, so their projections onto the line spanned by $(1, 1)$ are not the zero vector. Therefore, for such θ, $L(\mathcal{S}_\theta)$ is not the zero linear subspace. Since it is also not equal to all of $\mathbb{R}^2$ (it must be contained in the line spanned by $(1, 1)$), it is one-dimensional, and hence must equal the line spanned by $(1, 1)$. But when $\theta = 3\pi/4$, all the vectors in $\mathcal{S}_\theta$ are orthogonal to $(1, 1)$, so their projections onto the line spanned by $(1, 1)$ are all equal to the zero vector. Another way to see this is to calculate from the formula for L. Each $X \in \mathcal{S}_{3\pi/4}$ has the form $(v, -v)$ for some v, and for all v,

$$L((v, -v)) = \left(\frac{v - v}{2}, \frac{v - v}{2} \right) = \mathbf{0} .$$

Thus, $L(\mathcal{S}_{3\pi/4})$ is the zero-dimensional space $\{\mathbf{0}\}$.

5. Let us decide whether $(2, 1, -3)$ belongs to the image of L defined by

$$L((x_1, x_2, x_3)) = (x_1 - 2x_2 + x_3 , \, 2x_1 - 3x_2 - x_3 , \, x_1 - x_2 - 2x_3) .$$

Thus, we want to decide if the following linear system of equations has at least one solution:

$$
\begin{aligned}
x_1 - 2x_2 + x_3 &= 2 \\
2x_1 - 3x_2 - x_3 &= 1 \\
x_1 - x_2 - 2x_3 &= -3 .
\end{aligned}
$$

Solving this system involves a simple application of the Gauss-Jordan method, which all readers should be able to carry out for themselves. However, for reasons that will become apparent in the next example, we will briefly carry out the steps for the reader. We subtract twice the first equation from the second and also the first from the third to obtain the system:

$$
\begin{aligned}
x_1 - 2x_2 + x_3 &= 2 \\
x_2 - 3x_3 &= -3 \\
x_2 - 3x_3 &= -5 .
\end{aligned}
$$

We now add twice the second equation to the first equation and subtract the second equation from the third equation to obtain

$$
\begin{aligned}
x_1 - 5x_3 &= -4 \\
x_2 - 3x_3 &= -3 \\
0 &= -2 .
\end{aligned}
$$

The last of these three equations cannot be satisfied by any choice of x_1, x_2, and x_3. Hence, the original system has no solutions and therefore $(2, 1, -3) \notin \mathcal{I}(L)$.

6. Let us calculate $\mathcal{I}(L)$ for L as defined in the preceding example. Thus, for each $Y = (y_1, y_2, y_3)$ we want to decide whether $Y \in \mathcal{I}(L)$, that is, whether the following system of linear equations has at least one solution:

$$\begin{aligned}
x_1 - 2x_2 + x_3 &= y_1 \\
2x_1 - 3x_2 - x_3 &= y_2 \\
x_1 - x_2 - 2x_3 &= y_3 \, .
\end{aligned}$$

Mimicking the preceding example we first obtain the system

$$\begin{aligned}
x_1 - 2x_2 + x_3 &= y_1 \\
x_2 - 3x_3 &= y_2 - 2y_1 \\
x_2 - 3x_3 &= y_3 - y_1 \, ;
\end{aligned}$$

and then

$$\begin{aligned}
x_1 - 5x_3 &= 2y_2 - 3y_1 \\
x_2 - 3x_3 &= y_2 - 2y_1 \\
0 &= y_1 - y_2 + y_3 \, .
\end{aligned}$$

There are two cases. If $y_1 - y_2 + y_3 \neq 0$, then as in the previous example, this system has no solution and, as a consequence, $Y \notin \mathcal{I}(L)$. If $y_1 - y_2 + y_3 = 0$, then this system has infinitely many solutions; x_3 is the free variable, and x_2 and x_1 are determined by the second and first equations, respectively. The result is that $\mathcal{I}(L)$ equals the plane determined by the equation $y_1 - y_2 + y_3 = 0$, that is, the plane of vectors perpendicular to the vector $(1, -1, 1)$.

We already know that $L(\mathbf{e}_1)$, $L(\mathbf{e}_2)$, ..., and $L(\mathbf{e}_n)$ span the image of a linear map L with $\mathcal{D}(L) = \mathbb{R}^n$. Here is a generalization.

Theorem 3.2.2 If X_1, X_2, ..., and X_k span a linear subspace $\mathcal{S}$ of the domain of a linear function L, then $L(X_1)$, $L(X_2)$, ..., and $L(X_k)$ span $L(\mathcal{S})$.

PROOF: Let Y be an arbitrary member of $L(\mathcal{S})$. We plan to complete the proof by showing that Y is a linear combination of $L(X_1), \ldots, L(X_k)$. Since $Y \in L(\mathcal{S})$, we know that $Y = L(X)$ for some $X \in \mathcal{S}$. Since $X_1, \ldots,$ and X_k span $\mathcal{S}$, there exist constants $\alpha_1, \ldots, \alpha_k$ such that

$$X = \alpha_1 X_1 + \cdots + \alpha_k X_k \, .$$

Then, by the linearity of L,

$$Y = L(X) = \alpha_1 L(X_1) + \cdots + \alpha_k L(X_k) \, ,$$

a desired linear combination. $<<$

EXAMPLES:

7. For L as defined in Example 5 and S_1 equal to the linear subspace spanned by $(1, 1, 0)$ and $(0, 0, 1)$, let us calculate a basis for $L(S_1)$. The preceding theorem tells us that $L(S_1)$ is the linear subspace spanned by $L((1, 1, 0))$ and $L((0, 0, 1))$. These two vectors are $L((1, 1, 0)) = (-1, -1, 0)$ and $L((0, 0, 1)) = (1, -1, -2)$. Neither of these vectors is a multiple of the other. Therefore, they constitute a basis for the two-dimensional linear subspace $L(S_1)$.

8. Let us modify the preceding example by replacing S_1 by S_2, the linear subspace spanned by $(1, 1, 0)$ and $(2, 0, 1)$. Following the preceding example, we calculate $L((1, 1, 0)) = (-1, -1, 0)$ and $L((2, 0, 1)) = (3, 3, 0)$. These two vectors span $L(S_2)$. Taking account of the obvious linear dependence we conclude that $(-1, -1, 0)$ constitutes a basis for the one-dimensional linear subspace $L(S_2)$.

3.2B THE GRAPH OF A LINEAR MAP

Let $L : \mathbb{R}^n \to \mathbb{R}^q$ be a linear map. The **graph** of L is the set of those members of $\mathbb{R}^{n+q}$ that are of the form $(X, L(X))$ for some $X \in \mathbb{R}^n$. Keep in mind that $X, L(X)$, and $(X, L(X))$ are all vectors. The first n coordinates of the vector $(X, L(X))$ are the coordinates of X, and the last q coordinates of $(X, L(X))$ are the coordinates of $L(X)$. We denote the graph of L by $\mathcal{G}(L)$.

EXAMPLES:

9. Let $L : \mathbb{R} \to \mathbb{R}$ be the linear map defined by $L(x) = 2x$. The graph of L is the set of points in $\mathbb{R}^2$ of the form $(x, 2x)$ for $x \in \mathbb{R}$. This set is the line through the origin in $\mathbb{R}^2$ spanned by the vector $(1, 2)$. You should recognize this as the graph of the function $y = 2x$, familiar from single-variable calculus.

10. Let $L : \mathbb{R} \to \mathbb{R}^2$ be defined by $L(x) = (-x, 3x)$. Then the graph of L is the set of points in $\mathbb{R}^3$ of the form $(x, -x, 3x)$, which is the line through the origin in $\mathbb{R}^3$ spanned by the vector $(1, -1, 3)$.

11. In this example and in the next, $\mathcal{G}(L)$ will be a plane rather than a line. Let $L : \mathbb{R}^2 \to \mathbb{R}$ be defined by $L(x_1, x_2) = x_1 + 2x_2$. The graph of L is the set of vectors in $\mathbb{R}^3$ of the form $(x_1, x_2, x_1 + 2x_2)$, for real numbers x_1 and x_2. We may rewrite this set as the set of vectors of the form $x_1(1, 0, 1) + x_2(0, 1, 2)$, which is clearly the plane through the origin in $\mathbb{R}^3$ spanned by the vectors $(1, 0, 1)$ and $(0, 1, 2)$.

12. Let $L : \mathbb{R}^2 \to \mathbb{R}^3$ be defined by

$$L(x_1, x_2) = (x_2, \, 3x_1 - x_2, \, x_1 + x_2) .$$

Then $\mathcal{G}(L)$ is the set of vectors in $\mathbb{R}^5$ of the form

$$(x_1,\, x_2,\, x_2,\, 3x_1 - x_2, x_1 + x_2) = x_1(1,0,0,3,1) + x_2(0,1,1,-1,1)\,.$$

This set is the plane through the origin in $\mathbb{R}^5$ spanned by the vectors $(1,0,0,3,1)$ and $(0,1,1,-1,1)$.

REMARKS:

- We have difficulty picturing the graph of a linear map $L : \mathbb{R}^n \to \mathbb{R}^q$ if $n + q > 3$, since such a graph is a subset of a vector space with dimension higher than 3. Nevertheless, as Example 12 shows, we may still be able to describe $\mathcal{G}(L)$ in terms with which we have become familiar.

- One reason that single-variable calculus is easy in certain respects is that the graphs of functions from $\mathbb{R}$ to $\mathbb{R}$ are subsets of $\mathbb{R}^2$, and hence often easily drawn. In multivariable calculus, we will be able to draw the pictures of graphs of functions from $\mathbb{R}^2$ to $\mathbb{R}$, and also of functions from $\mathbb{R}$ to $\mathbb{R}^2$. For higher dimensional cases, pictures fail us, and we need to rely more on our algebraic skills.

- The graph of a linear map is *not* the same as the image of a linear map. In the examples, the images are: $\mathbb{R}$ in Examples 9 and 11; the line through the origin in $\mathbb{R}^2$ spanned by the vector $(-1,3)$ in Example 10; the plane through the origin in $\mathbb{R}^3$ spanned by the vectors $(0,3,1),(1,-1,1)$ in Example 12.

In each of the examples, $\mathcal{G}(L)$ is a linear subspace of $\mathbb{R}^{n+q}$. Look closely at Example 12. The graph of L in that example is the linear subspace spanned by the vectors $(\mathbf{e}_1, L(\mathbf{e}_1))$ and $(\mathbf{e}_2, L(\mathbf{e}_2))$, where $\mathbf{e}_1$ and $\mathbf{e}_2$ are the standard basis vectors in the domain of L. An analogous statement is true for the other examples. In general, we have the following result:

Theorem 3.2.3 The graph of a linear map from $\mathbb{R}^n$ into $\mathbb{R}^q$ is the n-dimensional linear subspace of $\mathbb{R}^{n+q}$ with basis

$$(\mathbf{e}_1, L(\mathbf{e}_1)), \ldots, (\mathbf{e}_n, L(\mathbf{e}_n))\,,$$

where $\mathbf{e}_1, \ldots,$ and $\mathbf{e}_n$ are the standard basis vectors of $\mathbb{R}^n$.

PROOF: Let X be a vector in the domain of L. As usual, write $X = (x_1, \ldots, x_n) = x_1\mathbf{e}_1 + \cdots + x_n\mathbf{e}_n$. Then

$$
\begin{aligned}
(X, L(X)) &= (x_1\mathbf{e}_1 + \cdots + x_n\mathbf{e}_n, L(x_1\mathbf{e}_1 + \cdots + x_n\mathbf{e}_n)) \\
&= (x_1\mathbf{e}_1 + \cdots + x_n\mathbf{e}_n, x_1 L(\mathbf{e}_1) + \cdots + x_n L(\mathbf{e}_n)) \\
&= x_1(\mathbf{e}_1, L(\mathbf{e}_1)) + \cdots + x_n(\mathbf{e}_n, L(\mathbf{e}_n))\,.
\end{aligned}
$$

Note how the linearity of L was used in the second line. Therefore, $\mathcal{G}(L)$ is the linear span of the vectors $(\mathbf{e}_1, L(\mathbf{e}_1)), \ldots, (\mathbf{e}_n, L(\mathbf{e}_n))$. (Compare this part of the proof to Example 12.) These vectors are linearly independent, since each of them has a 1 in a coordinate slot in which all the other vectors have 0's. Therefore they form a basis for $\mathcal{G}(L)$. Since there are n basis vectors, $\mathcal{G}(L)$ is an n-dimensional linear subspace of $\mathbb{R}^{n+q}$. Done. $<<$

To summarize: the graph of a linear map L is a linear subspace with the same dimension as the domain of L. If the domain of L is $\mathbb{R}$, then $\mathcal{G}(L)$ is a line; if the domain of L is $\mathbb{R}^2$, then $\mathcal{G}(L)$ is a plane, and so on.

3.2C THE NULL SPACE OF A LINEAR MAP

If $L : \mathbb{R}^n \to \mathbb{R}^q$ is a linear map, the set of vectors $X \in \mathbb{R}^n$ such that $L(X) = \mathbf{0}$ is called the **null space** of L (also called the **kernel** of L), and denoted by $\mathcal{N}(L)$. Note that the null space of L is a subset of the domain of L, in contrast to the image of L, which is a subset of the target of L. As we will see, the null space of a linear map is a linear subspace of its domain.

EXAMPLES:

13. If $L : \mathbb{R}^2 \to \mathbb{R}$ is given by $L(X) = x_1 - x_2$, then $\mathcal{N}(L)$ is the line through $\mathbf{0} \in \mathbb{R}^2$ given by $x_1 - x_2 = 0$. See Figure 3.12.

14. The null space of the linear map $L : \mathbb{R}^4 \to \mathbb{R}^3$ defined by

$$L((x_1, x_2, x_3, x_4)) =$$
$$(x_1 + 3x_2 - x_3 + 2x_4 , \, 4x_1 - x_2 - x_4 , \, 2x_1 - 3x_2 + 2x_3 + 3x_4)$$

is the set of vectors (x_1, x_2, x_3, x_4) such that all three coordinates of $L((x_1, x_2, x_3, x_4))$ are zero. In other words, it is the solution set of the homogeneous system

$$\begin{aligned} x_1 + 3x_2 - x_3 + 2x_4 &= 0 \\ 4x_1 - x_2 - x_4 &= 0 \\ 2x_1 - 3x_2 + 2x_3 + 3x_4 &= 0 . \end{aligned}$$

Thus we can find the null space of L by applying Gauss-Jordan elimination to this system. The answer, which you might wish to work out for yourself, is a linear subspace of $\mathbb{R}^4$.

Is the null space a linear subspace of the domain $\mathbb{R}^n$, as the name seems to imply? In both of the examples just given, the answer is "yes". The fact that the answer is always "yes" is stated in the following theorem, whose proof is left to you.

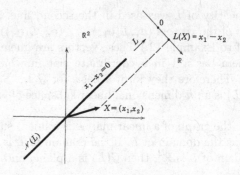

FIGURE 3.12.

Theorem 3.2.4 Let $L : \mathbb{R}^n \to \mathbb{R}^q$ be linear. Then the null space $\mathcal{N}(L)$ is a linear subspace of $\mathbb{R}^n$.

There is an important connection between the null space of L and the solution set of an equation of the form $L(X) = Y$. Suppose that $L(X_0) = \mathbf{0}$ and $L(X_1) = Y$, where $\mathbf{0}$ is the zero vector in the target space of L, and Y is an arbitrary vector in the target space. Then by the linearity of L,

$$L(X_0 + X_1) = L(X_0) + L(X_1) = \mathbf{0} + Y = Y .$$

Thus, if X_0 is in $\mathcal{N}(L)$ and if X_1 is a solution of the equation $L(X) = Y$, then the vector $X = X_0 + X_1$ is also a solution of $L(X) = Y$. Conversely, suppose that X_2 is any other solution of the equation $L(X) = Y$. Then $X_2 - X_1$ is in the null space of L, since

$$L(X_2 - X_1) = L(X_2) - L(X_1) = Y - Y = \mathbf{0} .$$

Since $X_2 = (X_2 - X_1) + X_1$, we have shown that every solution of $L(X) = Y$ is the sum of X_1 and some vector in $\mathcal{N}(L)$. Using the terminology of affine subspaces, we have proved the following:

Theorem 3.2.5 Let $L : \mathbb{R}^n \to \mathbb{R}^q$ be linear. Let $Y \in \mathbb{R}^q$ be given, and suppose that $L(X_1) = Y$ for a particular $X_1 \in \mathbb{R}^n$. Then the solution set of $L(X) = Y$ is the affine subspace $\mathcal{N}(L) + X_1$.

Corollary 3.2.1 If $\mathcal{N}(L) = \{\mathbf{0}\}$, then, for a given Y, there is either no solution or exactly one solution X of the equation $L(X) = Y$.

EXAMPLE:

15. For various choices of $Y \in \mathbb{R}^3$ let us solve the equation $L(X) = Y$, where $L : \mathbb{R}^2 \to \mathbb{R}^3$ is defined by

$$L((x_1, x_2)) = (3x_1 - x_2 , x_1 + 2x_2 , x_2) .$$

First we consider $Y = 0$, so that our task is to find the null space of L. So, we solve the system of homogeneous linear equations

$$
\begin{aligned}
3x_1 - x_2 &= 0 \\
x_1 + 2x_2 &= 0 \\
x_2 &= 0 .
\end{aligned}
$$

It is easily determined that this system has only the trivial solution $(x_1, x_2) = (0, 0)$, so $\mathcal{N}(L) = \mathbf{0}$. If $Y = (2, 3, 1)$, the unique solution of $L(X) = Y$ is $(1, 1)$, as may be found by solving the system

$$
\begin{aligned}
3x_1 - x_2 &= 2 \\
x_1 + 2x_2 &= 3 \\
x_2 &= 1 .
\end{aligned}
$$

If $Y = (2, 3, 2)$, there is no solution.

Here is a simple but useful consequence of Theorem 3.2.5. Suppose that we know, for two particular vectors, Y_1 and Y_2, that the solution sets of $L(X) = Y_1$ and $L(X) = Y_2$ are both nonempty. Then, according to Theorem 3.2.5, the two solution sets are of the form $\mathcal{N}(L) + X_1$ and $\mathcal{N}(L) + X_2$. The term "superposition principle" is used for the easy observation that a solution of $L(X) = Y_1 + Y_2$ is $X_1 + X_2$ and, thus, that the solution set of $L(X) = Y_1 + Y_2$ is $X_1 + X_2 + \mathcal{N}(L)$. In particular, the solution set of $L(X) = Y_1 + Y_2$ is nonempty.

Let us summarize the results of this section. The null space of a linear map is a linear subspace of its domain. The solution set of an equation like $L(X) = Y$ is, when solutions exist, an affine space parallel to $\mathcal{N}(L)$. The reader should be reminded of the material concerning the relationship between linear equations and affine subspaces presented in Section 1.6a. The similarity is no accident. When an equation involving a linear map is put into coordinate form, it becomes a system of linear equations. We will have more to say about this connection in Section 3.5.

3.2D $\dim(\mathcal{N}(L)) + \dim(\mathcal{I}(L)) = \dim(\mathcal{G}(L))$

We have introduced three vector spaces in connection with a linear map $L : \mathbb{R}^n \to \mathbb{R}^q$: its null space, its image, and its graph. The null space of L is a linear subspace of its domain $\mathbb{R}^n$, the image of L is a linear subspace of its target $\mathbb{R}^q$, and the graph is a linear subspace of the space $\mathbb{R}^{n+q}$. We found that the dimension of the graph was the same as the dimension of the domain of L, namely, $\dim(\mathcal{G}(L)) = n$. The dimension of the null space of L is called the **nullity** of L, and the dimension of the image of L is called the **rank** of L. In this section, we will discuss the remarkable result that is the title of this section: the sum of the nullity and rank of a linear map on $\mathbb{R}^n$ equals n. We start with a series of examples designed to highlight connections among various concepts.

EXAMPLES:

16. Let $L : \mathbb{R}^3 \to \mathbb{R}^4$ be defined by

$$\begin{aligned}
L(\mathbf{e}_1) &= (1, -1, -1, 1), \\
L(\mathbf{e}_2) &= (0, 3, 3, 0), \\
L(\mathbf{e}_3) &= (2, 1, 1, 2).
\end{aligned}$$

Let us determine the null space of L, which is a linear subspace of $\mathbb{R}^3$. Thus, we need to solve the following homogeneous system:

$$\begin{aligned}
x_1 + 2x_3 &= 0 \\
-x_1 + 3x_2 + x_3 &= 0 \\
-x_1 + 3x_2 + x_3 &= 0 \\
x_1 + 2x_3 &= 0.
\end{aligned}$$

Using Gauss-Jordan elimination, we can reduce this system to the equivalent system

$$\begin{aligned}
x_1 + 2x_3 &= 0 \\
3x_2 + 3x_3 &= 0 \\
0 &= 0 \\
0 &= 0.
\end{aligned}$$

There is one free variable, x_3, so the solution set is one-dimensional. The vector $(-2, -1, 1)$ constitutes a basis for the null space of L, and the nullity of L equals 1.

17. For L as in the preceding example, let us find $\mathcal{I}(L)$. Thus, we want to find those $Y \in \mathbb{R}^4$ for which the following system has a solution:

$$\begin{aligned}
x_1 + 2x_3 &= y_1 \\
-x_1 + 3x_2 + x_3 &= y_2 \\
-x_1 + 3x_2 + x_3 &= y_3 \\
x_1 + 2x_3 &= y_4.
\end{aligned}$$

This is the same as the system we solved to find the null space of L, except that the right-hand sides of the equation are not necessarily zero. If we carry out the same elimination steps as we did for the homogeneous system, we obtain

$$\begin{aligned}
x_1 + 2x_3 &= y_1 \\
3x_2 + 3x_3 &= y_1 + y_2 \\
0 &= y_3 - y_2 \\
0 &= y_4 - y_1.
\end{aligned}$$

For Y to belong to the image of L it is necessary and sufficient that the last two equations be satisfied:

$$\begin{aligned} y_3 - y_2 &= 0 \\ y_4 - y_1 &= 0 . \end{aligned}$$

We will treat y_1 and y_2 as the free variables and y_3 and y_4 as the basic variables. (When this method is used to calculate the image it is usually most efficient to treat the variables with the smaller subscripts as the free variables and those with the larger subscripts as the basic variables.) A basis for $\mathcal{I}(L)$ can be obtained by first letting $y_1 = 1$ and $y_2 = 0$ and then letting $y_1 = 0$ and $y_2 = 1$. The basis of $\mathcal{I}(L)$ thus obtained consists of the vectors: $(1, 0, 0, 1)$ and $(0, 1, 1, 0)$. The rank of L equals 2.

18. For L as defined for the preceding two examples let us find, for the graph, a basis that is closely related to the bases for the null space and image found in the preceding two examples.

Note that the nullity of L plus the rank of L equals the dimension of the domain of L: $1 + 2 = 3$. The dimension of the graph of L is the same as the dimension of the domain, so $\dim(\mathcal{N}(L)) + \dim(\mathcal{I}(L)) = \dim(\mathcal{G}(L))$. We will form a basis of three vectors for $\mathcal{G}(L)$ by choosing one vector in $\mathbb{R}^7$ that is closely related to the basis vector $(-2, -1, 1)$ found above for the null space and two vectors in $\mathbb{R}^7$ that are closely related to the two basis vectors $(1, 0, 0, 1)$ and $(0, 1, 1, 0)$ found above for the image.

To the vector $(-2, -1, 1)$ we adjoin the four coordinates of the vector $L((-2, -1, 1))$ —namely, four zeroes—to obtain

$$(-2, -1, 1, 0, 0, 0, 0) .$$

It is the first three coordinates that are missing from $(1, 0, 0, 1)$ and $(0, 1, 1, 0)$. Thus, we need to solve the equations

$$L(X) = (1, 0, 0, 1)$$

and

$$L(X) = (0, 1, 1, 0) .$$

From the work done in the preceding example we see that we can solve both problems at once by solving

$$\begin{aligned} x_1 + 2x_3 &= y_1 \\ 3x_2 + 3x_3 &= y_1 + y_2 . \end{aligned}$$

Gauss-Jordan elimination has already been completed. Since we are looking for only one solution we can set $x_3 = 0$ obtaining

$$x_1 = y_1 \quad \text{and} \quad x_2 = \frac{1}{3}y_1 + \frac{1}{3}y_2 \, .$$

Setting $y_1 = 1$ and $y_2 = 0$ we obtain the solution $(1, \frac{1}{3}, 0)$ of the equation $L(X) = (1, 0, 0, 1)$. Setting $y_1 = 0$ and $y_2 = 1$ we obtain the solution $(0, \frac{1}{3}, 0)$ of $L(X) = (0, 1, 1, 0)$. The resulting basis vectors are

$$\left(1, \frac{1}{3}, 0, 1, 0, 0, 1\right) \quad \text{and} \quad \left(0, \frac{1}{3}, 0, 0, 1, 1, 0\right) .$$

So it appears that we have a basis for $\mathcal{G}(L)$ consisting of the three vectors

$$(-2, -1, 1, 0, 0, 0, 0)', \quad \left(1, \frac{1}{3}, 0, 1, 0, 0, 1\right), \quad \text{and} \quad \left(0, \frac{1}{3}, 0, 0, 1, 1, 0\right) .$$

We certainly have the correct number of vectors, and by our calculations, we have arranged for them all to be members of $\mathcal{G}(L)$. We only have to check if they are linearly independent. Focusing on the third, fourth, and fifth coordinates, corresponding to the locations of the free variables in the two preceding exercises, we see that each vector has a 1 in the same position that all the other have a 0. Thus, we have the desired linear independence.

We now state the general result.

Theorem 3.2.6 Let L be a linear map on $\mathbb{R}^n$. Then the nullity of L plus the rank of L equals n (which equals the dimension of the graph of L or, equivalently, the dimension of the domain of L).

PROOF: In one way the example preceding the theorem fails to be a proof because it lacks generality. On the other hand it is more than a proof because it shows that, for the graph of a linear transformation, a basis can be constructed that is closely related to a given bases for the null space and image. The issue left unresolved by the example is whether it is coincidence that enough vectors were obtained to give a basis for the graph. Here is an argument which shows that it is not a coincidence.

After Gauss-Jordan elimination is carried out, there are a certain number of equations—say r equations—that do not have 0 as their left-hand sides. Each such equation corresponds to a basic variable. The number of basic variables is thus, the number of these equations, namely r. As a consequence, the number of free variables is $n - r$. Hence, the nullity equals $n - r$.

It is the equations that do have 0 on the left that give conditions on the image. There are $q - r$ such conditions, each of which corresponds to one basic variable. Thus, there are $q - (q - r) = r$ free variables and, hence, the rank equals r.

Therefore, the sum of the nullity and rank equals $(n - r) + r = n$, as desired. $<<$

REMARKS:

- It follows easily from the theorem that the image of L cannot have dimension higher than the dimension of the domain of L. A linear map cannot produce an image that is larger than its domain.

- Since the image is a linear subspace of the target, the rank r can be no larger than q. If j is the nullity, the theorem says that $j + r = n$, so $j = n - r \geq n - q$. Now suppose that $q < n$. Then $j > 0$. In other words, a linear map cannot squeeze $\mathbb{R}^n$ into the smaller $\mathbb{R}^q$ without mapping some nonzero vectors to $\mathbf{0}$.

- In the important special case $q = n$, we see that a linear map L is onto if and only if $r = q$ which is true if and only if $r = n$ which is true if and only if the nullity of L equals 0 which is true if and only if L is one-to-one. So, in the special case where the domain and the target are the same space, ontoness is equivalent to one-to-oneness.

- One-to-oneness is impossible for a linear map if its target has smaller dimension than its domain.

- Ontoness is impossible for a linear map if its target has larger dimension than its domain.

We now know that the dimension of the image of a linear transformation L equals the number of basic variables arising when Gauss-Jordan elimination is used on the corresponding system of homogeneous (or inhomogeneous) linear equations. Another fact we know is that each column (written as a vector) from $[L]$ belongs to the image. Taking a leap of faith based on count, one might suspect that the vectors corresponding to the basic variables constitutes a basis for $\mathcal{I}(L)$. There are the correct number of such vectors. We must prove that they are linearly independent. For convenience of notation, let us suppose that the basic variables are $x_1, x_2, \ldots, x_r$. Thus, we want to prove that $L(\mathbf{e}_1), L(\mathbf{e}_2), \ldots, L(\mathbf{e}_r)$ are linearly independent. Consider a linear combination that equals 0:

$$\alpha_1 L(\mathbf{e}_1) + \alpha_2 L(\mathbf{e}_2) + \cdots + \alpha_r L(\mathbf{e}_r) = 0 .$$

By linearity,

$$L(\alpha_1 \mathbf{e}_1 + \alpha_2 \mathbf{e}_2 + \cdots + \alpha_r \mathbf{e}_r) = 0 .$$

So,

$$(\alpha_1, \alpha_2, \ldots, \alpha_r, 0, 0, \ldots, 0) \in \mathcal{N}(L) .$$

Thus, we have a vector that both belongs to the null space and has all of its coordinates corresponding to free variables equal to 0. But *the* solution of the homogeneous system of equations that results from choosing the free variables all equal to 0 is the zero vector. Thus, $\alpha_1 = \alpha_2 = \cdots = \alpha_r = 0$, from which we conclude that we have the desired linear independence. We have proved the following:

Theorem 3.2.7 A basis for the image of a linear transformation L consists of those vectors obtained by writing as vectors the columns of $[L]$ corresponding to the basic variables.

EXAMPLE:

19. In Example 17 we obtained the basic variables x_1 and x_2. So, we obtain a basis for $\mathcal{I}(L)$ from the first two columns of $[L]$:

$$(1, -1, -1, 1) \quad \text{and} \quad (0, 3, 3, 0) \, .$$

3.2E ACCOMPLISHMENTS SO FAR

First, we have defined a vector space as a set in which addition of members and multiplication by real scalars are possible. A mathematician might say that a vector space is a set endowed with a "linear structure".

Second, we have discussed linear maps, namely those functions which *preserve* linear structure. It is because of this amicable relation between vector spaces and linear maps that the important Theorems 3.1.1, 3.2.1, and 3.2.2 are true.

Third, we have discovered striking connections among linear maps, linear subspaces, and their dimensions.

Exercises

Linear maps and related vector spaces.

1. For each of the following maps find bases for $\mathcal{N}(L)$ and $\mathcal{I}(L)$ and compute $\dim(\mathcal{N}(L))$ and $\dim(\mathcal{I}(L))$.

   (a) $L((x_1, x_2)) = (2x_1 - x_2, \, x_1 + x_2)$
   (b) $L((x_1, x_2, x_3)) = (2x_1 - x_2 + x_3, \, -2x_1 + x_2 - x_3)$
   (c) $L((x_1, x_2, x_3)) = (2x_1 - x_2, \, -2x_1 + x_2, \, 2x_1 - x_3)$.

2. Two quite different methods have been described in the text for finding a basis for the image of a linear map. Redo the "basis for image" portion of the preceding exercise using the method that you did not use there.

3. Let $L : \mathbb{R}^2 \to \mathbb{R}^2$ be the linear map given by $L(X) = (x_1, x_1 + x_2)$. Draw a picture of the target and locate the images of the x_1-axis, the x_2-axis, and the line $x_1 + x_2 = 0$.

4. Given: $L : \mathbb{R}^2 \to \mathbb{R}^2$ is linear, $L(\mathbf{e}_1) = (2, 3)$, and $L(\mathbf{e}_2) = (8, 12)$.

(a) If $Y = (1, 1)$, does there exist X such that $L(X) = Y$?

(b) Same question, with $Y = (4, 6)$?

(c) Draw a picture of the target, illustrating the linear subspace $\mathcal{I}(L)$.

(d) Is there a nonzero X such that $L(X) = 0$?

5. Draw pictures of the image space $\mathcal{I}(L)$ in $\mathbb{R}^2$ for the following L:

(a) $L : \mathbb{R} \to \mathbb{R}^2$, $L(t) = (2t, t)$

(b) $L : \mathbb{R}^2 \to \mathbb{R}^2$, $L(X) = (2x_1 - 6x_2, \; x_1 - 3x_2)$

(c) $L : \mathbb{R}^3 \to \mathbb{R}^2$, $L(X) = (2x_1 - 2x_2 + 6x_3, \; x_1 - x_2 + 3x_3)$.

6. Decide for which y_1 and y_2 the system

$$
\begin{aligned}
2x_1 - 6x_2 &= y_1 \\
x_1 - 3x_2 &= y_2
\end{aligned}
$$

has a solution. Note that you are not asked to find a solution if one exists. But can you?

7. For each of the following points Y and linear maps L decide whether $Y \in \mathcal{I}(L)$:

(a) $Y = (4, -6, 16)$, $L(x) = (-2x, 3x, -8x)$

(b) $Y = 4$, $L((x_1, x_2, x_3)) = 534x_1 - 351x_2 - 8174x_3$

(c) $Y = (4, 13, -38, 6)$,
$L((x_1, x_2)) = (x_1 - x_2, \; 3x_1 - 2x_2, \; -8x_1 + x_2, \; x_1 + x_2)$

(d) $Y = (0, 0, 0)$, $L((x_1, x_2)) = (-27x_1 + 45x_2, \; 53x_1 - 8x_2)$ (be careful)

(e) $Y = (3, 5)$, $L((x_1, x_2, x_3)) = (-6x_1 + 9x_2 - 21x_3, \; 8x_1 - 12x_2 + 28x_3)$

(f) $Y = (2, 3, 2)$, $L((x_1, x_2)) = (x_1 + x_2, \; 7x_1, \; x_1 + x_2)$

(g) $Y = (1, 0, 1, 0, 0)$, $L((x_1, x_2, x_3, x_4, x_5, x_6))$
$= (x_1 + x_6, \; x_2 + x_5, \; x_3 + x_4, \; x_1 + x_2 + x_3, \; x_4 + x_5 + x_6)$

(h) $Y = (1, 2, 3, \ldots, n)$, $L((x_1, x_2, x_3, \ldots, x_n))$
$= (x_1, \; x_1 + x_2, \; x_1 + x_2 + x_3, \ldots, \; x_1 + x_2 + x_3 + \cdots + x_n)$.

8. For each of the following points Z and linear maps L decide whether Z is a member of the graph of L:

(a) $Z = (2, 4, 2)$; $L((x_1, x_2)) = 3x_1 - x_2$

(b) $Z = (5, 7, 8, 3, 5, 6)$; $L((x_1, x_2, x_3, x_4)) = (x_1, \; -x_1 - 2x_2 + x_3 + 6x_4)$

(c) $Z = (1, 2, 3)$; $L((x_1, x_2)) = (x_1, \; x_1 + x_2, \; x_1 + 2x_2)$ (be careful).

9. Sketch the graph in $\mathbb{R}^2$ of the linear function $L(x) = (-2/5)x$ from $\mathbb{R}$ into $\mathbb{R}$. Find a basis for the graph of L. Convince yourself that your picture is indeed a picture of the linear subspace spanned by the basis.

10. Sketch the graph in $\mathbb{R}^3$ of the linear function $L(x) = (2x, 3x)$ from $\mathbb{R}$ into $\mathbb{R}^2$. Find a basis for the graph of L.

11. Sketch the graph in $\mathbb{R}^3$ of the linear function $L((x_1, x_2)) = x_1 + x_2$ from $\mathbb{R}^2$ into $\mathbb{R}$. Find a basis for the graph of L.

12. Find a basis for the graph of each of the following linear maps:

   (a) $L(x) = (x, x, x)$

   (b) $L(x) = (x, x, x, \ldots, x) \in \mathbb{R}^q$

   (c) $L((x_1, x_2, x_3)) = x_2$

   (d) $L((x_1, x_2, x_3)) = x_1 + x_2 + x_3$

   (e) $L((x_1, x_2)) = (3x_1 - 5x_2 \,,\, 5x_1 + 3x_2)$

   (f) $L((x_1, x_2, \ldots, x_n)) = (x_n, x_{n-1}, \ldots, x_1)$.

13. Let $L : \mathbb{R}^2 \to \mathbb{R}^2$. Since the null space $\mathcal{N}(L)$ is a linear subspace, it is $\{0\}$ or it is some line through 0 or it is all of $\mathbb{R}^2$. Decide which of the three possibilitites is appropriate for each of the following linear maps L:

   (a) $L(X) = (x_1, -x_2)$

   (b) $L(X) = (x_1, 0)$

   (c) $L(X) = (x_1, -x_1)$

   (d) $L(X) = (x_1 \,,\, x_1 - x_2)$

   (e) $L(X) = (2x_1 - 2x_2 \,,\, x_1 - x_2)$.

14. If L is linear, can $\mathcal{N}(L)$ be the empty set? Explain.

15. Let $L : \mathbb{R}^n \to \mathbb{R}^q$ be a linear map. Consider the equation $L(X) = B$, where B is a vector in the target space $\mathbb{R}^q$. For each of the following decide whether it is true or false, and, if it is false, give a counterexample.

   (a) If $n = q$, there is *at most one* solution of $L(X) = B$.

   (b) If $n > q$, there is *at least one* solution of $L(X) = B$.

   (c) If $n = q$, there is *at least one* solution of $L(X) = B$.

   (d) If $n < q$, there is *no* solution of $L(X) = B$.

   (e) If $n > q$, there is *no* solution of $L(X) = B$.

   (f) If $n < q$, the *only* solution of $L(X) = 0$ is $X = 0$.

16. Given that L is a linear map for which $L(X_1) = Y$ and $L(X_2) = Y$ for distinct X_1 and X_2. Explain why $\dim(\mathcal{N}(L)) \geq 1$.

17. Let L be a linear map for which $L(X_1) = L(X_2)$ implies $X_1 = X_2$. What can you say about X if $L(X) = 0$?

18. Given that L is linear and $\mathcal{N}(L) = \{0\}$. What can you say about X_1 and X_2 if $L(X_1) = L(X_2)$? *Hint:* Look at $X_1 - X_2$.

19. Let $L : \mathbb{R}^3 \to \mathbb{R}^3$ be defined by $L(X) = (x_1, 0, x_3)$. Geometrically describe $\mathcal{N}(L)$ and $\mathcal{I}(L)$. What are the dimensions of $\mathcal{N}(L)$ and $\mathcal{I}(L)$?

20. Prove Theorem 3.2.4.

3.3 A Special Case: Linear Functionals

A linear map $L : \mathbb{R}^n \to \mathbb{R}$ is commonly called a linear **functional**. We note that the values $L(X)$ of a functional are *real numbers*, that is, vectors in the one-dimensional space $\mathbb{R}$.

EXAMPLE:

1. Let $Z = (3, -1, 2)$. For $X \in \mathbb{R}^3$, define $L(X) = \langle Z, X \rangle$. For example, if $X = (1, -1, -1)$, then $L(X) = 2$. It is easy to see from the properties of the inner product (specifically homogeneity and the distributive law) that L is a linear map from $\mathbb{R}^3$ to $\mathbb{R}$, or in other words, a linear functional.

In a sense, the example says it all. Let Z be a fixed vector in $\mathbb{R}^n$. For $X \in \mathbb{R}^n$ define $L(X) = \langle Z, X \rangle$, the inner product of Z and X. Then $L : \mathbb{R}^n \to \mathbb{R}$, since $\langle Z, X \rangle$ is a number, and, as you should check, L is linear. Hence, we conclude that to each vector $Z \in \mathbb{R}^n$, there corresponds a linear functional L defined in terms of the inner product. An illustrative notation for this functional is $L = \langle Z, \cdot \rangle$, where the dot denotes the location at which the variable vector is to be inserted.

We have proved in Section 2.2b that if Y and Z are two vectors in $\mathbb{R}^n$, then $Y = Z$ if and only if $\langle Y, X \rangle = \langle Z, X \rangle$ for all $X \in \mathbb{R}^n$. Thus, different vectors correspond to different linear functionals. On the other hand, suppose L is a linear function defined on $\mathbb{R}^n$. Then the quantities $L(\mathbf{e}_1), \dots, L(\mathbf{e}_n)$ are all real numbers. Let Z be the vector whose coordinates are these numbers. Then

$$\langle Z, \mathbf{e}_1 \rangle = \langle (L(\mathbf{e}_1), \dots, L(\mathbf{e}_n)), (1, 0, \dots, 0) \rangle = L(\mathbf{e}_1) \, .$$

Similarly, $\langle Z; \mathbf{e}_j \rangle = L(\mathbf{e}_j)$ for all $j = 1, \dots, n$. It follows (see Section 3.1c) that $L(X) = \langle Z, X \rangle$ for all $X \in \mathbb{R}^n$. We have proved the following:

Theorem 3.3.1 (Representation Theorem) There is a one-to-one correspondence between $\mathbb{R}^n$ and the collection of all linear functionals on $\mathbb{R}^n$; the linear functional corresponding to the vector Z is $\langle Z, \cdot \rangle$.

EXAMPLE:

2. Let L be the linear function with matrix $[\ 3 \quad 0 \quad -2 \quad 8 \ ]$. Then

$$L(X) = 3x_1 - 2x_3 + 8x_4 = \langle (3, 0, -2, 8), X \rangle$$

for $X \in \mathbb{R}^4$.

REMARKS:

- The term "linear form" is sometimes used instead of "linear functional".

- Linear functionals are of basic importance in Chapter 7 because, roughly speaking, their graphs are the tangent planes to surfaces. We begin to consider these graphs in the exercises below.

Exercises
Linear functionals and related subspaces.

1. Which of the scalar-valued maps $L : \mathbb{R}^2 \to \mathbb{R}$ are linear functionals?

   (a) $L(X) = x_1$

   (b) $L(X) = \|X\|$

   (c) $L(X) = x_1 - x_2$

   (d) $L(X) = x_1 x_2$

   (e) $L(X) = \langle X, X \rangle$

   (f) $L(X) = |x_1|$

   (g) $L(X) = 1 - x_1 - x_2$

   (h) $L(X) = (\sin x_1 + \cos x_1)x_2$.

2. The image $\mathcal{I}(O)$ of the zero linear functional is the set consisting of the number 0 only. What is $\mathcal{I}(L)$ if the linear functional L is not the zero linear functional?

3. (a) Let $L : \mathbb{R}^2 \to \mathbb{R}$ be given by $L(X) = 2x_1 - 5x_2$. Find a vector Z such that $L(X) = \langle Z, X \rangle$ for all X.

   (b) Let $L : \mathbb{R}^2 \to \mathbb{R}$ be a linear functional for which $L(\mathbf{e}_1) = 2$ and $L(\mathbf{e}_2) = -5$. Compute a vector Z such that $L(X) = \langle Z, X \rangle$ for all X.

   (c) What is the matrix of the linear map L in the preceding part?

4. Let $L(X) = 2x_1 - 5x_2$, as in the preceding exercise.

   (a) Sketch $\mathcal{N}(L)$ as a linear subspace of $\mathbb{R}^2$.

   (b) What is the relation between $\mathcal{N}(L)$ and the vector Z found in the preceding exercise?

5. Does there exist a linear functional $L : \mathbb{R}^2 \to \mathbb{R}$ whose null space consists of $\mathbf{0}$ alone? Why?

6. True or false. The null space $\mathcal{N}(L)$ of a nonzero linear functional $L : \mathbb{R}^3 \to \mathbb{R}$ is a line through the origin in $\mathbb{R}^3$.

7. Find a linear functional on $\mathbb{R}^2$ whose graph is the plane $3x_1 - 5x_2 - 9y = 0$ in $\mathbb{R}^3$.

8. Find a linear functional on $\mathbb{R}^5$ whose graph is the hyperplane $x_1 + 6x_2 - 4x_3 - 5x_4 + 5x_5 + 3y = 0$ in $\mathbb{R}^6$.

9. Find a nonzero linear functional $L : \mathbb{R}^3 \to \mathbb{R}$ such that $L(1,1,0) = 0$ and $L(1,0,1) = 0$. Is there more than one correct answer?

3.4   The Algebra of Linear Maps

3.4A   A NEW DIRECTION

So far in this chapter we have considered linear maps one at a time. In this section we learn how to combine them. We will study the sum of linear maps, multiplication of a linear map by a scalar, and, finally, the composition of two linear maps. From this we will see that the collection of linear maps is itself a rich algebraic structure.

3.4B   ADDITION OF LINEAR MAPS

Let L and M be linear maps from $\mathbb{R}^n$ into $\mathbb{R}^q$; we may write this as $L, M : \mathbb{R}^n \to \mathbb{R}^q$. The **sum** of L and M, denoted by $L + M$, is the function from $\mathbb{R}^n$ into $\mathbb{R}^q$ defined by

$$(L + M)(X) = L(X) + M(X) .$$

This definition makes sense, because both $L(X)$ and $M(X)$ are in $\mathbb{R}^q$ and hence may be added by the usual addition of vectors.

EXAMPLE:

1. Let $L, M : \mathbb{R}^3 \to \mathbb{R}$ be functionals given by $L(X) = 3x_1 - x_2 + 12x_3$ and $M(X) = 16x_1 + x_2 + 17x_3$. Then, according to the definition of the sum of two linear maps,

$$(L+M)(X) = (3x_1 - x_2 + 12x_3) + (16x_1 + x_2 + 17x_3) = 19x_1 + 29x_3 .$$

Note here that $L + M$ is linear.

Theorem 3.4.1 Let $L, M, P : \mathbb{R}^n \to \mathbb{R}^q$ be linear maps. Then the following assertions are true:

1. The sum $L + M$ is also a linear map.

2. Associativity: $L + (M + P) = (L + M) + P$.

3. Commutativity: $L + M = M + L$.

4. Additive Identity: $L + O = L$. where O denotes the zero map (which is the map that assigns to each $X \in \mathbb{R}^n$ the zero vector in $\mathbb{R}^q$).

5. Additive Inverse: $L + (-L) = O$, where $-L$ denotes the map that assigns to each $X \in \mathbb{R}^n$ the vector $-(L(X))$ in $\mathbb{R}^q$.

PROOF: 1. We must show that $L+M$ has the additivity and homogeneity properties. To verify additivity, note that

$$
\begin{aligned}
(L+M)&(X+X') \\
&= L(X+X') + M(X+X') \quad \text{definition of } L+M \\
&= L(X) + L(X') + M(X) + M(X') \quad \text{additivity of both } L \text{ and } M \\
&= L(X) + M(X) + L(X') + M(X') \quad \text{commutativity in } \mathbb{R}^q \\
&= (L+M)(X) + (L+M)(X') \quad \text{definition of } L+M .
\end{aligned}
$$

We leave the verification of homogeneity to you.

Parts 2 through 5 of the theorem, which follow from the definition of the sum of linear maps and the analogous properties in $\mathbb{R}^q$, are left for the exercises. $<<$

3.4C MULTIPLICATION OF A LINEAR MAP BY A SCALAR

Just as for a vector, it is possible to multiply a linear map by a scalar. Let $L : \mathbb{R}^n \to \mathbb{R}^q$ be linear and let $\alpha \in \mathbb{R}$. We define αL to be the function given by

$$(\alpha L)(X) = \alpha(L(X)) .$$

EXAMPLE:

2. If $L : \mathbb{R}^3 \to \mathbb{R}$ is given by $L(X) = x_1 - 2x_2 + 4x_3$ and $\alpha = -5$, then $(-5L)(X) = -5x_1 + 10x_2 - 20x_3$.

The proof of the following theorem is left for the exercises.

Theorem 3.4.2 Suppose that $L, M : \mathbb{R}^n \to \mathbb{R}^q$ are linear maps and α and β are scalars. Then the following assertions are true:

1. $\alpha L : \mathbb{R}^n \to \mathbb{R}^q$ is a linear map.

2. $(\alpha\beta)L = \alpha(\beta L)$.

3. $1L = L$.

4. $(\alpha + \beta)L = \alpha L + \beta L$.

5. $\alpha(L+M) = \alpha L + \alpha M$.

REMARK:

- We digress to describe the content of Theorems 3.4.1 and 3.4.2 from the viewpoint of a vector space not previously discussed. If we let $\mathcal{L}$ denote the set of all linear maps from $\mathbb{R}^n$ into $\mathbb{R}^q$, that is,

$$\mathcal{L} = \{\text{linear maps } L : \mathbb{R}^n \to \mathbb{R}^q\} ,$$

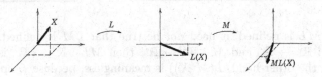

FIGURE 3.13. The composition of linear maps

and if we define $L + M$ and αL as above, then *the set $\mathcal{L}$ is itself a vector space*, because Theorems 3.4.1 and 3.4.2 assure us that the requirements for associativity, commutativity, zero, additive inverse, and so on, are all met. This is the first example we have encountered of a vector space that is not a Euclidean space $\mathbb{R}^n$ or a linear subspace of $\mathbb{R}^n$. The vector space $\mathcal{L}$ consists of functions and might therefore be called a "function space". The study of function spaces, using generalizations of the geometric notions of length, norm, orthogonality, and inner product, is an important and fruitful branch of modern higher mathematics.

3.4D Composition of linear maps

The third algebraic operation we consider for linear maps is composition. In single-variable algebra, if $f(u) = e^u$ and $u(x) = 2x^2 + 1$, then the composition is $f(u(x)) = e^{2x^2+1}$. We now extend this notion to linear maps in higher dimensions.

Let $L : \mathbb{R}^n \to \mathbb{R}^q$ and $M : \mathbb{R}^q \to \mathbb{R}^r$ be linear maps. We denote this $\mathbb{R}^n \xrightarrow{L} \mathbb{R}^q \xrightarrow{M} \mathbb{R}^r$. Note that if $X \in \mathbb{R}^n$, then $L(X) \in \mathbb{R}^q$, and $\mathbb{R}^q$, the target of the first map, is the domain of the second map M. Hence we may define the **composition**, denoted by $M \circ L$, of M on L as

$$(M \circ L)(X) = M(L(X)).$$

Thus $M \circ L : \mathbb{R}^n \to \mathbb{R}^r$ is a function given by the rule: (1) to $X \in \mathbb{R}^n$, apply L, obtaining $L(X) \in \mathbb{R}^q$; (2) to $L(X)$ apply the map M, obtaining $M(L(X)) \in \mathbb{R}^r$. See Figure 3.13.

The symbol $(M \circ L)(X)$ denotes the value of the function $M \circ L$ at the point X. Without creating ambiguity, the first pair of parentheses can be dropped; so, $M \circ L(X)$ is often written instead of $(M \circ L)(X)$. It is customary to streamline the notation even further and write ML instead of $M \circ L$, and $ML(X)$ for the value of the function $M \circ L$ at X. When we use the notation ML, it looks as if we are multiplying M and L, and, in fact, ML is even called the **product** of M and L. Thus, in the setting of linear maps, "composition" and "product" are actually synonyms. (It may seem to the reader as if we are making an effort to create confusion. What we are doing is preparing the reader for other books on linear maps which use established though somewhat confusing notation and terminology.)

REMARKS:

- If ML is defined, it need *not* be true that LM is defined. Thus, if $L : \mathbb{R}^3 \to \mathbb{R}^2$ and $M : \mathbb{R}^2 \to \mathbb{R}^7$, then $ML : \mathbb{R}^3 \to \mathbb{R}^7$ is defined; on the other hand, $L(M(X))$ is meaningless, because L operates on vectors in $\mathbb{R}^3$ only and $M(X) \in \mathbb{R}^7$.

- Even if LM and ML are both defined, say that $L : \mathbb{R}^2 \to \mathbb{R}^2$ and $M : \mathbb{R}^2 \to \mathbb{R}^2$, then it need not be true that $LM = ML$; in fact, this equality is rather rare. In general, the operation of composing two linear maps is *not* commutative.

EXAMPLES:

3. Let $L : \mathbb{R}^2 \to \mathbb{R}^3$ be determined by

$$L(\mathbf{e}_1) = \bar{\mathbf{e}}_1 - \bar{\mathbf{e}}_2 + 5\bar{\mathbf{e}}_3 , \quad L(\mathbf{e}_2) = 2\bar{\mathbf{e}}_1 + \bar{\mathbf{e}}_2 + \bar{\mathbf{e}}_3 ,$$

where $\mathbf{e}_1$ and $\mathbf{e}_2$ and $\bar{\mathbf{e}}_1$, $\bar{\mathbf{e}}_2$, and $\bar{\mathbf{e}}_3$ are the standard bases for $\mathbb{R}^2$ and $\mathbb{R}^3$, respectively. Also, let $M : \mathbb{R}^3 \to \mathbb{R}^3$ be determined by

$$M(\bar{\mathbf{e}}_1) = \bar{\mathbf{e}}_2 , \quad M(\bar{\mathbf{e}}_2) = \bar{\mathbf{e}}_3 , \quad M(\bar{\mathbf{e}}_3) = \bar{\mathbf{e}}_1 .$$

Then $ML : \mathbb{R}^2 \to \mathbb{R}^3$ operates on $\mathbf{e}_1$ and $\mathbf{e}_2$ as follows:

$$\begin{aligned} ML(\mathbf{e}_1) = M(L(\mathbf{e}_1)) &= M(\bar{\mathbf{e}}_1 - \bar{\mathbf{e}}_2 + 5\bar{\mathbf{e}}_3) \\ &= M(\bar{\mathbf{e}}_1) - M(\bar{\mathbf{e}}_2) + 5M(\bar{\mathbf{e}}_3) \\ &= 5\bar{\mathbf{e}}_1 + \bar{\mathbf{e}}_2 - \bar{\mathbf{e}}_3 . \end{aligned}$$

We leave it to you to compute $ML(\mathbf{e}_2)$. Note that LM is not defined.

4. Let $L, M : \mathbb{R}^2 \to \mathbb{R}^2$ be linear maps determined as follows:

$$\begin{aligned} L(\mathbf{e}_1) = \mathbf{e}_2 , \quad L(\mathbf{e}_2) = -\mathbf{e}_1 \quad &\text{(rotation } 90° \text{ counterclockwise)}, \\ M(\mathbf{e}_1) = \mathbf{e}_1 , \quad M(\mathbf{e}_2) = -\mathbf{e}_2 \quad &\text{(reflection across horizontal axis)}. \end{aligned}$$

We first note that LM and ML are both defined. From Figure 3.14 we see that

$$ML(\mathbf{e}_1) = -\mathbf{e}_2 , \quad ML(\mathbf{e}_2) = -\mathbf{e}_1 .$$

On the other hand, from Figure 3.15 we conclude that

$$LM(\mathbf{e}_1) = \mathbf{e}_2 , \quad LM(\mathbf{e}_2) = \mathbf{e}_1 .$$

Hence, $LM \neq ML$. (In fact, $LM = -ML$; ML is a reflection across the line $x_1 + x_2 = 0$, and LM is a reflection across the line $x_1 - x_2 = 0$.)

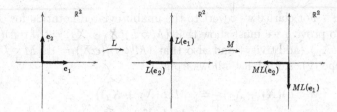

FIGURE 3.14. Rotation first, then reflection

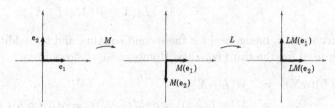

FIGURE 3.15. Reflection first, then rotation

5. Again let $L, M : \mathbb{R}^2 \to \mathbb{R}^2$, but this time let $L(X) = 7X$ for all $X \in \mathbb{R}^2$, and let M be a counterclockwise rotation through a fixed angle θ. Then LM and ML are both defined and, as you can verify by drawing pictures, $LM = ML$. The reason for commutativity in this case is the fact that the map L is very simple; it can be regarded as a relabelling of the scale on each axis, rather than as a moving of each point. A formal proof that $LM = ML$ follows:

$$LM(X) = 7M(X) = M(7(X)) = M(L(X)) = ML(X) .$$

You may have noted in the preceding examples that the composition ML, when defined, was linear. We now verify this in general and collect some other properties of composition.

Theorem 3.4.3 Let $\mathbb{R}^n \xrightarrow{L} \mathbb{R}^q \xrightarrow{M} \mathbb{R}^r \xrightarrow{P} \mathbb{R}^s$ be linear maps. Then the following assertions are true:

1. The composition $ML : \mathbb{R}^n \to \mathbb{R}^r$ is a linear map.

2. Associativity: $P(ML) = (PM)L$ as a map from $\mathbb{R}^n$ into $\mathbb{R}^s$.

3. Right Distributivity: $M(L_1 + L_2) = ML_1 + ML_2$, where $L_1, L_2 : \mathbb{R}^n \to \mathbb{R}^q$ are linear.

4. Left Distributivity: $(M_1 + M_2)L = M_1L + M_2L$, where $M_1, M_2 : \mathbb{R}^q \to \mathbb{R}^r$ are linear.

5. $I_q L = L = L I_n$, where $I_q(Y) = Y$ for $Y \in \mathbb{R}^q$ and $I_n(X) = X$ for $X \in \mathbb{R}^n$.

PROOF: For clarity we revert to the unabbreviated notation for composition. To prove 1 we must show that $(M \circ L)(X_1 + X_2) = (M \circ L)(X_1) + (M \circ L)(X_2)$ (additivity) and also that $(M \circ L)(\alpha X) = \alpha((M \circ L)(X))$. (homogeneity). Additivity follows from:

$$
\begin{aligned}
(M \circ L)(X_1 + X_2) &= M(L(X_1 + X_2)) \\
&= M(L(X_1) + L(X_2)) \\
&= M(L(X_1)) + M(L(X_2)) \\
&= (M \circ L)(X_1) + (M \circ L)(X_2) ,
\end{aligned}
$$

the additivity of L being used for the second equality and the additivity of M being used for the third equality. Homogeneity follows from:

$$
\begin{aligned}
(M \circ L)(\alpha X) &= M(L(\alpha X)) \\
&= M(\alpha(L(X))) = \alpha(M(L(X))) = \alpha((M \circ L)(X)) ,
\end{aligned}
$$

the homogeneity of L being used for the second equality and the homogeneity of M being used for the third equality.

The following string of equalities, which does not use the linearity of the maps, is a proof of Property 2:

$$
\begin{aligned}
(P \circ (M \circ L))(X) &= P((M \circ L)(X)) = P(M(L(X))) \\
&= (P \circ M)(L(X)) = ((P \circ M) \circ L)(X) .
\end{aligned}
$$

The proofs of the other three properties are left for the exercises. $<<$

REMARK:

- Linear maps whose target and domain are identical are particularly important. Consider the set $\mathcal{L}_n$ of all linear maps from $\mathbb{R}^n$ into $\mathbb{R}^n$. As mentioned earlier $\mathcal{L}_n$ is a vector space. In addition to the operation of addition and the operation of multiplication by scalars, a third operation is defined, namely, the composition of any two members of $\mathcal{L}_n$. Because of the Properties 1, 3, and 4 in the preceding theorem, $\mathcal{L}_n$ is called an "algebra". Because Property 5 also holds, it is an "algebra with identity". Because Property 2 also holds, it is an "associative algebra with identity". These terms from higher mathematics will not be needed throughout the remainder of the book.

Let $L \in \mathcal{L}_n$, that is, let L be a linear map from $\mathbb{R}^n$ into $\mathbb{R}^n$. Then the map LL is defined; it is denoted by L^2. By associativity,

$$
LL^2 = L(LL) = (LL)L = L^2 L .
$$

Thus, we use L^3 to denote the linear map LL^2 or, what is the same thing, the linear map $L^2 L$. Similarly L^m is the composition of m copies of L, with all groupings by parentheses resulting in the same linear map. Thus, $L^7 = L^2 L^5$, but it also equals $L^4 L^3$.

EXAMPLE:

6. If L denotes the counterclockwise rotation of $\mathbb{R}^2$ through an angle of $\pi/2$ radians, then L^2 is a rotation through an angle of π radians. Also, L^3 is a counterclockwise rotation through an angle of $3\pi/2$ radians, or, what is the same thing, a clockwise rotation through an angle of $\pi/2$ radians. The linear map L^4 is the identity map I, so, for example, we can easily calculate L^{87}:

$$L^{87} = L^{84} L^3 = (L^4)^{21} L^3 = I^{21} L^3 = I L^3 = L^3 \, ,$$

a clockwise rotation through an angle of $\pi/2$ radians. (Since only $\mathbb{R}^2$ is involved in this example, we have not placed any subscript on the identity map. If, say, both $\mathbb{R}^2$ and $\mathbb{R}^3$ had been involved, we would have wanted to distinguish between the identity map on $\mathbb{R}^2$ and the identity map on $\mathbb{R}^3$.)

REMARK:

- So far we have been careful to write double sets of parentheses in expressions like $L((2,3))$, to indicate that the vector $(2,3)$ is the argument of the function L. From now on, we will often drop one pair of parentheses, writing instead $L(2,3)$.

Exercises

Some calisthenics in the algebra of linear maps.

1. Let $L, M : \mathbb{R}^2 \to \mathbb{R}^2$ be linear maps determined by

$$L(\mathbf{e}_1) = \mathbf{e}_1 + \mathbf{e}_2 \, , \qquad L(\mathbf{e}_2) = \mathbf{e}_1 - \mathbf{e}_2 \, ;$$
$$M(\mathbf{e}_1) = 2\mathbf{e}_1 \, , \qquad M(\mathbf{e}_2) = \mathbf{e}_1 + 3\mathbf{e}_2 \, .$$

Compute the following in terms of $\mathbf{e}_1$ and $\mathbf{e}_2$ (don't use coordinates or matrices):

(a) $(L + M)(\mathbf{e}_1)$

(b) $(L + M)(\mathbf{e}_2)$

(c) $(L + M)(2\mathbf{e}_1 + \mathbf{e}_2)$

(d) $(4L)(\mathbf{e}_2)$

(e) $(4L - M)(\mathbf{e}_1)$

(f) $(ML)(\mathbf{e}_1)$

(g) $(ML)(\mathbf{e}_2)$

(h) $(ML)(2\mathbf{e}_1 + \mathbf{e}_2)$

(i) $(LM)(\mathbf{e}_1)$

   (j) $(LM)(\mathbf{e}_2)$

   (k) $L^2(\mathbf{e}_1)$.

2. For L as in the preceding exercise, find a formula for $L^m(\mathbf{e}_1)$ valid for all positive even integers m. Also find a formula valid for all positive odd integers m.

3.   (a) Find the 2-by-2 matrices $[L]$ and $[M]$ for the maps in Exercise 1.

    (b) Since $L+M$ is a linear map $(L+M): \mathbb{R}^2 \to \mathbb{R}^2$, it has a 2-by-2 matrix $[L+M]$. Find it by first calculating $(L+M)(\mathbf{e}_1)$ and $(L+M)(\mathbf{e}_2)$.

    (c) How is $[L+M]$ obtained from $[L]$ and $[M]$? We study the general version of this problem in the following section.

4. Complete Example 3 in Section 3.4d by calculating $ML(\mathbf{e}_2)$.

5. Let $L, M, P : \mathbb{R}^2 \to \mathbb{R}^2$ be linear maps given by

$$\begin{array}{lll} L(\mathbf{e}_1) = \mathbf{e}_2, & L(\mathbf{e}_2) = -\mathbf{e}_1 & \text{rotation } 90° \text{ counterclockwise;} \\ M(\mathbf{e}_1) = -\mathbf{e}_1, & M(\mathbf{e}_2) = \mathbf{e}_2 & \text{flip across } x_2\text{-axis;} \\ P(\mathbf{e}_1) = -\mathbf{e}_1, & P(\mathbf{e}_2) = -\mathbf{e}_2 & \text{reflection through the origin.} \end{array}$$

    (a) Draw pictures describing the actions of the maps L, M, and P and the compositions LM, ML, LP, PL, MP, and PM.

    (b) Which pairs of the maps L, M, and P commute?

6. Let L, M, and P be as in the preceding exercise and let I denote the identity map on $\mathbb{R}^2$. Identify each of the following as true or false:

    (a) $L^2 = P$

    (b) $P^2 = I$

    (c) $L^4 = I$

    (d) $L^5 = L$

    (e) $M^2 = I$

    (f) $M^3 = M$

    (g) $MPM = P$

    (h) $PMP = L$.

7. Let $L : \mathbb{R}^3 \to \mathbb{R}^2$ and $M : \mathbb{R}^2 \to \mathbb{R}^2$ be linear maps. Which of the following is a meaningful notation for a linear map?

    (a) $2L + M$

    (b) L^2

    (c) M^2

    (d) ML

    (e) LM

    (f) M^2L

    (g) $ML - 3M$.

8. Let $L, M : \mathbb{R}^n \to \mathbb{R}^n$. Use Theorem 3.4.3 to prove that $(L + M)^2 = L^2 + LM + ML + M^2$. Can this be simplified to $(L+M)^2 = L^2 + 2LM + M^2$? Why?

9. Let L and $M : \mathbb{R}^n \to \mathbb{R}^n$. Find an example to show that one of the following two statements is false:

$$(LM)^2 = L^2 M^2, \qquad (LM)^2 = LMLM.$$

10. Define $L, M : \mathbb{R}^2 \to \mathbb{R}^2$ by $L(x_1, x_2) = (0, x_1)$ and $M(x_1, x_2) = (x_1, 0)$. Let O denote the zero map from $\mathbb{R}^2$ into $\mathbb{R}^2$.

   (a) Compute $LM(X)$, $ML(X)$, $L^2(X)$, and $M^2(X)$ to deduce that $LM = L$, $L^2 = O$, $M^2 = M$, and $ML = O$.

   (b) Use the preceding part to express each of LML, MLM, $(LM)^2$, $(ML)^2$, M^{57}, $(L + M)^2$, $(2L + M)^2$, and $(L - 2M)^3$ in the form $\alpha L + \beta M$.

11. Let $L, M : \mathbb{R}^n \to \mathbb{R}^n$. For each of the following, give either a proof or a counterexample. *Hint:* See Exercises 6 and 10.

   (a) $L^2 = O \Rightarrow L = O$.

   (b) $L^2 = L \Rightarrow L = I$ or $L = O$.

   (c) $L^2 = I \Rightarrow L = I$ or $L = -I$.

12. Complete the proof of Theorem 3.4.1.

13. Prove Theorem 3.4.2.

14. Complete the proof of Theorem 3.4.3.

15. Let $L : \mathbb{R}^n \to \mathbb{R}^n$. Prove that $\mathcal{N}(L) \subseteq \mathcal{N}(L^2)$ and $\mathcal{I}(L^2) \subseteq \mathcal{I}(L)$.

16. Let $L : \mathbb{R}^n \to \mathbb{R}^n$. If, for every $X \in \mathbb{R}^n$, there is a number k (depending on X) such that $L^k(X) = \mathbf{0}$, prove that there is an m such that $L^m = O$. *Hint:* Prove that there is an m such that $L^m(\mathbf{e}_j) = \mathbf{0}$ for every standard basis vector $\mathbf{e}_j$.

17. Let $L : \mathbb{R}^n \to \mathbb{R}^n$. Prove that $\mathcal{I}(L) \subseteq \mathcal{N}(L) \Leftrightarrow L^2 = O$.

3.5   Matrices

3.5A   A SHORTHAND DEVICE

In Section 3.1d we used a rectangular array of numbers, namely a matrix, as a shorthand representation of a linear map. In this section we elaborate on the relation between linear maps and matrices and then go on to transfer the algebraic operations for linear maps—that is, addition, multiplication by scalars, and composition—to matrices. Having done this, we may use matrices not only to represent linear maps, as a mere shorthand device, but also to carry out numerical computations involving linear maps, a role for which matrices are especially well suited.

Let us recall how we associate a matrix $[L]$ to a linear map L. Say that $L : \mathbb{R}^3 \to \mathbb{R}^2$ is given by

$$L(\mathbf{e}_1) = a_1\bar{\mathbf{e}}_1 + a_2\bar{\mathbf{e}}_2 , \quad L(\mathbf{e}_2) = b_1\bar{\mathbf{e}}_1 + b_2\bar{\mathbf{e}}_2 , \quad L(\mathbf{e}_3) = c_1\bar{\mathbf{e}}_1 + c_2\bar{\mathbf{e}}_2,$$

where $\{\mathbf{e}_1, \mathbf{e}_2, \mathbf{e}_3\}$ and $\{\bar{\mathbf{e}}_1, \bar{\mathbf{e}}_2\}$ are the standard bases for $\mathbb{R}^3$ and $\mathbb{R}^2$, respectively. Then we associate to L the matrix

$$[L] = \begin{bmatrix} a_1 & b_1 & c_1 \\ a_2 & b_2 & c_2 \end{bmatrix} .$$

Note that the first (left-hand) column of $[L]$ gives information about $L(\mathbf{e}_1)$, the second column about $L(\mathbf{e}_2)$, and the third column about $L(\mathbf{e}_3)$.

REMARK:

- Though it may seem odd at first, we will become used to writing this information vertically. It will expedite computations later.

Now let us observe that we can reverse this process. Given a matrix such as

$$\begin{bmatrix} 3 & 2 & -2 \\ -1 & 0 & 1 \end{bmatrix}$$

we can use it to define a linear map $L : \mathbb{R}^3 \to \mathbb{R}^2$; define

$$L(\mathbf{e}_1) = 3\bar{\mathbf{e}}_1 - \bar{\mathbf{e}}_2 , \quad L(\mathbf{e}_2) = 2\bar{\mathbf{e}}_1 , \quad L(\mathbf{e}_3) = -2\bar{\mathbf{e}}_1 + \bar{\mathbf{e}}_2 .$$

Having defined the values of L on the basis vectors $\mathbf{e}_1$, $\mathbf{e}_2$, and $\mathbf{e}_3$, we know that there is precisely one linear map L from $\mathbb{R}^3$ into $\mathbb{R}^2$ that assumes these values at $\mathbf{e}_1$, $\mathbf{e}_2$, and $\mathbf{e}_3$. Further, we note that if we take the map L just defined and write $[L]$, we return to the array given just above.

In brief, we have set up a one-to-one correspondence between two sets: the set of all linear maps $L : \mathbb{R}^3 \to \mathbb{R}^2$ and the set of all matrices (rectangular arrays of numbers) having two horizontal rows and three vertical columns. To each linear map L there is precisely one matrix and vice versa.

3.5B MATRIX NOMENCLATURE

A **matrix** is any rectangular array of numbers, such as

$$A = \begin{bmatrix} a_{11} & a_{12} & \cdots & a_{1n} \\ a_{21} & a_{22} & \cdots & a_{2n} \\ \vdots & \vdots & & \vdots \\ a_{q1} & a_{q2} & \cdots & a_{qn} \end{bmatrix} .$$

The numbers $a_{11}, a_{12}, \ldots, a_{21}, a_{22}, \ldots, \ldots, a_{q(n-1)}$, and a_{qn} are the **entries** or **elements** of the matrix A. It is natural to call a_{ij} the $(i, j)^{\text{th}}$ **entry**.

The first index i in a_{ij} is called the **row index** and the second index j the **column index** (don't mix them up!). The ith **row** of the matrix A is

$$[a_{i1} \quad a_{i2} \quad \cdots \quad a_{in}] \, .$$

Note that $i = 1, 2, \cdots, q$ and $j = 1, 2, \cdots, n$; A has q rows and n columns and is thereby called a q-by-n matrix. (Remember: rows *before* columns!) We denote the set of all q-by-n matrices by $\mathcal{M}(q; n)$.

Note that a 1-by-1 matrix is the same as a real number. We call a 1-by-n matrix

$$[a_1 \quad a_2 \quad \cdots \quad a_n]$$

a **row matrix** and a q-by-1 matrix

$$\begin{bmatrix} a_1 \\ a_2 \\ \vdots \\ a_q \end{bmatrix}$$

a **column matrix**.

We typically use capital letters (A, B, and C, for example) from the beginning of the alphabet to denote matrices. Suppose the entries in a matrix A are a_{ij}, where it is understood that $i = 1, \ldots, q$ and $j = 1, \ldots, n$. Then it is customary to write $A = [a_{ij}]$. This notation can save space and writing and causes no confusion, provided the numbers of rows and columns have been established.

We remark that some authors would write (a_{ij}) instead of $[a_{ij}]$ and therefore

$$\begin{pmatrix} 1 & 7 \\ -3 & 4 \end{pmatrix} \quad \text{instead of} \quad \begin{bmatrix} 1 & 7 \\ -3 & 4 \end{bmatrix} \, .$$

3.5C   Addition of Matrices

Our aim now is to use the correspondence between linear maps and matrices to define the three algebraic operations: addition of matrices, multiplication of a matrix by a scalar, multiplication of matrices. We deal with addition first.

Let $L, M : \mathbb{R}^3 \to \mathbb{R}^2$ be linear maps. Suppose that

$$L(\mathbf{e}_1) = a_{11}\bar{\mathbf{e}}_1 + a_{21}\bar{\mathbf{e}}_2 \, , \quad L(\mathbf{e}_2) = a_{12}\bar{\mathbf{e}}_1 + a_{22}\bar{\mathbf{e}}_2 \, , \quad L(\mathbf{e}_3) = a_{13}\bar{\mathbf{e}}_1 + a_{23}\bar{\mathbf{e}}_2$$

and

$$M(\mathbf{e}_1) = b_{11}\bar{\mathbf{e}}_1 + b_{21}\bar{\mathbf{e}}_2 \, , \quad M(\mathbf{e}_2) = b_{12}\bar{\mathbf{e}}_1 + b_{22}\bar{\mathbf{e}}_2 \, , \quad M(\mathbf{e}_3) = b_{13}\bar{\mathbf{e}}_1 + b_{23}\bar{\mathbf{e}}_2 \, .$$

Thus we get the associated matrices

$$[L] = \begin{bmatrix} a_{11} & a_{12} & a_{13} \\ a_{21} & a_{22} & a_{23} \end{bmatrix} \, , \quad [M] = \begin{bmatrix} b_{11} & b_{12} & b_{13} \\ b_{21} & b_{22} & b_{23} \end{bmatrix} \, .$$

How should we define $[L] + [M]$? It is natural to do this as follows: Add the linear maps L and M, obtaining the linear map $L + M$, and then define $[L] + [M]$ to be the associated matrix $[L + M]$.

Adding the maps as in the preceding section, we obtain

$$
\begin{aligned}
(L + M)(\mathbf{e}_1) &= L(\mathbf{e}_1) + M(\mathbf{e}_1) \\
&= (a_{11}\bar{\mathbf{e}}_1 + a_{21}\bar{\mathbf{e}}_2) + (b_{11}\bar{\mathbf{e}}_1 + b_{21}\bar{\mathbf{e}}_2) \\
&= (a_{11} + b_{11})\bar{\mathbf{e}}_1 + (a_{21} + b_{21})\bar{\mathbf{e}}_2 .
\end{aligned}
$$

Likewise, we obtain

$$
\begin{aligned}
(L + M)(\mathbf{e}_2) &= (a_{12} + b_{12})\bar{\mathbf{e}}_1 + (a_{22} + b_{22})\bar{\mathbf{e}}_2 \\
(L + M)(\mathbf{e}_3) &= (a_{13} + b_{13})\bar{\mathbf{e}}_1 + (a_{23} + b_{23})\bar{\mathbf{e}}_2 .
\end{aligned}
$$

Hence we may immediately write $[L + M]$:

$$
[L + M] = \begin{bmatrix} a_{11} + b_{11} & a_{12} + b_{12} & a_{13} + b_{13} \\ a_{21} + b_{21} & a_{22} + b_{22} & a_{23} + b_{23} \end{bmatrix} .
$$

Now we *define* this matrix to be $[L] + [M]$.

Thus, motivated by the correspondence between maps and matrices, let us make our definition in full generality. Let A and B be matrices of the same shape, say $A = [a_{ij}]$ and $B = [b_{ij}]$, with $i = 1, \ldots, q$ and $j = 1, \ldots, n$. Then we define their **sum**, denoted $A + B$, to be the q-by-n matrix

$$
A + B = [a_{ij} + b_{ij}] .
$$

Thus, the $(i, j)^{\text{th}}$ entry of $A + B$ is the sum of the $(i, j)^{\text{th}}$ entries of A and B. Note that if A and B are of different shapes, their sum is undefined.

A particularly important matrix is that of the zero linear map. In this case, we use the same symbol for the matrix as for the linear map; thus, $[O] = O$. For A a matrix corresponding to a linear map L, the symbol $-A$ is used for the matrix corresponding to the linear map $-L$. It is easy to see that the $(i, j)^{\text{th}}$ entry of $-A$ is the negative of the $(i, j)^{\text{th}}$ entry of A.

EXAMPLES:

1.

$$
A = \begin{bmatrix} 5 & -1 \\ 2 & 0 \end{bmatrix}, \quad B = \begin{bmatrix} \frac{1}{2} & 4 \\ -3 & 3 \end{bmatrix}, \quad A + B = \begin{bmatrix} \frac{11}{2} & 3 \\ -1 & 3 \end{bmatrix} .
$$

2.

$$
A = \begin{bmatrix} 1 \\ 0 \\ -2 \end{bmatrix}, \quad B = \begin{bmatrix} 7 \\ -2 \\ -2 \end{bmatrix}, \quad A + B = \begin{bmatrix} 8 \\ -2 \\ -4 \end{bmatrix} .
$$

3.

$$\begin{bmatrix} 2 & 1 & 6 & -\frac{1}{2} \\ 3 & 0 & -1 & 0 \end{bmatrix} + \begin{bmatrix} 1 & 4 & 5 & 1 \\ 2 & 0 & -1 & -3 \end{bmatrix}$$

$$= \begin{bmatrix} 3 & 5 & 11 & \frac{1}{2} \\ & \text{left to you} & \end{bmatrix} .$$

4.

$$A = \begin{bmatrix} 3 & -5 \\ 4 & 2 \\ -2 & 0 \end{bmatrix} , \quad -A = \begin{bmatrix} -3 & 5 \\ -4 & -2 \\ 2 & 0 \end{bmatrix} .$$

$$A + (-A) = \begin{bmatrix} 0 & 0 \\ 0 & 0 \\ 0 & 0 \end{bmatrix} = O .$$

You should expect the following theorem, which does for matrices what Theorem 3.4.1 does for linear maps. A proof could be based on Theorem 3.4.1 and the connection between linear transformations and matrices. But a direct proof, using the above definition of addition of two matrices is so easy, except for the handling of notation, that we will neither give the proof nor ask for it as an exercise.

Theorem 3.5.1 Let A, B and C be q-by-n matrices. Then the following assertions are true:

1. $A + B$ is a q-by-n matrix.

2. $A + (B + C) = (A + B) + C$.

3. $A + B = B + A$.

4. $A + O = A$.

5. $A + (-A) = O$.

3.5D MULTIPLICATION OF A MATRIX BY A SCALAR

Let A be a q-by-n matrix and c any scalar. How will we define cA? Again we are guided by the correspondence between maps and matrices. Let $A = [a_{ij}]$, and suppose that $L : \mathbb{R}^n \to \mathbb{R}^q$ is a linear map whose associated matrix is A, that is, $[L] = A$. This means that $L(\mathbf{e}_1) = a_{11}\bar{\mathbf{e}}_1 + a_{21}\bar{\mathbf{e}}_2 + \cdots + a_{q1}\bar{\mathbf{e}}_q$, and so on, for $L(\mathbf{e}_2), \cdots, L(\mathbf{e}_n)$. Now we know that

$$\begin{aligned} (cL)(\mathbf{e}_1) = c(L(\mathbf{e}_1)) &= c(a_{11}\bar{\mathbf{e}}_1 + a_{21}\bar{\mathbf{e}}_2 + \cdots + a_{q1}\bar{\mathbf{e}}_q) \\ &= ca_{11}\bar{\mathbf{e}}_1 + ca_{21}\bar{\mathbf{e}}_2 + \cdots + ca_{q1}\bar{\mathbf{e}}_q , \end{aligned}$$

that is, the first column of $[L]$ is multiplied by c. The others are also. Since we want cA to equal $[cL]$, we *define*

$$cA = [ca_{ij}] .$$

EXAMPLES:

5.

$$A = \begin{bmatrix} 3 & 2 \\ -1 & 5 \end{bmatrix}, \quad 4A = 4 \begin{bmatrix} 3 & 2 \\ -1 & 5 \end{bmatrix} = \begin{bmatrix} 12 & 8 \\ -4 & 20 \end{bmatrix} .$$

6. If A is as above and $c = -1$, then

$$cA = -1A = -A = \begin{bmatrix} -3 & -2 \\ 1 & -5 \end{bmatrix} .$$

7. If A is any matrix and if the scalar c equals zero, then $cA = O$, the zero marix, that is, $0A = O$.

The following theorem does for matrices what Theorem 3.4.2 did for linear maps. The easy proof is left to the reader.

Theorem 3.5.2 Suppose that A and B be q-by-n matrices and α and β are scalars. Then the following assertions are true:

1. αA is a q-by-n matrix.

2. $(\alpha\beta)A = \alpha(\beta A)$.

3. $1A = A$.

4. $(\alpha + \beta)A = \alpha A + \beta A$.

5. $\alpha(A + B) = \alpha A + \alpha B$.

3.5E   MULTIPLICATION OF MATRICES

We have defined addition of matrices and multiplication of a matrix by a scalar, guided by the correspondence between matrices and linear maps. Now we ask: How should we define multiplication of matrices? Our answer is: We make it correspond to composition of linear maps. The reader might object, suggesting that we should be talking about "composition of matrices". We might agree in the abstract, but governing our choice of terms is a desire to be consistent with what has historically been used.

We proceed as follows. Let $L : \mathbb{R}^n \to \mathbb{R}^q$ and $M : \mathbb{R}^q \to \mathbb{R}^r$ be linear maps. In more succinct notation, we have $\mathbb{R}^n \xrightarrow{L} \mathbb{R}^q \xrightarrow{M} \mathbb{R}^r$. We know that the composition is a linear map $ML : \mathbb{R}^n \to \mathbb{R}^r$. As usual, we have the associated matrices $[L] \in \mathcal{M}(q; n)$, $[M] \in \mathcal{M}(r; q)$, and $[ML] \in \mathcal{M}(r; n)$, and we *define* $[M][L]$ to be the matrix $[ML]$. In order to make calculations based on this definition, we must obtain an expression for the entries of $[ML]$ in terms of the entries of $[M]$ and $[L]$.

First let us look at a special case, $\mathbb{R}^3 \xrightarrow{L} \mathbb{R}^2 \xrightarrow{M} \mathbb{R}^2$. Then the matrices $[L]$ and $[M]$ have the following forms:

$$[L] = \begin{bmatrix} a_{11} & a_{12} & a_{13} \\ a_{21} & a_{22} & a_{23} \end{bmatrix}, \quad [M] = \begin{bmatrix} b_{11} & b_{12} \\ b_{21} & b_{22} \end{bmatrix}.$$

What is $[ML]$?

To begin answering this, we focus on the first column of $[ML]$; it must contain the coefficients of $ML(\mathbf{e}_1)$ when this vector is written as a linear combination of $\bar{\mathbf{e}}_1$ and $\bar{\mathbf{e}}_2$. We have

$$\begin{aligned} ML(\mathbf{e}_1) &= M(L(\mathbf{e}_1)) \\ &= M(a_{11}\bar{\mathbf{e}}_1 + a_{21}\bar{\mathbf{e}}_2) \\ &= a_{11}M(\bar{\mathbf{e}}_1) + a_{21}M(\bar{\mathbf{e}}_2) \\ &= a_{11}(b_{11}\bar{\mathbf{e}}_1 + b_{21}\bar{\mathbf{e}}_2) + a_{21}(b_{12}\bar{\mathbf{e}}_1 + b_{22}\bar{\mathbf{e}}_2) \\ &= (b_{11}a_{11} + b_{12}a_{21})\bar{\mathbf{e}}_1 + (b_{21}a_{11} + b_{22}a_{21})\bar{\mathbf{e}}_2 \,. \end{aligned}$$

Hence, we have

$$[ML] = \begin{bmatrix} b_{11}a_{11} + b_{12}a_{21} & ? & ? \\ b_{21}a_{11} + b_{22}a_{21} & ? & ? \end{bmatrix}.$$

We obtain the second and third columns of $[ML]$ similarly, by computing $ML(\mathbf{e}_2)$ and $ML(\mathbf{e}_3)$ in terms of $\bar{\mathbf{e}}_1$ and $\bar{\mathbf{e}}_2$. The result is

$$[M][L] = [ML] = \begin{bmatrix} b_{11}a_{11} + b_{12}a_{21} & b_{11}a_{12} + b_{12}a_{22} & b_{11}a_{13} + b_{12}a_{23} \\ b_{21}a_{11} + b_{22}a_{21} & b_{21}a_{12} + b_{22}a_{22} & b_{21}a_{13} + b_{22}a_{23} \end{bmatrix},$$

that is,

$$\begin{bmatrix} b_{11} & b_{12} \\ b_{21} & b_{22} \end{bmatrix} \begin{bmatrix} a_{11} & a_{12} & a_{13} \\ a_{21} & a_{22} & a_{23} \end{bmatrix}$$

$$= \begin{bmatrix} b_{11}a_{11} + b_{12}a_{21} & b_{11}a_{12} + b_{12}a_{22} & b_{11}a_{13} + b_{12}a_{23} \\ b_{21}a_{11} + b_{22}a_{21} & b_{21}a_{12} + b_{22}a_{22} & b_{21}a_{13} + b_{22}a_{23} \end{bmatrix}.$$

Having this definition of $[M][L]$, we would like a means of remembering it, so that we need not repeat the entire process of computing $ML(\mathbf{e}_1)$, $ML(\mathbf{e}_2)$, and $ML(\mathbf{e}_3)$. If we examine $[M][L]$, we observe, for example, that its upper left-hand entry $b_{11}a_{11} + b_{12}a_{21}$ can be computed as follows. Take the first *row* of $[M]$ and first *column* of $[L]$ —namely,

$$[b_{11} b_{12}] \quad \text{and} \quad \begin{bmatrix} a_{11} \\ a_{21} \end{bmatrix}$$

—and define their product as

$$[b_{11} b_{12}] \begin{bmatrix} a_{11} \\ a_{21} \end{bmatrix} = b_{11}a_{11} + b_{12}a_{21} .$$

It may be helpful to view the right-hand side as the inner product of two vectors, one whose coordinates are the entries in the first row of B and one whose coordinates are the entries in the first column of A. This yields the upper left-hand entry of $[M][L]$.

More generally, the $(i,j)^{\text{th}}$ entry $b_{i1}a_{1j} + b_{i2}a_{2j}$ of $[M][L]$ is the inner product of the vector whose coordinates are the entries in the ith row of $[M]$ and the vector whose coordinates are the entries in the jth column of $[L]$:

$$\langle (b_{i1}, b_{i2}), (a_{1j}, a_{2j}) \rangle = b_{i1}a_{1j} + b_{i2}a_{2j} .$$

We turn to the general case. Let B and A be matrices and assume that B has as many columns as A has rows. Then the product BA is the matrix whose $(i,j)^{\text{th}}$ entry is the inner product of the vector whose coordinates are the entries in the ith row of B and the vector whose coordinates are the entries in the jth column of A.

REMARK:

- We do not even consider forming the product BA unless the number of columns of B equals the number of rows of A.

EXAMPLES:

8.

$$\begin{bmatrix} 5 & 2 \\ 1 & -3 \end{bmatrix} \begin{bmatrix} 2 & 0 & 1 \\ -1 & 4 & 2 \end{bmatrix} = \begin{bmatrix} 8 & 8 & 9 \\ 5 & -12 & -5 \end{bmatrix} .$$

Note, for instance, that the $(2,1)^{\text{th}}$ (lower left-hand) entry of the product is obtained from

$$[1 - 3] \begin{bmatrix} 2 \\ -1 \end{bmatrix} = 1 \cdot 2 + (-3) \cdot (-1) = 5 .$$

9.

$$\begin{bmatrix} 1 & 2 \\ 3 & 4 \end{bmatrix} \begin{bmatrix} 4 & 2 \\ 3 & 1 \end{bmatrix} = \begin{bmatrix} 10 & 4 \\ 24 & 10 \end{bmatrix} .$$

10.

$$\begin{bmatrix} 3 & 0 & 1 \\ 0 & 2 & -1 \\ 1 & 1 & 5 \end{bmatrix} \begin{bmatrix} 2 \\ -3 \\ 1 \end{bmatrix} = \begin{bmatrix} 7 \\ -7 \\ 4 \end{bmatrix} .$$

11.

$$\begin{bmatrix} 3 & 0 & 1 \\ 0 & 2 & -1 \\ 1 & 1 & 5 \end{bmatrix} \begin{bmatrix} 2 & 0 \\ -3 & 0 \\ 1 & 1 \end{bmatrix} = \begin{bmatrix} 7 & 1 \\ -7 & -1 \\ 4 & 5 \end{bmatrix}.$$

We suggest that you go back and perform some of these multiplications yourself (mentally, of course), keeping the right-hand side (product) covered and then comparing.

Just as for the zero map it is standard practice for the identity map to use the same symbol for both the map and the corresponding matrix. If I denotes the identity map on $\mathbb{R}^n$, then $I(\mathbf{e}_j) = \mathbf{e}_j$ for each j; so, for example, the corresponding matrices in $\mathbb{R}^3$ and $\mathbb{R}^4$ are

$$\begin{bmatrix} 1 & 0 & 0 \\ 0 & 1 & 0 \\ 0 & 0 & 1 \end{bmatrix} \quad \text{and} \quad \begin{bmatrix} 1 & 0 & 0 & 0 \\ 0 & 1 & 0 & 0 \\ 0 & 0 & 1 & 0 \\ 0 & 0 & 0 & 1 \end{bmatrix}.$$

Similarly, for a general n the identity matrix I is readily seen to have ones on the **main diagonal** (upper left to lower right) and zeroes elsewhere. As expected, the matrix I plays a role in matrix multiplication similar to that played by the identity map with respect to composition. Here is the matrix theorem analogous to Theorem 3.4.3 for linear maps.

Theorem 3.5.3 Let A be a q-by-n matrix, B an r-by-q matrix, and C an s-by-r matrix. Then the following assertions are true:

1. BA is an r-by-n matrix.

2. $C(BA) = (CB)A$.

3. $B(A_1 + A_2) = BA_1 + BA_2$, where A_1 and A_2 are q-by-n matrices.

4. $(B_1 + B_2)A = B_1A + B_2A$, where B_1 and B_2 are r-by-q matrices.

5. $AI_n = A$ and $I_qA = A$, where I_n denotes the n-by-n identity matrix and I_q denotes the q-by-q identity matrix.

REMARK:

- All properties of the preceding theorem, except the second, are easily proved using the formulas for adding two matrices, multiplying a matrix by a scalar, and multiplying two matrices. Property 2 can also be proved in that way, but not so easily. So, we will give a proof of this part that does not use formulas involving sums of products of scalars, but rather uses Theorem 3.4.3 and the connection between linear maps and matrices.

PROOF: (of Property 2 only) Let L, M, and P be the linear maps corresponding to A, B, and C, respectively. Then, since matrix multiplication is defined to correspond to composition of linear maps, $BA = [M][L] = [ML]$ and $CB = [PM]$. Therefore,

$$C(BA) = [P][ML] = [P(ML)] = [(PM)L] = [PM][L] = (CB)A \,,$$

where Theorem 3.4.3 has been used for the third equality. Done. $<<$

3.5F MATRICES AS FUNCTIONS

We have now learned how straightforward calculations involving matrices describe the linear maps obtained by multiplying given linear maps by scalars and combining them by addition and multiplication. In this section we will find that we can also use matrix arithmetic to decide what a particular linear map does to a particular vector.

For each vector $X = (x_1, x_2, \ldots, x_n)$ we use $[X]$ to denote the n-by-1 matrix whose entries are the coordinates of X:

$$[X] = \begin{bmatrix} x_1 \\ x_2 \\ \vdots \\ x_n \end{bmatrix}.$$

Thus $X = (1, 3, -2)$ in $\mathbb{R}^3$ and $Y = (0, 4)$ in $\mathbb{R}^2$ correspond to the column matrices

$$[X] = \begin{bmatrix} 1 \\ 3 \\ -2 \end{bmatrix} \quad \text{and} \quad [Y] = \begin{bmatrix} 0 \\ 4 \end{bmatrix},$$

respectively.

The row matrix corresponding to a vector $X = (x_1, \ldots, x_n)$ is the 1-by-n matrix

$$[\, x_1 \quad x_2 \quad \cdots \quad x_n \,].$$

Thus the row matrix corresponding to the vector $(1, 3, -2)$ is $[\, 1 \quad 3 \quad -2 \,]$.

REMARK:

- Because the most usual way for column and row matrices to arise is as matrices corresponding to vectors, these matrices are also called **column vectors** and **row vectors**, respectively. We will not use these terms however.

Let A be a q-by-n matrix and let $[X]$ denote the n-by-1 column matrix corresponding to an $X \in \mathbb{R}^n$. The product $A[X]$ is defined because the number of columns, n, of A equals the number of rows of $[X]$. The following theorem gives some significance to this product.

Theorem 3.5.4 Let $L : \mathbb{R}^n \to \mathbb{R}^q$ be a linear map and let $X \in \mathcal{D}(L)$. Then $L(X)$ is the vector corresponding to the q-by-1 matrix $[L][X]$, that is, $[L(X)] = [L][X]$.

PROOF: We will content ourselves with a proof in the case where $n = 2$ and $q = 3$. So,

$$[L] = \begin{bmatrix} a_{11} & a_{12} \\ a_{21} & a_{22} \\ a_{31} & a_{32} \end{bmatrix} \quad \text{and} \quad [X] = \begin{bmatrix} x_1 \\ x_2 \end{bmatrix}$$

for appropriate real numbers a_{ij} and x_j. We will first calculate $[L][X]$, then calculate $[L(X)]$, and complete the proof by observing that the results of these two calculations are equal.

The first calculation is merely the multiplication of matrices:

$$[L][X] = \begin{bmatrix} a_{11} & a_{12} \\ a_{21} & a_{22} \\ a_{31} & a_{32} \end{bmatrix} \begin{bmatrix} x_1 \\ x_2 \end{bmatrix} = \begin{bmatrix} a_{11}x_1 + a_{12}x_2 \\ a_{21}x_1 + a_{22}x_2 \\ a_{31}x_1 + a_{32}x_2 \end{bmatrix},$$

a 3-by-1 matrix (that is, a column matrix corresponding to a member of $\mathbb{R}^3$).

Now let us compute $L(X)$:

$$\begin{aligned} L(X) &= L(x_1\mathbf{e}_1 + x_2\mathbf{e}_2) \\ &= x_1(a_{11}\bar{\mathbf{e}}_1 + a_{21}\bar{\mathbf{e}}_2 + a_{31}\bar{\mathbf{e}}_3) + x_2(a_{12}\bar{\mathbf{e}}_1 + a_{22}\bar{\mathbf{e}}_2 + a_{32}\bar{\mathbf{e}}_3) \\ &= (a_{11}x_1 + a_{12}x_2)\bar{\mathbf{e}}_1 + (a_{21}x_1 + a_{22}x_2)\bar{\mathbf{e}}_2 + (a_{31}x_1 + a_{32}x_2)\bar{\mathbf{e}}_3 \\ &= (a_{11}x_1 + a_{12}x_2, \ a_{21}x_1 + a_{22}x_2, \ a_{31}x_1 + a_{32}x_2). \end{aligned}$$

Now we can write the column matrix $[L(X)]$:

$$[LX] = \begin{bmatrix} a_{11}x_1 + a_{12}x_2 \\ a_{21}x_1 + a_{22}x_2 \\ a_{31}x_1 + a_{32}x_2 \end{bmatrix}.$$

But this is just the product $[L][X]$ obtained earlier. Done. $<<$

EXAMPLES:

12. Let $L : \mathbb{R}^2 \to \mathbb{R}^3$ be given by the matrix

$$[L] = \begin{bmatrix} 2 & 12 \\ -1 & 8 \\ 3 & -17 \end{bmatrix}.$$

Now let $X = (x_1, x_2)$, and compute $[L(X)]$ as $[L][X]$:

$$\begin{bmatrix} 2 & 12 \\ -1 & 8 \\ 3 & -17 \end{bmatrix} \begin{bmatrix} x_1 \\ x_2 \end{bmatrix} = \begin{bmatrix} 2x_1 + 12x_2 \\ -x_1 + 8x_2 \\ 3x_1 - 17x_2 \end{bmatrix}.$$

Hence, we get the coordinate representation

$$L(X) = (2x_1 + 12x_2, -x_1 + 8x_2, 3x_1 - 17x_2).$$

If we want to calculate a particular value of L, say $L((4,-1))$, we can do as above for general X and then substitute obtaining

$$
\begin{aligned}
L((4,-1)) &= (2(4) + 12(-1), -4 + 8(-1), 3(4) - 17(-1)) \\
&= (-4, -12, 29),
\end{aligned}
$$

or we can use $(4,-1)$ from the beginning to obtain

$$
\begin{bmatrix} 2 & 12 \\ -1 & 8 \\ 3 & -17 \end{bmatrix}
\begin{bmatrix} 4 \\ -1 \end{bmatrix}
=
\begin{bmatrix} -4 \\ -12 \\ 29 \end{bmatrix},
$$

which corresponds to the member $(-4, -12, 29)$ of $\mathbb{R}^3$.

13. Let $L : \mathbb{R}^2 \to \mathbb{R}^2$ be a linear map with matrix

$$[L] = \begin{bmatrix} 2 & 5 \\ 4 & 9 \end{bmatrix}.$$

The problem we consider is that of finding those X (if any) for which $L(X) = (3, -7)$. This problem is equivalent to that of solving the matrix equation

$$
\begin{bmatrix} 2 & 5 \\ 4 & 9 \end{bmatrix}
\begin{bmatrix} x_1 \\ x_2 \end{bmatrix}
=
\begin{bmatrix} 3 \\ -7 \end{bmatrix}.
$$

The right-hand side and the product on the left-hand side are 2-by-1 matrices. For them to be equal the corresponding entries must be equal; thus the matrix equation is equivalent to the following system of two equations in two unknowns:

$$
\begin{aligned}
2x_1 + 5x_2 &= 3 \\
4x_1 + 9x_2 &= -7.
\end{aligned}
$$

We subtract the first equation from the second equation twice to obtain the equivalent system:

$$
\begin{aligned}
2x_1 + 5x_2 &= 3 \\
-x_2 &= -13.
\end{aligned}
$$

Then we add the second equation to the first five times:

$$
\begin{aligned}
2x_1 &= -62 \\
-x_2 &= -13.
\end{aligned}
$$

Thus, there is one and only one X for which $L(X) = (3, -7)$, namely $(-31, 13)$.

The preceding calculation can be performed with less writing by using matrix notation. We use the symbol "$\sim$" between two matrices when one can be obtained from the other in the process of Gauss-Jordan elimination. The symbol "$\sim$" can be read as "is equivalent to".

You can check that the following is merely a streamlined repeat of the steps performed above:

$$\left[\begin{array}{cc|c} 2 & 5 & 3 \\ 4 & 9 & -7 \end{array} \right]$$

$$\sim \left[\begin{array}{cc|c} 2 & 5 & 3 \\ 0 & -1 & -13 \end{array} \right] \quad \text{(subtracting twice first row from second)}$$

$$\sim \left[\begin{array}{cc|c} 2 & 0 & -62 \\ 0 & -1 & -13 \end{array} \right] \quad \text{(adding 5 times second row to first)}.$$

The matrix

$$\left[\begin{array}{cc} 2 & 5 \\ 4 & 9 \end{array} \right]$$

is called the **coefficient matrix** and the "matrix"

$$\left[\begin{array}{cc|c} 2 & 5 & 3 \\ 4 & 9 & -7 \end{array} \right]$$

is called the **augmented matrix** of the system of equations.

14. Let us solve the following system of three equations in four variables:

$$\begin{array}{rcl} 2x_1 - 4x_2 + 4x_3 + 5x_4 & = & 3 \\ -x_1 + 2x_2 - x_3 - 4x_4 & = & 0 \\ 6x_1 - 12x_2 + 10x_3 + 17x_4 & = & -2 . \end{array}$$

Even though neither matrices nor linear maps are mentioned in the problem we can keep our calculations organized and save writing by using matrix notation:

$$\left[\begin{array}{cccc|c} 2 & -4 & 4 & 5 & 3 \\ -1 & 2 & -1 & -4 & 0 \\ 6 & -12 & 10 & 17 & -2 \end{array} \right] \sim \left[\begin{array}{cccc|c} 2 & -4 & 4 & 5 & 3 \\ 0 & 0 & 1 & -\frac{3}{2} & \frac{3}{2} \\ 0 & 0 & -2 & 2 & -11 \end{array} \right].$$

(second row + half first row, third row $-$ thrice first row)

$$\sim \left[\begin{array}{cccc|c} 2 & -4 & 0 & 11 & -3 \\ 0 & 0 & 1 & -\frac{3}{2} & \frac{3}{2} \\ 0 & 0 & 0 & -1 & -8 \end{array} \right]$$

(first row − four times second row, third row + twice second row)

$$\sim \begin{bmatrix} 2 & -4 & 0 & 0 & \bigm| & -91 \\ 0 & 0 & 1 & 0 & \bigm| & \frac{27}{2} \\ 0 & 0 & 0 & -1 & \bigm| & -8 \end{bmatrix}$$

(first row + eleven times third row, second row $-\frac{3}{2}$ third row).

We now see that $x_4 = 8$, $x_3 = 27/2$, x_2 is arbitrary, and $x_1 = -(91/2) + 2x_2$. The solution set is the affine subspace described by

$$(-\frac{91}{2}, 0, \frac{27}{2}, 8) + t(2, 1, 0, 0), \quad t \in \mathbb{R}.$$

Before quitting let us use the preceding calculation to draw some quick conclusions about the linear map $L : \mathbb{R}^4 \to \mathbb{R}^3$ whose matrix is

$$[L] = \begin{bmatrix} 2 & -4 & 4 & 5 \\ -1 & 2 & -1 & -4 \\ 6 & -12 & 10 & 17 \end{bmatrix}.$$

The calculation shows that for one particular Y the solution set of the equation $L(X) = Y$ is an affine subspace of dimension 1. That affine subspace is the sum of any one of its members and a linear subspace of dimension 1. That linear subspace is the null space; so, the nullity equals 1. Since the sum of the nullity and the rank equals the dimension 4 of the domain, we conclude that the rank equals 3 and that, therefore, the image of L is its target $\mathbb{R}^3$.

The last paragraph in the preceding example illustrates a good study of technique. Once a problem is finished, it is useful to ask oneself about the various conclusions that one can reach about things that were not even mentioned in the problem.

3.5G SUMMARY

We began this chapter by giving a definition of a linear map, that is, a function that enjoys certain properties with respect to vector addition and multiplication by scalars. We then showed, through examples, that linear maps were discernible in a wide variety of mathematical situations: linear equations, rotations of the plane, and so on. This variety may have been a bit bewildering.

It is important to realize that our subsequent work has unified this picture somewhat. When we meet a linear map L in life (from some geometric, numerical, or physical problem), it is usually presented in one of three ways, namely, (1) in terms of standard bases, (2) as a matrix, or (3) in coordinates. Let us review these in turn:

1. Using the standard bases. We have learned that $L(\mathbf{e}_1) = a_{11}\bar{\mathbf{e}}_1 + \cdots + a_{q1}\bar{\mathbf{e}}_q$, and so on, and that the coefficients a_{ij} determine L precisely. If different coefficients are used, different maps result.

2. Using the matrix. We know that L is also determined by its matrix $[L] = [a_{ij}]$. In this representation, the standard bases are implicit. The matrix notation tidies up the symbolism, prompts the introduction of matrix multiplication, and thereby facilitates computing with L.

3. Using coordinates. We observe that *every* L may be written

$$L(X) = (a_{11}x_1 + \cdots + a_{1n}x_n, \ldots, a_{q1}x_1 + \cdots + a_{qn}x_n) ,$$

where $X = (x_1, \cdots, x_n)$. Thus every linear map looks like the "left-hand side" of a system of linear equations. This is the coordinate form of L.

Exercises

Practice in carrying out the basic algebraic operations on matrices, a skill that, like the multiplication table learned in grade school, will be taken for granted in subsequent sections.

1. Write down the 2-by-3 matrix $A = [a_{ij}]$ whose entries are $a_{11} = 2$, $a_{21} = -3$, $a_{12} = 0$, $a_{22} = -1$, $a_{13} = 4$, and $a_{23} = 144$.

2. Perform the additions indicated:

    (a) $\begin{bmatrix} 2 & 0 & 4 \\ -3 & -1 & 144 \end{bmatrix} + \begin{bmatrix} 1 & 1 & 2 \\ 0 & 1 & 2 \end{bmatrix}$

    (b) $\begin{bmatrix} 1 & 3 \\ 2 & 0 \end{bmatrix} + \begin{bmatrix} -1 & 2 \\ 3 & 1 \end{bmatrix}$

    (c) $\begin{bmatrix} 1 \\ 4 \\ 6 \\ 1 \end{bmatrix} + \begin{bmatrix} 0 \\ -4 \\ -6 \\ 0 \end{bmatrix}$

    (d) $\begin{bmatrix} 1 & 0 & -1 \\ 3 & 1 & -1 \\ 2 & 2 & 1 \end{bmatrix} + \begin{bmatrix} 4 & 0 & 0 \\ 0 & 4 & 0 \\ 0 & 0 & 4 \end{bmatrix} .$

3. Perform the operations indicated:

    (a) $\begin{bmatrix} 1 & 2 \\ 2 & 1 \end{bmatrix} + 4 \begin{bmatrix} 0 & -1 \\ 1 & 1 \end{bmatrix}$

    (b) $2 \begin{bmatrix} 1 \\ 0 \end{bmatrix} - 7 \begin{bmatrix} 0 \\ 1 \end{bmatrix}$

    (c) $7 \begin{bmatrix} 3 & 4 \\ 1 & -1 \\ 1 & 0 \end{bmatrix}$

(d) $2 \begin{bmatrix} 1 \\ 1 \\ 0 \end{bmatrix} + 4 \begin{bmatrix} -1 \\ 0 \\ -1 \end{bmatrix} - \begin{bmatrix} -2 \\ 2 \\ 4 \end{bmatrix}$.

4. Perform the following matrix multiplications:

(a) $\begin{bmatrix} 2 & 1 \\ 0 & 3 \end{bmatrix} \begin{bmatrix} -2 & 1 \\ 1 & 3 \end{bmatrix}$

(b) $\begin{bmatrix} 1 & 0 \\ 0 & 1 \end{bmatrix} \begin{bmatrix} 5 & 6 \\ 7 & 8 \end{bmatrix}$

(c) $\begin{bmatrix} 2 & 1 \\ 0 & 3 \end{bmatrix} \begin{bmatrix} 1 \\ 0 \end{bmatrix}$

(d) $\begin{bmatrix} -2 & 1 \\ 4 & -2 \end{bmatrix} \begin{bmatrix} 1 & 3 \\ 2 & 6 \end{bmatrix}$

(e) $\begin{bmatrix} 2 & 1 \\ 0 & 3 \end{bmatrix} \begin{bmatrix} 0 \\ 1 \end{bmatrix}$

(f) $\begin{bmatrix} 2 & 1 \\ 0 & 3 \end{bmatrix} \begin{bmatrix} 1 & 0 & 1 \\ 0 & 1 & 1 \end{bmatrix}$

(g) $\begin{bmatrix} 1 & 2 & 0 \\ 3 & 1 & 2 \\ -1 & 0 & 2 \end{bmatrix} \begin{bmatrix} -1 & 1 & 1 \\ 1 & -1 & 1 \\ 1 & -1 & -1 \end{bmatrix}$

(h) $\begin{bmatrix} 1 & 2 & 0 \\ 3 & 1 & 2 \\ -1 & 0 & 2 \end{bmatrix} \begin{bmatrix} 2 & -1 \\ 1 & 1 \\ -1 & 1 \end{bmatrix}$

(i) $\begin{bmatrix} 1 & 2 & 0 \\ 3 & 1 & 2 \end{bmatrix} \begin{bmatrix} 5 & 0 & 0 & 1 \\ 0 & 5 & 0 & 1 \\ 0 & 0 & 5 & 1 \end{bmatrix}$

(j) $\begin{bmatrix} 2 & 3 & 6 \end{bmatrix} \begin{bmatrix} -3 \\ 6 \\ -2 \end{bmatrix}$

(k) $\begin{bmatrix} -3 \\ 6 \\ -2 \end{bmatrix} \begin{bmatrix} 2 & 3 & 6 \end{bmatrix}$

(l) $\begin{bmatrix} -1 & 1 & 1 \\ 1 & -1 & 1 \\ 1 & -1 & -1 \end{bmatrix} \begin{bmatrix} 0 \\ 1 \\ 0 \end{bmatrix}$.

5. For each part, compute numbers a, b, c, and d (not all zero) so that the following equations hold:

(a) $\begin{bmatrix} 2 & 1 \\ 0 & 1 \end{bmatrix} \begin{bmatrix} a & b \\ c & d \end{bmatrix} = \begin{bmatrix} 1 & 0 \\ 0 & 1 \end{bmatrix}$

(b) $\begin{bmatrix} 3 & -1 \\ -6 & 2 \end{bmatrix} \begin{bmatrix} a & b \\ c & d \end{bmatrix} = \begin{bmatrix} 0 & 0 \\ 0 & 0 \end{bmatrix}$.

Hint: Solve linear equations.

6. Let $L : \mathbb{R}^2 \to \mathbb{R}^2$ be the linear map determined by

$$L(\mathbf{e}_1) = -2\mathbf{e}_1 + \mathbf{e}_2 \quad \text{and} \quad L(\mathbf{e}_2) = \mathbf{e}_1 + 3\mathbf{e}_2 \ .$$

(a) What is the matrix $[L]$?

(b) What is the relation between the image vector $L(\mathbf{e}_1)$ and the first column of the matrix $[L]$?

(c) What are the 2-by-1 matrices $[\mathbf{e}_1]$ and $[\mathbf{e}_2]$?

(d) What is the product of the matrix $[L]$ and the column matrix $[\mathbf{e}_1]$?

(e) Find $[X]$ if $X = 4\mathbf{e}_1 + 5\mathbf{e}_2$.

(f) Compute $L(X)$ in two ways: first, by calculating $L(\mathbf{e}_1)$ and $L(\mathbf{e}_2)$ and then using linearity and, second, by matrix multiplication. Which is more efficient? (Hopefully your answers agree.)

7. Let L be as in the preceding exercise, and let $M : \mathbb{R}^2 \to \mathbb{R}^2$ be linear and satisfy

$$M(\mathbf{e}_1) = 2\mathbf{e}_1 \quad \text{and} \quad M(\mathbf{e}_2) = \mathbf{e}_1 + 3\mathbf{e}_2 \ .$$

(a) What is the matrix $[M]$?

(b) Find $[ML]$.

(c) What is the relation between the image vector $ML(\mathbf{e}_1)$ and the first column of the matrix $[ML]$? What about $ML(\mathbf{e}_2)$?

8. Suppose that $L : \mathbb{R}^2 \to \mathbb{R}^2$ is a linear map for which

$$[L] = \begin{bmatrix} -2 & 1 \\ 1 & 3 \end{bmatrix} \ .$$

(a) Express $L(\mathbf{e}_1)$ and $L(\mathbf{e}_2)$ as linear combinations of $\mathbf{e}_1$ and $\mathbf{e}_2$.

(b) For $X \in \mathbb{R}^2$, write $L(X)$ in terms of coordinates, that is, in the form

$$(ax_1 + bx_2, \ cx_1 + dx_2)$$

for appropriate a, b, c, and d.

9. Let

$$A = \begin{bmatrix} 1 & -2 \\ 2 & 0 \\ 3 & 1 \end{bmatrix}, \quad B = \begin{bmatrix} -1 & 0 & -2 \\ 2 & 1 & 0 \end{bmatrix}, \quad C = \begin{bmatrix} 3 & 0 & 2 \\ 1 & 4 & -1 \\ 0 & -2 & 0 \end{bmatrix} \ .$$

Compute all the following products that make sense: AB, BA, AC, CA, BC, CB, A^2, B^2, C^2, ABC, and CAB.

10. If

$$A = \begin{bmatrix} 1 & 0 \\ 0 & 0 \end{bmatrix} \quad \text{and} \quad B = \begin{bmatrix} 0 & 0 \\ 1 & 0 \end{bmatrix},$$

express each of A^2, B^2, AB, BA, ABA, BAB, $(AB)^2$, $(BA)^2$, A^5, $(A + 2B)^2$, and $(3A - B)^2$ in the form $aA + bB$ for appropriate scalars a and b.

11. Given that A and B are two 2-by-2 matrices such that $AB = 0$. Which of the following statements is a consequence? For each that is, give a proof, and, for each that is not, give a counterexample.

    (a) $A = O$

    (b) $B = O$

    (c) Either $A = O$ or $B = O$

    (d) $BA = O$.

12. What can you say about the rightmost column of the augmented matrix corresponding to a homogeneous system of linear equations?

13. Use sequences of equivalent matrices to calculate the null spaces of the linear maps whose matrices are given.

    (a) $[L] = \begin{bmatrix} 4 & 3 \\ 3 & -4 \end{bmatrix}$

    (b) $[M] = \begin{bmatrix} 4 & 6 \\ -2 & -3 \end{bmatrix}$

    (c) $[P] = \begin{bmatrix} 1 & 3 & -4 \\ 3 & 9 & -12 \end{bmatrix}$

    (d) $[R] = \begin{bmatrix} 1 & 5 & -4 & 3 \\ -2 & -10 & 8 & 2 \end{bmatrix}$

    (e) $[S] = \begin{bmatrix} 1 & 5 & -4 & 3 \\ -2 & -10 & 8 & 2 \\ 2 & 10 & -8 & 14 \end{bmatrix}$

    (f) $[T] = \begin{bmatrix} 1 & 4 & -4 & -2 \\ 1 & 2 & 3 & -4 \\ 6 & -3 & 2 & 3 \\ 3 & -13 & 7 & 11 \end{bmatrix}$.

14. Calculate the nullities and ranks of the linear maps defined in the preceding exercise. Then, for those maps, use sequences of equivalent matrices to find the solutions sets of the following equations:

    (a) $L(X) = (2, 3)$

    (b) $M(X) = (2, 2)$

    (c) $P(X) = (-2, -6)$

    (d) $R(X) = (3, 5)$

    (e) $S(X) = (-3, 2, -10)$

    (f) $T(X) = (3, -2, 2, 1)$.

15. Find a 2-by-2 matrix A such that $A^2 = -I$.

3.6   Affine Maps

3.6A   DEFINITION

We now turn to maps that generalize linear maps in a way similar to the way in which affine subspaces generalize linear subspaces. These maps, which are built from linear maps, have graphs and images that are affine subspaces. (Recall that an affine subspace is obtained by translating a linear subspace.)

Definition 3.6.1 A function $T : \mathbb{R}^n \to \mathbb{R}^q$ is **affine** if and only if there is a fixed $Y_0 \in \mathbb{R}^q$ and a linear map $L : \mathbb{R}^n \to \mathbb{R}^q$ such that, for all $X \in \mathbb{R}^n$,

$$T(X) = L(X) + Y_0 .$$

An affine map T is calculated by first operating on X with a linear map L and then translating $L(X)$ in $\mathbb{R}^q$ by the vector Y_0. Since $T(\mathbf{0}) = L(\mathbf{0})+Y_0 = Y_0$, it is easy to see that the affine map T is linear if and only if the vector Y_0 is $\mathbf{0}$.

EXAMPLES:

1. In the case of $n = q = 1$, the vector X is a real number x, and an affine map is of the form $T(x) = ax + b$, where a and b are fixed scalars. The graph of T is a line.

2. Fix $Y_0 \in \mathbb{R}^n$, and define $T(X) = X + Y_0$. Here $T : \mathbb{R}^n \to \mathbb{R}^n$ translates every vector X by the fixed vector Y_0. Since $T(X) = I(X) + Y_0$, L is the identity map.

3. Suppose that $T : \mathbb{R}^2 \to \mathbb{R}^2$ is the affine map given by $T(X) = L(X) + Y_0$, where $L(X) = (x_1 - x_2 , 2x_1 + x_2)$ and $Y_0 = (6, -7)$. Then T is given in coordinate form by

$$T(X) = (x_1 - x_2 + 6 , 2x_1 + x_2 - 7) .$$

Each coordinate of $T(X)$ has a linear term—for example, $x_1 - x_2$ —and a constant term—for example, 6. This fact about coordinates is true for a general affine map $T : \mathbb{R}^n \to \mathbb{R}^q$.

3.6B   AFFINE MAPS AND AFFINE SUBSPACES

A basic property of affine maps is conveyed by Figure 3.16. Here $\mathcal{A}$ is an affine subspace (a plane) in $\mathbb{R}^3$, and T is an affine map $T : \mathbb{R}^3 \to \mathbb{R}^3$. Figure 3.16 indicates that the image of the set $\mathcal{A}$ under the map T, namely,

$$T(\mathcal{A}) = \{Y \in \mathbb{R}^3 : Y = T(X), X \in \mathcal{A}\} ,$$

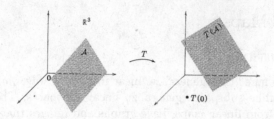

FIGURE 3.16.

is also an affine subspace of the target $\mathbb{R}^3$. In fact, $T(\mathcal{A})$ is here pictured to be a plane also, just as $\mathcal{A}$ was. The picture illustrates a particular case of the following theorem.

Theorem 3.6.1 The image of an affine subspace under an affine map is an affine subspace. In particular, $\mathcal{I}(T)$ is an affine subspace if T is an affine map.

PROOF: We will use the corresponding fact proved earlier for linear maps; that the image of a linear subspace under a linear map is a linear subspace. Consider an affine map $T(X) = L(X) + Y_0$, where $L : \mathbb{R}^n \to \mathbb{R}^q$ is linear and Y_0 is fixed in $\mathbb{R}^q$. Since $\mathcal{A}$ is affine, $\mathcal{A} = \mathcal{S} + Z$, where $\mathcal{S}$ is a linear subspace of $\mathbb{R}^n$ and Z is a fixed vector in $\mathcal{A}$. Thus, if $X \in \mathcal{A}$, we know that $X = X' + Z$, with $X' \in \mathcal{L}$. Applying the map T gives

$$T(X) = T(X' + Z) = L(X' + Z) + Y_0 = L(X') + [L(Z) + Y_0] \,,$$

which is a member of the affine subspace $L(\mathcal{S}) + [L(Z) + Y_0]$. (Note that $L(Z) + Y_0$ is a constant vector that does *not* depend on the variable X.)

Let us complete the proof by showing that every $Y \in L(\mathcal{S}) + [L(Z) + Y_0]$ equals $T(X)$ for some $X \in \mathcal{A}$. Consider such a Y. There exists some $W \in L(\mathcal{S})$ such that

$$Y = W + L(Z) + Y_0 \,.$$

Since $W \in L(\mathcal{S})$, by definition there exists $X_1 \in \mathcal{S}$ such that $W = L(X_1)$. Hence,

$$Y = L(X_1) + L(Z) + Y_0 = L(X_1 + Z) + Y_0 \,,$$

which equals $T(X)$ for $X = X_1 + Z \in \mathcal{S} + Z = \mathcal{A}$. <<

REMARKS:

- You might reflect on how much more cluttered the proof of Theorem 3.6.1 would be if, instead of using the coordinate-free vector notation, all the maps and linear subspaces were specified using the coordinates $x_1, \ldots, x_n$ and $y_1, \ldots, y_q$ and systems of linear equations in these coordinates. We would be lost in clouds of linear equations.

- In later chapters we will see examples of maps that "bend" or "fold" planes. They will not be affine maps, of course.

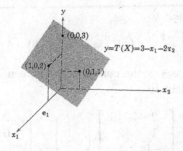

FIGURE 3.17. The graph of an affine map $T : \mathbb{R}^2 \to \mathbb{R}$

The graph of an affine map from $\mathbb{R}^n$ into $\mathbb{R}^q$ is a subset of $\mathbb{R}^{n+q}$. For example, let $T : \mathbb{R}^2 \to \mathbb{R}$ be defined by $T(X) = -x_1 - 2x_2 + 3$. Then $T(\mathbf{0}) = 3$, $T(\mathbf{e}_1) = 2$, and $T(\mathbf{e}_2) = 1$. The graph $y = T(X)$ is the subset of $\mathbb{R}^3$ shown in Figure 3.17. Of course, it must pass through the three points $(0,0,3)$, $(1,0,2)$, and $(0,1,1)$. That it is a plane, as shown, is a consequence of the following theorem.

Theorem 3.6.2 The graph of an affine map $T : \mathbb{R}^n \to \mathbb{R}^q$ is an n-dimensional affine subspace of $\mathbb{R}^{n+q}$.

The proof, which depends on the corresponding facts for linear maps, is left for the exercises. The proof is not very difficult conceptually, but organizing the notation is a bit of a challenge.

We have spoken of the image and the graph of an affine function. One does not usually speak of the "null space" of an affine map if it does not happen to be a linear map; indeed, $\mathbf{0}$ may not even belong to the image. There is a wide variety of questions one might ask about affine maps in general and also about particular affine maps. The standard way of approaching such questions is to use what one knows about linear maps, rather than to develop an entire theory for affine maps. An example will illustrate this point of view.

EXAMPLE:

4. Let $T(X) = L(X) + (4, 6, -3)$, where

$$[L] = \begin{bmatrix} 2 & -10 \\ 1 & 2 \\ -4 & 3 \end{bmatrix} .$$

Let us consider the problem of finding the solution set of $T(X) = (2, 0, 3)$. Thus, we want to solve

$$L(X) + (4, 6, -3) = (2, 0, 3) ,$$

which is equivalent to

$$L(X) = (-2, -6, 6) \ ,$$

which, on the basis of the connection between linear maps and matrices, is equivalent to

$$\begin{bmatrix} 2 & -10 \\ 1 & 2 \\ -4 & 3 \end{bmatrix} \begin{bmatrix} x_1 \\ x_2 \end{bmatrix} = \begin{bmatrix} -2 \\ -6 \\ 6 \end{bmatrix} \ .$$

Now, that we have reduced the problem to one that the reader can handle in expert fashion, we will stop.

Exercises

Practice in recognizing, constructing, and picturing affine maps.

1. Which of the following maps $T : \mathbb{R} \to \mathbb{R}$ are affine?

   (a) $T(x) = 1 + x^2$

   (b) $T(x) = 1 + 2x$

   (c) $T(x) = \frac{1}{2}x$.

2. Which of the following maps $T : \mathbb{R}^2 \to \mathbb{R}$ are affine?

   (a) $T(X) = x_1^2 + x_2^2$

   (b) $T(X) = x_1 - x_2$

   (c) $T(X) = 1 + x_1 + 3x_2$.

3. Which of the following maps $T : \mathbb{R}^2 \to \mathbb{R}^2$ are affine?

   (a) $T(X) = X$ (the identity map)

   (b) $T(X) = \langle X, X \rangle X$

   (c) $T(X) = X + Y_0$, Y_0 fixed

   (d) $T(X) = (x_1 , x_1 + x_2 + 2)$.

4. Which of the following maps $T : \mathbb{R} \to \mathbb{R}^2$ are affine?

   (a) $T(t) = (t, t^2)$

   (b) $T(t) = (-t + 1, 3t + 2)$

   (c) $T(t) = (t, t)$

   (d) $T(t) = (\langle t, t \rangle, 0)$

   (e) $T(t) = (0, 4)$ (a constant).

5. Which of the following maps $T : \mathbb{R}^3 \to \mathbb{R}^3$ are affine?

   (a) $T(X) = (x_3, x_1, x_2)$

   (b) $T(X) = (x_3, x_1, x_2) + \|X\|$

   (c) $T(X) = X \times Y_0$, Y_0 fixed

(d) $T(X) = (x_3 + 3, \, x_1 + 1, \, x_2 + 2)$.

6. Which of the following maps T on $\mathbb{R}^n$ are affine?

   (a) $T(X) = (x_n, x_{n-1}, x_{n-2}, \ldots, x_1)$

   (b) $T(X) = \langle Z, X + Y \rangle$, $\quad Y, Z$ fixed.

7. Let $T : \mathbb{R}^2 \to \mathbb{R}$ be the affine map $T(X) = -x_1 + 4x_2 + 7$.

   (a) Compute $T(\mathbf{e}_1)$.

   (b) Compute $T(\mathbf{e}_2)$.

   (c) Compute $T(\mathbf{0})$.

   (d) If we write this same map in the form $T(X) = y_0 + L(X)$, then what is the number y_0?

   (e) What is the coordinate expression for the linear map L?

   (f) What is the matrix $[L]$?

8. Let x_1, x_2, and y be coordinates in $\mathbb{R}^3$. The plane given by $3x_1 - x_2 + 6y = 7$ is the graph of an affine map $T : \mathbb{R}^2 \to \mathbb{R}$, $y = T(X)$. Find T.

9. Let x_1, x_2, and z be coordinates in $\mathbb{R}^3$. Find an affine function T whose graph is the plane $x_1 + x_2 - 2z = 1$.

10. (a) Find an affine map (coordinate form: $T(X) = ax_1 + bx_2 + c$) such that $T(\mathbf{0}) = 7$, $T(1, 0) = -1$, and $T(0, 1) = 4$.

    (b) Is the affine map you obtained for the preceding part the only correct response for that part?

    (c) Compute the equation of the plane in $\mathbb{R}^3$ (using the coordinates x_1, x_2, and z) through the three points

$$(0, 0, 7), \, (1, 0, -1), \, (0, 1, 4).$$

11. Find an affine map $T : \mathbb{R}^2 \to \mathbb{R}$, in the form of a vector added to a linear transformation given by its matrix, for which $T(1, 1) = 10$, $T(1, -1) = 2$, and $T(-1, 1) = 12$.

12. Decide whether the following are true or false:

    (a) Every affine map is linear.

    (b) Every constant map is affine.

    (c) If $T : \mathbb{R}^2 \to \mathbb{R}$ is affine, then it is uniquely determined by the two values $T(\mathbf{e}_1)$ and $T(\mathbf{e}_2)$.

    (d) If $T : \mathbb{R}^2 \to \mathbb{R}$ is affine, then its graph in $\mathbb{R}^3$ might be a line.

    (e) The graph of every affine map $T : \mathbb{R}^2 \to \mathbb{R}$ contains the origin of $\mathbb{R}^3$.

    (f) If $T : \mathbb{R}^2 \to \mathbb{R}$ is affine, then $T(X) = T(\mathbf{e}_1)x_1 + T(\mathbf{e}_2)x_2 + \mathrm{T}(\mathbf{0})$.

    (g) The map $T : \mathbb{R}^2 \to \mathbb{R}$ given by $T(X) = 3 + 4(x_1 - 1) + (x_2 + 1)$ is affine.

    (h) If $T((x_1, x_2)) = ax_1 + bx_2 + c$, then the graph of T in $\mathbb{R}^3$ is horizontal $\Leftrightarrow a = b = 0$.

(i) The graphs of the maps $T(X) = ax_1 + bx_2 + c$ and $T^*(X) = a^*x_1 + b^*x_2 + c^*$ are non-intersecting planes in $\mathbb{R}^3 \Leftrightarrow a = a^*$, $b = b^*$, and $c \neq c^*$.

(j) If T is affine, then $T(-X) = -T(X)$.

(k) Every affine map is a linear functional.

(l) Every linear map is affine.

13. Let $T : \mathbb{R}^n \to \mathbb{R}^q$ be affine. Prove that T is linear $\Leftrightarrow T(\mathbf{0}) = \mathbf{0}$.

14. Let $T : \mathbb{R}^2 \to \mathbb{R}^3$ be an affine map for which $T(\mathbf{e}_1) = (0, 1, 1)$, $T(\mathbf{e}_2) = (1, -1, 0)$, and $T(\mathbf{0}) = (1, -2, 3)$. Find a formula for $T(X)$. In particular, find $T((1, 2))$.

15. Let $T : \mathbb{R}^n \to \mathbb{R}^q$ be an affine map for which $T(\mathbf{e}_j) = Y_j$ for $j = 1, \ldots, n$ and $T(\mathbf{0}) = Y_0$.

   (a) Show that

$$T(X) = x_1(Y_1 - Y_0) + \cdots + x_n(Y_n - Y_0) + Y_0 \,.$$

   This shows that an affine map is uniquely determined by what it does to $\mathbf{0}$ and the standard basis vectors.

   (b) Let A denote the matrix whose j^{th} column is $[Y_j - Y_0]$. Show that $T(X) = L(X) + Y_0$, where L is the linear map with matrix A, that is show that

$$[T(X)] = A[X] + [Y_0] \,.$$

   (c) Redo Exercises 11 and 14 by using part (a) of this exercise and then again by using part (b) of this exercise.

16. Let $T : \mathbb{R}^n \to \mathbb{R}^q$ be an affine map, and let α and β be scalars such that $\alpha + \beta = 1$. Prove that $T(\alpha X + \beta Y) = \alpha T(X) + \beta T(Y)$ for all $X, Y \in \mathbb{R}^n$.

17. If $T : \mathbb{R}^n \to \mathbb{R}^q$ is an affine map, we define its **zero set** (a substitute for the "null space") as the set of all $X \in \mathbb{R}^n$ for which $T(X) = \mathbf{0}$. Prove that the zero set is either empty or an affine subspace of its domain.

18. Find the zero sets of the following affine maps.

   (a) $T(x_1, x_2, x_3) = (x_1 - 3x_2 + 4x_3 + 3, \, -3x_1 + 9x_2 - 2x_3 + 2)$

   (b) $T(X) = L(X) + (3, -2, 5)$, where

$$[L] = \begin{bmatrix} 3 & 1 \\ -2 & 0 \\ 1 & 1 \end{bmatrix} \,.$$

19. Let $T(X) = L(X) + Y_0$ be an affine map. If $T(X_1) = Y_1$, prove that $T(X) = L(X - X_1) + Y_1$.

20. Find the intersection of the graphs of the affine functions $T((x_1, x_2)) = -4x_1 + x_2 + 3$ and $U((x_1, x_2)) = 3x_1 + x_2 - 2$.

21. Prove Theorem 3.6.2. *Hint:* Use the correponding facts about linear maps.

3.7   Another Special Case: $L : \mathbb{R}^n \to \mathbb{R}^n$

3.7A   INVERSES OF LINEAR MAPS AND SQUARE MATRICES

When the domain and target space of a linear map $L : \mathbb{R}^n \to \mathbb{R}^q$ are the same, that is, when $n = q$, we say that L is a linear map **in** $\mathbb{R}^n$, rather than a linear map from $\mathbb{R}^n$ into $\mathbb{R}^n$. For such maps, it is sometimes convenient to visualize everything as happening in one copy of $\mathbb{R}^n$, rather than having one copy for the domain and another for the target. When this is done one might visualize the action of L as that of moving points around in $\mathbb{R}^n$.

As remarked at the end of Section 3.2d, the following four statements are equivalent for linear maps L in $\mathbb{R}^n$: (1) L is one-to-one; (2) L is onto; (3) the nullity of L is 0; (4) the rank of L is n. Suppose these conditions hold for a linear map L in $\mathbb{R}^n$. Then for each $Y \in \mathbb{R}^n$, there exists a unique $X \in \mathbb{R}^n$ such that $L(X) = Y$, since L is both one-to-one and onto. Therefore, we can define the **inverse** of L: for each $Y \in \mathbb{R}^n$, let $L^{-1}(Y) = X$, where X is the unique vector in $\mathbb{R}^n$ such that $L(X) = Y$. Note that, just as with L, the space $\mathbb{R}^n$ is both the domain and target of the function L^{-1}.

If the inverse of a function L exists, then we say that L is **invertible**. If L does not have an inverse, such as is the case when L is not one-to-one, then we say that L is non-invertible, or **singular**. The inverse of an invertible linear map is itself a linear map as stated in the following theorem:

Theorem 3.7.1 Let L be an invertible linear map in $\mathbb{R}^n$. Then L^{-1} is also a linear map in $\mathbb{R}^n$.

You may wish to prove this theorem for yourself, using the definition of linearity. Or, wait until we work out some examples, where it will become clear that inverses of linear maps are linear.

The map L^{-1} undoes the action of L (see Figure 3.18). If $L(X) = Y$, then

$$L^{-1}(L(X)) = L^{-1}(Y) = X .$$

Also,

$$L(L^{-1}(Y)) = L(X) = Y .$$

Thus,

$$L^{-1}L = I \quad \text{and} \quad LL^{-1} = I .$$

The first equality is simply a restatement of the definition of the inverse. You may use it to check whether a given linear map M is the inverse of L: if $ML = I$, then $M = L^{-1}$. If L is not invertible, then there does not exist a linear map M such that $ML = I$. The second equality says that the inverse of L^{-1} is L, or in symbols,

$$(L^{-1})^{-1} = L .$$

These observations are combined in the following theorem:

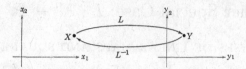

FIGURE 3.18. L^{-1} undoes the effect of L: $L^{-1}L(X) = X$ and $LL^{-1}(Y) = Y$

Theorem 3.7.2 Let L and M be linear transformations in $\mathbb{R}^n$ for which $ML = I$. Then L and M are invertible and each is the inverse of the other.

There is a corresponding terminology for matrices. If L is invertible and $A = [L]$, then we write A^{-1} for the matrix of L^{-1}. Thus,

$$A^{-1}A = AA^{-1} = I .$$

We call A^{-1} the **inverse** of A. The matrix of a linear map L is said to be invertible or singular, depending on whether L is **invertible** or **singular**.

REMARK:

- Since we are only concerned here with linear maps in $\mathbb{R}^n$, the only matrices of interest are those that are square. It is only for square matrices that the noun "inverse" and the adjectives "invertible" and "singular" are relevant.

EXAMPLE:

1. Let

$$A = \begin{bmatrix} 3 & 1 \\ 5 & 2 \end{bmatrix} .$$

The calculation

$$\begin{bmatrix} 2 & -1 \\ -5 & 3 \end{bmatrix} \begin{bmatrix} 3 & 1 \\ 5 & 2 \end{bmatrix} = \begin{bmatrix} 1 & 0 \\ 0 & 1 \end{bmatrix}$$

shows that

$$\begin{bmatrix} 2 & -1 \\ -5 & 3 \end{bmatrix} = \begin{bmatrix} 3 & 1 \\ 5 & 2 \end{bmatrix}^{-1} .$$

The following theorem collects in one place several equivalent statements that we have already used in this section, together with some results from earlier sections. Among other things, it provides us with a way to check whether a linear map is invertible.

Theorem 3.7.3 Let $L : \mathbb{R}^n \to \mathbb{R}^n$ be linear, and let $A = [L]$. Then the following assertions are equivalent (i.e. if one holds, then so do all the others):

1. L is invertible.

2. L is one-to-one.

3. L is onto.

4. The homogeneous system $AX = 0$ has only the trivial solution $X = 0$.

5. For every Y, the system $AX = Y$ has a solution X.

6. For every Y, the system $AX = Y$ has a unique solution X.

EXAMPLES:

2. Let us decide if the linear transformation L in $\mathbb{R}^4$ represented by

$$[L] = \begin{bmatrix} 0 & 2 & -3 & 1 \\ 3 & 6 & -2 & 2 \\ -1 & -2 & 2 & 3 \\ 2 & 8 & -6 & 7 \end{bmatrix}$$

is invertible. To do so we decide whether the nullity equals 0 by investigating whether the zero vector is the only member of the null space. As is typical when treating homogeneous systems, we work with the coefficient matrix rather than the augmented matrix, since the last column of the augmented matrix will be a zero column at every stage.

We interchange the first and third rows (not necessary but useful for organizational purposes) and then add the new first row three times to the second row and twice to the fourth row:

$$\begin{bmatrix} -1 & -2 & 2 & 3 \\ 3 & 6 & -2 & 2 \\ 0 & 2 & -3 & 1 \\ 2 & 8 & -6 & 7 \end{bmatrix} \sim \begin{bmatrix} -1 & -2 & 2 & 3 \\ 0 & 0 & 4 & 11 \\ 0 & 2 & -3 & 1 \\ 0 & 4 & -2 & 13 \end{bmatrix}.$$

Next we interchange the second and third rows and subtract twice the new second row from the fourth:

$$\begin{bmatrix} -1 & -2 & 2 & 3 \\ 0 & 2 & -3 & 1 \\ 0 & 0 & 4 & 11 \\ 0 & 4 & -2 & 13 \end{bmatrix} \sim \begin{bmatrix} -1 & -2 & 2 & 3 \\ 0 & 2 & -3 & 1 \\ 0 & 0 & 4 & 11 \\ 0 & 0 & 4 & 11 \end{bmatrix}.$$

After subtracting the third row from the fourth we obtain a row of zeroes. We see that x_4 is a free variable, the nullity equals one, and the linear transformation is singular. Since our only concern has been the invertibility or singularity of L we have not worked to obtain zeroes above the main diagonal.

3. Let us decide if the linear transformation L in $\mathbb{R}^3$ represented by

$$[L] = \begin{bmatrix} 0 & 3 & 1 \\ -1 & 3 & 0 \\ 1 & 0 & 2 \end{bmatrix}$$

is invertible. To do so we decide whether the nullity equals 0, as before.

We add the third row to the second and subtract the new second row from the first to obtain

$$\begin{bmatrix} 0 & 0 & -1 \\ 0 & 3 & 2 \\ 1 & 0 & 2 \end{bmatrix}.$$

It is now obvious that there is no free variable and so the nullity is 0. Therefore, L has an inverse map.

In the preceding example, we determined that L was invertible. How do we find L^{-1}? We turn to that problem now. Not surprisingly, the method is based on Gauss-Jordan elimination.

EXAMPLE:

4. Let L be as in Example 3, and let $A = [L]$. In order to find L^{-1}, it is enough to find the vectors $L^{-1}(e_1), L^{-1}(e_2)$, and $L^{-1}(e_3)$. These vectors will be the columns of A^{-1}.

First we find $L^{-1}(e_1)$. It is the solution of the equation $L(X) = e_1$. In order to find this solution, we solve the corresponding system. The augmented matrix is

$$\left[\begin{array}{ccc|c} 0 & 3 & 1 & 1 \\ -1 & 3 & 0 & 0 \\ 1 & 0 & 2 & 0 \end{array} \right].$$

Add the third row to the second, then subtract the new second row from the first to obtain the equivalent augmented matrix

$$\left[\begin{array}{ccc|c} 0 & 0 & -1 & 1 \\ 0 & 3 & 2 & 0 \\ 1 & 0 & 2 & 0 \end{array} \right].$$

Add the first row twice to each of the other two rows:

$$\left[\begin{array}{ccc|c} 0 & 0 & -1 & 1 \\ 0 & 3 & 0 & 2 \\ 1 & 0 & 0 & 2 \end{array} \right].$$

This completes the elimination part of the procedure. To solve for the unknowns x_1, x_2 and x_3, divide each row by its leading nonzero term:

$$\begin{bmatrix} 0 & 0 & 1 & | & -1 \\ 0 & 1 & 0 & | & \frac{2}{3} \\ 1 & 0 & 0 & | & 2 \end{bmatrix}.$$

We can read the solution off from this augmented matrix: $x_1 = 2$, $x_2 = 2/3$, and $x_3 = -1$. These are the three numbers that go into the first column of A^{-1}. However, it is useful to switch the first and last rows of the augmented matrix:

$$\begin{bmatrix} 1 & 0 & 0 & | & 2 \\ 0 & 1 & 0 & | & \frac{2}{3} \\ 0 & 0 & 1 & | & -1 \end{bmatrix}.$$

Now the coordinates of the solution appear in their correct order from top to bottom in the last column, so this column equals the first column of A^{-1}:

$$A^{-1} = \begin{bmatrix} 2 & ? & ? \\ \frac{2}{3} & ? & ? \\ -1 & ? & ? \end{bmatrix}.$$

In order to find the other two columns of A^{-1}, we solve the equations $L(X) = e_2$ and $L(X) = e_3$. The corresponding augmented matrices are

$$\begin{bmatrix} 0 & 3 & 1 & | & 0 \\ -1 & 3 & 0 & | & 1 \\ 1 & 0 & 2 & | & 0 \end{bmatrix} \quad \text{and} \quad \begin{bmatrix} 0 & 3 & 1 & | & 0 \\ -1 & 3 & 0 & | & 0 \\ 1 & 0 & 2 & | & 1 \end{bmatrix}.$$

Now apply the same Gauss-Jordan steps to these two matrices as we did to solve the first system, including the row switch at the end. The resulting pair of matrices is

$$\begin{bmatrix} 1 & 0 & 0 & | & -2 \\ 0 & 1 & 0 & | & -\frac{1}{3} \\ 0 & 0 & 1 & | & 1 \end{bmatrix} \quad \text{and} \quad \begin{bmatrix} 1 & 0 & 0 & | & -1 \\ 0 & 1 & 0 & | & -\frac{1}{3} \\ 0 & 0 & 1 & | & 1 \end{bmatrix}.$$

The rightmost columns of these two matrices are the second and third columns of A^{-1}. Therefore

$$A^{-1} = \begin{bmatrix} 2 & -2 & -1 \\ \frac{2}{3} & -\frac{1}{3} & -\frac{1}{3} \\ -1 & 1 & 1 \end{bmatrix}.$$

You should check that this is indeed the inverse of A by multiplying the two matrices.

It would have been more efficient to solve all three systems at the same time, by using the **double matrix** $A|I$:

$$\begin{bmatrix} 0 & 3 & 1 & 1 & 0 & 0 \\ -1 & 3 & 0 & 0 & 1 & 0 \\ 1 & 0 & 2 & 0 & 0 & 1 \end{bmatrix}.$$

You can think of this double matrix as a matrix that has been augmented three times. The three columns on the right are e_1, e_2, and e_3, that is, they are the right-hand sides of the three systems of equations that we solved to find A^{-1}. Applying the same Gauss-Jordan steps as before, we obtain the following series of equivalent double matrices:

$$\begin{bmatrix} 0 & 3 & 1 & 1 & 0 & 0 \\ -1 & 3 & 0 & 0 & 1 & 0 \\ 1 & 0 & 2 & 0 & 0 & 1 \end{bmatrix} \sim \begin{bmatrix} 0 & 3 & 1 & 1 & 0 & 0 \\ 0 & 3 & 2 & 0 & 1 & 1 \\ 1 & 0 & 2 & 0 & 0 & 1 \end{bmatrix}$$

$$\sim \begin{bmatrix} 0 & 0 & -1 & 1 & -1 & -1 \\ 0 & 3 & 2 & 0 & 1 & 1 \\ 1 & 0 & 2 & 0 & 0 & 1 \end{bmatrix} \sim \begin{bmatrix} 0 & 0 & -1 & 1 & -1 & -1 \\ 0 & 3 & 0 & 2 & -1 & -1 \\ 1 & 0 & 0 & 2 & -2 & -1 \end{bmatrix}.$$

At this point, we have completed the elimination part of the algorithm. Now we divide each row by its leading nonzero entry, and switch the first and third rows to complete the procedure.

$$\sim \begin{bmatrix} 1 & 0 & 0 & 2 & -2 & -1 \\ 0 & 3 & 0 & 2 & -1 & -1 \\ 0 & 0 & -1 & 1 & -1 & -1 \end{bmatrix} \sim \begin{bmatrix} 1 & 0 & 0 & 2 & -2 & -1 \\ 0 & 1 & 0 & \frac{2}{3} & -\frac{1}{3} & -\frac{1}{3} \\ 0 & 0 & 1 & -1 & 1 & 1 \end{bmatrix}.$$

The three columns on the right are the solutions of the three systems of equations. They are the columns of A^{-1}, in agreement with the earlier calculation.

Let us summarize. We started with the double matrix $A|I$. We carried out a series of Gauss-Jordan steps, including row switches and multiplication of rows by constants, if necessary, until finally we obtained a double matrix whose left half is the identity I. The right half of this matrix is A^{-1}. In symbols, $A|I \sim \cdots \sim I|A^{-1}$.

This method of the preceding example will work for any n-by-n matrix A in the sense that if A has an inverse, the method will find it, and if A does not have an inverse, that fact will be discovered when Gauss-Jordan elimination leads to a zero row on the left.

EXAMPLE:

5. Let us try to find the inverse of

$$A = \begin{bmatrix} 1 & 2 \\ -3 & -6 \end{bmatrix}.$$

We form the double matrix $A|I$ and start to carry out the elimination procedure:

$$\left[\begin{array}{cc|cc} 1 & 2 & 1 & 0 \\ -3 & -6 & 0 & 1 \end{array} \right] \sim \left[\begin{array}{cc|cc} 1 & 2 & 1 & 0 \\ 0 & 0 & 3 & 1 \end{array} \right].$$

We added the first row three times to the second row. The left half of the double matrix now has a row of zeroes. We conclude that A is singular.

The inverse of a linear map L can be used to quickly solve equations of the form $L(X) = Y$.

EXAMPLE:

6. Let L be as in Examples 3 and 4. Let us find all solutions of $L(X) = (3, 5, -8)$. Since L is invertible, there is a unique solution $X = L^{-1}((3, 5, -8))$. We can find X by using the matrix found in Example 4:

$$\begin{bmatrix} 2 & -2 & -1 \\ \frac{2}{3} & -\frac{1}{3} & -\frac{1}{3} \\ -1 & 1 & 1 \end{bmatrix} \begin{bmatrix} 3 \\ 5 \\ -8 \end{bmatrix} = \begin{bmatrix} 4 \\ 3 \\ -6 \end{bmatrix}.$$

The solution is $(4, 3, -6)$. You should check that $L((4, 3, -6)) = (3, 5, -8)$.

We conclude this section with a useful theorem about the inverse of the composition of two linear maps.

Theorem 3.7.4 If L and M are invertible linear maps in $\mathbb{R}^n$, then so is ML. Conversely, if ML is invertible, then so are L and M. In either case, $(ML)^{-1} = L^{-1}M^{-1}$.

PROOF: First assume that L and M are invertible. Then

$$(L^{-1}M^{-1})(ML) = L^{-1}(M^{-1}M)L = L^{-1}IL = L^{-1}L = I.$$

This calculation verifies that ML is invertible and it also verifies the formula for the inverse of ML.

For the proof of the converse half of the theorem, suppose that ML is invertible. Then the nullity of ML is 0. Any vector that is in the null space

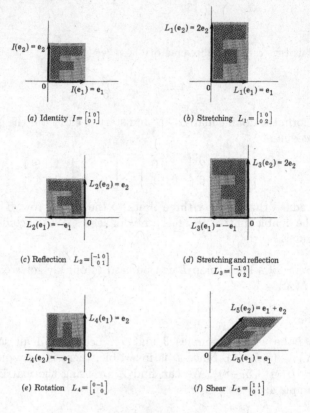

FIGURE 3.19.

of L is also in the null space of ML (can you see why?), so the nullity of L must also be 0. Thus, L is invertible. To show that M is also invertible, we use the first half of the theorem. Both ML and L^{-1} are invertible, so their composition $(ML)L^{-1}$ is invertible. Since $(ML)L^{-1} = M(LL^{-1}) = MI = M$, we are done. <<

Of course, there is a similar result for matrices. If A and B are invertible, then so is AB, and $(AB)^{-1} = B^{-1}A^{-1}$. If the product of two square matrices is invertible, then both matrices are invertible.

3.7B GEOMETRY IN $\mathbb{R}^2$

Let $L : \mathbb{R}^2 \to \mathbb{R}^2$ be a linear map. We know that L is determined by its effect on the standard basis $\mathbf{e}_1$ and $\mathbf{e}_2$ of $\mathbb{R}^2$. Figure 3.19 illustrates the geometric effect of some special linear maps in terms of their matrix representations. In each of the first ten cases the image of the region inside the unit square with corners at $(0,0)$, $(1,0)$, $(1,1)$, and $(0,1)$ is shown. We hope that you will ponder these pictures.

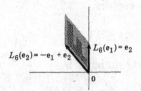

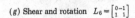

(g) Shear and rotation $L_6 = \begin{bmatrix} 0 & -1 \\ 1 & 1 \end{bmatrix}$

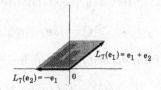

(h) Rotation and shear $L_7 = \begin{bmatrix} 1 & -1 \\ 1 & 0 \end{bmatrix}$

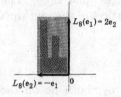

(i) Rotation and stretching
$L_8 = \begin{bmatrix} 0 & -1 \\ 2 & 0 \end{bmatrix}$

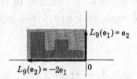

(j) Stretching and rotation
$L_9 = \begin{bmatrix} 0 & -2 \\ 1 & 0 \end{bmatrix}$

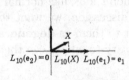

(k) Projection $L_{10} = \begin{bmatrix} 1 & 0 \\ 0 & 0 \end{bmatrix}$

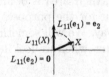

(l) Projection and rotation $L_{11} = \begin{bmatrix} 0 & 0 \\ 1 & 0 \end{bmatrix}$

FIGURE 3.19. (continued)

The region inside the unit square (boundary included) is the set of vectors in $\mathbb{R}^2$ of the form (s, t), where $0 \leq s, t \leq 1$. Let us use the symbol $\mathcal{Q}$ for this region. In terms of the standard basis vectors,

$$\mathcal{Q} = \{s\mathbf{e}_1 + t\mathbf{e}_2 : 0 \leq s, t \leq 1\}.$$

It follows from linearity that the image of $\mathcal{Q}$ under L is

$$L(\mathcal{Q}) = \{sL(\mathbf{e}_1) + tL(\mathbf{e}_2) : 0 \leq s, t \leq 1\}.$$

You should convince yourself that this set is the region inside the parallelogram in $\mathbb{R}^2$ generated by the vectors $L(\mathbf{e}_1)$ and $L(\mathbf{e}_2)$. The vertices of this parallelogram are the images of the vertices of the unit square, namely, $L(\mathbf{0})$, $L(\mathbf{e}_1)$, $L(\mathbf{e}_2)$, and $L(\mathbf{e}_1) + L(\mathbf{e}_2)$.

REMARK:

- If L is singular, then $L(\mathbf{e}_1)$ and $L(\mathbf{e}_2)$ both lie on the same line through the origin, so the "parallelogram" $L(\mathcal{Q})$ is just a line segment, or even just a single point in case L is the zero map. See for example Figure 3.19 (k) and (l). Such flattened parallelograms have no interior, their area is 0, and they don't look like normal parallelograms, but for the purposes of this discussion, we want to include them anyway. We will sometimes refer to them as **degenerate parallelograms**, when we feel it is necessary to emphasize their special nature.

We want to derive a formula for the area of $L(\mathcal{Q})$. Let $L(\mathbf{e}_1) = (a_{11}, a_{21})$ and $L(\mathbf{e}_2) = (a_{12}, a_{22})$. In Chapter 2, we learned that the area of the parallelogram generated by these two vectors is the norm of the cross product of the vectors $(a_{11}, a_{21}, 0)$ and $(a_{12}, a_{22}, 0)$:

$$
\begin{aligned}
\text{area of } L(\mathcal{Q}) &= \|(a_{11}, a_{12}, 0) \times (a_{21}, a_{22}, 0)\| \\
&= \|(a_{11}a_{22} - a_{12}a_{21})\mathbf{k}\| \\
&= |a_{11}a_{22} - a_{12}a_{21}|.
\end{aligned}
$$

This formula is correct even when the parallelogram is degenerate, since in that case, the two vectors lie on the same line through the origin and the corresponding cross product gives the zero vector.

The quantity $a_{11}a_{22} - a_{12}a_{21}$ (note the absence of an absolute value sign) is called the **determinant** of L and is denoted by $\det L$. If A is the matrix of L, then $\det L$ is also called the **determinant** of A and written $\det A$. We will also use the following notation for the determinant of A:

$$
\begin{vmatrix}
a_{11} & a_{12} \\
a_{21} & a_{22}
\end{vmatrix}.
$$

Don't be misled by this notation! Even though it looks similar to the notation for a matrix, it stands for the *number* $a_{11}a_{22} - a_{12}a_{21}$.

How does L affect the areas of figures other than Q?

Theorem 3.7.5 Let $\mathcal{R}$ be the region inside a parallelogram in $\mathbb{R}^2$. Then

$$\frac{\text{area of } L(\mathcal{R})}{\text{area of } \mathcal{R}} = |\det L|.$$

PROOF: First let us consider a special case. Assume that $\mathcal{R}$ is the region inside of the parallelogram in $\mathbb{R}^2$ generated by two vectors (b_{11}, b_{21}) and (b_{12}, b_{22}). Then, as noted above, $\mathcal{R}$ is the image of the unit square Q under the linear map M whose matrix is

$$\begin{bmatrix} b_{11} & b_{12} \\ b_{21} & b_{22} \end{bmatrix}.$$

It follows that $L(\mathcal{R})$ is the image of Q under the composition LM, that is, $L(\mathcal{R}) = LM(Q)$. We know how to compute the area of the image of Q under a linear map:

$$\text{area of } M(Q) = |\det M| \quad \text{and} \quad \text{area of } LM(Q) = |\det LM|.$$

It may be verified by direct (and slightly tedious) calculation with the corresponding matrices that the determinant of the composition of any two linear maps in $\mathbb{R}^2$ is the product of the determinants:

$$\det(LM) = (\det L)(\det M).$$

The desired result now follows for the special case under consideration.

In general, a parallelogram in $\mathbb{R}^2$ has vertices of the form X, $X + Y_1$, $X + Y_2$, and $X + Y_1 + Y_2$. Such a parallelogram is formed by taking the parallelogram generated by the vectors Y_1 and Y_2, and then shifting it along the vector X. The image under L of this parallelogram has vertices $L(X)$, $L(X) + L(Y_1)$, $L(X) + L(Y_1)$, and $L(X) + L(Y_1) + L(Y_2)$, which is the parallelogram generated by the vectors $L(Y_1)$ and $L(Y_2)$, shifted along the vector $L(X)$. Since shifting doesn't affect the areas of these parallelograms, we may assume that X is the zero vector. This puts us back into the special case with which we started the proof: $Y_1 = (b_{11}, b_{21})$ and $Y_2 = (b_{12}, b_{22})$. Done. $\cdot \ll$

REMARKS:

- The preceding theorem is valid for much more general regions $\mathcal{R}$. A more general treatment requires a rigorous definition of the area of such regions. Chapter 9 deals with such issues..

- It is natural to inquire about the significance of the sign of the determinant. Consider a counterclockwise walk around the unit square, starting at the origin, going then to $\mathbf{e}_1$, $\mathbf{e}_1 + \mathbf{e}_2$, and $\mathbf{e}_2$ in that order. As one walks counterclockwise around the image of the parallelogram beginning at $L(\mathbf{0})$, one either goes to the vertices $L(\mathbf{e}_1)$, $L(\mathbf{e}_1)+L(\mathbf{e}_2)$, and $L(\mathbf{e}_2)$ in the order listed or in the reverse order. The first of these possibilities occurs if the determinant is positive and the second occurs if the determinant is negative.

- If L is singular, it maps the unit square to a figure with area 0, and its determinant is 0. If L is not singular, then $\mathcal{I}(L) = \mathbb{R}^2$. Since $\mathcal{I}(L)$ is spanned by the vectors $L(\mathbf{e}_1)$ and $L(\mathbf{e}_2)$, it follows that those two vectors do not lie on the same line through the origin, hence they generate a parallelogram with positive area and, hence, the determinant of L is nonzero. The conclusion is that L is singular if and only if its determinant is zero.

EXAMPLES:

7. Here are some matrices whose determinants have absolute value equal to 1:

$$I, \; -I, \; \begin{bmatrix} -1 & 0 \\ 0 & 1 \end{bmatrix}, \; \begin{bmatrix} 0 & -1 \\ 1 & 0 \end{bmatrix}, \; \begin{bmatrix} \frac{1}{2} & 0 \\ 0 & 2 \end{bmatrix}, \; \begin{bmatrix} 1 & 1 \\ 0 & 1 \end{bmatrix}.$$

Theorem 3.7.5 says that if $\mathcal{R}$ is a parallelogram in $\mathbb{R}^2$ and if L is the linear map corresponding to one of the matrices given above, then $L(\mathcal{R})$ has the same area as $\mathcal{R}$. The first matrix is the identity, the second corresponds to a 180 degree rotation, the third to reflection about the vertical axis, and the fourth to a 90 degree counterclockwise rotation. It should be obvious to you that area is preserved for these first four maps, as well as for other rotations and reflections. We will have more to say about these types of linear maps in Section 3.8. The fifth matrix corresponds to shrinking in the horizontal direction and stretching in the vertical direction. The sixth is the shear map of Figure 3.18. You may have to think a bit to see why area is preserved for these last two cases. The fact that area is preserved by shear maps like the one given is a special case of "Cavalieri's principle" from geometry.

8. For any linear map L in $\mathbb{R}^2$ and any scalar a, the determinant of aL is a^2 times $\det L$, as you may easily check by direct computation using the formula for the determinant. Thus, for example, if $\mathcal{R}$ is a parallelogram in $\mathbb{R}^2$, then the area of $2L(\mathcal{R})$ is four times the area of $L(\mathcal{R})$.

3.7C   GEOMETRY IN $\mathbb{R}^3$

The situation in $\mathbb{R}^3$ is similar to that in $\mathbb{R}^2$. The images of cubes under a linear map are parallelepipeds, the ratio of volumes equals the absolute value of what is called the determinant of L, and L is singular if and only if its determinant equals 0. Similar statements hold in $\mathbb{R}^n$.

We will not give a general treatment of determinants and volume in $\mathbb{R}^n$, but we are equipped to treat $\mathbb{R}^3$. Let $\mathcal{Q}$ denote the region inside the unit cube generated by the standard basis vectors. Reasoning similar to that used in Section 3.7b shows that $L(\mathcal{Q})$ is the region inside the parallelepiped generated by $L(e_1)$, $L(e_2)$, and $L(e_3)$. In Section 2.4, we showed that the volume of this parallelepiped is given by

$$\text{volume of } L(\mathcal{Q}) \; = \; |[L(e_1), L(e_2), L(e_3)]| \, ,$$

that is, the volume is the absolute value of the triple product of the vectors that generate the parallelepiped. Thus, by analogy with what was done in $\mathbb{R}^2$, we drop the absolute value signs and define

$$\det L = [L(e_1), L(e_2), L(e_3)] \, .$$

Thus, we have defined $\det L$ so that the ratio of the volume of $L(\mathcal{Q})$ to the volume of $\mathcal{Q}$ is $|\det L|$. More generally, we have

Theorem 3.7.6 Let $\mathcal{R}$ be the region inside a parallelepiped in $\mathbb{R}^3$. Then

$$\frac{\text{volume of } L(\mathcal{R})}{\text{volume of } \mathcal{R}} = |\det L| \, .$$

The proof is similar to the proof of Theorem 3.7.5.

The determinant of a 3-by-3 matrix A is defined to be the determinant of the linear map L whose matrix is A. If

$$A = \begin{bmatrix} a_{11} & a_{12} & a_{13} \\ a_{21} & a_{22} & a_{23} \\ a_{31} & a_{32} & a_{33} \end{bmatrix} ,$$

then we use the following notation

$$\det A = \begin{vmatrix} a_{11} & a_{12} & a_{13} \\ a_{21} & a_{22} & a_{23} \\ a_{31} & a_{32} & a_{33} \end{vmatrix} .$$

Recall that when A is the matrix of L, the columns of A are the vectors $L(e_1)$, $L(e_2)$, and $L(e_3)$. Thus, for A given as above, the formula for the triple product yields the following formula for $\det A$:

$$\det A = (a_{21}a_{32} - a_{22}a_{31})a_{13} - (a_{11}a_{32} - a_{12}a_{31})a_{23} + (a_{11}a_{22} - a_{12}a_{21})a_{33} \, .$$

REMARKS:

- For linear maps L and M in $\mathbb{R}^3$, it can be verified by direct, very tedious, calculation that $\det(LM) = (\det L)(\det M)$. This fact is needed in the proof of Theorem 3.7.6. A similar fact holds for determinants of linear maps in $\mathbb{R}^n$.

- As in $\mathbb{R}^2$, the sign of the determinant has significance. If $\mathbf{e}_1$, $\mathbf{e}_2$, and $\mathbf{e}_3$ is a right-handed coordinate system, then $L(\mathbf{e}_1)$, $L(\mathbf{e}_2)$, and $L(\mathbf{e}_3)$ is a right- or left-handed coordinate system according to whether the determinant of L is positive or negative.

- Direct computation shows that

$$
\begin{vmatrix}
a_{11} & a_{12} & a_{13} \\
a_{21} & a_{22} & a_{23} \\
a_{31} & a_{32} & a_{33}
\end{vmatrix}
=
\begin{vmatrix}
a_{11} & a_{21} & a_{31} \\
a_{12} & a_{22} & a_{32} \\
a_{13} & a_{23} & a_{33}
\end{vmatrix} .
$$

In other words, if we use the columns of a matrix A to form the rows of a matrix B, then $\det A = \det B$. This comment will help you to reconcile our notation for the determinant with the notation found in Section 2.4, where, in the formula for the triple product, the coordinates of the three vectors were written in rows, rather than columns.

- You may also verify by direct calculation that

$$
\begin{vmatrix}
a_{11} & a_{12} & a_{13} \\
a_{21} & a_{22} & a_{23} \\
a_{31} & a_{32} & a_{33}
\end{vmatrix}
$$

$$
= a_{11}
\begin{vmatrix}
a_{22} & a_{23} \\
a_{32} & a_{33}
\end{vmatrix}
- a_{12}
\begin{vmatrix}
a_{21} & a_{23} \\
a_{31} & a_{33}
\end{vmatrix}
+ a_{13}
\begin{vmatrix}
a_{21} & a_{22} \\
a_{31} & a_{32}
\end{vmatrix} .
$$

For example,

$$
\begin{vmatrix}
1 & -2 & 0 \\
1 & 1 & 3 \\
2 & 0 & -1
\end{vmatrix}
= 1
\begin{vmatrix}
1 & 3 \\
0 & -1
\end{vmatrix}
- (-2)
\begin{vmatrix}
1 & 3 \\
2 & -1
\end{vmatrix}
+ 0 = -15 .
$$

This formula is called an "expansion by cofactors". The (i,j) **cofactor matrix** of an n-by-n matrix A is the $(n-1)$-by-$(n-1)$ matrix $A^{(i,j)}$ obtained by crossing out row i and column j of A. Using this notation, the above formula becomes

$$
\det A = a_{11} \det A^{(1,1)} - a_{12} \det A^{(1,2)} + a_{13} \det A^{(1,3)} .
$$

There are also other formulas for $\det A$ involving cofactors. We will not pursue the subject here. You should ask your teacher for the complete story if you are interested, or take a more advanced course in linear algebra. We will simply note that the cofactor expansion is used to obtain a formula for the determinants of general square matrices. If A is an n-by-n matrix, then it turns out that

$$\det A = a_{11} \det A^{(1,1)} - a_{12} \det A^{(1,2)} + \cdots + (-1)^{n-1} a_{1n} \det A^{(1,n)} .$$

For example, since you now know how to compute the determinants of 3-by-3 matrices, you could use this formula to compute the determinant of a 4-by-4 matrix.

Exercises

1. Let L be a linear map in $\mathbb{R}^2$, the matrix of which is

$$A = \begin{bmatrix} 5 & 0 \\ 0 & 6 \end{bmatrix} .$$

   (a) Calculate A^{-1} if it exists.

   (b) Solve the equation $L(X) = (12, -13)$.

   (c) Does the linear map L have an inverse map? If so, calculate its matrix, and also find $L^{-1}((12, -13))$.

2. For each of the following matrices compute its inverse or conclude that it does not have an inverse:

   (a) $A = \begin{bmatrix} -2 & 1 \\ 0 & 1 \end{bmatrix}$

   (b) $B = \begin{bmatrix} -2 & 1 \\ 1 & 1 \end{bmatrix}$

   (c) $C = \begin{bmatrix} -2 & 0 \\ 0 & 1 \end{bmatrix}$

   (d) $D = \begin{bmatrix} -2 & 1 \\ 1 & 3 \end{bmatrix}$

   (e) $E = \begin{bmatrix} 21 & -51 \\ -28 & 68 \end{bmatrix}$

   (f) $F = \begin{bmatrix} 0 & -7 \\ 12 & 19 \end{bmatrix} .$

3. (a) Solve the system

$$\begin{aligned} -2x_1 + x_2 &= 5 \\ x_1 + x_2 &= -7 . \end{aligned}$$

   (b) Solve for x_1 and x_2 in terms of y_1 and y_2 in the system

$$\begin{aligned} -2x_1 + x_2 &= y_1 \\ x_1 + x_2 &= y_2 . \end{aligned}$$

4. Use one of the answers of Exercise 2 to quickly redo Exercise 3.

5. Do this exercise to become a lightning-fast inverter of 2-by-2 matrices. Let

$$A = \begin{bmatrix} a & b \\ c & d \end{bmatrix}$$

and let

$$\Delta = ad - bc ,$$

the determinant of A. Show by direct computation that A is invertible $\Leftrightarrow$ $\Delta \neq 0$, and that

$$A^{-1} = \frac{1}{\Delta} \begin{bmatrix} d & -b \\ -c & a \end{bmatrix} = \begin{bmatrix} \frac{d}{\Delta} & -\frac{b}{\Delta} \\ -\frac{c}{\Delta} & \frac{a}{\Delta} \end{bmatrix}$$

if $\Delta \neq 0$.

6. Use the preceding exercise to do the following:

   (a) Decide if

$$\begin{bmatrix} 3 & 5 \\ 1 & 3 \end{bmatrix}$$

   has an inverse and to calculate it in case it does.

   (b) Redo Exercise 2.

7. Given $L: \mathbb{R}^2 \to \mathbb{R}^2$, compute L^{-1} and write it in terms of the standard basis, that is, write $L^{-1}(e_1)$ and $L^{-1}(e_2)$ in terms of e_1 and e_2:

   (a) $L(e_1) = e_2$, $L(e_2) = -e_1$

   (b) $L(e_1) = e_1$, $L(e_2) = e_1 + e_2$

   (c) $L(e_1) = ae_1 + ce_2$, $L(e_2) = be_1 + de_2$, under the assumption that $ad - bc \neq 0$.

8. For each part of the preceding exercise solve $L(X) = 1728e_1 - 1984e_2$ for X.

9. (a) What is the inverse of the 1-by-1 matrix $[5]$?

   (b) Solve $5x = 12$.

10. Which 1-by-1 matrices do not have inverses? *Hint:* most do.

11. Let A be an n-by-n (square) matrix and denote the corresponding linear transformation by L. Decide whether the following are true or false.

   (a) If A^{-1} exists and Y is given, then there exists precisely one X such that $L(X) = Y$.

   (b) If A^{-1} exists and $A[X] = [0]$, then $X = \mathbf{0}$.

   (c) If $L(X_1) = L(X_2)$ for $X_1 \neq X_2$, then there exists a nonzero X such that $L(X) = \mathbf{0}$.

   (d) If $L(X_1) = L(X_2)$ for $X_1 \neq X_2$, then A is singular.

   (e) If $A[X] = \mathbf{0}$ for some $X \neq \mathbf{0}$, then L does not have an inverse.

   (f) If $A[X_1] = A[X_2]$ with $X_1 \neq X_2$, then A does not have an inverse.

(g) If A does not have an inverse, then A is the zero matrix.

(h) If $A^{-1} = A$, then $A = I$ or $A = -I$.

12. Compute A^{-1} or conclude that it does not exist:

(a) $A = \begin{bmatrix} 1 & 2 & 0 \\ 0 & 1 & -1 \\ 0 & 0 & 1 \end{bmatrix}$

(b) $A = \begin{bmatrix} 0 & -3 & 2 \\ -2 & 1 & 3 \\ -4 & -1 & 8 \end{bmatrix}$

(c) $A = \begin{bmatrix} 2 & 0 & 3 \\ 0 & 1 & 0 \\ 1 & 0 & 2 \end{bmatrix}$

(d) $A = \begin{bmatrix} 1 & 1 & 0 & 0 \\ 0 & 1 & 1 & 0 \\ 0 & 0 & 1 & 1 \\ 0 & 0 & 0 & 1 \end{bmatrix}$

(e) $A = \begin{bmatrix} 0 & 0 & 0 & 0 & 0 & 1 \\ 1 & 0 & 0 & 0 & 0 & 0 \\ 0 & 1 & 0 & 0 & 0 & 0 \\ 0 & 0 & 1 & 0 & 0 & 0 \\ 0 & 0 & 0 & 1 & 0 & 0 \\ 0 & 0 & 0 & 0 & 1 & 0 \end{bmatrix}$.

13. Compute the determinants of the matrices in parts (a) through (c) of the preceding exercise. Check that you get the value 0 for exactly those matrices which you showed to be singular in the preceding exercise.

14. Let

$$A = \begin{bmatrix} \frac{1}{2} & -\frac{\sqrt{3}}{2} \\ \frac{\sqrt{3}}{2} & \frac{1}{2} \end{bmatrix} \quad \text{and} \quad B = \begin{bmatrix} 5 & \sqrt{3} \\ \sqrt{3} & 3 \end{bmatrix} .$$

(a) Compute A^{-1}.

(b) Compute ABA^{-1}.

(c) Compute $(ABA^{-1})^{99}$.

(d) Show that $(ABA^{-1})^3 = AB^3 A^{-1}$.

(e) Compute B^3.

(f) Compute B^{99}.

15. Given two 2-by-2 matrices A and B such that $AB = O$, which of the following statements are *necessarily* true? Give a proof or counterexample:

(a) $A = O$.

(b) $B = O$.

(c) Either $A = O$ or $B = O$.

(d) A is singular.

(e) If A^{-1} exists, then $B = O$.

(f) $BA = O$.

(g) If $A \neq O$ and $B \neq O$, then neither A nor B is invertible.

16. Here is an excursion in matrix equations. Often one meets the following kind of problem: Given matrices A and B, find C so that $AC = B$, assuming that A, B, and C are n-by-n and A is invertible. To solve this, we multiply both sides of $AC = B$ on the *left* by A^{-1}, to give $A^{-1}AC = A^{-1}B$, that is, $C = A^{-1}B$. Use this method and slight variations of it to solve the following equations for C, where

$$A = \begin{bmatrix} 2 & 3 \\ 1 & 2 \end{bmatrix} \quad \text{and} \quad B = \begin{bmatrix} 1 & -1 \\ 1 & 0 \end{bmatrix}.$$

(a) $AC = B$

(b) $CA = B$

(c) $BC = O$

(d) $CA = A$

(e) $ACA = I$

(f) $ACB = A + B$.

17. Find a 2-by-2 matrix A such that $A^2 = -I$.

18. If a square matrix A satisfies $A^2 - 2A = I$, find A^{-1} in terms of A. *Hint:* By inspection find B such that $BA = I$.

19. If a square matrix A satisfies $A^4 = O$, verify that $(I - A)^{-1} = I + A + A^2 + A^3$. Find $(I + A)^{-1}$ too.

20. Let L denote a linear transformation in $\mathbb{R}^n$. Decide whether the following are true or false.

(a) If L is invertible, then every equation $L(X) = Y$ has a unique solution given by $X = L^{-1}(Y)$.

(b) If L is invertible, then L^{-1} is invertible.

(c) If $L(X) = 5X$ for all X, then $L^{-1}(X) = \frac{1}{5}X$ for all X.

(d) If $n = 2$ and L rotates the plane $90°$ counterclockwise, then L^{-1} rotates the plane $90°$ clockwise.

(e) If $n = 2$ and L rotates the plane $90°$ counterclockwise, then L^{-1} rotates the plane $270°$ counterclockwise.

(f) The zero map O is invertible.

(g) If L is invertible, then $L^{-1} \neq O$.

(h) The identity map I is the only map that is its own inverse.

(i) If L is invertible, then L^{-1} is also linear.

21. Let A be an invertible matrix. Define $A^0 = I$. Why does law of exponents $A^{(r+s)} = A^r A^s$ hold for all integers r and s, positive, negative, and 0?

22. Let $T(X) = L(X) + Y_0$ be an affine map from $\mathbb{R}^n$ into $\mathbb{R}^n$ which we also call an affine map or affine transformation in $\mathbb{R}^n$. The transformation T is said to be **invertible** if and only if there is a map S such that for all X and Y, $ST(X) = X$ and $TS(Y) = Y$, in which case S is called the **inverse transformation** of T and we write $S = T^{-1}$. Prove the following facts. (*Hint:* Use both geometric and algebraic thinking. For instance, view T as the result of first using L and then translating by the constant vector Y_0.)

   (a) T is invertible $\Leftrightarrow$ L is invertible.

   (b) T is invertible $\Leftrightarrow$ $\mathcal{I}(T) = \mathbb{R}^n$.

   (c) If T is invertible, then $T^{-1}(Y) = L^{-1}(Y - Y_0)$ for every $Y \in \mathbb{R}^n$.

   (d) If T is invertible, then $T^{-1}(Y) = L^{-1}(Y) - L^{-1}(Y_0)$ for every $Y \in \mathbb{R}^n$.

   (e) If T is invertible, its inverse map is an affine map.

   (f) If there is a map $S : \mathbb{R}^n \to \mathbb{R}^n$ such that $ST = I$, then T is invertible.

   (g) If there is a map $S : \mathbb{R}^n \to \mathbb{R}^n$ such that $TS = I$, then T is invertible.

23. For each affine transformation T decide if it has an inverse and, if so, write the inverse in the form $T^{-1}(Y) = M(Y) + X_0$, where M is a linear transformation, and calculate $[M]$.

   (a) $T(X) = L(X) + (1, 0)$, where

   $$[L] = \begin{bmatrix} 1 & 5 \\ -3 & 4 \end{bmatrix}$$

   (b) $T(X) = L(X) + (1, -1)$, where

   $$[L] = \begin{bmatrix} -9 & 15 \\ 33 & -55 \end{bmatrix}$$

   (c)

   $$[T(X)] = \begin{bmatrix} 1 & -3 & 3 \\ 3 & -2 & 2 \\ 0 & 4 & -5 \end{bmatrix} [X] + \begin{bmatrix} 1 \\ 0 \\ -2 \end{bmatrix}$$

   (d) $T(X) = O(X) + e_1$

   (e) $T(X) = I(X) + e_1$

   (f) $T(X) = L(X) + (-1, 0, -1, 0)$, where

   $$[L] = \begin{bmatrix} 0 & 1 & 0 & 0 \\ 1 & 0 & 0 & 0 \\ 0 & 0 & 0 & 1 \\ 0 & 0 & 1 & 0 \end{bmatrix}.$$

24. For each part in the preceding exercise find those points which are not moved by T, that is, find those X for which $T(X) = X$.

25. Let T denote an affine transformation in $\mathbb{R}^n$. Those points X for which $T(X) = X$ are called **fixed points** of T.

   (a) Prove that if T is linear, then the set of fixed points of T is a linear subspace.

   (b) Prove that the set of fixed points of T is either empty or an affine subspace (whether or not T is linear).

26. There are three classes of affine transformations in $\mathbb{R}^1$: those that have no fixed points, those that have a single fixed point, and those for which every point is a fixed point. Identify which affine transformations belong to which of the three classes.

27. For each of the affine transformations in $\mathbb{R}^2$ described below sketch the image of the square region $\mathcal{R}$ with vertices at $(0,0)$, $(1,0)$, $(1,1)$, and $(0,1)$. Then use 2-by-2 determinants to calculate the area of the image of $\mathcal{R}$ and to also draw arrows on the boundary of that image to indicate the direction that corresponds to the counterclockwise direction around $\mathcal{R}$.

   (a) $T(\mathbf{0}) = -\mathbf{e}_2$, $T(\mathbf{e}_1) = \mathbf{e}_1$ $T(\mathbf{e}_2) = \mathbf{e}_2$

   (b) $T(\mathbf{0}) = \mathbf{0}$, $T(\mathbf{e}_1) = \mathbf{e}_2$, $T(\mathbf{e}_2) = \mathbf{e}_1$

   (c) $T(X) = \mathbf{e}_1 + 2X$ for all X

   (d) $T(\mathbf{0}) = \mathbf{0}$, $T(\mathbf{e}_1) = \mathbf{e}_1$ $T(\mathbf{e}_2) = 10\mathbf{e}_1 + \mathbf{e}_2$

   (e) $T((0,0)) = (1,0)$, $T((1,0)) = (1,1)$, $T((0,1)) = (0,0)$

   (f) $T(\mathbf{0}) = a\mathbf{e}_1$, $T(\mathbf{e}_1) = b\mathbf{e}_1$, $T(\mathbf{e}_2) = c\mathbf{e}_2$ where a, b, and c are positive constants with $a < b$

   (g) $T(\mathbf{0}) = a\mathbf{e}_1$, $T(\mathbf{e}_1) = b\mathbf{e}_1$, $T(\mathbf{e}_2) = c\mathbf{e}_2$ where a, b, and c are positive constants with $a > b$.

28. Use the determinant you have calculated for each affine transformation in the preceding exercise to calculate the area of the image of the region inside the parallelogram with vertices at $(1,2)$, $(5,7)$, $(3,8)$, and $(-1,3)$.

29. Fill in the blank. The affine transformation T in $\mathbb{R}^3$ given by $T(X) = cX + Y_0$, where c is a scalar, multiplies volumes by a factor of _____.

30. Let L denote the linear transformation in $\mathbb{R}^3$ with matrix

$$[L] = \begin{bmatrix} 3 & -2 & 0 \\ 1 & 2 & -2 \\ 4 & 1 & -2 \end{bmatrix}.$$

Calculate the volume of $L(\mathcal{R})$ for $\mathcal{R} =$

   (a) the cubical region, three of whose edges are the vectors $\mathbf{e}_1$, $\mathbf{e}_2$, and $\mathbf{e}_3$

   (b) the interior of the parallelepiped, three of whose edges are the line segments connecting $(0,0,0)$ to each of $(1,1,1)$, $(1,1,0)$, and $(1,0,0)$

   (c) the interior of the parallelepiped, three of whose edges are the segments connecting $\mathbf{0}$ to each of $\mathbf{e}_1 + \mathbf{e}_2$, $\mathbf{e}_1 + \mathbf{e}_3$, and $\mathbf{e}_2 + \mathbf{e}_3$

   (d) the interior of the parallelepiped whose vertices are $(-1,2,3)$, $(3,3,5)$, $(4,5,4)$, $(0,4,2)$, $(1,-1,0)$, $(5,0,2)$, $(6,2,1)$, and $(2,1,-1)$. *Hint:* First sort out which vertices of the parallelepiped are connected by edges.

31. Let $A = [a]$ be the 1-by-1 matrix whose single element is the number a. We define $\det A = a$. Why does this definition seem reasonable in light of the definitions of determinants of 2-by-2 and 3-by-3 matrices? Think in terms of the length of the image of the unit interval $[0, 1]$ under the linear transformation corresponding to A. What significance does the sign of the determinant have?

32. This exercise concerns properties of determinants. Prove the following for 2-by-2 matrices A and B. (If you have the interest and computational fortitude, prove them for 3-by-3 matrices as well.)

    (a) $\det I = 1$.

    (b) Let B be obtained from A by switching two rows of A. Then $\det B = -\det A$.

    (c) Let B be obtained from A by multiplying one of the rows of A by a scalar c. Then $\det B = c \det A$.

    (d) Let B be obtained from A by adding a scalar multiple of one of the rows of A to the another row of A. Then $\det B = \det A$. (Note that each of the preceding three properties corresponds to different types of changes in a coefficient matrix that are allowed in Gaussian elimination.)

    (e) $\det(AB) = (\det A)(\det B)$.

    (f) If A is invertible, then $\det A^{-1} = 1/(\det A)$.

These may all be done by direct computation. However, for the last one, you might try using the identity in the preceding part. Corresponding properties hold for the determinants of n-by-n matrices as well. They are quite important and all have geometric interpretations, which you may enjoy trying to discover.

3.8   Isometries

3.8A   FUNDAMENTAL CONCEPTS

An **isometry** in $\mathbb{R}^n$ is an affine map from $\mathbb{R}^n$ to $\mathbb{R}^n$ that preserves the distances between points. In other words, an affine map $T : \mathbb{R}^n \to \mathbb{R}^n$ is called an isometry if for all vectors $X, Y \in \mathbb{R}^n$,

$$\|X - Y\| = \|T(X) - T(Y)\| .$$

As we will see, it turns out that some isometries in $\mathbb{R}^n$ correspond to actual motions (such as translations and rotations) that can be carried out within $\mathbb{R}^n$. Such isometries are known as "rigid motions".

EXAMPLES:

1. Let Z be a fixed vector in $\mathbb{R}^n$, and define $T : \mathbb{R}^n \to \mathbb{R}^n$ by $T(X) = X + Z$. The function T is obviously an isometry, since

$$\|T(X) - T(Y)\| = \|(X + Z) - (Y + Z)\| = \|X - Y\| \,.$$

The isometry T corresponds to the action of shifting every vector in $\mathbb{R}^n$ by adding the vector Z. It is called **translation** by Z. Translations are rigid motions.

2. Let T be the linear map in $\mathbb{R}^2$ whose matrix is

$$\begin{bmatrix} 0 & -1 \\ 1 & 0 \end{bmatrix} \,.$$

We have seen this map before. It corresponds to the action of rotating every vector in $\mathbb{R}^2$ counterclockwise 90 degrees about the origin. For $X = (x_1, x_2)$ and $Y = (y_1, y_2)$, we compute

$$\begin{aligned} \|T(X) - T(Y)\| &= \|(-x_2, x_1) - (-y_2, y_1)\| \\ &= \sqrt{(x_2 - y_2)^2 + (x_1 - y_1)^2} = \|X - Y\| \,. \end{aligned}$$

Thus, T is an isometry. In general, all linear maps corresponding to rotations about the origin are isometries. They are also rigid motions, as we will see.

3. Let T be the linear map in $\mathbb{R}^2$ with matrix

$$\begin{bmatrix} -1 & 0 \\ 0 & 1 \end{bmatrix} \,.$$

We leave it to you to check that $\|T(X) - T(Y)\| = \|X - Y\|$ for all $X, Y \in \mathbb{R}^2$. The isometry T corresponds to the action of reflection about the vertical axis in $\mathbb{R}^2$. We will find that reflection is *not* a rigid motion. There is no motion that can occur *within* $\mathbb{R}^2$ that corresponds to the action of T.

It is rather clear that if $T(X) = L(X) + Z$ for every X, where L is a linear transformation in $\mathbb{R}^n$, then either both L and the affine transformation T are isometries or neither is, since

$$\|T(X) - T(Y)\| = \|(L(X) + Z) - (L(Y) + Z)\| = \|L(X) - L(Y)\| \,.$$

Therefore, in the sections that follow, we will focus on linear isometries rather than general affine isometries, although a few exercises will treat the affine case.

We conclude this introductory section with an important criterion for linear isometries.

Theorem 3.8.1 Let L be a linear map in $\mathbb{R}^n$. Then the following three conditions are equivalent:

1. L is an isometry in $\mathbb{R}^n$.

2. For all $X \in \mathbb{R}^n$, $\|L(X)\| = \|X\|$.

3. For all $X, Y \in \mathbb{R}^n$, $\langle L(X), L(Y) \rangle = \langle X, Y \rangle$.

PROOF: Let L be a linear isometry in $\mathbb{R}^n$. Then for $X \in \mathbb{R}^n$,

$$\|L(X)\| = \|L(X - 0)\| = \|L(X) - L(0)\| = \|X - 0\| = \|X\|.$$

On the other hand, suppose that L is a linear map in $\mathbb{R}^n$ such that $\|L(X)\| = \|X\|$ for all $X \in \mathbb{R}^n$. Then

$$\|L(X) - L(Y)\| = \|L(X - Y)\| = \|X - Y\|,$$

so L is a linear isometry. We have thus far shown that the first two conditions are equivalent to each other. We will complete the proof by showing that the second and third conditions are equivalent to each other.

Since the norm of a vector is determined by its inner product with itself, the third condition obviously implies the second condition. It remains therefore to show that if L satisfies the second condition, then it also satisfies the third condition. Suppose that L satisfies the second condition. Fix $X, Y \in \mathbb{R}^n$. Then

$$\begin{aligned}
\|X - Y\|^2 &= \|L(X - Y)\|^2 \\
&= \langle L(X - Y), L(X - Y) \rangle \\
&= \langle L(X) - L(Y), L(X) - L(Y) \rangle \\
&= \langle L(X), L(X) \rangle - 2\langle L(X), L(Y) \rangle + \langle L(Y), L(Y) \rangle \\
&= \|L(X)\|^2 + \|L(Y)\|^2 - 2\langle L(X), L(Y) \rangle \\
&= \|X\|^2 + \|Y\|^2 - 2\langle L(X), L(Y) \rangle.
\end{aligned}$$

We also have

$$\|X - Y\|^2 = \langle X - Y, X - Y \rangle = \|X\|^2 + \|Y\|^2 - 2\langle X, Y \rangle.$$

Upon equating the two expressions, we find that $\langle L(X), L(Y) \rangle = \langle X, Y \rangle$, as desired. Done. $\ll$

REMARKS:

- Since the angle between two vectors in $\mathbb{R}^n$ is defined in terms of the inner product and the norms of the two vectors, it follows from the theorem that linear isometries preserve angles. Since translations also preserve angles, all isometries preserve angles.

- In the preceding proof we showed that the first two conditions are equivalent and that the second and third conditions are equivalent. This means that the first two conditions are either both true or both false, and similarly for the second and third conditions. Therefore, the first and third conditions are either both true or both false, that is, they are equivalent. This reasoning explains the lack of a direct argument in the proof that the first and third conditions are equivalent.

3.8B ROTATIONS AND REFLECTIONS OF $\mathbb{R}^2$

In this section we will find and classify all the linear isometries in $\mathbb{R}^2$. Let L be a linear isometry in $\mathbb{R}^2$. By Theorem 3.8.1, $\langle L(\mathbf{e}_1), L(\mathbf{e}_2) \rangle = \langle \mathbf{e}_1, \mathbf{e}_2 \rangle = 0$. Also, $\|L(\mathbf{e}_1)\| = \|\mathbf{e}_1\| = 1$, and similarly for $L(\mathbf{e}_2)$. Thus, $L(\mathbf{e}_1)$ and $L(\mathbf{e}_2)$ are perpendicular unit vectors. Let θ be the angle between $\mathbf{e}_1$ and $L(\mathbf{e}_1)$, measured counterclockwise from $\mathbf{e}_1$. Then

$$L(\mathbf{e}_1) = (\cos\theta, \sin\theta) \ .$$

It follows, from the requirements that $L(\mathbf{e}_1)$ and $L(\mathbf{e}_2)$ be perpendicular and that $L(\mathbf{e}_2)$ be a unit vector, that

$$L(\mathbf{e}_2) = (\cos(\theta \pm \pi/2), \sin(\theta \pm \pi/2)) = \pm(-\sin\theta, \cos\theta) \ .$$

Thus, the matrix of L takes one of the following two forms:

$$\begin{bmatrix} \cos\theta & -\sin\theta \\ \sin\theta & \cos\theta \end{bmatrix}, \quad \begin{bmatrix} \cos\theta & \sin\theta \\ \sin\theta & -\cos\theta \end{bmatrix} \ .$$

You may check by direct computation that for any angle θ, if L is the linear map whose matrix is one of the two matrices above, then for all $X, Y \in \mathbb{R}^2$, $\|L(X) - L(Y)\| = \|X - Y\|$. (This computation relies chiefly on the fact that $\cos^2\theta + \sin^2\theta = 1$.) We have thus found that all linear isometries in $\mathbb{R}^2$ are of one of two types, corresponding to the two types of matrices given above. In the discussion that follows, we will informally and temporarily refer to these two types of linear isometries as the "first type" and the "second type", respectively.

Let L_θ be an isometry of the first type, with matrix given in terms of the angle θ as above. As already noted, θ is the angle between $\mathbf{e}_1$ and $L_\theta(\mathbf{e}_1)$, measured counterclockwise from $\mathbf{e}_1$. Since $L(\mathbf{e}_2) = (\cos(\theta + \pi/2), \sin(\theta + \pi/2))$, it is easily seen that θ is also the angle between $\mathbf{e}_2$, which can be written as $(\cos(\pi/2), \sin(\pi/2))$, and $L(\mathbf{e}_2)$, measured counterclockwise from $\mathbf{e}_2$. Since each of the basis vectors is rotated counterclockwise through the angle θ, it follows that L_θ rotates every vector in $\mathbb{R}^2$ counterclockwise about the origin through the angle θ.

A counterclockwise rotation about the origin through an angle θ_1 followed by a counterclockwise rotation about the origin through an angle θ_2

results in a counterclockwise rotation about the origin through an angle $\theta_1 + \theta_2$. Thus,

$$\begin{bmatrix} \cos\theta_2 & -\sin\theta_2 \\ \sin\theta_2 & \cos\theta_2 \end{bmatrix} \begin{bmatrix} \cos\theta_1 & -\sin\theta_1 \\ \sin\theta_1 & \cos\theta_1 \end{bmatrix}$$
$$= \begin{bmatrix} \cos(\theta_1 + \theta_2) & -\sin(\theta_1 + \theta_2) \\ \sin(\theta_1 + \theta_2) & \cos(\theta_1 + \theta_2) \end{bmatrix}.$$

You may want to carry out the multiplication in order to obtain quick proofs of the standard formulas for $\sin(\theta_1 + \theta_2)$ and $\cos(\theta_1 + \theta_2)$.

EXAMPLES:

4. We are already familiar with three isometries of the first type, namely, $L_0, L_{\pi/2}$, and L_π. As you may easily check, $L_0 = I, L_\pi = -I$, and $L_{\pi/2}$ is the 90 degrees counterclockwise rotation that we have seen before in several examples.

5. Let $\theta_1 = \pi/3$ and $\theta_2 = \pi/4$. Then the composition $L_{\theta_2} \circ L_{\theta_1}$ corresponds to a counterclockwise rotation through $\pi/3$ radians, followed by a counterclockwise rotation through $\pi/4$ radians, which is a counterclockwise rotation through $7\pi/12$ radians. The corresponding matrix product relation is

$$\begin{bmatrix} \frac{\sqrt{2}}{2} & -\frac{\sqrt{2}}{2} \\ \frac{\sqrt{2}}{2} & \frac{\sqrt{2}}{2} \end{bmatrix} \begin{bmatrix} \frac{1}{2} & -\frac{\sqrt{3}}{2} \\ \frac{\sqrt{3}}{2} & \frac{1}{2} \end{bmatrix} = \begin{bmatrix} \frac{\sqrt{2}-\sqrt{6}}{4} & \frac{-\sqrt{6}-\sqrt{2}}{4} \\ \frac{\sqrt{6}+\sqrt{2}}{4} & \frac{\sqrt{2}-\sqrt{6}}{4} \end{bmatrix}.$$

The entries in the first column of the matrix on the right are, of course, the cosine and sine of $7\pi/12$, and the entries in the second column also are the correct quantities for the matrix $L_{7\pi/12}$. We have obtained in the example a couple of formulas with which you may have not been familiar:

$$\cos\frac{7\pi}{12} = -\frac{\sqrt{6} - \sqrt{2}}{4} \quad \text{and} \quad \sin\frac{7\pi}{12} = \frac{\sqrt{6} + \sqrt{2}}{4}.$$

6. A particular instance of the sum of two angles is $\theta + (-\theta) = 0$. Hence, $L_{-\theta} \circ L_\theta = L_0 = I$. That is,

$$\begin{bmatrix} \cos\theta & -\sin\theta \\ \sin\theta & \cos\theta \end{bmatrix}^{-1} = \begin{bmatrix} \cos(-\theta) & -\sin(-\theta) \\ \sin(-\theta) & \cos(-\theta) \end{bmatrix}$$
$$= \begin{bmatrix} \cos\theta & \sin\theta \\ -\sin\theta & \cos\theta \end{bmatrix}.$$

The inverse of a counterclockwise rotation about the origin through an angle θ is a clockwise rotation about the origin through an angle θ. Note that $L_0 = L_{-0}$ and $L_\pi = L_{-\pi}$, so each of these two matrices (which we also know as I and $-I$) is its own inverse.

Now let us consider linear isometries of the second type. As we now show, these correspond to reflections in lines through the origin. Let S_φ be the line through the origin spanned by the vector $(\cos\varphi, \sin\varphi)$, where φ is some angle. Note that S_φ can be obtained by rotating the horizontal axis counterclockwise about the origin through the angle φ. Let M_φ be the linear map corresponding to reflection about S_φ. Then it can be shown using elementary trigonometry that the matrix of M_φ is

$$[M_\varphi] = \begin{bmatrix} \cos(2\varphi) & \sin(2\varphi) \\ \sin(2\varphi) & -\cos(2\varphi) \end{bmatrix}.$$

(To check this, draw a picture, and then determine the results of reflecting the standard basis vectors about the line S_φ.) Now note that M_φ is the matrix of a linear isometry in $\mathbb{R}^2$ of the second type, with $\theta = 2\varphi$:

$$[M_{\theta/2}] = \begin{bmatrix} \cos(\theta) & \sin(\theta) \\ \sin(\theta) & -\cos(\theta) \end{bmatrix}.$$

It is easily checked that $M_\varphi^{-1} = M_\varphi$. That is, each linear isometry of the second type in $\mathbb{R}^2$ is its own inverse. This fact is to be expected, since reflecting twice about the same line puts every point back where it started.

We no longer have need of the terms "first type" and "second type". Instead we now have the descriptive terms "rotation" and "reflection", where, if necessary for clarity, we adjoin phrases such as "about the origin" and "in a line through the origin", respectively.

EXAMPLES:

7. Let us calculate the result of a counterclockwise rotation through the angle $\pi/7$ followed by a reflection about the line $x_2 = x_1$ followed by a counterclockwise rotation through the angle $6\pi/7$. The reflection is characterized by $\varphi = \pi/4$. We only have to calculate the product of the three corresponding matrices, possibly simplify the answer, and then check our list of isometries to geometrically describe the result. Remembering that the matrix corresponding to the last transformation comes first, we calculate:

$$\begin{bmatrix} \cos(\frac{6\pi}{7}) & -\sin(\frac{6\pi}{7}) \\ \sin(\frac{6\pi}{7}) & \cos(\frac{6\pi}{7}) \end{bmatrix} \begin{bmatrix} \cos(\frac{2\pi}{4}) & \sin(\frac{2\pi}{4}) \\ \sin(\frac{2\pi}{4}) & -\cos(\frac{2\pi}{4}) \end{bmatrix} \begin{bmatrix} \cos(\frac{\pi}{7}) & -\sin(\frac{\pi}{7}) \\ \sin(\frac{\pi}{7}) & \cos(\frac{\pi}{7}) \end{bmatrix}$$

$$= \begin{bmatrix} \cos(\frac{6\pi}{7}) & -\sin(\frac{6\pi}{7}) \\ \sin(\frac{6\pi}{7}) & \cos(\frac{6\pi}{7}) \end{bmatrix} \begin{bmatrix} 0 & 1 \\ 1 & 0 \end{bmatrix} \begin{bmatrix} \cos(\frac{\pi}{7}) & -\sin(\frac{\pi}{7}) \\ \sin(\frac{\pi}{7}) & \cos(\frac{\pi}{7}) \end{bmatrix}$$

$$= \begin{bmatrix} \cos(\frac{6\pi}{7}) & -\sin(\frac{6\pi}{7}) \\ \sin(\frac{6\pi}{7}) & \cos(\frac{6\pi}{7}) \end{bmatrix} \begin{bmatrix} \sin(\frac{\pi}{7}) & \cos(\frac{\pi}{7}) \\ \cos(\frac{\pi}{7}) & -\sin(\frac{\pi}{7}) \end{bmatrix} =$$

$$\begin{bmatrix} \cos(\frac{6\pi}{7})\sin(\frac{\pi}{7}) - \sin(\frac{6\pi}{7})\cos(\frac{\pi}{7}) & \cos(\frac{6\pi}{7})\cos(\frac{\pi}{7}) + \sin(\frac{6\pi}{7})\sin(\frac{\pi}{7}) \\ \sin(\frac{6\pi}{7})\sin(\frac{\pi}{7}) + \cos(\frac{6\pi}{7})\cos(\frac{\pi}{7}) & \sin(\frac{6\pi}{7})\cos(\frac{\pi}{7}) - \cos(\frac{6\pi}{7})\sin(\frac{\pi}{7}) \end{bmatrix}$$

$$= \begin{bmatrix} -\sin(\frac{5\pi}{7}) & \cos(\frac{5\pi}{7}) \\ \cos(\frac{5\pi}{7}) & \sin(\frac{5\pi}{7}) \end{bmatrix} = \begin{bmatrix} \cos(\frac{17\pi}{14}) & \sin(\frac{17\pi}{14}) \\ \sin(\frac{17\pi}{14}) & -\cos(\frac{17\pi}{14}) \end{bmatrix} = [M_{17\pi/28}],$$

where the switch from $5\pi/7$ to $17\pi/14$ was made in order to put the matrix in a form suitable for an immediate geometric interpretation based on the earlier discussion in this section. Thus, the result of the three successive transformations is a reflection in the line $S_{17\pi/28}$.

8. Consider the square $\mathcal{R}$ with vertices at $(1,1)$, $(-1,1)$, $(-1,-1)$, and $(1,-1)$. Let us find those isometries L for which $\mathcal{R} = L(\mathcal{R})$. Since isometries preserve distances and angles, you should be able to convince youself that the image of any vertex of the square under L is again a vertex, although not necessarily the same vertex.

Consider two adjacent vertices on the square, say $(1,1)$ and $(-1,1)$. Those two vertices can be moved by L to any of the four pairs of adjacent vertices on the square: $(1,1)$ *and* $(-1,1)$, $(-1,1)$ *and* $(-1,-1)$, $(-1,-1)$ *and* $(1,-1)$, or $(1,-1)$ *and* $(1,1)$. For each such pair of points there are two ways of moving the two vertices $(1,1)$ and $(-1,1)$ to them. Once these two vertices are moved, the final positions of the other two vertices are determined, and so the final positions of all the points in the square are determined. Thus, there are at most $4 \times 2 = 8$ isometries that transform the given square to itself. Some thought, possibly in conjunction with experiments with a piece of square paper, convinces one that there are, in fact, exactly eight isometries, namely: $I = L_0$, $-I = L_\pi$, $L_{\pi/2}$, $L_{-\pi/2}$, M_0, $M_{\pi/4}$, $M_{\pi/2}$, and $M_{3\pi/4}$. These eight isometries are called the "symmetries" of the square $\mathcal{R}$.

If one wanted to see the net result of several of these isometries acting in succession, one could, as in the preceding example, carry out the appropriate sequence of matrix multiplications, obtaining, of course, one of the isometries in the above list of eight, since the result will certainly be a symmetry of $\mathcal{R}$. If one were anticipating making many such calculations, one could make an 8-by-8 multiplication table involving these symmetries, to be used as a reference. (Since multiplication is not commutative, it is *not* sufficient to make, say, only the upper half of the table.) For instance, the calculation

$$[M_{\pi/2}][M_{3\pi/4}] = \begin{bmatrix} -1 & 0 \\ 0 & 1 \end{bmatrix} \begin{bmatrix} 0 & -1 \\ -1 & 0 \end{bmatrix} = \begin{bmatrix} 0 & 1 \\ -1 & 0 \end{bmatrix}$$

gives the entry $M_{\pi/2} M_{3\pi/4} = L_{-\pi/2}$ for such a table.

The **symmetries** of a figure $\mathcal{R}$ in $\mathbb{R}^2$, and more generally in $\mathbb{R}^n$, are those isometries L such that $L(\mathcal{R}) = \mathcal{R}$. The preceding example may be used as a pattern for finding the symmetries of many geometrical figures. By following it, one may show that a regular n-gon (an n-sided figure with

n equal sides and n equal interior angles) has $2n$ symmetries, n of which are rotations, and n of which are reflections.

REMARK:

- Often, when determining the symmetries of a figure in $\mathbb{R}^n$, it is convenient to locate the figure so that all its symmetries are *linear* isometries, as in the example above. This can be accomplished by placing the center (or more generally the centroid) of the figure at the origin, since any isometry that maps a figure to itself must leave the center (centroid) of that figure fixed, and isometries that map the origin to itself are always linear isometries (can you see why?).

3.8C ISOMETRIES OF $\mathbb{R}^n$

Since we have some geometrical background in $\mathbb{R}^2$, let us here take a more systematic algebraic viewpoint. We begin with some symbolism and terminology for (not necessarily square) matrices. Let $A = (a_{ij})$ be a q-by-n matrix and $B = (b_{kl})$ be an n-by-q matrix. The matrices A and B are said to be **transposes** of each other if $a_{ij} = b_{ji}$ for $i = 1, 2, \ldots, q$ and $j = 1, 2, \ldots, n$. Starting with a matrix A one can get its transpose by turning the columns into rows and the rows into columns, making sure that one does not change the order of the rows or columns while doing so. If A and B are transposes of each other, we write $B = A^T$ or, equivalently, $A = B^T$. For instance,

$$\begin{bmatrix} 2 & 4 & 9 \\ 3 & 5 & 1 \end{bmatrix} = \begin{bmatrix} 2 & 3 \\ 4 & 5 \\ 9 & 1 \end{bmatrix}^T .$$

Let X be a vector. Recall that $[X]$ denotes the corresponding column matrix. Now we also have a notation for the corresponding row matrix, namely $[X^T]$. To illustrate how different notations can be used we consider the expression

$$[X]^T[Y],$$

where X and Y are members of $\mathbb{R}^n$. This is a matrix product—a product of a 1-by-n matrix and a n-by-1 matrix. The product is a 1-by-1 matrix, which we can regard as a real number. What real number is it? It is the inner product of X and Y. That is,

$$[X]^T[Y] = [x_1 y_1 + \cdots + x_n y_n] = x_1 y_1 + \cdots + x_n y_n = \langle X, Y \rangle$$

where the second equality reflects a slight abuse of notation, confusing a 1-by-1 matrix with the number which is its single entry.

Here is a theorem that is useful for making calculations with transposes.

Theorem 3.8.2 Let A and B be matrices and c a scalar. Then the following assertions are true:

1. $(cA)^T = cA^T$.

2. $(A + B)^T = A^T + B^T$ if A and B have the same number of rows and the same number of columns.

3. $(A^T)^{-1} = (A^{-1})^T$ if A is a square matrix whose inverse exists.

4. $(BA)^T = A^T B^T$ if A has the same number of rows as B has columns.

We will omit the proof but illustrate the last conclusion with an example.

EXAMPLE:

9. Let

$$A = \begin{bmatrix} 2 & 1 & 5 \\ 3 & -1 & -3 \end{bmatrix} \quad \text{and} \quad B = \begin{bmatrix} -4 & 0 \\ -2 & 1 \\ -1 & -5 \\ 2 & 4 \end{bmatrix}.$$

Then

$$BA = \begin{bmatrix} -8 & -4 & -20 \\ -1 & -3 & -13 \\ -17 & 4 & 10 \\ 16 & -2 & -2 \end{bmatrix};$$

so

$$(BA)^T = \begin{bmatrix} -8 & -1 & -17 & 16 \\ -4 & -3 & 4 & -2 \\ -20 & -13 & 10 & -2 \end{bmatrix}.$$

We want to compare this with $A^T B^T$, where

$$A^T = \begin{bmatrix} 2 & 3 \\ 1 & -1 \\ 5 & -3 \end{bmatrix} \quad \text{and} \quad B^T = \begin{bmatrix} -4 & -2 & -1 & 2 \\ 0 & 1 & -5 & 4 \end{bmatrix}.$$

Notice that A^T has the same number of columns as B^T has rows, so the product $A^T B^T$ is meaningful; the product has the same number of rows as does A^T, namely 3, and the same number of columns as does B^T, namely 4. We leave it to the reader to calculate $A^T B^T$ and compare the result with $(BA)^T$. It is also instructive to compare step-by-step the calculations leading to $A^T B^T$ with those leading to $(BA)^T$.

A square matrix A is said to be **orthogonal** if and only if each column vector has norm 1 and the inner product of any two distinct column vectors is 0. Notice that these two conditions are encompassed in the condition $A^T A = I$ (can you see why?). Thus A is orthogonal if and only if $A^T = A^{-1}$; so, in particular, orthogonal matrices are invertible.

Suppose A is orthogonal. By Theorem 3.8.2, A^T is invertible, and

$$(A^T)^{-1} = (A^{-1})^T = (A^T)^T = A .$$

Thus, like the matrix A, A^T is a matrix whose inverse and transpose are identical, that is, A^T is orthogonal. So, the transpose of an orthogonal matrix is orthogonal; it follows that the row vectors of an orthogonal matrix have norm one and distinct row vectors are orthogonal to each other.

Theorem 3.8.3 A linear map in $\mathbb{R}^n$ is an isometry if and only if its matrix is orthogonal.

PROOF: Consider the matrix of an isometry. Its columns, being that into which the standard bases vectors are transformed, must have the same norms as the standard bases vectors, namely 1. Since the standard bases vectors are orthogonal to each other, the same must be true of the column vectors; thus the inner products of distinct column vectors must be 0.

We have proved that the matrix of an isometry is orthogonal. Now let us begin with an orthogonal matrix A and prove that it represents an isometry. We will do so by proving that the corresponding linear transformation L preserves inner products. For any two members W and X of $\mathbb{R}^n$ we have:

$$\langle L(W), L(X) \rangle = [L(W)]^T [L(X)] = (A[W])^T A[X]$$
$$= [W]^T A^T A[X] = [W]^T I[X] = [W]^T [X] = \langle W, X \rangle .$$

Done. <<

EXAMPLES:

10. Here are some 3-by-3 orthogonal matrices:

$$\begin{bmatrix} 1 & 0 & 0 \\ 0 & -1 & 0 \\ 0 & 0 & 1 \end{bmatrix}, \quad \begin{bmatrix} 0 & 0 & 1 \\ 1 & 0 & 0 \\ 0 & 1 & 0 \end{bmatrix}, \quad \begin{bmatrix} \cos\theta & -\sin\theta & 0 \\ \sin\theta & \cos\theta & 0 \\ 0 & 0 & 1 \end{bmatrix},$$

$$\begin{bmatrix} \cos\theta & 0 & \sin\theta \\ 0 & 1 & 0 \\ \sin\theta & 0 & -\cos\theta \end{bmatrix}, \quad \begin{bmatrix} \frac{1}{2} & \frac{-\sqrt{2}}{2} & \frac{1}{2} \\ \frac{1}{2} & \frac{\sqrt{2}}{2} & \frac{1}{2} \\ \frac{\sqrt{2}}{2} & 0 & \frac{-\sqrt{2}}{2} \end{bmatrix}.$$

The first matrix corresponds to a reflection about the $x_1 x_3$-plane. The second one corresponds to a rotation that moves the x_3-axis to the position originally occupied by the x_1-axis, the x_1-axis to the original location of the x_2-axis, and the x_2-axis to the position vacated by the x_3-axis. This is a rotation of $2\pi/3$ radians about the line through the origin in $\mathbb{R}^3$ spanned by the vector $(1, 1, 1)$. The third orthogonal matrix corresponds to a rotation through the angle θ about the x_3-axis. The direction of the rotation is counterclockwise if viewed from the positive x_3 direction in a right-handed coordinate system. Compare this rotation to a rotation in $\mathbb{R}^2$. The fourth orthogonal matrix is like a reflection involving the x_1 and x_3 coordinates, with the x_2 coordinate left unchanged. The third and fourth matrices both illustrate one way to produce linear isometries in $\mathbb{R}^3$: apply any one of the linear isometries in $\mathbb{R}^2$ to two of the coordinates in $\mathbb{R}^3$, and leave the remaining coordinate fixed. The fifth orthogonal matrix above was constructed by multiplying the third and fourth matrices, with $\theta = \pi/4$.

11. If you multiply any column (or row) of an orthogonal matrix A by -1, you get another orthogonal matrix. Can you give a geometric interpretation of this fact? Think in terms of the composition of the isometry corresponding to A with a reflection. Try to find a matrix B corresponding to a reflection, such that the product AB is the matrix you get by multiplying one of the columns of A by -1. Also try to find a matrix C such that CA is the matrix you get by multiplying one of the rows of A by -1.

12. Let us find those isometries in $\mathbb{R}^3$ that are the symmetries of the regular pentagonal prism with vertices at

$$(1, 0, 2) \qquad\qquad (1, 0, -2)$$
$$(\cos(2\pi/5), \sin(2\pi/5), 2) \qquad (\cos(2\pi/5), \sin(2\pi/5), -2)$$
$$(\cos(4\pi/5), \sin(4\pi/5), 2) \qquad (\cos(4\pi/5), \sin(4\pi/5), -2)\;.$$
$$(\cos(6\pi/5), \sin(6\pi/5), 2) \qquad (\cos(6\pi/5), \sin(6\pi/5), -2)$$
$$(\cos(8\pi/5), \sin(8\pi/5), 2) \qquad (\cos(8\pi/5), \sin(8\pi/5), -2)$$

From our study of isometries in $\mathbb{R}^2$ we conclude that there are ten symmetries that do not interchange the top and the bottom of the prism. Their matrices are

$$\begin{bmatrix} \cos\theta & -\sin\theta & 0 \\ \sin\theta & \cos\theta & 0 \\ 0 & 0 & 1 \end{bmatrix} \quad \text{and} \quad \begin{bmatrix} \cos\theta & \sin\theta & 0 \\ \sin\theta & -\cos\theta & 0 \\ 0 & 0 & 1 \end{bmatrix}$$

for $\theta = 0, \frac{2\pi}{5}, \frac{4\pi}{5}, \frac{6\pi}{5}$, and $\frac{8\pi}{5}$. There are ten other symmetries that can be obtained by first reflecting in the plane $x_3 = 0$ and then following with one of the above ten symmetries. The matrix describing the reflection about $x_3 = 0$ is

$$\begin{bmatrix} 1 & 0 & 0 \\ 0 & 1 & 0 \\ 0 & 0 & -1 \end{bmatrix}.$$

Thus, we multiply the ten matrices identified above by this matrix to obtain the other ten symmetries:

$$\begin{bmatrix} \cos\theta & -\sin\theta & 0 \\ \sin\theta & \cos\theta & 0 \\ 0 & 0 & -1 \end{bmatrix} \quad \text{and} \quad \begin{bmatrix} \cos\theta & \sin\theta & 0 \\ \sin\theta & -\cos\theta & 0 \\ 0 & 0 & -1 \end{bmatrix}$$

for $\theta = 0$, $\frac{2\pi}{5}$, $\frac{4\pi}{5}$, $\frac{6\pi}{5}$, and $\frac{8\pi}{5}$.

3.8D RIGID MOTIONS

The rotations about the origin in $\mathbb{R}^2$ can be labelled, as in Section 3.8b, by L_θ. In particular, $L_0 = I$ and $L_\pi = -I$. Fix a value θ of interest and consider the isometries $L_{t\theta}$ for $0 \leq t \leq 1$. When $t = 0$, we get the identity map. When $t = 1$, we get L_θ. As t varies from 0 to 1, there is a continuous change, involving only isometries, that begins with the identity and ends with the map of interest. We may think of this continuous change as a motion occuring within $\mathbb{R}^2$. As t moves from 0 to 1, $\mathbb{R}^2$ is rotated counterclockwise continuously (clockwise if $\theta < 0$) until the rotation through the angle θ is completed. The end result L_θ of such a continuous change is what we mean by a **rigid motion**. Thus, L_θ is a rigid motion. Is the word "continuous" that we have used really appropriate? Yes it is, because the entries in the matrix, $\cos t\theta$ and $\pm\sin t\theta$, are continuous functions of t.

There does not seem to be a way of carrying out the same procedure for the reflections M_φ. In fact, here is a way of seeing that it cannot be done. If we were able to get from I to M_φ continuously with isometries, the determinant would change continuously. But, as may be checked by direct computation, the isometries L_θ have determinant 1, and the isometries M_φ have determinant -1. There is no way to change continuously from the determinant of the identity matrix $+1$ to the determinant of M_φ, which is -1.

A similar story holds in $\mathbb{R}^3$. The isometries with determinant -1 are not rigid motions and those that have determinant $+1$ are rigid motions. The first half of this statement is proved just as it was for $\mathbb{R}^2$. The second half is more difficult, and we will omit it. The same story is valid for linear isometries in $\mathbb{R}^n$.

EXAMPLES:

13. Let us identify which of the twenty symmetries of the prism described in the example in Section 3.8c are rigid motions. All we need do is calculate the determinants of their matrices. There are ten matrices having positive determinants (that is, equal to 1 rather than -1):

$$\begin{bmatrix} \cos\theta & -\sin\theta & 0 \\ \sin\theta & \cos\theta & 0 \\ 0 & 0 & 1 \end{bmatrix} \quad \text{and} \quad \begin{bmatrix} \cos\theta & \sin\theta & 0 \\ \sin\theta & -\cos\theta & 0 \\ 0 & 0 & -1 \end{bmatrix}$$

for $\theta = 0$, $\frac{2\pi}{5}$, $\frac{4\pi}{5}$, $\frac{6\pi}{5}$, and $\frac{8\pi}{5}$.

14. Consider the matrix

$$A = \begin{pmatrix} 0 & 1 & 0 \\ 1 & 0 & 0 \\ 0 & 0 & -1 \end{pmatrix}.$$

The effect of this matrix on the first two coordinates x_1 and x_2 is the same as that of the linear isometry $M_{\pi/4}$ in $\mathbb{R}^2$. In particular, the x_1x_2-plane is reflected about the line $x_1 = x_2$. Recall that $M_{\pi/4}$ is not a rigid motion in $\mathbb{R}^2$. But the determinant of the matrix A is $+1$, so A corresponds to a rigid motion in $\mathbb{R}^3$. How is this possible? It is possible because A represents a rotation through π radians about the line through the origin spanned by the vector $(1,1,0)$. Such a rotation can be carried out continuously in $\mathbb{R}^3$, but there is no corresponding motion that can occur within $\mathbb{R}^2$. Note that the reflection about the plane $x_1 = x_2$ in $\mathbb{R}^3$, which has the matrix

$$\begin{pmatrix} 0 & 1 & 0 \\ 1 & 0 & 0 \\ 0 & 0 & 1 \end{pmatrix},$$

also agrees with $M_{\pi/4}$ as far as the first two coordinates are concerned, but that this reflection is not a rigid motion in $\mathbb{R}^3$, since its determinant is -1.

Exercises

1. Let

$$A = \begin{bmatrix} 2 & -3 \\ -1 & 2 \end{bmatrix}, \quad B = \begin{bmatrix} 2 & 3 & -3 & 1 \\ -1 & 2 & 0 & 1 \end{bmatrix},$$

$$C = \begin{bmatrix} -1 \\ 0 \\ 5 \\ 2 \end{bmatrix}, \quad D = \begin{bmatrix} 6 & -2 \end{bmatrix}.$$

Decide which of the following are defined and calculate them:

(a) B^T

(b) $B^T D^T$

(c) DA^T

(d) DA

(e) DC

(f) CD

(g) $(5CD)^T$

(h) $5D^T C^T$

(i) $DA^2 + (A^{-1}D^T)^T$

(j) $C^T B^T$

(k) $(CB)^T$

(l) B^{-1} .

2. Decide which of the following matrices represent isometries:

(a) $\begin{bmatrix} \frac{3}{5} & \frac{4}{5} \\ -\frac{4}{5} & \frac{3}{5} \end{bmatrix}$

(b) $\begin{bmatrix} \frac{5}{13} & -\frac{12}{13} \\ -\frac{12}{13} & -\frac{5}{13} \end{bmatrix}$

(c) $\begin{bmatrix} \frac{3}{5} & \frac{3}{5} \\ -\frac{4}{5} & \frac{4}{5} \end{bmatrix}$

(d) $\begin{bmatrix} -1 & 0 \\ 0 & -1 \end{bmatrix}$

(e) $\begin{bmatrix} -1 & 0 \\ 0 & 1 \end{bmatrix}$

(f) $\begin{bmatrix} \sin\gamma & -\cos\gamma \\ \cos\gamma & \sin\gamma \end{bmatrix}$ where $\frac{-\pi}{2} < \gamma \le \frac{3\pi}{2}$

(g) $\begin{bmatrix} 1/2 & -\sqrt{3}/2 \\ \sqrt{3}/2 & 1/2 \end{bmatrix}$

(h) $\begin{bmatrix} 3 & -4 \\ 4 & 3 \end{bmatrix}$.

3. For each of the matrices in the preceding exercise that represents an isometry decide whether it represents a rigid motion, find its inverse matrix, and decide whether its inverse represents a rigid motion. Identify each isometry as L_θ for some θ having absolute value belonging to the interval $[0, \pi]$, or M_φ for some $\varphi \in [0, \pi)$.

4. Write down the matrices that represent the symmetries of the triangle with vertices at $(1,0)$, $(-1/2, 1/2)$, and $(-1/2, -1/2)$.

5. Write down the matrices that represent the symmetries of the triangle with vertices at $(1,0)$, $(-1/2, \sqrt{3}/2)$, and $(-1/2, -\sqrt{3}/2)$.

6. Consider the hexagon with vertices at

$$(2,0),\ (1, \sqrt{3}),\ (-1, \sqrt{3}),\ (-2, 0),\ (-1, -\sqrt{3}),\ (1, -\sqrt{3}) .$$

(a) Prove that the hexagon is regular, that is, that all its edges are congruent *and* all its angles are congruent.

(b) Write the matrices that represent all the symmetries of the hexagon.

(c) Calculate the matrix that represents the result of first reflecting in the vertical axis, then rotating $2\pi/3$ radians in the clockwise direction, then reflecting in the horizontal axis, and finally rotating $\pi/3$ radians in the clockwise direction. Is your result one of the symmetries of the hexagon?

7. Consider the hexagon with vertices at

$$(4,0),\ (2,3),\ (-2,3),\ (-4,0),\ (-2,-3),\ (2,-3).$$

(a) Write the matrices that represent all the symmetries of this hexagon.

(b) Complete a multiplication table for the symmetries of this hexagon.

8. Let L_θ be a rotation in $\mathbb{R}^2$, as discussed in Section 3.8b. Find two linear isometries in $\mathbb{R}^3$, one a rigid motion and the other not, whose effect on the coordinates x_1 and x_2 is the same as that of L_θ. Do the same for M_φ.

9. Decide which of the following matrices represent isometries:

(a) $\begin{bmatrix} \frac{2}{7} & \frac{3}{7} & \frac{6}{7} \\ \frac{3}{7} & -\frac{6}{7} & \frac{2}{7} \\ \frac{6}{7} & \frac{2}{7} & -\frac{3}{7} \end{bmatrix}$

(b) $\begin{bmatrix} \frac{2}{3} & -\frac{1}{3} & \frac{2}{3} \\ -\frac{2}{3} & -\frac{2}{3} & \frac{1}{3} \\ -\frac{1}{3} & \frac{1}{3} & \frac{1}{3} \end{bmatrix}$

(c) $\begin{bmatrix} \frac{2}{3} & -\frac{1}{3} & \frac{2}{3} \\ -\frac{2}{3} & -\frac{2}{3} & \frac{1}{3} \\ -\frac{1}{3} & \frac{2}{3} & \frac{2}{3} \end{bmatrix}$

(d) $\frac{1}{2}\begin{bmatrix} \sqrt{2} & -\sqrt{2} & 0 \\ 1 & 1 & -\sqrt{2} \\ 1 & 1 & \sqrt{2} \end{bmatrix}$

10. For each of the matrices in the preceding exercise that represents an isometry decide whether it represents a rigid motion, calculate its inverse, and decide whether its inverse represents a rigid motion.

11. Consider the cube with vertices at $(1,1,1)$, $(1,1,-1)$, $(1,-1,1)$, $(-1,1,1)$, $(1,-1,-1)$, $(-1,1,-1)$, $(-1,-1,1)$, and $(-1,-1,-1)$.

(a) Give an argument that shows that the cube has 48 symmetries.

(b) Give an argument that shows that the corresponding matrices are those whose entries are only zeroes, ones, and negative ones and which have exactly two zeroes in each row and each column.

(c) Identify the 24 matrices that represent rigid motions.

(d) Identify each of the 23 rigid motions different from I as a rotation about some axis. For instance, the two matrices

$$\begin{bmatrix} 0 & 1 & 0 \\ 0 & 0 & 1 \\ 1 & 0 & 0 \end{bmatrix} \quad \text{and} \quad \begin{bmatrix} 0 & 0 & 1 \\ 1 & 0 & 0 \\ 0 & 1 & 0 \end{bmatrix}$$

each represent $120°$ rotations about the line passing through $(1,1,1)$ and $(-1,-1,-1)$. *Hint:* The two isometries corresponding to these two matrices transform the two points $(1,1,1)$ and $(-1,-1,-1)$ to themselves. It is for that reason that we conclude that they represent rotations about the axis indicated. As a check we note that each of these matrices is the square of the other.

(e) Use matrix multiplication to find the result of a 90 degree rotation about an axis through the centers of two opposite faces, followed by a 180 degree rotation about an axis through the centers of two opposite edges, each of which is an edge of one of the two faces involved in the first rotation.

12. Visualize the ten rigid motions described in Example 13 in Section 3.8d, or, better, carry them out with a real prism.

13. Decide whether

(a)

$$\frac{1}{2}\begin{bmatrix} 1 & -1 & -1 & 1 \\ -1 & 1 & -1 & 1 \\ -1 & -1 & 1 & 1 \\ -1 & -1 & -1 & -1 \end{bmatrix}$$

represents an isometry.

(b) Decide whether the transpose of this matrix represents an isometry.

Extra: Linear Maps on Function Spaces

In the extra section of Chapter 1 we observed that sets of functions may be vector spaces. It turns out that some linear maps on these spaces are well known to every student of calculus. These maps are the familiar derivative and integral.

To be more precise, let $\mathcal{C}(\mathbb{R})$ denote the set of all real-valued continuous functions on $\mathbb{R}$, and let $\mathcal{C}^1(\mathbb{R})$ denote the subset of $\mathcal{C}(\mathbb{R})$ consisting of those functions whose first derivative exists and is continuous. Some elements of $\mathcal{C}^1(\mathbb{R})$ are: the trigonometric functions sin and cos and polynomials such as $x^2 + 1$. The absolute value function $|x|$ is a member of $\mathcal{C}(\mathbb{R})$ but not of $\mathcal{C}^1(\mathbb{R})$. Consider the map $D : \mathcal{C}^1(\mathbb{R}) \to \mathcal{C}(\mathbb{R})$ defined by

$$D(f) = f' \, ,$$

where f' is the derivative of f. Basic theorems about derivatives tell us that D is a linear map. For example, $D(\sin) = \cos$, $D(\cos) = -\sin$,

$$D(\sin + \cos) = \cos - \sin \, ,$$

and $D(3\sin) = 3\cos$; and, generally,

$$D(f + g) = D(f) + D(g), \quad D(cf) = c(Df) \, .$$

Similarly, the map $L : \mathcal{C}(\mathbb{R}) \to \mathcal{C}(\mathbb{R})$ defined for $f \in \mathcal{C}(\mathbf{R})$ by

$$(L(f))(x) = \int_0^x f(t)dt$$

is a linear map. Two values of this map are $L(x^2) = x^3/3$ and $L(e^x) = e^x - 1$. Note that $L(\mathbf{0}) = \mathbf{0}$, as required for linear maps. (Here, $\mathbf{0}$ denotes the function $f(x) = 0$.)

Because both the derivative and the integral are linear maps, we see that all calculus can be viewed as a study of these two special linear maps on appropriate vector spaces. For instance, one version of the Fundamental Theorem of Calculus states that

$$\frac{d}{dx} \int_0^x f(t)\, dt = f(x)\,,$$

which can be written in the simple symbolic form

$$DL = I\,,$$

where I is the identity map in $\mathcal{C}(\mathbb{R})$; that is, $I(f) = f$ for all functions $f \in \mathcal{C}(\mathbb{R})$.

Other linear maps can now be constructed. For instance, define $T : \mathcal{C}^1(\mathbf{R}) \to \mathcal{C}(\mathbf{R})$ by the rule

$$T(u) = u' - u\,,$$

so that, for instance, $T(\sin) = \cos - \sin$ and $T(7e^x) = \mathbf{0}$. Since T is a linear combination of two linear maps,

$$T = D - I\,,$$

it too is a linear map. The problem of solving the differential equation

$$u' - u = f\,,$$

where f is a given continuous function, is exactly that of solving $T(u) = f$. The question of whether there are any solutions at all is that of whether f is in $\mathcal{I}(T)$. Just as in the case of maps on $\mathbb{R}^n$, and with identical proofs, one sees that the image $\mathcal{I}(T)$ and null space $\mathcal{N}(T)$ are linear subspaces of $\mathcal{C}(\mathbb{R})$ and $\mathcal{C}^1(\mathbb{R})$, respectively.

We whet our appetite for future courses by making calculations that show that $\mathcal{N}(T)$ *is one-dimensional*, and that, therefore, the solution set of $u' - u = f$ is either empty or else an affine subspace of dimension 1. That the dimension of $\mathcal{N}(T)$ is at least 1 follows from the calculation

$$D(e^x) - I(e^x) = e^x - e^x = \mathbf{0}\,.$$

To show that the dimension is no more than 1 we consider an arbitrary $v(x) \in \mathcal{N}(T)$ and plan to show that $v(x) = ce^x$ for some c. We define a new function $\psi(x) = v(x)e^{-x}$. Thus, our plan is to show that ψ is a constant function. Since $v' - v = \mathbf{0}$, we easily obtain

$$\psi'(x) = v'(x)e^{-x} - v(x)e^{-x} = (v'(x) - v(x))e^{-x} = 0$$

for each x. We know from single-variable calculus that functions whose derivative is the zero function are constant functions. We conclude, as desired, that ψ is a constant function.

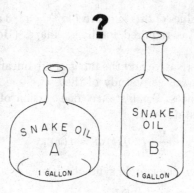

FIGURE 3.20.

Extra: Linear Programming

You are a traveling snake-oil salesman, selling brand A at 5 dollars per gallon and brand B at 3 dollars per gallon. You ply your trade in a territory that craves snake oil, so you can be sure of selling all you carry at the prices mentioned. Things are not quite so simple, however. Snake oil requires the magical ingredient M. More precisely, brand A requires 4 ounces of ingredient M per 100 gallons of oil, and brand B requires 3 ounces per 100 gallons. And you have only 12 ounces of ingredient M. See Figure 3.20.

If the scarcity of ingredient M were the only constraint, it would be easy to see that you would maximize gross income by producing and selling only brand A. There is, however, one more fly in the ointment. Brand A weighs 15 pounds per gallon and brand B weighs one-third this much. You are capable of transporting only 3,000 pounds of snake oil. How much should be brand A and how much brand B?

We proceed as follows. Let $x =$ amount of brand A in hundreds of gallons and $y =$ amount of brand B in hundreds of gallons. We wish to compute x and y so that the gross income is maximized. Thus, we consider the linear functional

$$\text{gross income } = L(x,y) = 500x + 300y \, ,$$

since, for example, brand A sells for $500 per 100 gallons.

We may not use more than 12 ounces of ingredient M, that is,

$$4x + 3y \le 12 \, .$$

Also brands A and B weigh 1,500 and 500 pounds per 100 gallons, respectively, so that we must satisfy

$$1,500x + 500y \le 3,000 \, ,$$

which is equivalent to

$$3x + y \le 6 \, .$$

In summary, the problem may be stated in mathematical terms as follows: maximize $L(x, y) = 500x + 300y$ subject to the constraints

$$x \geq 0, \quad y \geq 0 \quad \text{(the amounts must be nonnegative)},$$

$$4x + 3y \leq 12 \quad \text{(ingredient M is limited)},$$

$$3x + y \leq 6 \quad \text{(the total weight is limited)}.$$

It is not hard to see that the set $\mathcal{K}$ of points (x, y) satisfying all four inequalities above looks as shown in Figure 3.21. This set is **convex** in that if (x_1, y_1) and (x_2, y_2) are two points of $\mathcal{K}$, then the line segment connecting them is also in $\mathcal{K}$.

For certain numbers c, the line $L(x, y) = 500x + 300y = c$ actually intersects the set $\mathcal{K}$, for example, $c = 0$ and $(x, y) = (0, 0)$. What we seek is the *largest* number c having this property and the point (or points) (x, y) of intersection. The lines $500x + 300y = c$ all have the same slope; they are parallel. Examination of the picture should convince you that the maximum value of $L(x, y)$ for $(x, y) \in \mathcal{K}$ occurs at the point $X^* = (x^*, y^*) = (\frac{6}{5}, \frac{12}{5})$, which is one of the corners of the convex set $\mathcal{K}$. Therefore, in order to maximize gross income, you should produce 120 gallons of brand A and 240 gallons of brand B, an action which will yield an income of 1,320.

For the general linear programming problem, we have a linear functional $L : \mathbb{R}^n \to \mathbb{R}$, like $L(x, y) = 500x + 300y$ above, and some convex subset (not subspace) $\mathcal{K}$ of $\mathbb{R}^n$. This constraint set $\mathcal{K}$ of admissible values is usually

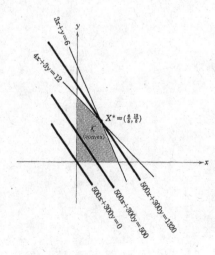

FIGURE 3.21.

defined by many inequalities (four in the example). The problem is to find X^* in $\mathcal{K}$ that maximizes (or minimizes) L, so that $L(X^*) \geq L(X)$ for all X in the set $\mathcal{K}$ of admissible values. In searching for such a vector X^*, we look at the solution sets of $L(X) = c$. For different values of c, these are parallel affine subspaces of $\mathbb{R}^n$. The maximum value of L on $\mathcal{K}$ is the largest value of c such that the corresponding affine subspace $\mathcal{A}$ intersects $\mathcal{K}$. An important theorem asserts that this affine subspace $\mathcal{A}$ intersects $\mathcal{K}$ at one of the corners of $\mathcal{K}$ so the desired "best point" X^* is always one of these corners. Computers are essential in solving these problems when, as is typical, many variables and constraints are involved.

REMARK:

- We have just discussed maximizing a *linear* functional. The problem of maximizing a *nonlinear* function is discussed in Chapter 6.

4

Curves: Mappings $F : \mathbb{R} \to \mathbb{R}^q$

4.0 Introduction

Now we begin calculus proper. In single-variable calculus we considered a function $y = f(x)$ with $x \in \mathbb{R}$. Such a function assigns to each real x another real number y. Thus we may write

$$f : \mathbb{R} \to \mathbb{R} .$$

What are the simplest scalar-valued functions of a scalar variable? Linear functions, of course:

$$f(x) = ax ,$$

and the closely related affine functions

$$g(x) = ax + b .$$

Note that a linear function is a special kind of affine function where $b = 0$.

It is tempting to call both these functions "linear", because the graphs of both f and g are lines. But if $b \neq 0$, then $g(0) \neq 0$, and so, according to our definition, g is not linear when $b \neq 0$. See Figure 4.1.

At an early age we enlarged our vocabulary of functions to a more general class that includes as examples

$$f_1(x) = x^2 + 1 , \quad f_2(x) = \sin x , \quad f_3(x) = \sqrt{x} , \quad f_4(x) = \frac{1}{x} .$$

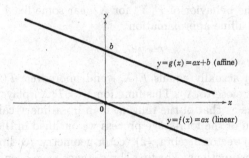

FIGURE 4.1.

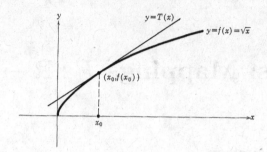

FIGURE 4.2. The tangent line at $(x_0, f(x_0))$

None of these functions is linear. Recall, for example, that $\sin(x_1 + x_2) \neq \sin x_1 + \sin x_2$ for most values of x_1 and x_2. Each maps some subset of the reals into the reals. As discussed in Section 0.3a, the subset of the reals for which a function is defined is its domain or domain of definition, and denoted by $\mathcal{D}(f)$. Thus

$$\begin{aligned}
\mathcal{D}(f_1) &= \mathbb{R}, & \mathcal{D}(f_2) &= \mathbb{R}, \\
\mathcal{D}(f_3) &= \{x \in \mathbb{R} : x \geq 0\}, & \mathcal{D}(f_4) &= \{x \in \mathbb{R} : x \neq 0\}.
\end{aligned}$$

The main problem of differential calculus is this: Given a function, such as $y = f(x) = \sqrt{x}$ (see Figure 4.2), and given a point $x_0 \in \mathcal{D}(f)$, define and construct the tangent line to the graph at the point x_0. This tangent line is the graph of an affine function $y = T(x)$. We call the function T the "tangent map". When x is close to x_0, $T(x)$ is close to $f(x)$, that is, $T(x)$ approximates $f(x)$. Later on we will speak of the tangent map $y = T(x)$ as the "best affine approximation" to $f(x)$ near the point x_0.

In this chapter and in chapters to come, we parallel the study above for nonlinear functions

$$F : \mathbb{R}^n \to \mathbb{R}^q,$$

with $n \geq 1$ and $q \geq 1$. Writing $Y = F(X)$ with $X \in \mathbb{R}^n$ and $Y \in \mathbb{R}^q$, we say that F is a vector-valued function of a vector variable. Then, in order to understand the behavior of $F(X)$ for X near some fixed vector X_0, we construct a best affine approximation

$$T : \mathbb{R}^n \to \mathbb{R}^q$$

such that $T(X_0)$ actually equals $F(X_0)$ and, moreover, $T(X)$ is close to $F(X)$ when X is close to X_0. This function $Y = T(X)$ plays the same role in vector calculus as that of the tangent map in ordinary calculus.

We remind you of the four-stage process we outlined in the introduction to Chapter 1: (1) vector algebra, (2) vector geometry, (3) linear functions, and (4) nonlinear functions. The first three stages have been accomplished. In carrying out the fourth step, it is helpful to consider three different classes of nonlinear functions, as follows:

1. $F : \mathbb{R} \to \mathbb{R}^q$. These are vector-valued functions of a scalar variable. They constitute the central topic of this chapter in which curves in $\mathbb{R}^q$ are viewed as the images of such functions.

2. $f : \mathbb{R}^n \to \mathbb{R}$. These are scalar-valued functions of several variables $x_1, \ldots, x_n$, or, more compactly, of the vector X. They are discussed in Chapters 6 and 7.

3. $F : \mathbb{R}^n \to \mathbb{R}^q$ with $n \geq 1$, $q \geq 1$. These are the vector-valued functions of a vector variable. In the important special case $n = q$, the term **vector field** is used for F. They are discussed in Chapter 8.

Even though the discussion is broken into three cases, it is very important to appreciate the similarity of approaches. Given a function $Y = F(X)$ defined near a point X_0, we construct another, simpler function $Y = T(X)$, the best affine approximation to F at X_0. This is just what we did in ordinary calculus by constructing the tangent line to the graph of $y = f(x)$.

REMARK:

- The description of coordinate forms of linear vector-valued functions in Example 10 in Section 3.1c carries over to the present setting. Suppose that F is a function with $\mathcal{D}(F) \subseteq \mathbb{R}^n$ and target $\mathbb{R}^q$. Then, for each X, $F(X)$ is a *vector* with q coordinates. We can write either $F(X) = Y$, where $Y \in \mathbb{R}^q$, or

$$F(X) = (y_1, \ldots, y_q) .$$

As X varies, each of the coordinates $y_1, \ldots, y_q$ may vary, that is, each y_j is scalar-valued function of $X : y_j = f_j(X)$. Hence, we may write

$$F(X) = (f_1(X), \ldots, f_q(X))$$

or

$$F = (f_1, \ldots, f_q) .$$

We specify F by specifying the q **coordinate functions** $f_1, \ldots, f_q$. The target space of each f_j is $\mathbb{R}$ and the domain of each is $\mathcal{D}(F)$.

Exercises
A few linear and nonlinear functions.

1. Is each of the following linear, affine, or neither? Note that most of these maps are given in terms of coordinates:

   (a) $f : \mathbb{R} \to \mathbb{R}$, $f(x) = 1 + 5x + \cos x$

   (b) $f : \mathbb{R}^2 \to \mathbb{R}$, $X = (x_1, x_2)$, $f(X) = 1 + x_1 + x_2$

   (c) $g : \mathbb{R}^2 \to \mathbb{R}$, $X = (x_1, x_2)$, $g(X) = x_1 x_2$

   (d) $F : \mathbb{R}^2 \to \mathbb{R}^2$, $X = (x_1, x_2)$, $F(X) = (x_1 x_2, 0)$

    (e) $G : \mathbb{R}^2 \to \mathbb{R}^2$, $X = (x_1, x_2)$, $G(X) = (1 + x_1 + x_2, 2 - x_1)$

    (f) $F : \mathbb{R}^2 \to \mathbb{R}^3$, $X = (x_1, x_2)$, $F(X) = (0, x_1, x_1 x_2)$

    (g) $f : \mathbb{R}^2 \to \mathbb{R}$, $X = (x_1, x_2)$, $f(X) = x_1$

    (h) $f : \mathbb{R}^3 \to \mathbb{R}$, $f(X) = \langle X, X \rangle$

    (i) $f : \mathbb{R}^4 \to \mathbb{R}$, $f(X) = \langle X, X \rangle$

    (j) $f : \mathbb{R} \to \mathbb{R}$, $f(X) = \langle X, X \rangle$

    (k) $g : \mathbb{R}^2 \to \mathbb{R}$, $g(x, y) = e^{xy}$

    (l) $F : \mathbb{R} \to \mathbb{R}^2$, $F(t) = (t, t^2)$

    (m) $G : \mathbb{R} \to \mathbb{R}^2$, $G(t) = (t, t)$

    (n) $F : \mathbb{R} \to \mathbb{R}^2$, $F(t) = (1 - t, 3 + 5t)$

    (o) $f : \mathbb{R}^n \to \mathbb{R}$, $f(X) = \langle Z, X \rangle$, Z a fixed member of $\mathbb{R}^n$.

2. For the functions in Exercise 1, parts (d), (e), (f), (l), (m), and (n), write the coordinate functions.

3. Which of the functions in Exercise 1 is a vector field?

4. How many coordinate functions do the functions in Exercise 1, parts (a) and (b), have? Identify them.

4.1   Limits, Continuity, and Curves

In this section we generalize to *vector-valued* functions of a *scalar* variable two topics from single-variable calculus: limits and continuity.

4.1A   LIMITS

We consider a function F whose domain $\mathcal{D}(F)$ is an open interval (a, b) of $\mathbb{R}$ and whose target space is $\mathbb{R}^q$. We wish to define

$$\lim_{x \to x_0} F(x),$$

"the limit as x approaches x_0 of $F(x)$", where x_0 may either belong to $\mathcal{D}(F)$ or equal one of the endpoints a or b. We want the limit to be a vector $L \in \mathbb{R}^q$.

Definition 4.1.1 Let F be a function with domain $(a, b) \subseteq \mathbb{R}$ and target space $\mathbb{R}^q$ and let $L \in \mathbb{R}^q$. Also let x_0 be a real number such that $a \le x_0 \le b$. The **limit** of $F(x)$ as x approaches x_0 is equal to L, written as

$$\lim_{x \to x_0} F(x) = L,$$

if and only if

$$\lim_{x \to x_0} \|F(x) - L\| = 0;$$

that is, if and only if for every positive number ε there is a positive number δ such that $\|F(x) - L\| < \varepsilon$ whenever $x \in \mathcal{D}(F)$ and $0 < |x - x_0| < \delta$.

This definition is a precise mathematical way of expressing the notion that as x gets close to x_0, the vector $F(x)$ gets close to the vector L. Note how it states the concept of the convergence of a vector-valued function $F(x)$ to a vector L in terms of the convergence of the *scalar-valued* function $\|F(x) - L\|$ to the scalar 0.

The assertion that

$$\lim_{x \to x_0} F(x) = L \,,$$

can also be written as $F(x) \to L$ as $x \to x_0$, which is read: "$F(x)$ approaches L as x approaches x_0." When the condition in the definition is satisfied, the interval $(x_0 - \delta, x_0 + \delta)$ might be called the "confidence interval" because when x belongs to that interval and is different from x_0, one is confident that the distance from $F(x)$ to L is less than ε. In case x_0 is equal to a, the left-hand endpoint of the interval of definition, one might, for emphasis, write $x \downarrow a$ rather than $x \to a$. The use of this notation may eliminate the necessity of explicitly discussing the domain being considered. The notation $x \uparrow b$ is, of course, also useful. One sometimes speaks of a "limit existing". The limit of $F(x)$ **exists** as x approaches x_0 if and only if there is some $L \in \mathbb{R}^q$ for which

$$\lim_{x \to x_0} F(x) = L \,.$$

In Definition 4.1.1 $\mathcal{D}(F)$ was assumed to equal (a, b). Nothing would change were the domain equal to $[a, b)$ or $(a, b]$ or $[a, b]$. The reason is that the limit as $x \downarrow a$, for example, does not involve the value of the function at a.

It is sometimes useful to consider the limit of a function as x approaches ∞ or $-\infty$. If the domain of F is an interval of the form (a, ∞), then

$$\lim_{x \to \infty} F(x) = L$$

if and only if for every positive number ε, there is a positive number δ such that $\|F(x) - L\| < \varepsilon$ whenever $x > 1/\delta$. We leave it to the reader to formulate a similar definition for the limit as x approaches $-\infty$.

Before giving some examples, it will be helpful to work out several of the properties of limits. These properties are quite similar to those found in single-variable calculus for limits of scalar-valued functions. Our first property says that a function F cannot have two different limits as x approaches x_0.

Theorem 4.1.1 If F is an $\mathbb{R}^q$-valued function of a real variable for which $F(x) \to K$ and $F(x) \to L$ as $x \to x_0$, then $K = L$.

PROOF: The plan is to prove that $\|K - L\| = 0$, for then it will immediately follow that $K = L$. We perform this task by using the triangle inequality:

$$\|K - L\| = \|(K - F(x)) + (F(x) - L)\| \le \|K - F(x)\| + \|F(x) - L\| ,$$

which approaches $0 + 0 = 0$ as $x \to x_0$, since $F(x) \to K$ and $F(x) \to L$ as $x \to x_0$. Thus $\|K - L\| \le 0$. Since the norm of a vector is always nonnegative, it follows that $\|K - L\| = 0$. <<

The next theorem says that when a function approaches a limit then so does its norm. Before stating this theorem, we need to prove the subtraction form of the triangle inequality.

Lemma 4.1.1 Let X and Y be vectors in $\mathbb{R}^q$. Then

$$| \|X\| - \|Y\| | \le \|X - Y\| .$$

PROOF: The usual triangle inequality says $\|P+Q\| \le \|P\| + \|Q\|$, where P and Q are any vectors in $\mathbb{R}^q$. Let $P = Y$ and $Q = X - Y$, so $P + Q = X$. Then $\|X\| \le \|Y\| + \|X - Y\|$, so $\|X\| - \|Y\| \le \|X - Y\|$. Interchanging the roles of X and Y, we also see that $\|Y\| - \|X\| \le \|Y - X\| = \|X - Y\|$. Putting these last two inequalities together, we obtain the lemma. <<

Theorem 4.1.2 If F is an $\mathbb{R}^q$-valued function of a real variable for which $F(x) \to L$ as $x \to x_0$, then

$$\|F(x)\| \to \|L\|$$

as $x \to x_0$.

PROOF: The desired conclusion can be rephrased as

$$\|F(x)\| - \|L\| \to 0 .$$

To show this, we will "pinch" $\|F(x)\| - \|L\|$ between two functions each of which approaches 0. By the lemma,

$$-\|F(x) - L\| \le \|F(x)\| - \|L\| \le \|F(x) - L\| .$$

Since $F(x) \to L$, both the leftmost function and the rightmost function approach 0. By the "pinching theorem" (also called the "sandwich theorem" in some single-variable calculus books),

$$\|F(x)\| - \|L\| \to 0$$

as desired. <<

Let $Y = (y_1, y_2, \ldots, y_q)$ and $Z = (z_1, z_2, \ldots, z_q)$ be two members of $\mathbb{R}^q$. Then

$$\max\{|y_1 - z_1|, |y_2 - z_2|, \ldots, |y_q - z_q|\}$$
$$\le \sqrt{[y_1 - z_1]^2 + [y_2 - z_2]^2 + \cdots + [y_q - z_q]^2} = \|Y - Z\| .$$

Thus, the distance between any two coordinates of Y and Z is no larger than the distance between Y and Z themselves. So, if $F(x) \to L$ as $x \to x_0$, then each coordinate function of F approaches the corresponding coordinate of the vector L.

On the other hand,

$$
\begin{aligned}
\|Y - Z\| &= \sqrt{[y_1 - z_1]^2 + [y_2 - z_2]^2 + \cdots + [y_q - z_q]^2} \\
&\leq \sqrt{q \max\{[y_1 - z_1]^2, [y_2 - z_2]^2, \ldots, [y_q - z_q]^2\}} \\
&= \sqrt{q} \max\{[y_1 - z_1], [y_2 - z_2], \ldots, [y_q - z_q]\} .
\end{aligned}
$$

So, the distance between the vectors Y and Z is no larger than the constant $\sqrt{q}$ multiplied by the largest distance between their corresponding coordinates. Therefore, if each coordinate function of F approaches a limit as $x \to x_0$, then the same is true for the vector-valued function F itself, and, moreover, the coordinates of its limit are the limits of its coordinate functions. Let us summarize our conclusions.

Theorem 4.1.3 Let F be an $\mathbb{R}^q$-valued function of a real variable and let $L \in \mathbb{R}^q$. Denote the coordinate functions of F by f_j and the coordinates of L by l_j, $j = 1, 2, \ldots, q$. Then $F(x) \to L$ as $x \to x_0$ if and only if, for each j, $f_j(x) \to l_j$ as $x \to x_0$.

The preceding theorem is very useful for obtaining other properties of limits, since it enables us to apply what we know from single-variable calculus to coordinate functions and thence to draw conclusions about vector-valued functions. It is left for the reader to use this approach to provide proofs for the following two theorems.

Theorem 4.1.4 Suppose that F and G are $\mathbb{R}^q$-valued functions defined on the same interval in $\mathbb{R}$. If $F(x) \to K$ and $G(x) \to L$ as $x \to x_0$, then

$$
\lim_{x \to x_0} (F(x) + G(x)) = K + L .
$$

It should be realized that part of the preceding theorem is the assertion that if $\lim_{x \to x_0} F(x)$ and $\lim_{x \to x_0} G(x)$ exist, then so does $\lim_{x \to x_0} (F(x) + G(x))$. The assumption that F and G have a common domain is made chiefly to ensure that the sum $F + G$ has an appropriate domain. It should be obvious to the reader how to relax this assumption here and in subsequent theorems of a similar nature.

Theorem 4.1.5 Let F be an $\mathbb{R}^q$-valued function and let h be an $\mathbb{R}$-valued function, with both functions being defined on the same interval in $\mathbb{R}$. If $F(x) \to L$ and $h(x) \to b$ as $x \to x_0$, then

$$
\lim_{x \to x_0} h(x) F(x) = bL .
$$

Once one sees for herself or himself how easy it is to prove the preceding two theorems using Theorem 4.1.3, one is tempted to *always* think of a vector-valued function in terms of its coordinate functions. It is wise to avoid that temptation, especially when geometrical intuition is relevant. If the values of a function are points in three-dimensional space, it is often useful to regard them as exactly that, rather than as ordered triples of real numbers.

The preceding theorem deals with the product of a scalar-valued function and a vector-valued function. The next result is concerned with the inner product of two vector-valued functions.

Theorem 4.1.6 Let F and G be $\mathbb{R}^q$-valued functions defined on the same interval in $\mathbb{R}$. If $F(x) \to K$ and $G(x) \to L$ as $x \to x_0$, then

$$\lim_{x \to x_0} \langle F(x), G(x) \rangle = \langle K, L \rangle .$$

PROOF: Denote the coordinate functions of F and G by f_j and g_j, respectively, $j = 1, 2, \ldots, q$. Write $K = (k_1, k_2, \ldots, k_q)$ and $L = (l_1, l_2, \ldots, l_q)$ We use the single-variable versions of Theorems 4.1.4 and 4.1.5:

$$\begin{aligned}
\lim_{x \to x_0} \langle F(x), G(x) \rangle &= \lim_{x \to x_0} [f_1(x)g_1(x) + f_2(x)g_2(x) + \cdots + f_q(x)g_q(x)] \\
&= k_1 l_1 + k_2 l_2 + \cdots + k_q l_q \\
&= \langle K, L \rangle .
\end{aligned}$$

Done. <<

We also have a theorem concerning the composition of two functions. Let F be an $\mathbb{R}^q$-valued function defined on an interval in $\mathbb{R}$, and let g be a scalar-valued function, also defined on an interval in $\mathbb{R}$. In order for $F \circ g$ to be defined, we need to assume that the image of g is contained in the domain of F. The coordinate functions of $F \circ g$ are the functions $f_1 \circ g, \ldots, f_q \circ g$, where $f_1, \ldots, f_q$ are the coordinate functions of F. From single-variable calculus, we know that if $g(x) \to t$ as $x \to x_0$, and if f is a function which is continuous at t, then $(f \circ g)(x) \to f(t)$ as $x \to x_0$. Thus we have the following, whose proof is an immediate consequence of Theorem 4.1.3:

Theorem 4.1.7 Let F be an $\mathbb{R}^q$-valued function defined on an interval in $\mathbb{R}$, and let g be a scalar-valued function, also defined on an interval in $\mathbb{R}$. Assume that the image of g is contained in the domain of F. Suppose further that

$$\lim_{x \to x_0} g(x) = t ,$$

and that each of the coordinate functions of F is continuous at t. Then

$$\lim_{x \to x_0} (F \circ g)(x) = F(t) .$$

Theorems 4.1.3 through 4.1.7 are tools for calculating many different limits.

EXAMPLES:

1. In many applications of Theorems 4.1.4, 4.1.5, and 4.1.6, one of the two functions is a constant. For instance, if $C \in \mathbb{R}^q$ and F is an $\mathbb{R}^q$-valued function for which $F(x) \to L$ as $x \to x_0$, then

$$C + F(x) \to C + L \quad \text{and} \quad \langle C, F(x) \rangle \to \langle C, L \rangle$$

as $x \to x_0$.

2. If the coordinate functions of F are continuous, then limits involving F are easy to compute by using Theorem 4.1.3. Thus, for example, if $F : \mathbb{R} \to \mathbb{R}^2$ is defined by $F(x) = (e^{-x} \cos x, e^{-x} \sin x)$, then

$$\lim_{x \to 0} F(x) = (1, 0) .$$

If we let $g(x) = \sqrt{x}$ for $x \geq 0$, then by Theorem 4.1.7,

$$\lim_{x \downarrow 0} (F \circ g)(x) = F(0) = (1, 0) .$$

Note that the function F of this example could also be written as $F(x) = h(x)G(x)$ where $h(x) = e^{-x}$ and $G(x) = (\cos x, \sin x)$.

3. If $\|F(x)\| \to 0$ as $x \to x_0$, then $F(x) \to \mathbf{0}$ as $x \to x_0$. PROOF: let $L = \mathbf{0}$ in the definition of limit, and then use the fact that $\|F(x) - \mathbf{0}\| = \|F(x)\|$.

As an application consider the previous example, where

$$\|F(x)\| = |e^{-x}| \sqrt{\cos^2 x + \sin^2 x} = e^{-x} .$$

Since e^{-x} approaches 0 as x approaches ∞, it follows that

$$\lim_{x \to \infty} F(x) = \mathbf{0} .$$

4.1B CONTINUITY

The definition of continuity for vector-valued functions of a scalar variable is the same as for scalar-valued functions.

Definition 4.1.2 Let F be an $\mathbb{R}^q$-valued function whose domain is an interval. The function F is **continuous at** a point x_0 in its domain if and only if $F(x) \to F(x_0)$ as $x \to x_0$.

If F is not continuous at some point x_0 in its domain, then we say that F is **discontinuous** at x_0. A function F that is continuous at every point in its domain is said to be **continuous**. If F is not continuous, then F is **discontinuous**. Thus, a discontinuous function might be continuous at some points in its domain.

Since continuity is defined in terms of limits and we know several theorems about limits, we immediately obtain some facts about continuity which we summarize in the following theorem.

Theorem 4.1.8 Let F and G be $\mathbb{R}^q$-valued functions, and let h be a scalar-valued function. Assume that the functions F, G and h are defined on the same interval in $\mathbb{R}$. Also, let g be a scalar-valued function that is defined on an interval and whose image is a subset of the domain of F.

1. $G = (g_1, g_2, \ldots, g_q)$ is continuous at x if and only if each of its coordinate functions g_j is continuous at x.

2. If F and G are continuous at x, then so are $F + G$ and $\langle F, G \rangle$.

3. If h and F are continuous at x, then so is hF.

4. If F is continuous at x, then so is $\|F\|$.

5. If g is continuous at x and F is continuous at $g(x)$, then $F \circ g$ is continuous at x.

Subtraction and division are not mentioned in the theorem because they can be treated in terms of addition and multiplication. For example, replace h by $1/h$ in Property 3 and assume that $h(x) \neq 0$ to obtain a result about F/h. Note that when $q > 1$ the function F/G is *not* defined, since division by vectors is *not* defined.

The preceding theorem enables us to create continuous vector-valued functions from continuous $\mathbb{R}$-valued functions and new continuous vector-valued functions from vector-valued functions that we already know to be continuous. The following theorem states that affine functions with real domains are continuous.

Theorem 4.1.9 Every affine map $T : \mathbb{R} \to \mathbb{R}^q$ is continuous.

PROOF: The domain of T equals the entire real line $\mathbb{R}$ and a formula for $T(x)$ is given by $Ax + B$ where $A, B \in \mathbb{R}^q$. We may think of both A and B as $\mathbb{R}^q$-valued constant functions. By Property 3 of Theorem 4.1.8, Ax is a continuous $\mathbb{R}^q$-valued function (let $h(x) = x$). By Property 2 of that same theorem, $Ax + B$ is continuous. $\ll$

The main object of study in this chapter will now be defined.

Definition 4.1.3 A **parametrized curve** in $\mathbb{R}^q$ is a continuous function $F : I \to \mathbb{R}^q$, where I is some interval of real numbers.

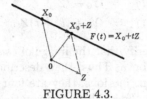

FIGURE 4.3.

Definition 4.1.4 A **curve** in $\mathbb{R}^q$ is the image of some parametrized curve F in $\mathbb{R}^q$. The function F is called a **parametrization** of the curve.

When discussing parametrized curves, we will usually denote the domain variable by t rather than x. The variable t, called a **parameter** in the context of curves, encourages the reader to think of "time". A parametrized curve can be regarded as a precise and detailed schedule for the motion of some object in $\mathbb{R}^q$; $F(t)$ denotes the location of this object at time t.

It is important that one become very familiar with the following examples so that one feels comfortable when they arise in the middle of other discussions and problems. The first example consists of a line and some related topics. The second and third also focus on new points of view in relation to familiar objects, for example, circles and graphs of $\mathbb{R}$-valued functions of a real variable.

EXAMPLES:

4. *Lines and subsets of lines.* We have previously defined a line in $\mathbb{R}^q$ to be

$$\{tZ + Y_0 : t \in \mathbb{R}\}$$

for some Y_0 and Z belonging to $\mathbb{R}^q$ with $Z \neq \mathbf{0}$. We say the same thing by describing it as the image of the function $F(t) = Y_0 + tZ$. Thus, it is a *curve*, since the affine function F is a continuous function, according to Theorem 4.1.9. (It may not seem appropriate to call a line a curve, but this is one of those cases where a name which seems appropriate for a general class of objects does not fit well for certain special objects within that class.) See Figure 4.3 in which the image of the function F is illustrated for the case where Z and Y_0 are members of $\mathbb{R}^2$. The parametrized curve F represents motion along the line shown, with $F(t) = tZ + Y_0$ being the location at the instant t. This is a slightly different point of view from that taken in Chapters 1, 2, and 3, where a rather static view is taken.

Now let us compare F to the following parametrized curves:

$$\begin{aligned} G(t) &= Y_0 + (t\sqrt{2})Z \,, \\ H(t) &= Y_0 + (\log t)Z \,, \quad t > 0 \,, \end{aligned}$$

$$J(t) = Y_0 - tZ ,$$
$$K(t) = Y_0 + (t^3 - t)Z .$$

The reader should verify that the image of each is the same, namely, the line described above. Thus, they describe the same curve, but they describe different motions; they are four different *parametrized* curves. The function G describes a motion that goes faster than that described by F, by a factor of $\sqrt{2}$. The motion described by H is sometimes faster than that described by F and sometimes slower, and, moreover, the entire line is traversed even though only positive times are used. All three of the functions F, G, and H describe motion in the same direction with no turning around. There is no turning around because the functions are one-to-one functions, as is J. With J the line is traversed in the opposite direction from that indicated by F, G, and H. Since $K(0) = K(1) \neq K(1/2)$, we conclude that the function K describes motion in which there is some turning around.

5. *Rays and line segments.* The curves described by each of the following are subsets of the line of the previous example:

$$M(t) = tZ + Y_0 , \quad t \geq 0 ;$$
$$N(t) = tZ + Y_0 , \quad 0 \leq t \leq 1 ;$$
$$P(t) = tZ + Y_0 , \quad 0 < t < 1 .$$

The curve described by M is a **ray** emanating from Y_0 and going in the same direction as does Z. The curves described by N and P are almost identical—a **line segment** for N, and for P, the set obtained by removing the endpoints Y_0 and $Z + Y_0$ from that segment.

6. *Circles in the plane.* Let a denote a fixed positive constant. The functions we now consider all have the same formula (but different domains), so we will abuse notation somewhat by using the symbol F for all of them:

$$F(t) = (a \cos t, a \sin t) .$$

Alternatively, we could write

$$F(t) = a(\cos t, \sin t)$$

or $F = a(f_1, f_2)$, where $f_1 = \cos$ and $f_2 = \sin$. The reader can check, using the trigonometric identity

$$\cos^2 t + \sin^2 t = 1 ,$$

that every point in the image of F belongs to the circle of radius a centered at $(0,0)$. What will distinguish one function from the other in our discussion will be their various domains. For the domain $[0, 2\pi)$,

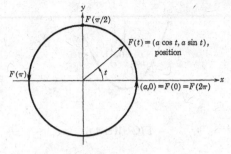

FIGURE 4.4.

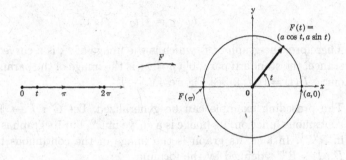

FIGURE 4.5. $F(t)$ traces out the circle

the motion consists of one revolution around the circle in a counterclockwise direction. In particular, $F(0) = (a, 0)$, $F(\pi/2) = (0, a)$, $F(\pi) = (-a, 0)$, $F(3\pi/2) = (0, -a)$, and $F(t) \to (a, 0)$ as $t \uparrow 2\pi$. If $F(t)$ is viewed as an arrow emanating from the origin, then its head traces out the circle and t is the measure (in radians) of the angle from the positive x-axis counterclockwise to the arrow $F(t)$. If we enlarge the domain to $[0, 2\pi]$, then the parametrized curve returns to its starting point at time $t = 2\pi$. See Figure 4.4 for a picture of the image of F. An appropriate, although not widely used, name for Figure 4.5 is **domain-image picture**. If we change the domain to $[0, 4\pi)$, then F indicates two revolutions around the circle. The domain $[0, \infty)$ is appropriate for a continuing revolution around the circle, starting from time $t = 0$. The domain $(-\infty, \infty)$ continues this motion into the infinite past before time 0. The graph of F requires three dimensions, one for its domain and two for its target space. It is a good exercise, for the various domains, to draw or at least visualize the graph of F.

7. *Graphs as curves.* If $y = f(x)$ is a continuous function defined on an interval I, then we may regard its graph as the image of the mapping $F : \mathbb{R} \to \mathbb{R}^2$, where

$$F(t) = (t, f(t)) .$$

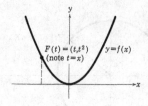

FIGURE 4.6.

In terms of coordinates

$$x = t , \quad y = f(t) = f(x) .$$

Therefore, the graph of f, which is the image of F, is a curve. For instance, the standard parabola $y = x^2$ is the image of the parametrized curve $F(t) = (t, t^2)$. See Figure 4.6.

8. The preceding example can be generalized. Let $G : I \to \mathbb{R}^q$ be a continuous function. Its image is a curve in $\mathbb{R}^q$, but its graph is a curve in $\mathbb{R}^{q+1}$. In fact, its graph is the image of the continuous function $F : I \to \mathbb{R}^{q+1}$ defined by the formula

$$F(t) = (t, G(t)) ,$$

where $(t, G(t))$ represents a member of $\mathbb{R}^{q+1}$, of which the first coordinate is t, and the remaining coordinates are the q coordinates of G. The image of the function $F = (\cos, \sin)$ of Example 6 is the same for the two domains $[0, 2\pi)$ and $\mathbb{R}$, namely, a circle, but the two graphs are quite different. A limited pictorial representation of the graph is shown in the next example for the case where the domain is $\mathbb{R}$; however, in that example t is the third coordinate rather than the first.

9. *A helix.* This curve may be given by $F : \mathbb{R} \to \mathbb{R}^3$,

$$F(t) = (\cos t, \sin t, t) .$$

At $t = 0$, $F(0) = (1, 0, 0)$. See Figure 4.7. As t increases from 0, the first two coordinates indicate that $F(t)$ moves in a circle with respect to a horizontal plane. Meanwhile the vertical z-coordinate is increasing, so that the point $F(t)$ is constantly rising. At $t = 2\pi$, the point $F(2\pi) = (1, 0, 2\pi)$ is 2π units from and directly above $F(0) = (1, 0, 0)$.

We finish this section with a theorem that is an analogue of the intermediate value theorem of single-variable calculus.

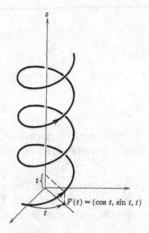

FIGURE 4.7. A helix

Theorem 4.1.10 Suppose that the points Y and Z are on a curve in $\mathbb{R}^q$ and that r is a positive number less than $\|Y - Z\|$. Then there exists a point P on the curve such that $\|Y - P\| = r$.

PROOF: Let F denote a parametrization of the curve. Then $F - Y$ (with Y viewed as a constant function) is a continuous function, and, by Theorem 4.1.8, so is the function $\|F - Y\|$. This latter scalar-valued function takes on the value 0 (since Y belongs to the image of F) and also the value $\|Z - Y\|$. Since r is between these two values and the domain of $\|F - Y\|$ is an interval, the intermediate value theorem applies. We conclude there exists t such that $\|F(t) - Y\| = r$, as desired. $<<$

Exercises

1. Each of the following is a modification of the function in Example 6 of Section 4.1b. For each, describe differences from and similarities to Example 6.

   (a) $F(t) = (\cos t, \sin t)$, $t \in [0, 4\pi)$

   (b) $F(t) = (\cos t, \sin t)$, $t \in [0, 2\pi]$

   (c) $F(t) = (\sin t, \cos t)$, $t \in [0, 2\pi)$

   (d) $F(t) = (3\cos t, -3\sin t)$, $t \in [0, 4\pi)$

   (e) $F(t) = (\cos(t/2), \sin(t/2))$, $t \in [0, 4\pi)$

   (f) $F(t) = (\cos t, -\sin t)$, $t \in [0, 2\pi)$

   (g) $F(t) = (\cos t, \sin t)$, $t \in (-\pi, \pi]$

   (h) $F(t) = (\cos(2\pi t), \sin(2\pi t))$, $t \in [0, 1)$

   (i) $F(t) = (3\cos t, 7, 3\sin t)$, $t \in [0, 2\pi)$.

2. Sketch the following curves and indicate with arrows the direction in which the parameter increases:

(a) $F(t) = (2t, 3 - t)$

(b) $G(t) = (2t, |3 - t|)$

(c) $\Phi(s) = (s^2, 1 + s^2)$

(d) $H(\theta) = (\theta, \sin \theta)$

(e) $V(t) = (t^2, 1 + t^4)$

(f) $F(t) = (t^2 \cos t, t^2 \sin t)$.

3. The helix in Example 9 moves counterclockwise as it moves upward, when viewed from above. Find a parametrized equation of a helix that moves clockwise.

4. For $t \in [0, 2\pi)$, let $F(t) = (3 \cos t, 2 \sin t)$. Describe the image of F. Is F a one-to-one function? (Recall that a function F is one-to-one if for each point Y in its image there is only one point t in its domain for which $F(t) = Y$.)

5. Parametrically describe the ray:

   (a) emanating from $(0, 0, 0)$ and passing through $(2, 5, -3)$

   (b) emanating from $(4, -4)$ and passing through $(-4, 4)$

   (c) emanating from $(0, 1, 0, 1)$ and passing through $(1, 0, 1, 0)$.

6. Parametrically describe the line segment:

   (a) with endpoints $(1, 3)$ and $(3, 1)$

   (b) with endpoints $(0, 0, 0)$ and $(5, 5, 5)$

   (c) with endpoint $(1, 3)$ and midpoint $(5, 6)$

   (d) with endpoints $(0, 0, \ldots, 0)$ and $(1, 1, \ldots, 1)$

   (e) with endpoints $(-1, -1, \ldots, -1)$ and $(1, 1, \ldots, 1)$

   (f) with endpoints $(0, 1, 1, \ldots, 1)$ and $(1, 1, 1, \ldots, 1)$.

7. A particle moves along some parabola $y = cx^2$ in such a way that for $t \geq 0$ the y-coordinate is proportional to the cube of the time elapsed, and so $y = kt^3$ for some constant k. If the particle is at $(2, 1)$ when $t = 1$, find the function $F(t) = (x(t), y(t))$. Where is the particle when $t = 4$?

8. Repeat the preceding exercise, but this time with the particle moving along the cubic $y = x^3 + c$. Where is the particle when $t = 4$?

9. Show that the curve $F(t) = (\cos t, \cos 2t)$, $0 \leq t \leq \pi$, lies on a parabola, find a (nonparametric) equation for that parabola, and sketch that portion of the parabola that is the image of F.

10. Show that the curve $F(t) = (a \cos t, b \sin t)$, $0 \leq t \leq 2\pi$, $a > 0$, $b > 0$, lies on the ellipse $x^2/a^2 + y^2/b^2 = 1$.

11. Sketch the planar curve $F(t) = (t(t^2 - 1), t^2 - 1)$. Note that the curve crosses itself. *Hint:* plot some points, and examine, roughly, the nature of the curve for $|t|$ large.

12. Sketch the planar curve $F(t) = (e^{-t}\cos t, e^{-t}\sin t)$. Evaluate whichever of the following exist:
$$\lim_{t\to\infty} F(t),$$
$$\lim_{t\to-\infty} F(t).$$

13. Explain why the function
$$F(t) = \begin{cases} (t+1, 2t-3) & \text{if} \quad 0 \le t < 1 \\ (t^2+1, t) & \text{if} \quad 1 \le t < 2 \\ (7-t, |t-4|) & \text{if} \quad 2 \le t \end{cases}$$
is discontinuous. At which points is it discontinuous? Sketch its image.

14. True or false:

   (a) If $F, G, H : \mathbb{R} \to \mathbb{R}^3$ are continuous, then so is $[F, G, H]$.

   (b) If $F, G : \mathbb{R} \to \mathbb{R}^q$ are functions such that $\langle F, G \rangle$ is continuous, then F and G are continuous.

   (c) If $\lim_{x\to x_0} \|F(x)\|$ exists, then $\lim_{x\to x_0} F(x)$ exists.

   (d) If $\lim_{x\to x_0} h(x)F(x)$ exists and is not equal to 0, for $h : \mathbb{R} \to \mathbb{R}$, $F : \mathbb{R} \to \mathbb{R}^q$, then $\lim_{x\to x_0} F(x)$ exists.

   (e) If $\lim_{x\to x_0} F(x)$ exists, then F is continuous at x_0.

   (f) If F is continuous at x_0, then $\lim_{x\to x_0} e^x F(x)$ exists.

   (g) If $F : (a, b) \to \mathbb{R}^q$ is continuous, then $\lim_{x\downarrow a} F(x)$ exists.

   (h) If $F, G : \mathbb{R} \to \mathbb{R}^q$ have the same images, and if F is continuous, then so is G.

   (i) If the image of $F : \mathbb{R} \to \mathbb{R}^q$ is a single point, then F is continuous.

   (j) If $F : [a, b] \to \mathbb{R}^q$ is continuous, then $F(a) = F(b)$.

15. Discuss how one might define continuity for functions whose domain is not an interval. Then consider the function
$$F(t) = (t, \sqrt{t^2 - 4}).$$
Explain why it is continuous, identify its domain, and sketch its image.

16. Show that the curve $G(s) = (s\cos s, s\sin s, s^2 + 1)$ lies on the paraboloid $x^2 + y^2 = z - 1$. Draw a rough sketch.

17. (a) Find a parametric representation of a line that passes through $P = (-3, 1, -2)$ at $t = -1$ and $Q = (0, 1, 1)$ at $t = 2$.

   (b) Let P and Q be distinct points in $\mathbb{R}^n$. Find a parametric equation of a line that passes through the point P at $t = t_1$ and Q at $t = t_2$.

18. Prove the following:

   (a) Theorem 4.1.4

   (b) Theorem 4.1.5.

19. Prove Theorem 4.1.7 by focusing on the coordinate functions of F and using a theorem from single-variable calculus about limits of compositions of functions.

20. Give an example of an $\mathbb{R}^2$-valued function F with $\mathcal{D}(F)$ equal to an interval (a, b) and having the following properties:

$$\lim_{x \uparrow b} F(x) \text{ does not exist ;}$$
$$\lim_{x \uparrow b} \|F(x)\| \text{ does exist .}$$

Is such an example possible in the $\mathbb{R}^1$-setting?

21. Let $q > 1$. A **circle** in $\mathbb{R}^q$ is defined to be the set of all points in some two-dimensional affine subspace of $\mathbb{R}^q$ that are a fixed positive distance from some fixed point in that affine subspace. The fixed positive distance is the **radius** of the circle and the fixed point is its **center**. Let Y, Z_1, and Z_2 be members of $\mathbb{R}^q$ and suppose that Z_1 and Z_2 have the same norm and are perpendicular to each other. Let

$$F(t) = Y + Z_1 \cos t + Z_2 \sin t , \quad 0 \le t < 2\pi .$$

Prove that the image of F is a circle by first showing that every member of the image of F is on some circle and then showing that every member of that circle belongs to the image of F. What are the center and radius of the circle?

22. Consider a particular case of the circles defined in the preceding exercise, namely, assume that $\langle Y, Z_1 \rangle = 0$ and $\langle Y, Z_2 \rangle = 0$. Prove that all the points on the circle are on a sphere centered at the origin and find the radius of that sphere.

23. Consider the parametrized curve

$$(\cos^2 s, \cos s \sin s, \cos s \sin s, \sin^2 s) , \quad s \in [0, \pi) .$$

Show that the image is a circle in $\mathbb{R}^4$ by parametrizing the image as in Exercise 21. Then apply Exercise 22 if it is relevant.

24. Parametrize the square whose vertices are $(-1, -1)$, $(-1, 1)$, $(1, 1)$, and $(1, -1)$.

25. Find the points of intersection of the two ellipses in $\mathbb{R}^2$ given parametrically by the following:

$$(\cos s, 3 \sin s) , \quad 0 \le s < 2\pi ;$$
$$(2 \cos t, \sin t) , \quad 0 \le t < 2\pi .$$

Hint: It is possible for the two ellipses to intersect at a point without the corresponding values of the two parameters being equal.

26. Let

$$F(t) = \left(\frac{1 - t^2}{1 + t^2}, \frac{2t}{1 + t^2} \right) , \quad t \in [0, 1] .$$

(a) Calculate $F(1/4)$, $F(1/3)$, $F(1/2)$, and $F(2/3)$.

(b) Find the image of F.

(c) Explain why the coordinates of $F(t)$ are rational if t is rational.

(d) Show that F is a one-to-one function.

(e) Show that if the coordinates of $F(t)$ are rational, then t is rational.

(f) Describe how the function F can be used to make an infinite list of all pairs of nonnegative rational numbers that have the property that the sum of their squares equals 1.

(g) Describe a method for listing all **primitive Pythagorean triples**—that is, triples (a, b, c) such that a, b, and c are nonnegative integers which have no common positive integral divisors other than 1 and which satisfy $a^2 + b^2 = c^2$.

27. For fixed positive integers n, consider the parametrized curves

$$F(t) = (\cos t \cos nt, \sin t \cos nt, \sin nt), \ t \in [0, 2\pi] \,.$$

Show that their images lie on the sphere of radius 1 about the origin in $\mathbb{R}^3$. In other words, show that every point on these curves is 1 unit from the origin. For various values of n, draw the curves on the surface of a ball.

28. Follow the instructions of the preceding exercise for the parametrized curves

$$F(t) = (\cos nt \cos t, \sin nt \cos t, \sin t), \ t \in [0, 2\pi].$$

4.2   The Tangent Map

4.2A   WHAT DIFFERENTIAL CALCULUS IS ALL ABOUT

The chief triumph of differential calculus of one real variable, due to Barrow, Newton, Leibnitz, and their contemporaries, is illustrated in Figure 4.8. Given a function f, it is possible to define, compute, and interpret the tangent line to the graph $y = f(x)$ above each point x_0, provided the function f is differentiable at x_0. Recall that f is said to be differentiable at the point x_0 if the limit

$$f'(x_0) = \lim_{x \to x_0} \frac{f(x) - f(x_0)}{x - x_0}$$

exists as a finite number. This limit is known as the first derivative of f at x_0.

If f is differentiable at x_0, then the tangent line there is itself the graph of a function, the affine function

$$y = T(x) = f(x_0) + f'(x_0)(x - x_0).$$

To convince yourself that this formula for $y = T(x)$ does give the tangent line to $y = f(x)$ at x_0, you should check three things:

1. The graph $y = T(x)$ is a line in the xy-plane.

2. The line $y = T(x)$ goes through $(x_0, f(x_0))$, the point of tangency.

3. The line $y = T(x)$ has the correct slope $f'(x_0)$.

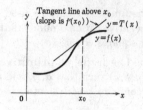

FIGURE 4.8.

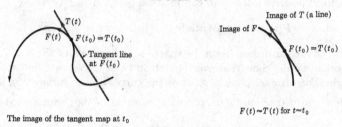

The image of the tangent map at t_0

$F(t) \approx T(t)$ for $t \approx t_0$

FIGURE 4.9. $F(t) \approx T(t)$ for $t \approx t_0$

If these three criteria are satisfied, then the graph of $y = T(x)$ clearly deserves to be called the tangent line to $y = f(x)$ at x_0. We urge you to verify these criteria; they will appear again.

There is an important connection that we now describe between the formula for the derivative and the formula for the tangent line. Subtract $f'(x_0)$ from both sides of the defining relation for the derivative to obtain

$$
\begin{aligned}
0 &= \lim_{x \to x_0} \frac{f(x) - f(x_0)}{x - x_0} - f'(x_0) \\
&= \lim_{x \to x_0} \frac{f(x) - f(x_0) - f'(x_0)(x - x_0)}{x - x_0} \\
&= \lim_{x \to x_0} \frac{f(x) - T(x)}{x - x_0} .
\end{aligned}
$$

Thus, we see that $T(x)$ is a very good approximation to $f(x)$ when x is close to x_0, so good that the difference between it and $f(x)$ is small compared to the difference between x and x_0. See Figure 4.8.

We have seen that if f is differentiable at x_0, then there is an affine function T that approximates f nicely near x_0. Moreover, we have a formula for T in terms of the derivative of f. The pictorial representation of the approximation property of T is that the graph of T is tangent to the graph of f at x_0.

Now consider the converse point of view. Suppose that T_1 is some affine function that approximates f near x_0 in the sense that $T_1(x_0) = f(x_0)$ and

$$
\lim_{x \to x_0} \frac{f(x) - T_1(x)}{x - x_0} = 0 .
$$

Does it follow that f is differentiable at x_0 and that $T_1(x) = f(x_0) + f'(x_0)(x - x_0)$? We will now give an affirmative answer to this question, and on that basis, if f is differentiable at x_0, call the affine function T *the best affine approximation* to f near x_0.

Consider T_1 as described above. Since T_1 is affine there exist scalars y_0 and m for which

$$T_1(x) = y_0 + m(x - x_0) .$$

Therefore, $y_0 = T_1(x_0)$, which, by definition of T_1, equals $f(x_0)$. We thus conclude that $y_0 = f(x_0)$ and that

$$\frac{f(x) - T_1(x)}{x - x_0} = \frac{f(x) - f(x_0)}{x - x_0} - m .$$

Since the limit of the left-hand side equals 0, so does the limit of the right-hand side. Therefore,

$$\lim_{x \to x_0} \frac{f(x) - f(x_0)}{x - x_0} = m .$$

That is, $f'(x_0)$ exists and equals m as desired.

As we generalize from scalars to vectors it is the two requirements that

$$f(x_0) = T(x_0) \quad \text{and} \quad \frac{f(x) - T(x)}{x - x_0} \to 0$$

that we will generalize.

So much for the philosophy of the best affine approximation T, also called the "tangent map", and its graph, the tangent line. Let us now recall some of its uses:

1. In seeking the local maxima and minima of $f(x)$, we know that we need only consider those points x_0 at which the tangent line is horizontal, that is, where its slope $f'(x_0) = 0$.

2. In computing the values of complicated functions, such as $f(x) = \sqrt{x}$, it often suffices to compute an approximate value $T(x)$ where $y = T(x)$ is the tangent map of $y = f(x)$ at a point x_0 near x.

3. If $y = f(x)$ is regarded as a position on the y–axis at time x, then the slope $f'(x_0)$ of the tangent line is the velocity or rate of change at the instant x_0.

Thus, an answer to the question, "With what is differential calculus concerned?" is "the best affine approximation" or, alternatively, "the tangent map."

4.2B   THE BASIC DEFINITIONS

Let F be a parametrized curve. We want to mimic ideas from the preceding section in order to define the following objects and illuminate their interconnections: the "tangent line to the graph" of F at a particular point $F(t_0)$, the "derivative" $F'(t_0)$, and the "tangent line to the image" of F at the point $F(t_0)$. It will develop that, by definition, the first of these three objects exists if and only if the second exists, but that the third may fail to exist if $F'(t_0)$ equals $\mathbf{0}$.

Definition 4.2.1 Let F be a parametrized curve with target space $\mathbb{R}^q$ and let $t_0 \in \mathcal{D}(F)$. The graph of an affine map $T : \mathbb{R} \to \mathbb{R}^q$ is **tangent** to the graph F at the point corresponding to t_0 if and only if

$$F(t_0) = T(t_0) \qquad \text{and} \qquad \lim_{t \to t_0} \frac{F(t) - T(t)}{t - t_0} = 0 \,.$$

If this is the case, we say that T is the **best affine approximation** to F near t_0, and that T is the **tangent map** of the parametrized curve F at t_0.

The following theorem, the proof of which is left for Exercise 17, says that the use of the word "the" in the preceding definition is valid.

Theorem 4.2.1 If T_1 and T_2 are tangent maps of the the parametrized curve F at t_0, then $T_1 = T_2$.

Thus, there are only two possibilities: no tangent map or one tangent map. Suppose there is one. It can be written in the form

$$T(t) = Y + tZ \,,$$

with $Y, Z \in \mathbb{R}^q$. Now we have $T(t_0) = Y + t_0 Z$, whence $Y = T(t_0) - t_0 Z$.
    Thus, we may write

$$
\begin{aligned}
T(t) &= T(t_0) - t_0 Z + tZ \\
&= F(t_0) + Z(t - t_0) \,.
\end{aligned}
$$

Here $Z(t - t_0)$ denotes multiplication of the vector Z by the scalar $t - t_0$. This form is especially relevant to studying $T(t)$ for t near t_0. See Figure 4.9.

Definition 4.2.2 If $T(t) = F(t_0) + Z(t - t_0)$ is the best affine approximation to $F : \mathbb{R} \to \mathbb{R}^q$ at t_0 for some Z, then we say that F is **differentiable at** t_0 and the vector $Z \in \mathbb{R}^q$ is called the **derivative** (also **total derivative**, **first derivative**, **velocity vector**) of F at t_0. We write

$$Z = F'(t_0) \,.$$

If F is differentiable at each point in its domain we say that F is **differentiable**. Some other notations for the total derivative are

$$\frac{dF}{dt}(t_0), \ (DF)(t_0), \ DF(t_0), \ \dot{F}(t_0).$$

The length $\|F'(t_0)\|$ is called the **speed** of F at t_0 in case F refers to particle motion.

When F is differentiable at t_0, we have the relation

$$T(t) = F(t_0) + F'(t_0)(t - t_0).$$

All the calculations connected with this relation are straightforward except that of $F'(t_0)$. The following theorem reduces the problem of calculating the derivative of a vector-valued function to that of calculating the derivatives of scalar-valued functions.

Theorem 4.2.2 Let F be a parametrized curve. Then F is differentiable at a point $t_0 \in \mathcal{D}(F)$ if and only if each coordinate function f_j of F is differentiable at t_0, in which case

$$F'(t_0) = (f_1'(t_0), f_2'(t_0), \ldots, f_q'(t_0)).$$

PROOF: We first assume that each coordinate function is differentiable at t_0. Let

$$T(t) = F(t_0) + (f_1'(t_0), f_2'(t_0), \ldots, f_q'(t_0))(t - t_0).$$

We use the fact that the limit of a vector function is the vector whose coordinates are the limits of the corresponding coordinates functions:

$$\lim_{t \to t_0} \frac{F(t) - T(t)}{t - t_0}$$

$$= \left(\lim_{t \to t_0} \frac{f_1(t) - [f_1(t_0) + f_1'(t_0)(t - t_0)]}{t - t_0}, \lim_{t \to t_0} \frac{f_2(t) - [f_2(t_0) + f_2'(t_0)(t - t_0)}{t - t_0}, \right.$$

$$\left. \ldots, \lim_{t \to t_0} \frac{f_q(t) - [f_q(t_0) + f_q'(t_0)(t - t_0)}{t - t_0} \right)$$

$$= (0, 0, \ldots, 0).$$

Therefore, T is the tangent map of F at t_0.

For the proof in the opposite direction assume that $F'(t_0)$ exists. Denote its coordinates by z_j and the coordinate functions of F by $f_j, j = 1, 2, \ldots, q$. Then

$$\lim_{t \to t_0} \frac{(f_1(t), \ldots, f_q(t)) - [(f_1(t_0), \ldots, f_q(t_0)) + (z_1, \ldots, z_q)(t - t_0)]}{t - t_0} = 0.$$

Hence, for each j,

$$\lim_{t \to t_0} \frac{f_j(t) - [f_j(t_0) + z_j(t - t_0)]}{t - t_0} = 0 \, ,$$

from which it follows that $z_j = f_j'(t_0)$. $<<$

REMARK:

- The preceding theorem says that the derivative of $F = (f_1, f_2, \ldots, f_q)$ is the most natural candidate, namely,

$$F' = (f_1', f_2', \ldots, f_q') \, .$$

We could have *defined* F' to be this, rather than defining it in terms of affine maps. Had we done this, however, the burden of proof would have been on us to show that $F'(t_0)$, defined computationally as $(f_1'(t_0), \ldots, f_q'(t_0))$, is *interesting* in that it helps us understand the parametrized curve $F(t)$. Our point of view is that tangent maps are clearly interesting and relevant from a geometric point of view, and therefore we should prove a theorem that tells us how to compute them. This we have just done.

EXAMPLE:

1. The function

$$F(t) = (a \cos t, a \sin t) \, , \quad t \in \mathbb{R} \, ,$$

parametrizes the circle of radius a centered at $(0, 0)$. Note that this parametrization describes the motion of a particle moving repeatedly around the circle. By Theorem 4.2.2, differentiation of $a \cos t$ and $a \sin t$ gives

$$F'(t_0) = (-a \sin t_0, a \cos t_0)$$

for each t_0. For instance, if $t_0 = \pi/6$, then $F'(\pi/6) = (-a/2, \sqrt{3}\, a/2)$. Let us find the best affine approximation to $F(t)$ at $t_0 = \pi/6$.

We know that tangent map has the form $T(t) = F(t_0) + F'(t_0)(t - t_0)$. Now

$$F(t_0) = F\left(\frac{\pi}{6}\right) = \left(\frac{\sqrt{3}\, a}{2}, \frac{a}{2}\right) \, .$$

Using the velocity vector $F'(\pi/6)$ computed above, we have

$$T(t) = \left(\frac{\sqrt{3}\, a}{2}, \frac{a}{2}\right) + \left(\frac{-a}{2}, \frac{\sqrt{3}\, a}{2}\right)\left(t - \frac{\pi}{6}\right) \, .$$

This may also be written

$$T(t) = \left(\frac{\sqrt{3}\, a}{2} - \frac{a}{2}\left(t - \frac{\pi}{6}\right), \frac{a}{2} + \frac{\sqrt{3}\, a}{2}\left(t - \frac{\pi}{6}\right)\right) \, .$$

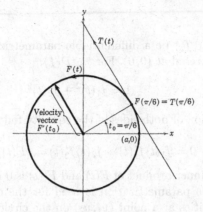

FIGURE 4.10.

The graph of F is a helix, and the graph of T is a tangent line to the helix in $\mathbb{R}^3$. More important for many purposes than this three-dimensional point of view is the two-dimensional point of view (see Figure 4.10) that focuses on the image of F and does *not* include an axis for the domain variable t. The tangent line shown is the *image* of T, not the *graph* of T, and is tangent to the *image* of F, not to the *graph* of F. Note that the tangent line is parallel to the velocity vector. The length of the velocity vector is the speed of motion at time t_0. For this example,

$$\|F'(t)\| = a$$

for all t, so the speed does not depend on t. The velocity does depend on t, since the direction of the velocity vector changes, but the speed does not. Intuitively, it is not surprising that in our example, the speed equals a, once it is known that the speed is constant. The perimeter of the circle of radius a equals $2\pi a$ units of length, and the particle traverses this circle in 2π units of time (why?). Thus the constant speed is $2\pi a/2\pi = a$ units of length per unit of time.

The preceding example illustrates how the derivative may be used to obtain the tangent line to a curve at a particular point. If the affine function $T(t) = Y + Zt$ is the tangent map of the parametrized curve F at t_0 and if $Z \neq \mathbf{0}$, we then also say that the *image* of T is the **tangent line** to the *image* of F at $F(t_0)$. The following examples will illustrate various issues concerned with this concept.

EXAMPLES:

2. Let $F = (f_1, f_2)$ be a differentiable parametrization of the circle of radius 1 centered at $(0,0)$. For $t \in \mathcal{D}(f)$,

$$1 = f_1(t)^2 + f_2(t)^2 .$$

Differentiation of both sides of the equality followed with division by 2 gives

$$0 = f_1(t)f_1'(t) + f_2(t)f_2'(t) = \langle F(t), F'(t) \rangle ,$$

that is, the inner product of $F(t)$ and $F'(t)$ is 0 for all t. So, whatever differentiable parametrization is used for the circle, there are only two possibilities at a point (y_1, y_2) on the circle: (i) the tangent line obtained by applying the definition is the one and only line that is perpendicular to the radius at (y_1, y_2); or (ii) the definition does not apply because $F'(t) = \mathbf{0}$ for t chosen so that $F(t) = (y_1, y_2)$.

We are comforted that these are the only two possibilities, but somewhat dismayed that the second of the two possibilities can actually occur. Suppose that

$$F(t) = (-\sin \pi t^3, \cos \pi t^3) , \quad t \in (-1, 1] .$$

It is easily checked that F represents a journey that goes once around the circle of radius 1 centered at $(0,0)$ in the counterclockwise direction and terminates at $(0, -1)$. The velocity when $t = 0$ is $(0,0)$ so the speed at that time is 0. Thus the tangent map is the constant function $(0, 1)$, a function whose image is a point, not a line. Thus, we are unable to use this tangent map to find the line tangent to the circle at the point $(0, 1)$ or even to conclude that there is a tangent line to the circle there. What we have seen is that the above particular parametrization of the circle is unsuitable for finding a tangent line to the circle at the point $(0, 1)$.

3. Let

$$F(t) = (t^2, t|t|) .$$

The image of F is a curve passing through $(0,0)$. Since $F(0) = (0,0)$, we look for a tangent line to the curve at $(0,0)$ by calculating $F'(0)$. To make this calculation we use Theorem 4.2.2. The first coordinate is easy to treat; its derivative at 0 is $2t$ evaluated at 0, that is, 0. To treat the second coordinate one can use the single-variable calculus definition of derivative; the result is 0. Therefore, $F'(0) = \mathbf{0}$. So we cannot use this calculation to obtain a tangent line to the image of F. (It is worth noting that the tangent *map* does exist at 0 and that the constant function $(0,0)$ is the best affine approximation to F near 0.)

In the two preceding examples we have faced the issue of not being able
to calculate a tangent line to a curve at a particular point because the
derivative is $(0,0)$ at the corresponding value of t. However, there is an essential difference between the two situations. In the case of the circle, there
was another parametrization that would do the job. On the other hand, it
develops that in the example just completed, there is no parametrization
of the image of F that will give a nonzero velocity at a t corresponding to
the point $(0,0)$. That is a good thing. The image of F is a right angle with
vertex at $(0,0)$, so it is appropriate that there be no tangent line defined
at $(0,0)$. It is a good exercise to draw the graph of F (in $\mathbb{R}^3$) and notice
that at each point there is a tangent line *to the graph*.

EXAMPLES:

4. Let
$$F(t) = (\cos t, \sin 2t), \quad t \in [0, 2\pi).$$
Notice that $F(\pi/2) = F(3\pi/2) = \mathbf{0}$. It is easy to calculate the tangent
map T_1 at $\pi/2$ and the tangent map T_2 at $3\pi/2$:
$$T_1(t) = (-1, -2)(t - \tfrac{\pi}{2}),$$
$$T_2(t) = (1, -2)(t - \tfrac{3\pi}{2}).$$
The images of T_1 and T_2 are *different* since the slope of the image
of T_1 is positive and that of the image of T_2 is negative, so there are
two tangent lines at the same point on the image of F. This image
resembles the numeral "8" rotated $\pi/2$ radians; at the point $(0,0)$,
the point of crossing, it is natural that there would be two tangent
lines.

5. *The graph of $y = f(x)$.* Say that $y = f(x) = x^2$. We realize its
graph as a parametrized curve in $\mathbb{R}^2$ by defining $F : \mathbb{R} \to \mathbb{R}^2$, $F(t) =$
$(x, y) = (t, t^2)$. Let us compute the tangent map of this curve at
$t_0 = 1$. See Figure 4.11.

We have, first of all, $F(1) = (1, 1)$. Further, $F'(t) = (1, 2t)$, whence
$F'(1) = (1, 2)$. It follows that the tangent map or best affine approximation is
$$\begin{aligned}
T(t) &= F(1) + F'(1)(t - 1) \\
&= (1, 1) + (1, 2)(t - 1) \\
&= (1 + (t - 1), 1 + 2(t - 1)) \\
&= (t, 1 + 2(t - 1)).
\end{aligned}$$

This is familiar. Let $g(x) = 1 + 2(x - 1)$. Then $T(t)$ traces out the
graph of $y = g(x)$, for
$$x = t, \quad y = g(t) = 1 + 2(t - 1).$$

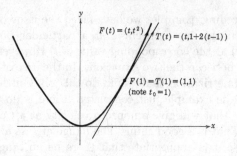

FIGURE 4.11.

But also $y = g(x) = f(x_0) + f'(x_0)(x - x_0)$ in the case of $f(x) = x^2$ and $x_0 = 1$, as you may check. This is the ordinary tangent line in one-variable calculus. Hence both senses of tangent line—the image of the tangent map of a parametrized curve $T(t)$ and the graph of the tangent map $g(x)$ to $y = f(x)$—coincide. (Note that the graph of T is a subset of $\mathbb{R}^3$ and has not been shown.)

The preceding examples are intended in part to help the reader get used to the difference between graphs and images, and between their corresponding tangent lines. The distinction between graph and image is not new to this section; it already arises for affine functions. If $T : \mathbb{R} \to \mathbb{R}^q$ is an affine function, its image is a subset of $\mathbb{R}^q$ and its graph is a subset of $\mathbb{R}^{q+1}$. It is easy to check using the definition that T is its own best affine approximation.

EXAMPLE:

6. *How to fly off on a tangent.* Here is a problem we are now able to solve. A particle is traveling up a helical path in space, its position at time t given by $F(t) = (\cos t, \sin t, t)$. At time $t_0 = 13\pi/6$ it flies off on a tangent, leaving the helical path and traveling in a straight line at constant velocity. Where is the particle at time $t_1 = 5\pi/2$? See Figure 4.12.

(a) At the start, $t = 0$ and the particle is at $F(0) = (1,0,0)$.

(b) At the instant the particle flies off on the tangent line, it is at the point

$$F(t_0) = (\cos 13\pi/6, \sin 13\pi/6, 13\pi/6) = (\sqrt{3}/2, 1/2, 13\pi/6) ,$$

because $\cos 13\pi/6 = \cos(2\pi + \pi/6) = \cos \pi/6$, and so on.

(c) Now we compute the tangent map of the helix $F(t)$ at the point t_0. This, of course, traces out the straight line path taken by the particle as it flies away "on the tangent".

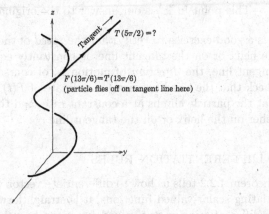

FIGURE 4.12.

The tangent map is $T(t) = F(t_0) + F'(t_0)(t - t_0)$. We have just obtained $F(t_0)$. Now we get $F'(t_0)$. We have

$$F'(t) = (-\sin t, \cos t, 1),$$

so that, when $t_0 = 13\pi/6$,

$$F'(t_0) = \left(-\sin \frac{13\pi}{6}, \cos \frac{13\pi}{6}, 1\right)$$

$$= \left(-\frac{1}{2}, \frac{\sqrt{3}}{2}, 1\right).$$

Thus the tangent map at t_0 is given by

$$T(t) = F(t_0) + F'(t_0)(t - t_0)$$

$$= \left(\frac{\sqrt{3}}{2}, \frac{1}{2}, \frac{13\pi}{6}\right) + \left(-\frac{1}{2}, \frac{\sqrt{3}}{2}, 1\right)\left(t - \frac{13\pi}{6}\right).$$

(d) Now let us locate the particle at time $t_1 = 5\pi/2$. We note that $5\pi/2 = 15\pi/6$, and so

$$t_1 - t_0 = \frac{\pi}{3}.$$

Thus at time $t_1 = 5\pi/2$, the particle is located on the tangent line $T(t)$ at the point

$$T(t_1) = F(t_0) + F'(t_0)(t_1 - t_0)$$

$$= \left(\frac{\sqrt{3}}{2}, \frac{1}{2}, \frac{13\pi}{6}\right) + \left(-\frac{1}{2}, \frac{\sqrt{3}}{2}, 1\right)\frac{\pi}{3}$$

$$= \frac{1}{6}(3\sqrt{3} - \pi, 3 + \pi\sqrt{3}, 15\pi).$$

This point in $\mathbb{R}^3$ is our answer to the original question.

It is a good exercise to verify that the *speed* of the particle, either on the helix or on the tangent line, is constantly equal to $\sqrt{2}$. On the tangent line, the *direction* of motion is, of course, constant as well. Check that the third (vertical) coordinate of $T(t)$ equals t. Conclude that the particle climbs at a constant rate, and this rate is the same either on the helix or on the tangent line.

4.2C DIFFERENTIATION RULES

Since Theorem 4.2.2 tells us how to differentiate vector-valued functions by differentiating scalar-valued functions, it is straightforward to prove some rules for differentiating vector-valued functions by using known rules for scalar-valued functions. The proofs of the following facts are left for the exercises.

Theorem 4.2.3 Let F and G be parametrized curves with common domain and target space and let h be a scalar-valued function having the same domain as F and G. Assume that F, G, and h are each differentiable. Then the following properties hold:

1. $(F + G)' = F' + G'$;

2. $(hF)' = h'F + hF'$;

3. $\langle F, G \rangle' = \langle F', G \rangle + \langle F, G' \rangle$;

4. $(F \times G)' = F' \times G + F \times G'$ if the target space is $\mathbb{R}^3$.

EXAMPLE:

7. Suppose that G describes the motion of a particle on a sphere of radius λ centered at $\mathbf{0} \in \mathbb{R}^q$, and further, suppose that G is differentiable. We have $\langle G(t), G(t) \rangle = \|G(t)\|^2 = \lambda^2$. Let us differentiate both sides of this relation (a method used in Example 2). On the left-hand side we have $(d/dt)\langle G(t), G(t) \rangle = 2\langle G'(t), G(t) \rangle$, as you may verify using Theorem 4.2.3. On the right-hand side we have $(d/dt)\lambda^2 = 0$, since λ is constant. Now these derivatives are equal, so

$$\langle G'(t), G(t) \rangle = 0 .$$

But this says that $G(t)$ is orthogonal to its derivative $G'(t)$ for all t. See Figure 4.13.

Let us make an official statement of what we have just proved.

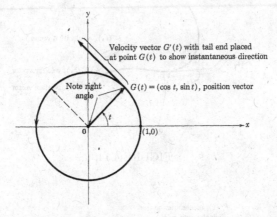

FIGURE 4.13. Showing $G(t) \perp G'(t)$ in the case $\|G(t)\| = 1$ and $q = 2$

Theorem 4.2.4 Let G be a differentiable parametrized curve and suppose that $\|G(t)\| = \lambda$ for each $t \in \mathcal{D}(G)$. Then $\langle G(t), G'(t) \rangle = 0$ for each t.

There is also a chain rule, the proof of which is left for the reader.

Theorem 4.2.5 Let h be a differentiable scalar-valued function defined on an interval and let F be a parametrized curve whose domain contains the image of h. Then $(F \circ h)' = (F' \circ h)h'$.

4.2D ACCELERATION

If the first derivative F' of a parametrized curve $F = (f_1, f_2, \ldots, f_q)$ is differentiable, then the derivative of F' is called the **second derivative** of F. If one is thinking of motion, then one might speak of **acceleration** rather than "second derivative". The acceleration at a particular instant t_0 is

$$(f_1''(t_0), f_2''(t_0), \ldots, f_q''(t_0)) .$$

The acceleration vector $F''(t)$ is often pictured as extending from the tip of the velocity vector $F'(t)$. See Figure 4.14. Pictured there it indicates the tendency of the velocity vector to vary, that is, to turn in some direction and to alter its length (speed of the moving particle). For since the acceleration vector $F''(t_0)$ is the first derivative at $t = t_0$ of $F'(t)$, it carries information about the change of $F'(t)$ at $t = t_0$.

EXAMPLES:

8. Let $F(t) = (a \cos t, a \sin t)$ trace out the circle of radius $a > 0$ in $\mathbb{R}^2$. See Figure 4.15. We then have

$$F'(t) = (-a \sin t, a \cos t),$$

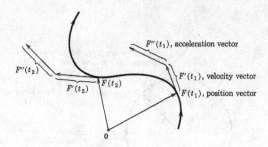

FIGURE 4.14.

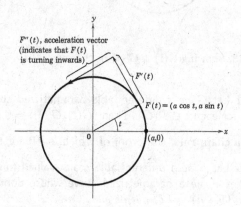

FIGURE 4.15.

whence
$$F''(t) = (-a \cos t, -a \sin t) = -F(t).$$

The vector $F''(t)$ points exactly opposite to $F(t)$, because the particle at $F(t)$ is always turning to maintain its constant distance from the origin; it is turning *inward*, so that its acceleration is toward the center of the circle. Since $\|F'(t)\| = 1$, we could have used Theorem 4.2.4 to predict $F'' \perp F'$ without computation.

9. Let $G(t) = (t, t^3)$ trace out the graph of $y = x^3$ in the xy-plane. See Figure 4.16. Then

$$G'(t) = (1, 3t^2), \quad G''(t) = (0, 6t).$$

Note that $G''(t_0)$ points downward if $t_0 < 0$, in accordance with the fact that, as t increases from $-\infty$ to 0, the velocity vector $G'(t)$ changes from almost vertical to horizontal, $G'(0) = (1, 0)$. Likewise, the acceleration vector at $t_1 > 0$ points upward, because the velocity vector $G'(t)$ points more and more toward the vertical as $t > 0$ increases.

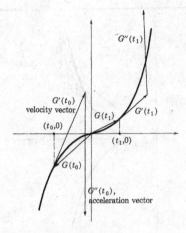

FIGURE 4.16.

4.2E CURVES WITH PRESCRIBED TANGENTS OR ACCELERATIONS

So far we have been given $F(t)$ and asked to compute $F'(t_0)$ and $F''(t_0)$ for a particular t_0. Now we reverse the question: Given the velocity vector $F'(t)$ for each t, or given $F''(t)$, plus some other information, compute $F(t)$. Here are some examples.

EXAMPLE:

10. *Given tangent and starting point.* Say that $F'(t) = (2t, 3t^2)$ and $F(0) = (1, 0)$. What is $F(t)$?

    To get $F(t) = (x(t), y(t))$, note that $F'(t) = (x'(t), y'(t))$, and so

    $$x'(t) = 2t , \quad y'(t) = 3t^2 .$$

    Ordinary integration gives

    $$x(t) = t^2 + c_1 , \quad y(t) = t^3 + c_2 ,$$

    where the constants c_1 and c_2 must be found. But we know that $x(0) = 1$ (why?), and so $c_1 = 1$. Likewise $c_2 = 0$. Thus we have

    $$F(t) = (x(t), y(t)) = (t^2 + 1, t^3) .$$

REMARKS:

- You may sketch $F(t)$ for the preceding example by plotting points $F(0)$, $F(1)$, and so on. But also a direct relation between x and y will help. Since, $x - 1 = t^2$, we obtain $(x-1)^3 = y^2$, that is, $y = \pm(x-1)^{3/2}$. See Figure 4.17.

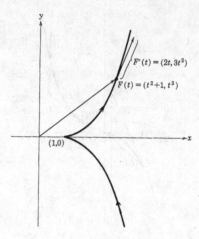

FIGURE 4.17.

- If we know the velocity $F'(t)$ for all t and the position at one instant ($t = 0$ in the above), then we can find the position for *all* times t. Why is this reasonable?

- In the preceding example the speed equals zero at time 0, raising the possibility of getting a curve with a corner. Figure 4.17 indicates that this has happened. The moving particle instantaneously changes its direction from westerly to easterly.

EXAMPLE:

11. *Given acceleration, starting point, and starting velocity.* Say that $F''(t) = (3\pi^2 \cos \pi t, 3\pi^2 \sin \pi t)$, $F'(0) = (0, -3\pi)$, and $F(0) = (-3, 0)$. We compute $F(t) = (f_1(t), f_2(t))$.

We have

$$f_1''(t) = 3\pi^2 \cos \pi t , \quad f_2''(t) = 3\pi^2 \sin \pi t ,$$

so that, by integrating once, we obtain

$$f_1'(t) = 3\pi \sin \pi t + c_1 , \quad f_2'(t) = -3\pi \cos \pi t + c_2 .$$

Since $F'(0) = (f_1'(0), f_2'(0)) = (0, -3\pi)$, we conclude that $c_1 = c_2 = 0$. A second integration (the second derivative necessitates two integrations) yields

$$f_1(t) = -3 \cos \pi t + c_3 , \quad f_2(t) = -3 \sin \pi t + c_4 .$$

Since $F(0) = (-3, 0)$, we conclude that $c_3 = c_4 = 0$. Finally

$$F(t) = (f_1(t), f_2(t)) = -3(\cos \pi t, \sin \pi t) .$$

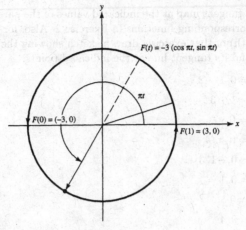

FIGURE 4.18.

You might check that this $F(t)$ has the prescribed starting point $F(0)$ and starting velocity $F'(0)$. See Figure 4.18.

Exercises

1. Let $f(x) = 1 - x^2$. Sketch the graph $y = f(x)$ and compute the slope $f'(x_0)$ and best affine approximation $y = T(x) = f(x_0) + f'(x_0)(x - x_0)$ at the following points x_0:

   (a) $x_0 = 0$

   (b) $x_0 = 1$

   (c) $x_0 = -2$

   (d) $x_0 = 10$.

   Sketch the graphs of these functions as tangent lines to $y = f(x)$.

2. Find the derivative of each of the following functions:

   (a) $F(t) = (1 - 7t, 5 + 9t)$

   (b) $G(t) = (3t, t^2 - 7)$

   (c) $H(\theta) = (2 + e^\theta, 3\theta - 7\theta^2)$

   (d) $P(s) = (\sin 2s, \cos 2s, s)$

   (e) $\Phi(t) = (3 - t, 1 + 4t, -9t)$

   (f) $V(t) = (\sin \pi(1 + 3t), 4, \log \sqrt{1 + t^4})$

   (g) $Q(r) = r^2(\sin r, e^{3r})$

   (h) $K(t) = (3, t)/\|(3, t)\|$

   (i) $M(t) = (e^{-t} \cos \pi t, e^{-t} \sin \pi t, \cos 2\pi t, \sin 2\pi t)$.

3. Find the tangent map at the indicated values of the parameter listed below for the corresponding functions in Exercise 2. Also, for parts (a), (b), (d), (e), and (h) of that exercise, draw a sketch showing the curve (that is, the image) and its tangent line at the indicated points:

  (a) $t_0 = 0, 1, -2$

  (b) $t_0 = 1, -3$

  (c) $\theta_0 = 0$

  (d) $s_0 = 0, \pi/4$

  (e) $t_0 = 0, -1, 3$

  (f) $t_0 = 0, 1$

  (g) $r_0 = 0$

  (h) $t_0 = 0, 4$

  (i) $t_0 = 0$.

4. Check directly that the tangent maps found in Exercise 3 satisfy the conditions of Definition 4.2.1 in Section 4.2b.

5. If each of the functions F, G, Φ, V, K and M in Exercise 2 describes the motion of a particle at time t, find the position, velocity, and speed at $t = 0$ and $t = 1$.

6. Find the second derivative of each of the functions in Exercise 2.

7. Find functions $F(t) = (x(t), y(t))$ or $F(t) = (f_1(t), f_2(t), f_3(t))$, such that

  (a) $F'(t) = (1 + 2t, 3)$, $F(0) = (1, 3)$

  (b) $F'(t) = (2 \cos t, 3 \sin t)$, $F(0) = (0, 1)$

  (c) $F'(t) = (1 - 4t, 3 + 2t, 1 + 3t^2)$, $F(1) = (1, 4, 0)$

  (d) $F'(t) = (1 + e^{2t}, t - t^2, 5 + t^4)$, $F(0) = (0, 1, -2)$

  (e) $F'(t) = (1, t^2, t^3)$, $F(0) = (1, 1, 1)$.

8. (a) A particle travels on a curve in $\mathbb{R}^3$ in such a way that its velocity is $V(t) = (2, 5 + 6t, 0)$ and its position at $t = 0$ is $(1, 0, 3)$. Find its position at $t = 1$.

   (b) Repeat with $V(t) = (3t^2, -2t + 7, \cos \pi t)$ and the position at $t = 0$ equal to $(1, 5, 0)$.

9. Find functions $F(t)$ such that

  (a) $F''(t) = (1 + 2t, 3)$, $F(0) = (0, 0)$, $F'(0) = (1, 3)$

  (b) $F''(t) = (-\cos t, -\sin t)$, $F(0) = (0, 1)$, $F'(0) = (-1, 0)$

  (c) $F''(t) = (1 - 4t, 3 + 2t, 1 + 3t^2)$, $F(1) = (0, 0, 1)$, $F'(1) = (1, 4, 0)$.

10. (a) A particle moves on a curve in $\mathbb{R}^3$ in such a way that its acceleration is $F''(t) = (-4, 6t, -2 + 4t^2)$. If its initial position is $F(0) = (2, 0, 0)$ and its initial velocity is $F'(0) = (0, 0, 1)$, find its position at $t = 3$.

   (b) Repeat with $F''(t) = (e^t, 0, 1 + 6t)$.

11. (a) Let $F(t) = (1-7t, 5+9t)$, $G(t) = (3t, t^2-7)$, and $\varphi = \langle F, G \rangle$. Compute $\varphi'(t)$ in two ways: by using Theorem 4.2.3 and by first substituting for $F(t)$ and $G(t)$ to find a formula for $\varphi(t)$ and then differentiating as in single-variable calculus.

    (b) Repeat with $F(t) = (3 - t^2, 1 + 4t, -9t)$ and $G(t) = (1, e^t, 2/(1+t^2))$.

12. Let $F(t)$ be a differentiable parametrized curve. If $F(t)$ is perpendicular to $F'(t)$ for all t, prove that $\|F(t)\|$ is a constant.

13. (a) If the speed of a particle is identically constant, prove that the acceleration vector is perpendicular to the velocity vector.

    (b) Prove the converse that if the acceleration vector is perpendicular to the velocity vector, then the speed is constant.

14. (a) Let $F : \mathbb{R} \to \mathbb{R}^q$ define a differentiable curve that does not pass through the origin. If the point $X_0 = F(t_0)$ is a point on the curve closest to the origin and not an endpoint of the curve, show that $F'(t_0) \perp F(t_0)$, that is, the velocity vector is orthogonal to the position vector. *Hint:* consider $\varphi(t) = \|F(t)\|^2$.

    (b) Apply this to prove anew the well-known fact that the radius vector to any point on a circle is perpendicular to the tangent vector at that point.

    (c) What can you say about a point $X_1 = F(t_1)$ farthest from the origin?

    (d) Modify, if necessary, your answers to the preceding parts to accommodate curves that do pass through the origin.

15. (a) Sketch the curve $F(t) = (t(t^2 - 1), t^2 - 1)$, observing that it crosses itself at the origin, which corresponds to $t = \pm 1$.

    (b) Find the equation of the tangent line for $t = +1$ and $t = -1$. Sketch these tangent lines.

16. Fill in the blank. The image of a parametrized curve in $\mathbb{R}^1$ is an _____.

17. Prove Theorem 4.2.1.

18. Prove Theorem 4.2.3.

19. Prove Theorem 4.2.5.

20. This is a continuation of Exercise 25 of Section 4.1. At each point of intersection, calculate the angle from the first ellipse counterclockwise to the second ellipse.

21. What can you conclude about a vector-valued function F whose derivative is $\mathbf{0}$ at every point in the interval $\mathcal{D}(F)$? Prove your assertion. Discuss the situation in case it is not assumed that $\mathcal{D}(F)$ is an interval.

22. What can you conclude about a vector-valued function F for which $F''(t) = \mathbf{0}$ at all t in the interval $\mathcal{D}(F)$? Prove your assertion.

23. If $F'(t) = V_0$ for all $t \in \mathbb{R}$ and $F(0) = X_0$, where X_0 and V_0 are fixed vectors, what can you conclude about $F(t)$? Proof?

24. Let $F : \mathbb{R} \to \mathbb{R}^q$. If $F''(t) = C$, where C is a constant vector, show that F has the form

$$F(t) = \frac{1}{2}Ct^2 + V_0 t + X_0 ,$$

where X_0 and V_0 are fixed vectors. Show how X_0 and V_0 are determined by $F(0)$ and $F'(0)$. (This is precisely the situation for projectiles, such as baseballs, moving near the surface of the earth. Then the acceleration $F''(t) = -gN$, where N is a unit vector directed upward from the surface of the earth and $g = 32\text{ft/sec}^2$ is the acceleration due to gravity.)

25. A particle is traveling on a curve in outer space in such a way that its position at time t is given by $F(t) = (1+t, t^2, -2t)$. At time $t = 2$, however, the particle flies off the curve and moves on the tangent line with constant velocity. Where is the particle at $t = 4$?

26. Let $F : \mathbb{R} \to \mathbb{R}^q$ describe a curve, and let $Z \in \mathbb{R}^q$ be a fixed vector. If $F'(t) \perp Z$ for all t and if $F(0) \perp Z$, prove that $F(t) \perp Z$ for all t. *Hint:* Consider $\varphi(t) = \langle Z, F(t) \rangle$ and write an equation for a $(q-1)$-dimensional linear subspace in which the image of F lies.

27. Drop the assumption that $F(0) \perp Z$ in the preceding problem. Show that the image of F lies in some $(q-1)$-dimensional affine subspace of $\mathbb{R}^q$ and find an equation for that affine subspace.

28. Let $F : \mathbb{R} \to \mathbb{R}^3$ describe the position of a particle, and let $Z \in \mathbb{R}^3$ be a fixed vector. If the acceleration $F''(t)$ is perpendicular to Z for all t and if the initial position and initial velocity vectors $F(0)$ and $F'(0)$ are perpendicular to Z, show that the position $F(t)$ is perpendicular to Z for all t.

29. A particle moves on a straight line so that its position at time $t \geq 0$ is $F(t) = (\sqrt{t}, \sqrt{2} - \sqrt{t})$, until, at a certain time, it meets the circle $x^2 + y^2 = 1$. It is then trapped into a circular orbit on that circle, with constant speed equal to the speed it had when it met the circle. Show that the direction of motion on the circle may be chosen so that the velocity of the particle is defined for all t. When is the particle at the point $(0, 1)$ on the circle?

30. A particle moves so that its position $F(t)$ is a constant distance r from the origin. If you know that $F''(t) = -a^2 F(t)$, where a is a nonnegative constant, prove that its speed is a constant. In fact, show that $\|F'(t)\| = ar$. Does this check with your intuition in the special case where $a = 0$?

4.3   Arc Length and Curvature

4.3A   Least Upper Bounds

Let S denote a set of real numbers. A real number u is an **upper bound** of S if $u \geq s$ for every $s \in S$. Notice that an upper bound of S may, but need not, be a member of S. It is an important convention that the symbol ∞ is said to be an **upper bound** of every set of real numbers. A not so important convention, especially for our purposes, is that $-\infty$ is an **upper bound** of the empty set, but of no other set of real numbers.

EXAMPLES:

1. For the set $S = \{s : s^2 < 2\}$ let us calculate whichever of the following exist: three upper bounds, an upper bound that belongs to S, three members of S that are not upper bounds, two things that neither belong to S nor are upper bounds of S, and the smallest of all the upper bounds.

   For $s > 0$, $s^2 < 2$ if and only if $s < \sqrt{2}$. Thus, 2, 17, and ∞ are upper bounds.

   Here is an argument that shows that no member of S is an upper bound. Consider a fixed member x of S, and let ε denote a positive number yet to be chosen. Whatever the choice for ε, $x < x + \varepsilon$. So, if we can choose ε so that $x + \varepsilon \in S$, it will follow that x is not an upper bound of S. Since $(x + \varepsilon)^2 = x^2 + 2x\varepsilon + \varepsilon^2$, what we want is

   $$x^2 + 2x\varepsilon + \varepsilon^2 < 2 ,$$

   or, equivalently,

   $$2x\varepsilon + \varepsilon^2 < 2 - x^2 .$$

   Since $x \in S$, the number $2 - x^2$ on the right-hand side is positive. The function of ε on the left-hand side is continuous and has the value 0 when $\varepsilon = 0$ and, thus, it is less than the positive number $2 - x^2$ when ε is sufficiently close to 0. Such a positive ε is our choice, and, for that choice, $x + \varepsilon \in S$. Hence, x is not an upper bound. Since x was an arbitrary member of S, it follows that no member of S is an upper bound of S.

   From the preceding paragraph we conclude that no member of S is an upper bound. By calculating the squares of -1, 0, and $41/29$ we see that each of these is a member of S, and therefore, that each is a member of S that is not an upper bound of S.

   The number -3 does not belong to S, since its square 9 is larger than 2; nor is it an upper bound of S, since it is less than the member 0 of S. Also $-\infty$ neither belongs to S nor is an upper bound of S.

   The number $\sqrt{2}$ is an upper bound of S. To see that it is the smallest upper bound, consider an arbitrary $x < \sqrt{2}$. Then

   $$x^2 < \sqrt{2}^2 = 2 ;$$

   so, $x \in S$. We have already seen that, in this example, no member of S is an upper bound of S. Therefore, x is not an upper bound of S. Since x is an arbitrary number less than $\sqrt{2}$ we conclude that no number less than $\sqrt{2}$ is an upper bound. Consequently, $\sqrt{2}$ is the least of all upper bounds of S.

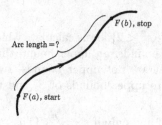

FIGURE 4.19.

2. For the set Z of all integers we will calculate whichever of the following exist: three upper bounds, an upper bound that belongs to S, three members of S that are not upper bounds, a number that neither belongs to S nor is an upper bound of S, and the smallest of all the upper bounds.

Every real number lies between two integers. Thus no real number is an upper bound of Z. So, there are not three upper bounds and no member of Z is an upper bound. In particular, none of the three members 5, -1899, and 10^{10} of Z are upper bounds of Z. The number $1/2$, being neither an integer nor a number as large as the integer 5, is neither a member of Z nor an upper bound of Z. The symbol ∞, being the only upper bound, is the smallest of all the upper bounds.

In both of the preceding examples we discovered that the set of upper bounds is itself a set that has a smallest member. Not all sets have smallest members. For instance, the set of all positive numbers has no smallest member. The question arises: Does the set $\mathcal{U}$ of upper bounds of a set S of real numbers necessarily have a smallest member? According to the following theorem, a theorem typically proved in an advanced calculus or analysis course, the answer is "yes".

Theorem 4.3.1 Every set of real numbers has a least upper bound.

Even though we do not prove this theorem, you should find it in accordance with your mental picture of the real numbers. We will use the theorem in an essential way to define the arc length of a parametrized curve in the next section.

4.3B   DEFINITION OF ARC LENGTH

Let $F : [a, b] \to \mathbb{R}^q$ be a parametrized curve in $\mathbb{R}^q$. How can we measure arc length along the curve from $F(a)$ to $F(b)$? See Figure 4.19. This is the same as asking for the distance traveled by a particle whose position at time t is $F(t)$ for $t \in [a, b]$. If the curve is actually a line segment from a point X_0 to a point X, then we know its arc length; it is simply $\|X - X_0\|$.

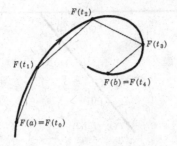

FIGURE 4.20. Approximating a curve by segments

If the curve is *not* a line segment, then we have not yet defined in a precise way what we mean by its arc length. And certainly we must define arc length before we can hope to discover a formula to compute it. Let us attend to this definition now. The idea is to approximate this curve by line segments, whose lengths we know. See Figure 4.20.

A **partition** of the interval $[a, b]$ is any finite set

$$\mathcal{P} = \{t_0, t_1, \ldots, t_m\}$$

consisting of numbers that satisfy

$$a = t_0 < t_1 < \cdots < t_m = b.$$

The points $F(a) = F(t_0)$, $F(t_1), \ldots, F(t_m) = F(b)$ are on the curve from $F(a)$ to $F(b)$. We connect $F(t_0)$ to $F(t_1)$ by a line segment of length $\|F(t_1) - F(t_0)\|$, then connect $F(t_1)$ to $F(t_2)$ by another segment, of length $\|F(t_2) - F(t_1)\|$, then continue on to $F(t_3)$, and so on, until we reach the endpoint $F(t_m)$, that is, $F(b)$.

The total distance traveled along these line segments from $F(t_0)$ to $F(t_1)$, $F(t_1)$ to $F(t_2), \ldots, F(t_{m-1})$ to $F(t_m)$ is the sum

$$S(\mathcal{P}) = \|F(t_1) - F(t_0)\| + \cdots + \|F(t_m) - F(t_{m-1})\|.$$

In Figure 4.20, we have $\mathcal{P} = \{t_0, t_1, t_2, t_3, t_4\}$, and so

$$S(\mathcal{P}) = \|F(t_1) - F(t_0)\| + \|F(t_2) - F(_1)\| + \cdots + \|F(t_4) - F(t_3)\|.$$

It would appear that this number $S(\mathcal{P})$ is somewhat less than the "true arc length" of the curve, because, in brief, a line segment gives the *shortest* distance between two points.

These ideas are made precise in the following definition, whose meaningfulness is assured by Theorem 4.3.1.

Definition 4.3.1 Let F denote a parametrized curve, and let $a < b$ be two real numbers in $\mathcal{D}(F)$. The **arc length** of F on $[a, b]$, denoted $\mathcal{L}(F; a, b)$, is the least upper bound of the set of numbers

$$\{S(\mathcal{P}) : \mathcal{P} \text{ a partition of } [a, b]\}.$$

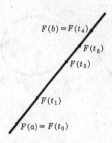

FIGURE 4.21. Partition for a line segment

The **arc length** of F is the least upper bound of the set

$$\{\mathcal{L}(F; a, b) : a, b \in \mathcal{D}(F)\} \, .$$

The preceding definition says that the arc length of a parametrized curve is the smallest value that is at least as large as the arc length of any polygonal path with corner points on the curve, where the corner points on the polygonal path are ordered in the same way as they are ordered on the given parametrized curve. If $\mathcal{D}(F)$ is a closed interval $[a, b]$, then the arc length of F is the same as the arc length of F on $[a, b]$. The notation $\mathcal{L}(F; a, b)$ is sometimes used for the arc length of F when a and b are the endpoints of $\mathcal{D}(F)$, even if either a or b is not in $\mathcal{D}(F)$.

You may check that if the curve from $F(a)$ to $F(b)$ is actually a line segment, then the arc length as defined above is equal to $\|F(b) - F(a)\|$. Note that, for instance, on the segment pictured in Figure 4.21 $\mathcal{L}(F; a, b) = \|F(b) - F(a)\|$ does equal

$$\|F(t_1) - F(a)\| \quad + \quad \|F(t_2) - F(t_1)\|$$
$$+ \quad \|F(t_3) - F(t_2)\| + \|F(b) - F(t_3)\|.$$

Some curves, even some that lie in a bounded region, have arc length equal to ∞. In many discussions one wants to restrict consideration to those that have finite arc length; we say a parametrized curve is **rectifiable** if and only if it has finite arc length.

People will say that two travelers following the same route will have traveled the same distance even though, due to differences in their speeds, their journeys are represented by *different parametrizations*, that is, by *different parametrized* curves. Based on this real-world experience we suspect that there is a theorem which says that the arc lengths of two parametrized curves are the same if they represent the same journey. The following theorem is such a result. We, but possibly not your teacher, omit its proof, not because of conceptual difficulty, but because of notational cumbersomeness. The reader is advised, for the statement of the theorem, to satisfy himself or herself that the notation in the theorem corresponds to the description of the two travelers just given.

Theorem 4.3.2 Let γ denote a strictly monotone continuous function with domain $[a, b]$ and image $[c, d]$. Let F denote a parametrized curve whose domain contains $[c, d]$. Then the arc length of the parametrized curve $F \circ \gamma$ on $[a, b]$ equals the arc length of the parametrized curve F on $[c, d]$.

The definition of the arc length $\mathcal{L}(F; a, b)$ is geometrically convincing but computationally not very useful. We certainly cannot form *all* partitions $\mathcal{P}$ and the corresponding sums $S(\mathcal{P})$; the number of such partitions is infinite. In the following section we derive a formula for the arc length $\mathcal{L}(F; a, b)$.

4.3C   A FORMULA FOR ARC LENGTH

Here is a formula for the arc length of a parametrized curve in $\mathbb{R}^q$ in terms of the integral of a scalar-valued function of a scalar.

Theorem 4.3.3 Let F be a parametrized curve in $\mathbb{R}^q$, and let $a < b$ be real numbers which either lie in $\mathcal{D}(F)$ or are endpoints of $\mathcal{D}(F)$. Suppose that F' is continuous. Then

$$\mathcal{L}(F; a, b) = \int_a^b \|F'(t)\| \, dt \, .$$

(If either a or b is an endpoint of $\mathcal{D}(F)$ but does not actually lie in $\mathcal{D}(F)$, then the integral is improper.)

PROOF: (an outline only.) For a partition $\mathcal{P}$ of $[a, b]$, a typical term in the approximate arc length $S(\mathcal{P})$ is of the form $\|F(t_{k+1}) - F(t_k)\|$. Let us write $\Delta_k t = t_{k+1} - t_k$. Then the typical term equals

$$\|F(t_k + \Delta_k t) - F(t_k)\| = \left\| \frac{F(t_k + \Delta_k t) - F(t_k)}{\Delta_k t} \right\| \Delta_k t \, .$$

Hence the approximating sum $S(\mathcal{P})$ has been put in the form

$$S(\mathcal{P}) = \sum_{k=0}^{m-1} \frac{\|F(t_k + \Delta_k t) - F(t_k)\|}{\Delta_k t} \Delta_k t \, .$$

Recall that we obtain the arc length $\mathcal{L}(F; a, b)$ from $S(\mathcal{P})$ by considering various partitions $\mathcal{P}$. It can be shown that we only need to consider partitions with larger and larger numbers of points $t_0, t_1, \ldots, t_m$, that is, $m \to \infty$, and, moreover, that we may insist that all $\Delta_k t = t_{k+1} - t_k$ approach zero. If we do this, two things happen (these are plausible, but we will omit details):

1. The expression $(F(t_k + \Delta_k t) - F(t_k))/\Delta_k t$ is very close to $F'(t_k)$, since $\Delta_k t$ is approaching 0. Hence $\|(F(t_k + \Delta_k t) - F(t_k))/\Delta_k t\|$ is close to $\|F'(t_k)\|$.

2. The sum

$$\sum_{k=0}^{m-1} \|F'(t_k)\| \Delta_k t$$

approaches the integral

$$\int_a^b \|F'(t)\| \, dt \, .$$

This integral exists as a finite number, because the integrand $\|F'(t)\|$ is a continuous function. Thus the limit of the approximating sums $S(\mathcal{P})$ over all $\mathcal{P}$ exists and equals the integral as claimed. $<<$

REMARKS:

- If $F(t) = (f_1(t), \dots, f_q(t))$, then

$$\|F'(t)\| = \sqrt{f_1'(t)^2 + \cdots + f_q'(t)^2} \, .$$

Suppose $F(t)$ traces out a curve in the plane, $F : \mathbb{R} \to \mathbb{R}^2$. Then we might write $x = f_1(t)$ and $y = f_2(t)$. Thus $f_1'(t) = dx/dt$, and $f_2'(t) = dy/dt$. So, the arc length may be written as

$$\mathcal{L}(F; a, b) = \int_a^b \sqrt{\left(\frac{dx}{dt}\right)^2 + \left(\frac{dy}{dt}\right)^2} \, dt \, .$$

- If we think of $\|F'(t)\|$ as speed and dt as a measure of time, then the integrand in the formula is the product of speed and time, which gives distance or length, as expected.

EXAMPLES:

3. *The arc length of a circle.* Let $F(t) = (a \cos t, a \sin t)$. This traces out the circle of radius $a > 0$ centered at the origin as t varies in the interval $[0, 2\pi]$. What is the distance around this circle?

   Since $F'(t) = (-a \sin t, a \cos t)$, we have $\|F'(t)\| = a$. Thus, the arc length is

$$\mathcal{L}(F; 0, 2\pi) = \int_0^{2\pi} \|F'(t)\| \, dt = a \int_0^{2\pi} dt = 2\pi a \, ,$$

   as expected.

4. *The circle revisited.* Some may object to the inclusion of the point $(a, 0)$ at both ends of the parametrized curve in the preceding example. So, let us modify that example so that the domain is $[0, 2\pi)$. The

calculations in that example leading to the integral

$$a \int_0^{2\pi} dt$$

are relevant here also, but now this integral is viewed as being improper at its upper limit 2π. Thus, we get a slightly different calculation leading to the same result $2\pi a$ that was attained in the preceding example:

$$\lim_{b\uparrow 2\pi} \left(a \int_0^b dt\right) = \lim_{b\uparrow 2\pi} ab = 2\pi a .$$

5. *The circle revisited again.* Let us calculate the arc length of the parametrized curve F given by

$$F(t) = (a\cos t, a\sin t) , \quad 0 \le t \le 4\pi .$$

The initial calculations made in Example 3 are applicable here also, but the endpoints of integration change. We get

$$a \int_0^{4\pi} dt = 4\pi a ,$$

which is twice the answer obtained in Example 3, even though the image of the parametrized curves in Example 3 and this example are identical. The doubling of the answer arises from the fact that the parametrization in this example represents a journey twice around the circle. We thus learn the lesson that if we actually want the arc length of a curve rather than the length of a journey, then we should parametrize the curve in a manner that corresponds to a journey that does not overlap itself.

6. *A spiral.* We will find the arc length of the parametrized curve:

$$F(t) = e^t(\cos 2\pi t, \sin 2\pi t) , \quad -\infty < t \le 0 .$$

A straightforward calculation gives

$$F'(t) = e^t(-2\pi \sin 2\pi t + \cos 2\pi t, 2\pi \cos 2\pi t + \sin 2\pi t) ,$$

from which it follows that

$$\|F'(t)\| = e^t\sqrt{4\pi^2 + 1} .$$

We then get the arc length by evaluating an improper integral:

$$\int_{-\infty}^0 e^t\sqrt{4\pi^2 + 1}\, dt = \sqrt{4\pi^2 + 1} .$$

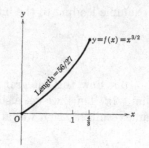

FIGURE 4.22.

7. *The arc length of a graph.* The graph of $y = f(x)$ in the xy–plane is traced out by the mapping $F(t) = (t, f(t))$. We have $F'(t) = (1, f'(t))$, whence $\|F'(t)\| = \sqrt{1 + f'(t)^2}$. Using $x = t$ and $y = f(x) = f(t)$, we conclude that the arc length $\mathcal{L}$ of the graph of $y = f(x)$ for $a \le x \le b$ is given by

$$\mathcal{L} = \int_a^b \sqrt{1 + f'(x)^2} \, dx \, .$$

8. Let us use the formula in the preceding example to find the arc length of the graph $y = f(x) = x^{3/2}$ from $x = 0$ to $x = \frac{4}{3}$. We have $f'(x) = \frac{3}{2}x^{1/2}$, so the arc length is

$$\mathcal{L} = \int_0^{4/3} \sqrt{1 + \frac{9}{4}x} \, dx = \frac{8}{27}\left(1 + \frac{9}{4}x\right)^{3/2} \Bigg]_{x=0}^{x=4/3} = \frac{56}{27} \, .$$

Thus $\mathcal{L} = \frac{56}{27}$; the curve is slightly more than 2 units long. See Figure 4.22.

9. *A helix.* We will calculate the arc length of one turn of the helix

$$F(\theta) = (\cos\theta, \sin\theta, \theta) \, ,$$

where $0 \le \theta \le 2\pi$.

We readily compute that $\|F'(\theta)\| = \sqrt{2}$, a constant, whence the arc length is given by

$$\mathcal{L}(F; 0, 2\pi) = \sqrt{2} \int_0^{2\pi} d\theta = 2\sqrt{2}\,\pi \, .$$

In this example we have used θ, not t, as the parameter. Needless to say, the concept of arc length does not depend on the name we give the parameter. Notice, as expected, that the answer is larger than 2π, the circumference of a circle of radius 1.

10. *How to concoct some rectifiable plane curves.* This example is more
a curiosity then an essential part of the development. The integrand
$\sqrt{(dx/dt)^2 + (dy/dt)^2}$ may be formidable, in great part because of
the square root. Here is a method for constructing curves $F(t) =$
$(x(t), y(t))$ such that $(dx/dt)^2 + (dy/dt)^2$ is a perfect square. (Compare
with Exercise 26 of Section 4.1.) This will enable you to make up some
reasonable practice problems.

If u and v are numbers, and if we define α, β, and γ by

$$\alpha = u^2 - v^2 \; , \quad \beta = 2uv \; , \quad \gamma = u^2 + v^2 \; ,$$

then it is easy to see that $\alpha^2 + \beta^2 = \gamma^2$. (These are classical formulas
for right triangles with sides α, β, and γ.) Now let $u = u(t)$ and
$v = v(t)$ be ordinary functions such that the indefinite integrals

$$\int u^2(t) \, dt \; , \quad \int v^2(t) \, dt \; , \quad \int u(t)v(t) \, dt$$

are not difficult. For example, let $u(t)$ and $v(t)$ be any polynomials.
If we now define

$$x(t) = \int [u^2(t) - v^2(t)] \, dt \; , \quad y(t) = 2 \int u(t)v(t) \, dt \; ,$$

then we note that

$$\left(\frac{dx}{dt}\right)^2 + \left(\frac{dy}{dt}\right)^2 = (u^2 + v^2)^2.$$

This is because $dx/dt = u^2 - v^2$ and $dy/dt = 2uv$ by the Fundamental
Theorem of Calculus. Thus the problem of finding the arc length of
the curve given by $F(t) = (x(t), y(t))$ as defined here reduces to the
integration of $u^2(t) + v^2(t)$ (no square root), and we *chose* u and v
so that this integration would not be difficult. You might pick two
functions $u(t)$, and $v(t)$ and try this procedure for yourself.

4.3D   THE NATURAL PARAMETRIZATION OF A RECTIFIABLE CURVE

When watching a race, we do not usually concern ourselves with the coordinates of the various racers. Rather we think of the distance along the
race track that each has traveled. That is, we label points on the track by
numbers indicating distance from the start, rather than ordered pairs of
numbers indicating distances from perpendicular axes.

To formalize these ideas, let F be a rectifiable parametrized curve defined
on an interval $[a, b]$. For $t \in [a, b]$, let $\mathcal{L}(F; a, t)$ be the arc length of the

curve parametrized by F on the interval $[a, t]$. From the fact that the curve is rectifiable, it follows (but we will omit the proof) that $\mathcal{L}(F; a, t)$ is a continuous function of t. We will also assume that this function is strictly increasing, or, what is the same thing, that the motion described by F is never stationary over a period of time. It then follows that this function has a strictly increasing continuous inverse function which we call γ. According to Theorem 4.3.2 the arc length of the parametrized curve F on $[a, t]$ and the arc length of the parametrized curve $F \circ \gamma$ on $[0, \mathcal{L}(F; a, t)]$ are equal. That is, the arc length of the curve traced by F as its domain variable ranges from a to t is equal to the arc length of the curve traced by $F \circ \gamma$ as its domain variable ranges from 0 to $\mathcal{L}(F; a, t)$.

Let us use τ for the parameter of $F \circ \gamma$. Then we have just shown that on the interval $[0, \tau]$, the parametrized curve $F \circ \gamma$ travels distance τ. Thus we have a new parametrization of the curve corresponding to F, and this new parametrization has the property that its parameter τ and the arc length of the corresponding initial portion of the curve are the *same*. The function $F \circ \gamma$ is called the **natural parametrization** of the curve originally defined by F.

In case $\mathcal{L}(F; a, t)$ is not a strictly increasing function, it is still possible to parametrize the rectifiable curve naturally, but for the proof, an argument somewhat different from that just given is required. The result is that there is an increasing function γ such that $F \circ \gamma$ is natural. The function γ will not be continuous in a case where the increasing function $\mathcal{L}(F; a, t)$ is not strictly increasing.

In most cases it is easy to recognize whether a parametrization G is natural. Suppose that G' is continuous. Then

$$\mathcal{L}(G; 0, \tau) = \int_0^\tau \|G'(u)\| \, du \, .$$

If the integrand equals 1 for every u, then $\mathcal{L}(G; 0, \tau) = \tau$ for every τ and so G is natural. On the other hand if G is natural, then $\mathcal{L}(G; 0, \tau) = \tau$ and

$$\tau = \int_0^\tau \|G'(u)\| \, du \, ,$$

for every τ. Differentiation of both sides and an application of one version of the Fundamental Theorem of Calculus gives

$$1 = \|G'\| \, .$$

Hence, once G is known to have a continuous derivative one can decide whether G is natural by checking whether the norm of its derivative is identically equal to 1.

Let us explore consequences of the relation $F \circ \gamma = G$ when G is a natural parametrization, γ is increasing, and F and γ, and therefore G, have continuous derivatives. By the chain rule,

$$G'(\tau) = F'(\gamma(\tau))\gamma'(\tau) ,$$

so

$$1 = \|G'(\tau)\| = \|F'(\gamma(\tau))\gamma'(\tau)\| = \|F'(\gamma(\tau))\|\gamma'(\tau) .$$

We conclude therefore that

$$\gamma'(\tau) = \frac{1}{\|(F' \circ \gamma)(\tau)\|}$$

and

$$G'(\tau) = \frac{F'(\gamma(\tau))}{\|F'(\gamma(\tau))\|} .$$

We see that the two vectors $G'(\tau)$ and $F'(\gamma(\tau))$ point in the same direction at the point $G(\tau) = F(\gamma(\tau))$ on the curve, namely, in the direction of the tangent ray indicating the direction of motion.

In this section we have assumed the domain of the original parametrization to be a *closed* interval. Appropriate modifications are required when either endpoint of the domain does not belong to the domain. Rather than describe the modifications in general, we illustrate with the particular case of a spiral.

EXAMPLE:

11. Consider the parametrized curve (a spiral) of Example 6 of Section 4.3c. For any real number $t \leq 0$,

$$\mathcal{L}(F; -\infty, t) = \int_{-\infty}^{t} e^u \sqrt{4\pi^2 + 1} \, du = e^t \sqrt{4\pi^2 + 1} .$$

Set the expression on the right-hand side equal to τ and then solve for t to calculate the inverse function γ explicitly. The result is

$$\gamma(\tau) = -\ln \frac{\sqrt{4\pi^2 + 1}}{\tau} \quad \tau \in (0, \sqrt{4\pi^2 + 1}] .$$

Therefore, a formula for the natural parametrization is

$$(F \circ \gamma)(\tau) =$$
$$\frac{\tau}{\sqrt{4\pi^2 + 1}}(\cos(-2\pi \ln[\sqrt{4\pi^2 + 1}/\tau]), \sin(-2\pi \ln[\sqrt{4\pi^2 + 1}/\tau])),$$
$$0 < \tau \leq \sqrt{4\pi^2 + 1} .$$

The curve has no initial point and thus no point for which the natural parameter τ is equal to 0.

Let us find the "midpoint of a journey" along the curve. We do this by setting $\tau = \sqrt{4\pi^2 + 1}/2$. For this value of τ, we have

$$(F \circ \gamma)(\tau) = \frac{1}{2}(\cos(-2\pi \ln 2), \sin(-2\pi \ln 2)) .$$

4.3E CURVATURE

In this section we want to discuss a geometrical aspect of curves that does not depend on the parametrization. Thus, we begin our discussion with a rectifiable curve G parametrized naturally. We focus on the structure of the curve near a particular point $G(\tau)$ on it, where τ is an arbitrary but fixed point in $\mathcal{D}(G)$. Let us assume that G' is continuous at τ. Since G is natural, $\|G'(\tau)\| = 1$ at each τ. So the norm of $G'(\tau)$ gives us no information about the curve near $G(\tau)$; rather its value is 1 just because we are working with the natural parametrization. But the direction of $G'(\tau)$ does give us information about the direction of the curve. From it we can calculate the tangent line to the curve at $G(\tau)$. This tangent line is quite a good approximation to the curve near the point $G(\tau)$, and it would actually lie on a portion of the curve were G'' equal to $\mathbf{0}$ at and near τ.

We now ask: Can we get a better approximation to the curve near the point $G(\tau)$, an approximation that takes into account $G''(\tau)$ in order to describe the manner in which the direction of the curve is changing? To answer this question we consider a special case, the simplest curve for which the tangent line changes direction as one moves along the curve, namely, a circle in $\mathbb{R}^2$. So, consider a circle of radius r centered at $\mathbf{0} \in \mathbb{R}^2$ parametrized by

$$G(\tau) = (r\cos(\tau/r), r\sin(\tau/r)) \,.$$

Notice that we have arranged for $\|G'(\tau)\| = 1$ for each τ, so that G is the natural parametrization. Easy calculations show that for each τ, $\mathbf{0}$ is on the ray $\{G(\tau) + vG''(\tau) : v \geq 0\}$ and the radius r of the circle is the reciprocal of $\|G''(\tau)\|$. That is the center of the circle is in the direction of $G''(\tau)$ from $G(\tau)$ and at a distance $\|G''(\tau)\|^{-1}$ from $G(\tau)$.

Now return to the general case, in which G is a natural parametrization of some curve in $\mathbb{R}^q$. If, for some value of the parameter τ, $G''(\tau)$ exists, its norm is called the **curvature** at the point $G(\tau)$. If the curvature exists and is not 0, its reciprocal is the **radius of curvature**. The **center of curvature** is the point in the direction of $G''(\tau)$ from $G(\tau)$ and whose distance from $G(\tau)$ equals the radius of curvature. That is, the center of curvature is the point

$$G(\tau) + \frac{G''(\tau)}{\|G''(\tau)\|^2} \,.$$

Both the center of curvature and the tangent line lie in the **osculating plane** defined as

$$\{G(\tau) + v_1 G'(\tau) + v_2 G''(\tau) : (v_1, v_2) \in \mathbb{R}^2\} \,.$$

The **osculating circle** is the circle in the osculating plane whose center and radius are the center of curvature and radius of curvature. All these numbers, points, and sets depend on the parameter τ so if several values of τ are being considered, then one must adjoin appropriate modifying phrases to avoid any ambiguity in communication.

The osculating circle is an improvement of the tangent-line approximation of a curve; except, of course, in case the curvature equals 0, a case in
which we are particularly happy with our tangent-line approximation. On
the other hand, a large value for the curvature indicates that the tangent-
line approximation is poor and that the curve is making large changes of
direction for small distances moved along the curve.

Given a curve via a parametrization F and the task of finding its osculating circles at points on the curve, we can follow the program of first
finding the natural parametrization G of the curve and then calculating
G''. To avoid always having to make an explicit calculation of the natural
parametrization, we derive a relation between the second derivatives of F
and G. Suppose that $G = F \circ \gamma$, as in the preceding section, in which an
application of the chain rule gave

$$G' = \frac{F' \circ \gamma}{\|F' \circ \gamma\|} .$$

Let us differentiate this formula. We will need the chain rule again and the
formula obtained earlier for the derivative of the norm of a vector-valued
function. We use the facts proved earlier that $\|F' \circ \gamma\| \gamma' = \|G'\| = 1$ and
$\gamma' = 1/\|F' \circ \gamma\|$. In working with the coordinate functions on the right-hand
side of the formula, we will also need the quotient rule (from single-variable
calculus). Try working this carefully through for yourself. You should get

$$G'' = \frac{\|F' \circ \gamma\|^2 (F'' \circ \gamma) - \langle F' \circ \gamma, F'' \circ \gamma \rangle (F' \circ \gamma)}{\|F' \circ \gamma\|^4} .$$

The following equivalent relation is more useful for the purpose of making
calculations without requiring an explicit formula for G composed on γ:

$$G''(\mathcal{L}(F; a, t)) = \frac{\|F'(t)\|^2 F''(t) - \langle F'(t), F''(t) \rangle F'(t)}{\|F'(t)\|^4} .$$

Example:

12. Let us apply the preceding formula to the spiral described in Example 6 of Section 4.3c. Formulas for F and F' are given there. Straightforward (somewhat tedious) calculations give

$$F''(t) =$$
$$e^t(-4\pi \sin 2\pi t - (4\pi^2 - 1)\cos 2\pi t, 4\pi \cos 2\pi t - (4\pi^2 - 1)\sin 2\pi t) ;$$
$$\|F'(t)\|^2(t) = e^{2t}(4\pi^2 + 1) ;$$
$$\langle F'(t), F''(t) \rangle = e^{2t}(4\pi^2 + 1) ;$$
$$\|F'(t)\|^2 F''(t) - \langle F'(t), F''(t) \rangle F'(t) =$$
$$e^{3t}(4\pi^2 + 1)(-2\pi \sin 2\pi t - 4\pi^2 \cos 2\pi t, 2\pi \cos 2\pi t - 4\pi^2 \sin 2\pi t) .$$

Therefore, for $\tau = \mathcal{L}(F; -\infty, t)$, we have

$$G''(\tau) = \frac{1}{e^t(4\pi^2 + 1)}(-2\pi \sin 2\pi t - 4\pi^2 \cos 2\pi t, 2\pi \cos 2\pi t - 4\pi^2 \sin 2\pi t)$$

and

$$\|G''(\tau)\| = \frac{2\pi}{e^t\sqrt{4\pi^2 + 1}}.$$

As might be expected the curvature gets large and the radius of curvature small when $t \to -\infty$, that is, at points on the spiral near $(0, 0)$. The center of curvature is the point

$$e^t(\cos 2\pi t, \sin 2\pi t)$$
$$+ \frac{1}{4\pi^2} e^t(-2\pi \sin 2\pi t - 4\pi^2 \cos 2\pi t, 2\pi \cos 2\pi t - 4\pi^2 \sin 2\pi t)$$
$$= \frac{1}{2\pi} e^t(-\sin 2\pi t, \cos 2\pi t) .$$

Thus, we see that as the given spiral is traced another spiral is traced by the centers of curvature. The center of curvature is $\pi/2$ radians "in front" of the corresponding point on the given spiral and closer to the origin by a factor of 2π.

Exercises

A table of integrals may be helpful in some of the arc length computations.

1. Find the arc lengths of the following curves:

   (a) $F(t) = (1 - 3t, 5 + 4t)$, $-1 \le t \le 2$

   (b) $G(r) = (r^3, 2 - 6r^2)$, $0 \le r \le 3$

   (c) $H(\tau) = (2 - \tau, 7 + 2\tau, -3 + 2\tau)$, $0 \le \tau \le 4$

   (d) Graph of $y = 7 + (x^2 + \frac{2}{3})^{3/2}$, $1 \le x \le 2$

   (e) Graph of $y = \frac{x^3}{6} + \frac{1}{2x}$, $1 \le x \le 3$

   (f) $F(\theta) = (\cos^3 \theta, 7, \sin^3 \theta)$, $0 \le \theta \le \pi/2$

   (g) $\Phi(t) = (e^{2t} \sin t, e^{2t} \cos t, e^{2t})$, $0 \le t \le \log 5$

   (h) $R(t) = (t \sin t, \sqrt{8t}, t \cos t, \frac{2}{3}\sqrt{6}t^{3/2})$, $4 \le t \le 8$.

2. (a) The position $R(t)$ of a particle at time t is given by

$$R(t) = (5 + 3t^2, 2(1 + 2t)^{3/2}) .$$

   How far does the particle travel between $t = 0$ and $t = 4$?

   (b) Repeat this for

$$R(t) = (\frac{2}{3}t, 5 - t^2, 4 + t^3), \quad 1 \le t \le 2 .$$

3. Let P and Q be distinct points in $\mathbb{R}^n$, and let

$$F(t) = \frac{(t-a)P + (b-t)Q}{b-a} \ .$$

What is the arc length for $a \leq t \leq b$? Could you have guessed the answer?

4. If $r(\theta)$ is a nonnegative continuous function of θ,

$$F(\theta) = (r(\theta) \cos \theta, r(\theta) \sin \theta)$$

is a parametric representation of a curve in $\mathbb{R}^2$ for which the parameter θ is the angle between the vector $F(\theta)$ and the horizontal axis. Note that $r(\theta) = \|F(\theta)\|$.

   (a) Show that the arc length of such a curve for $\alpha \leq \theta \leq \beta$ is

$$\mathcal{L} = \int_\alpha^\beta \sqrt{r'^2 + r^2} \, d\theta \ ,$$

   where $r' = dr/d\theta$.

   (b) Find the arc length of a circle, for which $r(\theta) = 1$.

   (c) Sketch and find the arc length of the curve determined by $r(\theta) = \theta^2$, $-2\sqrt{3} \leq \theta \leq 0$.

5. This is a standard example of a continuous curve that is not rectifiable. Let $y = x \sin(\pi/x)$ for $0 < x \leq 1$ and $y = 0$ for $x = 0$.

   (a) Show geometrically that the arc length of the portion of the curve for $1/(n+1) \leq x \leq 1/n$, that is, for one arch, is at least $2/(n+\frac{1}{2})$.

   (b) Use this to show that the arc length of the curve from $x = 1/N$ to $x = 1$ tends to infinity as $N \to \infty$ (you will need the fact that $1 + \frac{1}{2} + \frac{1}{3} + \cdots + \frac{1}{N} \to \infty$ as $N \to \infty$).

6. Find a natural parametrization G for a journey once around the square with vertices at $(1,0)$, $(0,1)$, $(-1,0)$, and $(0,-1)$. Show that G is not differentiable, but that at points where it has a derivative the norm of the derivative equals 1.

7. For F as in Example 6 of Section 4.3c calculate $\lim_{t \to -\infty} F(t)$.

8. Let $G = F \circ \gamma$ as defined in Example 11 of Section 4.3d. Calculate $\lim_{\tau \downarrow 0} G(\tau)$. Use this answer to modify G so that its domain is a closed interval, but that otherwise the example is changed as little as possible. What does this modification do to the midpoint calculation in Example 11?

9. Let F be a rectifiable curve with domain (a, b), where it is possible that $a = -\infty$ or $b = \infty$. Make a conjecture about the existence of limits at a and b.

10. Let

$$F(t) = \frac{t^2}{1+t^2}(\cos t, \sin t) \ , \quad -\infty < t < \infty \ .$$

Show that this curve is not rectifiable without involving yourself in a messy differentiation or complicated integration.

11. Draw the spiral defined in Example 6 of Section 4.3c.

12. Recalculate the curvature and center of curvature of Example 12 of Section 4.3e by using the natural parametrization of Example 11 of Section 4.3d rather than the parametrization of Example 6 of Section 4.3c.

13. Sharpen your intuition by trying to decide, without calculation, whether the radius of curvature of the helix $(\cos t, \sin t, t)$ is less than 1, equal to 1, or larger than 1. Then check your intuition by doing the appropriate calculation with the given parametrization.

14. Naturally parametrize the piece of the helix in the preceding exercise for which $0 \le t \le 2\pi$. Then use this parametrization to calculate the radius of curvature.

15. For k positive let

$$F(t) = e^t(\cos kt, \sin kt), \quad -\infty < t \le 0 .$$

Make a conjecture about how the arc length of the curve depends on k. In particular, what happens to the arc length as $k \to 0$? Then mimic Examples 6 of Section 4.3c, Example 11 of Section 4.3d, and Example 12 of Section 4.3e, and while doing so, check the correctness of your assertions about the dependence of the arc length on k.

16. Derive a formula for the curvature at an arbitrary point on the graph of a twice differentiable real-valued function of a real-variable.

17. Calculate the equation of the osculating circle at the point $(0,0)$ on the graph of the parabola $y = cx^2$, where c is a nonzero constant.

18. Calculate an equation in x and y whose solution set is the set of centers of curvature for the parabola of the preceding exercise.

19. For the graph of the function $f(x) = \cosh x$ decide whether every osculating circle intersects the axis of the dependent variable.

20. Consider the motion described by

$$F(t) = (t \cos t, t \sin t, 3t) \ t \ge 0 .$$

Make a conjecture about the general features of the function $\mathcal{L}(F; 0, t)$. Is it a constant multiplied by t or does it behave like a constant multiplied by t^2 or does it grow like an exponential function or, alternatively, a logarithmic function? Then check your conjecture by calculating the distance traveled by time t. Find the natural parametrization of the curve. Comment on the fact that you are able to find a natural parametrization even though the curve is not rectifiable.

21. Fill in the blanks. I could draw the curve in the preceding exercise if someone would only buy me an ____ ____ ____, although I would have to eat the ____ ____ first in order to prevent it from melting while I was doing the drawing.

22. Generalize Exercise 20 by considering the motion described by

$$F(t) = (t \cos t, t \sin t, kt) \ \ t \ge 0 ,$$

where k is a constant.

23. For a continuous $\mathbb{R}^q$-valued function H with domain $[a, b]$, define

$$\int_a^b H(u)\, du = \left(\int_a^b h_1(u)\, du, \int_a^b h_2(u)\, du, \ldots, \int_a^b h_q(u)\, du \right),$$

where $h_1, \ldots, h_q$ are the coordinate functions of H. Note that this expression defines a member of $\mathbb{R}^q$. For $t \in [a, b]$, let

$$F(t) = \int_a^t H(u)\, du.$$

(a) Why does the distance between $F(a)$ and $F(b)$ equal

$$\left\| \int_a^b H(u)\, du \right\| ?$$

(b) Why does the arc length of the parametrized curve F with domain $[a, b]$ equal

$$\int_a^b \|H(t)\|\, dt ?$$

(c) Show that

$$\left\| \int_a^b H(u)\, du \right\| \le \int_a^b \|H(t)\|\, dt.$$

24. Let $F(t) = (t(1 - t), t(1 - t))$ for $0 \le t \le 1$. Calculate $F(0)$, $F(1)$, and $\mathcal{L}(F; 0, 1)$. Make some descriptive comments describing the situation.

25. Find the arc lengths of the given curves:

(a) $F(t) = (6t, 3t^2, t^3) \quad -2 \le t \le 2$

(b) $F(t) = (3t^2, 2t^3, (3/4)t^4) \quad -2 \le t \le 2$

(c) $F(t) = (4t, 2\sqrt{3}t^2, (4/\sqrt{3})t^3, t^4) \quad -2 \le t \le 2.$

26. Find approximations to the arc length of the curve

$$F(t) = (t, t^2, t^3, t^4), \quad 0 \le t \le 1,$$

both by calculating an approximation to a certain integral and by using the definition of arc length. For each method subdivide the interval $[0, 1]$ into 10 subintervals of equal length.

5

Topics for Review and Preview

5.0   Introduction

The common thread in this chapter is that each section relies heavily on concepts presented in earlier chapters. In Sections 5.1 and 5.2 there is no one major theme; rather, the focus is on a variety of problems requiring techniques from earlier chapters, with the problems in Section 5.2 being somewhat more involved. Section 5.3 treats applications to motion under the influence of gravity. The chapter closes with some geometry of important subsets in $\mathbb{R}^n$. This geometry is important for the remainder of the book; for it provides examples and techniques that one can use as an aid in understanding general theory. Sections 5.2 and 5.3 are not essential prerequisites for the remainder of the book.

5.1   Further Concepts and Problems

We will introduce some new ideas in this section that are very closely related to concepts presented in the first four chapters. Working with old ideas in conjunction with a few new ideas provides an opportunity for honing previously developed skills.

5.1A   ANGLES IN $\mathbb{R}^3$

In Chapter 2 we obtained the following formula for the angle θ between two nonzero vectors X and Y:

$$\theta = \arccos \frac{\langle X, Y \rangle}{\|X\| \, \|Y\|}.$$

A similar formula,

$$\theta = \arccos \frac{|\langle X, Y \rangle|}{\|X\| \, \|Y\|},$$

is valid for the angle between two intersecting lines given parametrically by $F(t) = U + tX$ and $G(t) = V + tY$. This second formula gives, as is customary, an angle in the closed interval $[0, \pi/2]$.

The question of whether the two lines intersect is rather simple in $\mathbb{R}^2$. If the formula gives a value of θ other than 0, then the two lines do intersect.

If it gives the value 0, then either the two lines do not meet or else they are the same line.

In $\mathbb{R}^3$ (and $\mathbb{R}^n$ for n larger than 3) one has to solve a system of linear equations in two unknowns to decide if the two lines intersect; and, if they do, the above formula gives the angle between them.

EXAMPLE:

1. Let us decide if the lines given parametrically by

$$F(t) = (1, 2, 0) + t(-3, -4, 7) \quad \text{and} \quad G(t) = (4, 16, 1) + t(1, -2, -5)$$

intersect; and, if they do, let us find the angle of intersection.

We are *not* asking that the two lines have a common point for a common value of the parameter, but only that they have a common point. Therefore, let us use different names for the parameters of the two lines: s for the first line and t for the second. That the two lines intersect is equivalent to the following equation having a solution for s and t:

$$(1, 2, 0) + s(-3, -4, 7) = (4, 16, 1) + t(1, -2, -5) \,.$$

(If this equation has infinitely many solutions, then the two lines are identical.) We can rewrite this one vector equation as three scalar equations:

$$\begin{aligned} -t - 3s &= 3 \\ 2t - 4s &= 14 \\ 5t + 7s &= 1 \,. \end{aligned}$$

We use Gauss-Jordan elimination

$$\begin{bmatrix} -1 & -3 & | & 3 \\ 2 & -4 & | & 14 \\ 5 & 7 & | & 1 \end{bmatrix} \sim \begin{bmatrix} -1 & -3 & | & 3 \\ 0 & -10 & | & 20 \\ 0 & -8 & | & 16 \end{bmatrix} \sim \begin{bmatrix} -1 & 0 & | & -3 \\ 0 & -10 & | & 20 \\ 0 & 0 & | & 0 \end{bmatrix},$$

from which we conclude that $t = 3$ and $s = -2$ constitute the one and only solution.

The angle of intersection equals

$$\arccos \frac{|\langle (-3, -4, 7), (1, -2, -5) \rangle|}{\|(-3, -4, 7)\| \, \|(1, -2, -5)\|} = \arccos \frac{|-30|}{\sqrt{74}\sqrt{30}}$$

$$= \arccos \sqrt{15/37} \,,$$

which is approximately equal to 0.88 radians.

We now consider the intersection of two planes. There are three possibilities: they do not intersect, they are the same plane, or they intersect in a line. In the latter case we would like to speak of and evaluate the "angle" at which the planes meet. We will focus on planes in $\mathbb{R}^3$, but even there we need a definition.

Let us start with the case of two planes through the origin in $\mathbb{R}^3$. Suppose that an equation for the first plane is $a_1 x_1 + a_2 x_2 + a_3 x_3 = 0$ and that an equation for the second plane is $b_1 x_1 + b_2 x_2 + b_3 x_3 = 0$. The first plane is the set of vectors orthogonal to the vector (a_1, a_2, a_3) (see Section 2.3). Any multiple of (a_1, a_2, a_3) is also orthogonal to every vector in the first plane. In other words, every vector on the line spanned by (a_1, a_2, a_3) is orthogonal to every vector in the first plane. This line is called the **normal line** of the first plane. Similarly, the line spanned by the vector (b_1, b_2, b_3) is the normal line of the second plane. We define the **angle between two planes** through the origin in $\mathbb{R}^3$ to be the angle between their normal lines.

Now suppose that $\mathcal{A}_1$ and $\mathcal{A}_2$ are two planes in $\mathbb{R}^3$ that do not necessarily contain the origin. Let $\mathcal{P}_1$ and $\mathcal{P}_2$ be planes through the origin in $\mathbb{R}^3$ that are parallel to $\mathcal{A}_1$ and $\mathcal{A}_2$. If $\mathcal{A}_1$ and $\mathcal{A}_2$ intersect, then we define the angle between $\mathcal{A}_1$ and $\mathcal{A}_2$ to be the same as the angle between $\mathcal{P}_1$ and $\mathcal{P}_2$.

Without checking whether the two planes intersect, one may proceed with the "calculation of the angle". If the result is 0, the two planes either are identical or never meet. Otherwise, the intersection of the two planes is a line and the calculation gives the size of the angle of intersection of the two planes.

EXAMPLE:

2. Let us decide if the intersection of the two planes

$$x_1 - 3x_2 + x_3 = 3 \quad \text{and} \quad 2x_1 - 3x_2 - x_3 = 0$$

is a line; and, if it is, let us calculate the angle at which they meet. The corresponding planes through the origin have the equations

$$x_1 - 3x_2 + x_3 = 0 \quad \text{and} \quad 2x_1 - 3x_2 - x_3 = 0 \,.$$

The normal lines of these planes are given parametrically by

$$s(1, -3, 1) \quad \text{and} \quad t(2, -3, -1) \,,$$

so the angle of intersection equals

$$\arccos \frac{|\langle (1, -3, 1), (2, -3, -1) \rangle|}{\|(1, -3, 1)\| \, \|(2, -3, -1)\|} = \arccos \frac{10}{\sqrt{154}} \,.$$

The fact that we have obtained a nonzero answer implies that the intersection of the two given planes is a line.

There are three possibilities for a line S and a plane A in $\mathbb{R}^3$: the line lies in the plane, the line never meets the plane, or the line meets the plane in a single point P. In the latter case we define the angle between the line and the plane as follows. Let $\mathcal{L}$ be a line through the origin parallel to S and let $\mathcal{P}$ be a plane through the origin parallel to A. Then the angle between S and A is defined to be the same as the angle between $\mathcal{L}$ and $\mathcal{P}$, which is defined to be the complement of the angle between $\mathcal{L}$ and the normal line of $\mathcal{P}$. In other words, if θ is the angle between the normal line of $\mathcal{P}$ and the line $\mathcal{L}$, then $\pi/2 - \theta$ is the angle (in radians) between $\mathcal{L}$ and $\mathcal{P}$, which is also the angle between S and A. As in the case with the two planes, one need not give attention (in $\mathbb{R}^3$) to whether the line intersects the plane. Rather, it is enough to carry out the computation of the angle as if they did intersect. If the result is nonzero, then they do in fact intersect in a point, otherwise the line either never meets the plane or is a subset of the plane.

EXAMPLE:

3. Let us calculate the angle between the line given parametrically by

$$(0, 2, -5) + t(1, 3, 0)$$

and the plane $\mathcal{P} + (2, -1, 3)$, where $\mathcal{P}$ is the linear span of $(1, 0, 2)$ and $(-2, -3, 4)$. According to the definition, we calculate the angle between $\mathcal{P}$ and the line spanned by the vector $(1, 3, 0)$.

We first need to find the normal line of $\mathcal{P}$. This line is spanned by the vector

$$(1, 0, 2) \times (-2, -3, 4) = (6, -8, -3) .$$

The angle between the two lines is

$$\operatorname{arccos} \frac{|\langle (1, 3, 0), (6, -8, -3) \rangle|}{\|(1, 3, 0)\| \|(6, -8, -3)\|} = \operatorname{arccos} \frac{18}{\sqrt{1090}} .$$

Subtract this angle from $\pi/2$ to obtain

$$\operatorname{arcsin} \frac{18}{\sqrt{1090}} \approx .577 \text{ radians.}$$

REMARK:

- There are other, equivalent ways to define the angle between a plane and a line. Assuming for simplicity that both the plane and the line contain the origin, we may define the angle between them to be the minimum angle between the line and any line that is contained in the plane and passes through the origin. Or we may define the angle between the line and the plane to be the angle between the line and the orthogonal projection of the line onto the plane. These alternate definitions have the advantage that they make sense in dimensions higher than 3. They have the disadvantage in $\mathbb{R}^3$ that they require somewhat more computation.

5.1B   DISTANCE BETWEEN A POINT AND A LINE

In Example 9 of Section 2.2e we calculated the distance from a particular point to a particular line. We return to the ideas used in Chapter 2 in order to obtain a general formula.

Theorem 5.1.1 The distance between a point $Y \in \mathbb{R}^n$ and a line in $\mathbb{R}^n$ given parametrically by $Z + tX$, $t \in \mathbb{R}$, equals

$$\frac{\sqrt{\|Y - Z\|^2 \|X\|^2 - \langle Y - Z, X \rangle^2}}{\|X\|}.$$

PROOF: We first shift everything so that the line contains the origin. We accomplish this by subtracting Z from Y and from every point on the line. Our new point is $Y - Z$, which we shall call W, and our new line is given parametrically by tX, $t \in \mathbb{R}$. The distance between the new point and the new line is the same as the distance between the original point and the original line. We need to show that this distance, given in terms of W and X, is

$$\frac{\sqrt{\|W\|^2 \|X\|^2 - \langle W, X \rangle^2}}{\|X\|}.$$

In Chapter 2 we did obtain a formula for the distance between a point W and the line through the origin spanned by a vector X, namely $\|W - P\|$, where P denotes the projection of W onto X. Using the formula given there for P, we obtain the following formula for the distance between W and the line tX:

$$\left\| W - \frac{\langle W, X \rangle}{\|X\|^2} X \right\|$$

$$= \sqrt{\left\langle W - \frac{\langle W, X \rangle}{\|X\|^2} X, \, W - \frac{\langle W, X \rangle}{\|X\|^2} X \right\rangle}$$

$$= \sqrt{\|W\|^2 - 2\frac{\langle W, X \rangle^2}{\|X\|^2} + \frac{\langle W, X \rangle^2}{\|X\|^4} \langle X, X \rangle}$$

$$= \sqrt{\|W\|^2 - \frac{\langle W, X \rangle^2}{\|X\|^2}},$$

which equals

$$\frac{\sqrt{\|W\|^2 \|X\|^2 - \langle W, X \rangle^2}}{\|X\|}.$$

as desired. $<<$

EXAMPLE:

4. We will calculate the distance between the point $(3,5)$ and the line given parametrically by $(-1,3) + t(3,4)$. To apply the preceding theorem we first calculate $(3,5) - (-1,3) = (4,2)$. The distance equals

$$\frac{\sqrt{\|(4,2)\|^2 \, \|(3,4)\|^2 - \langle (4,2), (3,4) \rangle^2}}{\|(3,4)\|}$$

$$= \frac{\sqrt{(20)(25) - (20)^2}}{5} = \frac{\sqrt{100}}{5} = 2 \,,$$

in agreement with Example 9 of Section 2.2e.

Consider the special case where the point of interest is the origin $(0,0)$ in $\mathbb{R}^2$. Suppose that the line is not given parametrically, but instead by an equation. The distance between $(0,0)$ and the line equals the absolute value of the x_2-intercept if the line is horizontal and the absolute value of the x_1-intercept if the line is vertical. If the line is neither horizontal or vertical it can be represented by an equation in terms of its x_1- and x_2-intercepts b_1 and b_2 as follows:

$$\frac{x_1}{b_1} + \frac{x_2}{b_2} = 1 \,.$$

We look for a formula for the distance between the origin and this line, without first obtaining a parametric representation for the line.

Consider the triangle with vertices at $(0,0)$, $(b_1,0)$, and $(0,b_2)$. There is a right angle at $(0,0)$, so the area of the triangle can be expressed in terms of the two legs of the right triangle as $\frac{1}{2}|b_1| \, |b_2|$. But there is another way of calculating the area. We regard the hypotenuse as the base, and the distance between the origin and the line as the height of the triangle. The length of the hypotenuse is $\sqrt{b_1^2 + b_2^2}$. Let h denote the distance between the line and the origin. Then the area of the triangle equals $\frac{1}{2}h\sqrt{b_1^2 + b_2^2}$. Setting the two expressions for the area equal to each other and then solving for h, we obtain

$$h = \frac{|b_1| \, |b_2|}{\sqrt{b_1^2 + b_2^2}} \,.$$

The reader may want to use a parametrization of the line to derive another formula for h, and then compare the two formulas (they should be equal). A three-dimensional version of the reasoning used in this paragraph plays a central role in Section 5.1c.

EXAMPLE:

5. Let us find the distance between $(0,0)$ and the line $-\frac{2}{5}x_1 + \frac{3}{10}x_2 = 1$. The two intercepts are $(-\frac{5}{2}, 0)$ and $(0, \frac{10}{3})$ and the distance between

$(0,0)$ and the given line is

$$\frac{(5/2)(10/3)}{\sqrt{(5/2)^2 + (10/3)^2}} = \frac{5/3}{\sqrt{(1/4) + (4/9)}} = 2 .$$

The reader should consider whether the agreement between this answer and the answer to Example 4 is a coincidence.

5.1C Distance between two lines in $\mathbb{R}^3$

Consider two lines in $\mathbb{R}^3$ given parametrically by

$$U + rX \quad \text{and} \quad V + sY .$$

We look for a formula for the "distance between the two lines". There are two cases: if the two lines lie in a common plane and are parallel, we expect that there will be infinitely many line segments that are perpendicular to both the given lines, with the length of any one of the segments equalling the distance between the two lines. Otherwise, we expect that there will be one line segment which is perpendicular to both lines, and the length of that line segment is the distance between the two lines. We will not give a formal definition of the distance between two lines, but rather take the description in this paragraph as constituting the definition. The formulas for the two cases just described are given in the following two theorems.

Theorem 5.1.2 In $\mathbb{R}^3$: if $X \times Y = 0$ and neither X nor Y equals 0, then the distance between the lines described parametrically by $U + rX$ and $V + sY$ equals

$$\frac{\|(V-U) \times X\|}{\|X\|} .$$

PROOF: If $X \times Y = 0$, then X and Y are multiples of each other, so the two given lines lie in a common plane and are parallel. Consider the parallelogram with vertices $U, V, U + X, V + X$. (If the two lines are identical, this parallelogram will be degenerate. Our proof is still valid in this case.) We compute the area of this parallelogram in two ways. The first way involves the distance between the two parallel lines $U + rX$ and $V + sY$. One edge of the parallelogram, which we call the base, lies on the line $U + rX$ and has length $\|X\|$. The height of the parallelogram is the distance h between the line $U + rX$ and the line $V + sY$. Thus, a formula for the area of the parallelogram is $h\|X\|$. For a second computation of the area, shift the parallelogram by subtracting U from each of its vertices. The shifted figure is the parallelogram generated by the vectors X and $V - U$, so its area is

$$\|(V - U) \times X\| .$$

Setting the two expressions for the area equal to each other gives the desired formula for h. $<<$

Theorem 5.1.3 In $\mathbb{R}^3$: if $X \times Y \neq 0$, then the distance between the lines described parametrically by $U + rX$ and $V + sY$ equals

$$\frac{|[(V - U), X, Y]|}{\|X \times Y\|} .$$

(Recall: the notation $[\cdot, \cdot, \cdot]$ in the numerator represents the triple product.)

PROOF: Consider the parallelepiped with vertices $U, U + X, U + Y, U + X + Y, V, V + X, V + Y, V + X + Y$. We compute the volume of this parallelepiped in two ways (compare the proof of the previous theorem). The first method involves the distance between the two lines. The line $U + rX$ is contained in the plane which contains the points $U, U + X, U + Y$. The line $V + sY$ is contained in the plane that contains the points $V, V + X, V + Y$. These two planes contain opposite faces of the parallelepiped, so they are parallel. One of these planes contains the line $U + rX$ and is parallel to the line $V + sY$, while the other plane contains the line $V + sY$ and is parallel to the line $U + rX$. It follows that any line segment which is perpendicular to both lines is also perpendicular to both planes, so the distance between the planes is the distance between the two lines. We call this distance h. Consider one of the two faces, say, the one with vertices $U, U + X, U + Y, U + X + Y$. Think of this face as the base of the parallelepiped. The volume of the parallelepiped is h times the area of the base. The area of the base is $|X \times Y|$, as may be seen by shifting the base by subtracting U from each of its vertices, so that it becomes the parallelogram generated by the vectors X and Y. Thus, one formula for the area of the parallelepiped is $h|X \times Y|$.

To obtain a second formula, subtract U from all the vertices of the parallelepiped. The shifted figure is the parallelepiped generated by the vectors $X, Y, V - U$, whose volume is $|[(V - U, X, Y]|$. Setting the two formulas for the volume equal to each other and solving for h, we obtain the desired formula for the distance between the two lines. $<<$

EXAMPLES:

6. Let us find the distance between the lines given parametrically by

$$(3, 1, -4) + r(4, 10, -2) \quad \text{and} \quad (2, 2, 0) + s(-6, 0, 3) .$$

We first calculate the appropriate cross-product:

$$(4, 10, -2) \times (-6, 0, 3) = (30, 0, 60) .$$

Since this result is different from 0, it is Theorem 5.1.3 that is applicable. The denominator in the expression for the distance is

$$\|(30, 0, 60)\| = 30\sqrt{5} .$$

The triple product in the numerator is

$$\langle (30, 0, 60), (2, 2, 0) - (3, 1, -4) \rangle = \langle (30, 0, 60), (-1, 1, 4) \rangle = 210,$$

the absolute value of which equals 210. Thus, the distance between the two lines is $\frac{210}{30\sqrt{5}} = \frac{7}{\sqrt{5}}$.

7. Let us find the distance between the lines given parametrically by

$$(3, 1, -4) + r(4, 10, -2) \quad \text{and} \quad (2, 2, 0) + s(-6, -15, 3).$$

We first calculate the appropriate cross-product:

$$(4, 10, -2) \times (-6, -15, 3) = (0, 0, 0) = \mathbf{0}.$$

Thus, the two lines are parallel and Theorem 5.1.2 is applicable. The distance between them equals

$$\frac{\|((2, 2, 0) - (3, 1, -4)) \times (4, 10, -2)\|}{\|(4, 10, -2)\|}$$

$$= \frac{\|(-1, 1, 4) \times (4, 10, -2)\|}{\sqrt{120}} = \frac{\|(-42, 14, -14)\|}{2\sqrt{30}}$$

$$= \frac{7\|(-3, 1, -1)\|}{\sqrt{30}} = 7\sqrt{11/30}.$$

5.1D DISTANCE BETWEEN A POINT AND A CURVE

We start by reconsidering the problem of finding the distance between a point and a line. Let $f(t)$ denote the distance between the points Y and $Z + tX$, that is

$$f(t) = \|(Z + tX) - Y\|.$$

The function f records distances between pairs of points. As t varies, this distance varies. It is the case that the function f has a global (some use the term "absolute") minimum. This minimum is the distance between point and the line.

EXAMPLE:

8. Let us find the distance between the point $(2, 3)$ and the line given parametrically by $(5, -1) + t(1, 2)$ for $t \in \mathbb{R}$. For fixed t we calculate the distance $f(t)$ between the corresponding point on the line and the point $(2, 3)$ in the usual way:

$$f(t) = \sqrt{((5 + t) - 2)^2 + ((-1 + 2t) - 3)^2} = \sqrt{5t^2 - 10t + 25}.$$

To minimize we calculate the derivative:

$$f'(t) = \frac{5t - 5}{\sqrt{5t^2 - 10t + 25}},$$

which is negative for $t < 1$, is equal to 0 for $t = 1$, and is positive for $t > 1$. Thus, f has a global minimum at 1, the value of which is $\sqrt{5 - 10 + 25} = 2\sqrt{5}$.

As a check let us see if the line segment with endpoints $(2, 3)$ and $(5 + 1, -1 + 2)$ is perpendicular to the given line. It is, since

$$\langle (6 - 2, 1 - 3), (1, 2) \rangle = \langle (4, -2), (1, 2) \rangle = 4 - 4 = 0 \,.$$

Instead of working with f we could have worked with the function g defined by $g(t) = [f(t)]^2$, since g has a global minimum at a point t if and only if the nonnegative function f does and the value of the global minimum of f is the square root of the global minimum of g. Doing so would not essentially change the calculations, but would give formulas a more streamlined appearance.

When lines are replaced by more general curves it is not always the case that the distance function— f in the preceding example—has a global minimum. Even so, in this situation we would like to have a reasonable definition of the distance between a point and a curve. We need some preparatory work for this definition.

In analogy with the definition of an upper bound, we define a **lower bound** of a nonempty set S of numbers to be any real number or $-\infty$ that is less than or equal to every member of S. The largest of the lower bounds, which always exists, is the **greatest lower bound** of S. Consider a point Y and a curve C in $\mathbb{R}^q$. The **distance** between Y and C is the greatest lower bound of the set consisting of distances between Y and points in C.

EXAMPLE:

9. Let us calculate the distance between the point $(0, 5)$ and the parabola $x_2 = x_1^2$. We begin by parametrizing the parabola: $x_1(t) = t$ and $x_2(t) = t^2$. The distance $f(t)$ between the point (t, t^2) and $(0, 5)$ equals

$$f(t) = \sqrt{t^2 + (t^2 - 5)^2} \,.$$

Instead we work with the function $g(t) = f(t)^2$, that is,

$$g(t) = t^4 - 9t^2 + 25 \,.$$

For its derivative we obtain

$$g'(t) = 4t^3 - 18t \,,$$

which is negative for $t < -\frac{3}{\sqrt{2}}$, positive for $-\frac{3}{\sqrt{2}} < t < 0$, negative for $0 < t < \frac{3}{\sqrt{2}}$, and positive for $\frac{3}{\sqrt{2}} < t$. Thus g has a global minimum at $-\frac{3}{\sqrt{2}}$ or $\frac{3}{\sqrt{2}}$ (or both). The value of g at each of these two points is $\frac{81}{4} - 9(\frac{9}{2}) + 25 = \frac{19}{4}$. So the distance between the given point and given parabola is $\frac{\sqrt{19}}{2}$.

Exercise 14 of Section 4.2 tells us that at the point on a curve closest to the origin, the tangent line, if it exists, is perpendicular to the line through that point and the origin. The same is true if the origin is replaced by some other point. This fact gives us a method of checking our above calculation. The difference between the point $(0, 5)$ and the point $(\frac{3}{\sqrt{2}}, \frac{9}{2})$ equals $(-\frac{3}{\sqrt{2}}, \frac{1}{2})$ and the tangent line at the point $(\frac{3}{\sqrt{2}}, \frac{9}{2})$ has direction $(1, 2\frac{3}{\sqrt{2}}) = (1, \frac{6}{\sqrt{2}})$. We check the inner product

$$\langle (-\frac{3}{\sqrt{2}}, \frac{1}{2}), (1, \frac{6}{\sqrt{2}}) \rangle = 0 .$$

Thus, we have an affirmative check. (Note that it is possible for this check to be affirmative·even when the answer is wrong.)

The next example is similar to but more general than the last example. The point $(0, 5)$ is replaced by an arbitrary point on the vertical axis and an arbitrary, but fixed, constant is introduced into the parabola.

EXAMPLES:

10. Let us consider the parabola $x_2 = bx_1^2$, where b is a fixed positive constant; and let us calculate the distance between it and an arbitrary point $(0, c)$ on the vertical axis.

To do the calculation we parametrize the parabola, for example by $x_1(t) = t$ and $x_2(t) = bt^2$. The distance between $(0, c)$ and the point on the curve corresponding to t is

$$f(t) = \sqrt{(t - 0)^2 + (bt^2 - c)^2} = \sqrt{t^2 + (bt^2 - c)^2} .$$

To slightly simplify the calculus we work with the square of the distance function: $g(t) = [f(t)]^2$. Differentiating we get

$$g'(t) = 2t + 2(bt^2 - c)2bt = 2t[2b^2t^2 - (2bc - 1)] .$$

If $c \leq \frac{1}{2b}$, then $g'(t)$ is negative, zero, or positive according as t is negative, zero, or positive. So g and therefore f have absolute minima at $t = 0$. The value of the minimum value of f is $\sqrt{c^2} = |c|$.

If $c > \frac{1}{2b}$, then $g'(t)$ is negative for $t < -\frac{\sqrt{bc-(1/2)}}{b}$, positive for $-\frac{\sqrt{bc-(1/2)}}{b} < t < 0$, negative for $0 < t < \frac{\sqrt{bc-(1/2)}}{b}$, and positive for $\frac{\sqrt{bc-(1/2)}}{b} < t$. Therefore, g has an absolute minimum at $-\frac{\sqrt{bc-(1/2)}}{b}$ and/or $\frac{\sqrt{bc-(1/2)}}{b}$; and the same is true for f. The value of f is the same at these two points and so that common value is the distance between the point $(0, c)$ and the parabola. The distance is

$$\frac{1}{2b}\sqrt{4bc - 1}.$$

We conclude that the distance $h(c)$ between the point $(0, c)$ and the parabola equals

$$h(c) = \begin{cases} |c| & \text{if} \quad c \le \frac{1}{2b} \\ \frac{\sqrt{4bc-1}}{2b} & \text{if} \quad c > \frac{1}{2b} . \end{cases}$$

Let us do some checking. Our intuition based on the statement of the problem is that h should be continuous. It is obvious that the function given by the above formula is continuous at points different from $\frac{1}{2b}$, and that it is left-continuous at $\frac{1}{2b}$ where its value is $\frac{1}{2b}$. The limit from the right at $\frac{1}{2b}$ equals

$$\frac{\sqrt{4b(1/2b) - 1}}{2b} = \frac{1}{2b} ,$$

as desired. As a second check let us see if the inequality

$$\frac{\sqrt{4bc - 1}}{2b} \le c$$

holds for $c > \frac{1}{2b}$; it should since the distance from the point $(0, c)$ to the point $(0, 0)$ on the parabola equals $|c|$. Clearing the denominator $2b$ and squaring both sides gives the equivalent inequality

$$4bc - 1 \le 4(bc)^2$$

or

$$0 \le 4(bc)^2 - 4(bc) + 1 ,$$

which is true since the right-hand side is the square of $(2bc - 1)^2$.

11. Consider a "spiraling helix" parametrized by

$$F(t) = (e^t \cos t, e^t \sin t, -e^t), \quad -\infty < t < \infty .$$

The reader is advised to draw a picture. Let us calculate the distance between the spiraling helix and the point $(1, 0, 1)$.

The distance between the points $F(t)$ and $(1, 0, 1)$ equals

$$f(t) = \sqrt{(e^t \cos t - 1)^2 + (e^t \sin t)^2 + (-e^t - 1)^2} .$$

As is common we work with the square $g(t) = [f(t)]^2$ of the distance. Thus,

$$
\begin{aligned}
g(t) &= 2e^{2t} + 2e^t[1 - \cos t] + 2 \\
&= 2e^t \left[e^t + 1 - \cos t\right] + 2 \ .
\end{aligned}
$$

The distance between the spiraling helix and $(1, 0, 1)$ is the square root of the greatest lower bound of the image of g.

Rather than differentiate we first observe that

$$
\lim_{t \to -\infty} g(t) = 2 \ ,
$$

so the distance between the curve and the point is no larger than $\sqrt{2}$. To see whether it is smaller than $\sqrt{2}$ we need to ascertain whether $e^t + 1 - \cos t$ can ever be negative. Obviously, it cannot be. So, the distance between $(1, 0, 1)$ and the spiraling helix equals $\sqrt{2}$. The distance between $(1, 0, 1)$ and any particular point on the spiraling helix is larger than $\sqrt{2}$, but not much larger when the point on the spiraling helix is close to $(0, 0, 0)$.

5.1E   COMPLEX ARITHMETIC AND LINEAR MAPS IN $\mathbb{R}^2$

Recall that a complex number z is of the form $x + yi$, where x and y are real numbers and i is a square root of -1. Since each complex number corresponds to a unique ordered pair of real numbers and conversely, it follows that $\mathbb{R}^2$ can be regarded as a picture of the set of complex numbers.

One adds and subtracts complex numbers as members of $\mathbb{R}^2$. Thus $(u + vi) + (x + yi) = (u + x) + (v + y)i$ for real numbers u, v, x, and y. The fact that multiplication of complex numbers is defined distinguishes complex numbers from the vectors in $\mathbb{R}^2$ discussed in preceding sections of this book. This definition, based on the equality $i^2 = -1$, is

$$
(u + vi)(x + yi) = (ux - vy) + (vx + uy)i \ .
$$

Thus, for example, $(2 + 3i)(4 - i) = 11 + 10i$. With the preceding definitions of addition and multiplication, the usual properties of arithmetic hold, namely, both commutative laws, both associative laws, the distributive law, and $0 = 0 + 0i$ and $1 = 1 + 0i$ are the additive and multiplicative identities, respectively, and there are additive and multiplicative inverses so that subtraction and division, except by 0, are defined. To carry out division, a standard trick is needed.

EXAMPLE:

12. Let us divide $11 + 10i$ by $4 - i$, that is, let us find a complex number whose product with $4 - i$ is $11 + 10i$. We have

$$
\frac{11 + 10i}{4 - i} = \frac{(11 + 10i)(4 + i)}{(4 - i)(4 + i)} = \frac{34 + 51i}{17} = 2 + 3i \ .
$$

Fix a complex number $c = a + bi$, where a and b are real numbers. Consider the function $L(z) = cz$ defined for complex numbers z. Notice that L can be regarded as a function from $\mathbb{R}^2$ into $\mathbb{R}^2$ since both z and cz are complex numbers. For example, if $c = i$, then $L(1 + 2i) = -2 + i$; so $L(1, 2) = (-2, 1)$. For general $z = x + yi$ we have

$$L(z) = (a + bi)(x + yi) = (ax - by) + (bx + ay)i \,,$$

or, in notation appropriate for $\mathbb{R}^2$,

$$L(x, y) = (ax - by \,, \ bx + ay) \,.$$

We now recognize that L is a *linear* map for which

$$[L] = \begin{bmatrix} a & -b \\ b & a \end{bmatrix} .$$

The converse is also clear. A matrix of the form

$$\begin{bmatrix} a & -b \\ b & a \end{bmatrix}$$

represents a linear map that can be expressed in terms of complex multiplication: $L(z) = (a + bi)z$.

EXAMPLE:

13. Let us decide if a counterclockwise rotation of $\mathbb{R}^2$ through $\pi/3$ radians can be represented by complex multiplication. We know from Section 3.8 that the matrix representing this rotation is

$$\begin{bmatrix} \frac{1}{2} & -\frac{\sqrt{3}}{2} \\ \frac{\sqrt{3}}{2} & \frac{1}{2} \end{bmatrix} .$$

From the above discussion we see that this rotation can be represented by complex multiplication, namely $L(z) = (\frac{1}{2} + \frac{\sqrt{3}}{2} i)z$.

We leave it for the reader to show that a counterclockwise rotation of $\mathbb{R}^2$ through an angle θ can be represented by the complex multiplication

$$L(z) = (\cos\theta + i\sin\theta)\, z \,.$$

The effect of a counterclockwise rotation through an angle θ_1 followed by a counterclockwise rotation through an angle θ_2 can be represented in two ways involving complex multiplication:

$$(\cos(\theta_1 + \theta_2) + i\sin(\theta_1 + \theta_2))\, z$$

and
$$(\cos\theta_2 + i\sin\theta_2)(\cos\theta_1 + i\sin\theta_1)\,z\;.$$

Thus, it must be that
$$(\cos(\theta_1 + \theta_2) + i\sin(\theta_1 + \theta_2)) = (\cos\theta_2 + i\sin\theta_2)(\cos\theta_1 + i\sin\theta_1)\;.$$

As a consequence, we obtain the familiar addition formulas for cosine and sine:
$$\cos(\theta_1 + \theta_2) = \cos\theta_1 \cos\theta_2 - \sin\theta_1 \sin\theta_2\;,$$
$$\sin(\theta_1 + \theta_2) = \sin\theta_1 \cos\theta_2 + \sin\theta_2 \cos\theta_2\;.$$

Because the property
$$(\cos(\theta_1 + \theta_2) + i\sin(\theta_1 + \theta_2)) = (\cos\theta_2 + i\sin\theta_2)(\cos\theta_1 + i\sin\theta_1)$$

is so similar to a law of exponents for exponential functions of a real variable, we now define a new symbol:
$$e^{i\theta} = \cos\theta + i\sin\theta\;,$$

for all real numbers θ. Thus, the above relation becomes
$$e^{i(\theta_1 + \theta_2)} = e^{i\theta_1} e^{i\theta_2}\;.$$

We can use our new notation to write the complex multiplication that represents the rotation L through an angle θ:
$$L(z) = e^{i\theta}z\;.$$

EXAMPLES:

14. Let us decide if multiplication $L(X) = rX$ by a fixed real constant r can be represented by complex multiplication. Since $L(e_1) = re_1$ and $L(e_2) = re_2$, we conclude that
$$[L] = \begin{bmatrix} r & 0 \\ 0 & r \end{bmatrix}\;.$$

Our discussion in this section tells us that L can be viewed as complex multiplication by the real constant $r = r + 0i$, that is, $L(z) = rz$.

15. Let us calculate the result of rotating $\mathbb{R}^2$ counterclockwise about the origin through an angle of $\pi/5$ radians followed by multiplying each vector by 2 followed by a clockwise rotation through an angle of $7\pi/10$ radians and then multiplication of each vector by $1/2$. The following calculation indicates what happens to a point z (written as a complex number):
$$\frac{1}{2}e^{-7\pi i/10}2e^{\pi i/5}\,z = e^{-\pi i/2}\,z\;,$$

a clockwise rotation through a right angle. Since complex multiplication is commutative, we would obtain the same result for the same four operations performed in any order. The two factors of 2 and 1/2 cancel each other and the two rotations combine according to the law of exponents:

$$e^{-7\pi i/10} e^{\pi i/5} = e^{-(7\pi i/10)+(\pi i/5)} = e^{-\pi i/2} .$$

REMARK:

- As we noticed in the preceding example, composition of linear transformations representable by complex multiplication is commutative. This is because multiplication of complex numbers is commutative. In terms of matices, multiplication of matrices of the form

$$\begin{bmatrix} a & -b \\ b & a \end{bmatrix}$$

is commutative.

Consider the linear transformation represented by the complex number $re^{i\theta}$, where $r \geq 0$. It rotates by an angle θ and then multiplies by r. In particular, the point $(1, 0)$ goes first to $(\cos\theta, \sin\theta)$ and then to $(r\cos\theta, r\sin\theta)$, which is a point that has the complex representation $re^{i\theta}$. Since it is clear that every point in $\mathbb{R}^2$ can be obtained by rotating the vector $(1, 0)$ about the origin and then multiplying by the appropriate factor, any complex number $c = a + bi$ can be written in the form

$$c = re^{i\theta} = r\cos\theta + ir\sin\theta ,$$

called the **polar form** of c. Given r and θ one can find a and b by the formulas

$$a = r\cos\theta \quad \text{and} \quad b = r\sin\theta .$$

Alternatively, if a and b are given, these equations can be solved for r and θ. In particular, $r = \sqrt{a^2 + b^2}$ is the distance from the complex number $a + bi$ to the origin.

It is useful to write complex numbers in polar form when multiplications and divisions are anticipated. For instance,

$$(re^{i\theta})(se^{i\varphi}) = (rs)e^{i(\theta+\varphi)}$$

and, if $s \neq 0$,

$$(re^{i\theta}) \div (se^{i\varphi}) = \frac{r}{s}e^{i(\theta-\varphi)} .$$

In some sense the polar form is nothing new at all. When one writes $r\cos\theta + ir\sin\theta$ one is really writing $a + bi$, with $a = r\cos\theta$ and $b = r\sin\theta$.

But the way is open for a new point of view. Points in the plane can be identified by ordered pairs (r, θ) with r denoting the distance from a point to the origin and θ denoting its counterclockwise angle about the origin measured from the positive x-axis. When this point of view is taken, the quantities in the ordered pair (r, θ) are called the **polar coordinates** of the corresponding point $(a, b) = (r \cos \theta, r \sin \theta)$.

EXAMPLE:

16. Let us write the complex number $6 - 6\sqrt{3}\,i$ in polar form, and let us also represent in polar coordinates the point whose rectangular coordinates are $(6, -6\sqrt{3})$. We have

$$r = \sqrt{6^2 + (-6\sqrt{3})^2} = 12$$

and

$$\theta = \pm \arccos(6/12) = \pm\pi/3 \,.$$

Since $-6\sqrt{3} < 0$ we choose the negative sign in the formula for θ; thus, $\theta = -\pi/3$.

We conclude that

$$6 - 6\sqrt{3} = 12(\cos(-\pi/3) + i\sin(-\pi/3))$$

and that $(12, -\pi/3)$ is a polar coordinates representation of the point whose rectangular coordinates are $(6, -6\sqrt{3})$.

REMARKS:

- Polar forms are not unique. For instance, the complex number treated in the preceding example can be written as

$$12(\cos(5\pi/3) + i\sin(5\pi/3)) \quad \text{or} \quad 12(\cos(11\pi/3) + i\sin(11\pi/3)) \,.$$

The numbers are the same:

$$12\cos(-\pi/3) = 12\cos(5\pi/3) = 12\cos(11\pi/3) = 6$$

and

$$12\sin(-\pi/3) = 12\sin(5\pi/3) = 12\sin(11\pi/3) = -6\sqrt{3} \,;$$

only the appearances have changed.

- If we were planning to make extensive use of both polar coordinates and rectangular coordinates we would create notation to distinguish them. Without such notation, we must use words to avoid confusion.

The fact that an arbitrary complex number can be written in polar form implies that the linear map corresponding to multiplication by a fixed complex number can always be regarded as the composition of two particularly nice linear maps: rotation about the origin and multiplication of all distances from the origin by a factor. If this factor is not 0, then the map is one-to-one; and geometrically we see that angles are not changed. For this reason such linear maps are called **similarities** (in the same spirit as the adjective "similar" is used for two triangles that have equal angles, but not necessarily congruent edges).

Do all linear similarities have the form $L(z) = cz$ for some complex number $c \neq 0$? The answer is "no", but all linear similarities that *have positive determinant* are of this form. The linear similarities that have negative determinant reverse the orientation of angles, that is, if an angle is measured in the counterclockwise direction from one ray to a second, then the direction from the image of the first ray to the image of the second is clockwise.

Affine maps that preserve angles and orientation are called **orientation-preserving** similarities. Such similarities have the form $T(z) = cz + w_0$ for appropriate complex numbers $c \neq 0$ and w_0.

EXAMPLE:

17. Let us find the result of the similarity $2iz + (1 + i)$ followed by the similarity $(-1 + i)z + 2$. We have

$$(-1 + i)[2iz + (1 + i)] + 2 = (-2 - 2i)z ,$$

a counterclockwise rotation about the origin of $5\pi/4$ radians followed by a multiplication of all distances from the origin by $2\sqrt{2}$.

Exercises

1. Decide if the two lines are identical, never meet, or meet in exactly one point. In the latter case calculate the angle at the point of intersection.

   (a) $(1, 2) + t(-3, 2)$ and $(3, -3) + t(2, -1)$

   (b) $S_1 + (3, 5)$ and $S_2 + (11, -1)$, where S_1 is the linear span of $(4, -3)$ and S_2 is the linear span of $(-8, 6)$

   (c) $(2, 3, -1) + t(3, 0, 2)$ and $(1, -2, 5) + t(1, 4, -2)$

   (d) $S_1 + (2, 3, 0)$ and $S_2 + (1, -2, 3)$, where S_1 is the linear span of $(3, 0, 2)$ and S_2 is the linear span of $(1, 4, -2)$

   (e) $2x_1 - 3x_2 = 3$ and $5x_1 + 2x_2 = -4$ in $\mathbb{R}^2$

   (f) one line in $\mathbb{R}^3$ determined by $x_1 - x_2 - x_3 = 0$ and $2x_1 + x_2 + 3x_3 = 0$ and the second line determined by $x_1 - 2x_2 - 5x_3 = 0$ and $3x_1 + 2x_2 - 4x_3 = 0$

(g) one line in $\mathbb{R}^3$ determined by $x_1 - x_2 - x_3 = -2$ and $2x_1 + x_2 + 3x_3 = 9$ and the second line determined by $x_1 - 2x_2 - 5x_3 = -11$ and $3x_1 + 2x_2 - 4x_3 = -3$

(h) one line in $\mathbb{R}^3$ determined by $x_1 - x_2 - x_3 = -2$ and $2x_1 + x_2 + 3x_3 = 9$ and the second line determined by $x_1 - 2x_2 - 5x_3 = 0$ and $3x_1 + 2x_2 - 4x_3 = 0$

(i) $(2,3,0,-1) + t(3,0,2,0)$ and $(2,-3,1,2) + t(3,2,1,-1)$.

2. Decide if the two planes in $\mathbb{R}^3$ are identical, never meet, or have an intersection that is a line. In the latter case calculate the angle at which the two planes meet.

(a) $x_1 - x_2 - x_3 = 0$ and $x_1 - 2x_2 - 5x_3 = 0$

(b) $x_1 - x_2 - x_3 = -2$ and $x_1 - 2x_2 - 5x_3 = -11$

(c) $x_1 - x_2 - x_3 = -2$ and $-2x_1 + 2x_2 + 2x_3 = 4$

(d) $x_1 - x_2 - x_3 = -2$ and $-2x_1 + 2x_2 + 2x_3 = 5$

(e) $2x_1 - 3x_2 = 3$ and $5x_1 + 2x_2 = -4$

(f) span of $(3,-1,4)$ and $(1,1,0)$ and span of $(2,1,0)$ and $(-3,0,-1)$

(g) $\mathcal{S}_1 + (1,0,0)$ and $\mathcal{S}_2 + (0,1,0)$, where $\mathcal{S}_1$ is the span of $(3,-1,4)$ and $(1,1,0)$ and $\mathcal{S}_2$ is the span of $(2,1,0)$ and $(-3,0,-1)$

(h) $\mathcal{S}_1 + (1,0,0)$ and $\mathcal{S}_2 + (2,-3,4)$, where $\mathcal{S}_1$ is the span of $(3,-1,4)$ and $(1,1,0)$ and $\mathcal{S}_2$ is the span of $(1,0,1)$ and $(0,-1,1)$.

3. Compare parts (e) of the preceding two exercises.

4. For each part of Exercise 1, except the last, calculate the distance between the two lines.

5. For each part of Exercise 2 for which the given planes are parallel, calculate the distance between the two planes.

6. Decide if the line lies in the plane, does not meet the plane, or intersects the plane in one point. In the latter case calculate the angle at which the line meets the plane.

(a) $t(1,2,3)$ and $x_1 - 3x_2 + 2x_3 = 2$

(b) $t(1,-3,2)$ and $x_1 - 3x_2 + 2x_3 = 2$

(c) span of $(1,1,1)$, and the plane $x_1 - 3x_2 + 2x_3 = 2$

(d) $(1,1,2) + t(1,1,1)$ and $x_1 - 3x_2 + 2x_3 = 2$

(e) line determined by $x_1 + x_2 - 4x_3 = 1$ and $2x_1 + x_2 + 3x_3 = -1$, and the plane $x_1 + 2x_2 - 2x_3 = 0$

(f) line determined by $x_1 + x_2 - 4x_3 = 1$ and $2x_1 + x_2 + 3x_3 = -1$ and the plane spanned by $(1,2,0)$ and $(-1,0,3)$.

7. Consider a diagonal of a cube that passes through the center of the cube and ends at two opposite vertices.

(a) At what angle does the diagonal meet each face?

(b) The diagonal also meets six of the edges. At what angle does it meet each of these six edges?

8. Find the distance in $\mathbb{R}^2$ between the point and line.

   (a) $(0,0)$ and $(1,0) + t(0,1)$

   (b) $(-1,2)$ and $(1,0) + t(-1,1)$

   (c) $(2,2)$ and $(1,0) + t(-1,1)$

   (d) $(0,0)$ and $(-\sin\theta, \cos\theta) + t(\cos\theta, \sin\theta)$, where θ is a fixed constant

   (e) $(0,0)$ and the line whose intercepts are $(2,0)$ and $(0,3)$

   (f) $(0,0)$ and $2x_1 + 3x_2 = 0$

   (g) $(1,1)$ and $2x_1 + 3x_2 = 0$.

9. Consider a point Y and a hyperplane

$$\{X : \langle X, Z \rangle = c\}$$

in $\mathbb{R}^n$. Show that the distance between the hyperplane and the point Y equals

$$\frac{|\langle Y, Z \rangle - c|}{\|Z\|}.$$

   Hint: First prove the formula in the special case where $c = 0$.

10. The preceding exercise gives a formula for the distance between a point and a line in $\mathbb{R}^2$. Two other formulas have been given in the text. Show consistency among the three formulas.

11. Calculate the distance between the point and the hyperplane.

   (a) $(0,0)$ and $x_1 + x_2 = 1$

   (b) $(0,0,0)$ and $x_1 + x_2 + x_3 = 1$

   (c) $(0,0,0,0)$ and $x_1 + x_2 + x_3 + x_4 = 1$

   (d) $\mathbf{0} \in \mathbb{R}^n$ and $x_1 + x_2 + \cdots + x_n = 1$

   (e) $(0) \in \mathbb{R}^1$ and the hyperplane (that is, point) $x_1 = 1$.

   (f) $(0,0,1)$ and $x_3 = 0$

   (g) $(1,1,1)$ and $x_1 + 2x_2 + 3x_3 = 0$.

12. Consider the union of eight triangles in $\mathbb{R}^3$, where each triangle has one vertex on each of the coordinate axes at a point one unit away from the origin. This figure is called a "regular octahedron". It is called an **octahedron** because it has eight faces. It is said to be **regular** because given any two vertices there is an isometry that maps the first of them to the second and for which the image of the octahedron is itself.

   (a) Write the coordinates of each of the six vertices.

   (b) Give parametric representations of each of the twelve edges. In particular, describe the values of the parameter that give points on an edge, rather than just on the line containing the edge.

(c) Find equations of the eight planes containing the eight triangular faces.

(d) Calculate, using two different methods, the angle at which two edges of a common face meet.

(e) At each vertex there are edges that meet even though they are not edges of a common face. At what angle do they meet?

(f) Consider the twelve lines that contain the twelve edges. What are the distances between various pairs of such lines? Do these distances deserve to be called the distances between the edges themselves?

(g) At one of its endpoints, an edge is in contact with four faces, two of which have it as an edge. What is the angle at which this edge meets each of the other two faces?

(h) What is the distance between the origin and each of the edges?

(i) What is the distance between the origin and each of the faces?

13. A **regular tetrahedron** in $\mathbb{R}^3$ has four faces, each of which is an equilateral triangle. There are four vertices. At each of them three faces and three edges meet. Consider the regular tetrahedron obtained by placing one vertex at $(0, 0, 1)$ and arranging that the other three vertices have the same third coordinate as each other and that all four vertices are equidistant from the origin. Find the coordinates of the other three vertices. Then do an analysis similar to that which is requested in the preceding exercise for the octahedron.

14. Ignoring the fact that $(1, 0, 0) + t(1, -1, 1)$ is a line, but rather treating it as one would treat a curve in general, calculate the distance between it and the point $(0, 0, 1)$.

15. Calculate the distance between the point $(0, 0, 5)$ and the curve given parametrically by (t, t, t^2).

16. For Example 10 of Section 5.1d verify that the line passing through $(0, 1)$ and a point on the parabola closest to it is perpendicular to the parabola at the point of intersection.

17. For arbitrary k calculate the distance between the point $(k, 1/2b)$ and the parabola of Example 10 of Section 5.1d.

18. Consider a piece of glass in the shape of a square of edge length 10 feet. Protective material is to be placed in the plane of the glass everywhere outside the glass region within one foot of the glass. Draw a picture indicating the location of the protective material.

19. Calculate the distance between the point $\mathbf{0} \in \mathbb{R}^2$ and the curve given parametrically by

$$F(t) = e^t (\cos t, \sin t), \quad -\infty < t < \infty.$$

20. Calculate, as a function of b, the distance between the spiraling helix of Example 11 in Section 5.1d and the point $(0, 0, b)$.

21. Let C denote the curve given parametrically by

$$F(t) = \frac{t+1}{t+2}(\cos t, \sin t) \quad 0 \le t.$$

Find the distance from C to each of the following points:

(a) $(0,0)$

(b) $(1/2, 0)$

(c) $(1, 0)$

(d) $(2, 0)$.

22. Write each of the following complex numbers in the form $x + yi$ for appropriate real numbers x and y:

(a) $(2 - 3i) - (3 - 2i)$

(b) $(2 - 3i)(3 - 2i)$

(c) $(2 - 3i) \div (3 - 2i)$

(d) $(1 + i)^4$

(e) $(5 + 3i) \div (7 - 4i)$.

23. By what complex number should one multiply a complex variable z in order to move every point closer to the origin by a factor of 3?

24. By what complex number should one multiply a complex variable z in order to effect a counterclockwise rotation about the origin through an angle of $4\pi/3$ radians?

25. By what complex number should one multiply a complex variable z in order to effect a clockwise rotation about the origin through an angle of $\pi/2$ radians? Express your answer both in polar form and in rectangular coordinates.

26. What is the image of the point $(1, -2)$ under the given (in terms of complex arithmetic) affine map?

(a) $T(z) = iz$

(b) $T(z) = -iz$

(c) $T(z) = (1 + 2i)z$

(d) $T(z) = iz - 2$

(e) $T(z) = -iz + (3 - i)$.

27. Which of the affine maps in the preceding exercise are linear?

28. Each of the affine maps in Exercise 26 can be regarded as the composition of two maps: first a linear map, then translation by the addition of an appropriate member of $\mathbb{R}^2$. For the maps given in that exercise identify the matrix of the linear transformation and also identify the vector being added to effect the translation.

29. Find a formula, in terms of complex arithmetic, of some compositions of the affine maps in Exercise 26:

(a) the affine map of part (b) on that of part (a)

(b) the affine map of part (c) on that of part (a)

(c) the affine map of part (a) on that of part (c)

(d) the affine map of part (c) on that of part (e)

(e) the affine map of part (e) on that of part (c)

(f) the affine map of part (b) on that of part (c) on that of part (a).

30. In the text it was shown, by combining geometric reasoning with the fact that every complex number can be written in polar form, that multiplication by a nonzero complex number preserves angles. Give another proof, this one being algebraic using inner products and matrices.

31. Write the following complex numbers in polar form:

(a) $5 - 12i$

(b) $-6 + 7i$

(c) $24 - 7i$

(d) $-17i$.

32. Find polar coordinate representations of the points whose rectangular coordinate representations are given.

(a) $(5, -12)$

(b) $(-6, 7)$

(c) $(24, -7)$

(d) $(0, -17)$.

33. Find an orientation-preserving similarity that fixes the origin and moves $(1, 2)$ to

(a) $(-1, -2)$

(b) $(-1, 2)$.

34. Find an orientation-preserving similarity that interchanges e_1 and e_2 (but does not necessarily fix the origin).

35. Find an orientation-preserving similarity that changes the sign of e_1 and fixes e_2.

5.2   Some Challenging Problems

This section contains no new concepts, but focuses on some rather elaborate problems whose solutions require the use of several ideas. Sections 5.2a, 5.2b, and 5.2c, each of which treats essentially one problem, are independent of each other.

5.2A   SLICING A CUBE

Let us consider the cube $\mathcal{C}$ in $\mathbb{R}^3$ whose vertices are the eight points $(\pm 1, \pm 1, \pm 1)$. Thus, $\mathcal{C}$ is the union of portions of six planes:

$$\{X = (x_1, x_2, x_3) : x_3 = 1, |x_1| \leq 1, |x_2| \leq 1\}$$
$$\cup \ \{X : x_3 = -1, |x_1| \leq 1, |x_2| \leq 1\}$$
$$\cup \ \{X : x_2 = 1, |x_3| \leq 1, |x_1| \leq 1\}$$
$$\cup \ \{X : x_2 = -1, |x_3| \leq 1, |x_1| \leq 1\}$$
$$\cup \ \{X : x_1 = 1, |x_2| \leq 1, |x_3| \leq 1\}$$
$$\cup \ \{X : x_1 = -1, |x_2| \leq 1, |x_3| \leq 1\}\,.$$

These portions of the six planes are the six faces of the cube. Notice that we use the word "cube" for the "box", and the term "cubical region" to describe the box together with the space inside.

Diagonals of a cube connect two vertices that are not connected by an edge. There are two types of diagonals: those which connect two vertices of a common face and the longer diagonals which do not. We will focus on one of the longer diagonals. Let $\mathcal{L}$ denote the diagonal connecting $(-1, -1, -1)$ and $(1, 1, 1)$. We set the task of naturally parametrizing the intersection of the cube with each perpendicular trisector of $\mathcal{L}$. (Note that the perpendicular trisectors of a line segment in $\mathbb{R}^3$ are planes.)

Let us first naturally parametrize the indicated diagonal:

$$(-1, -1, -1) + t(1/\sqrt{3}, 1/\sqrt{3}, 1/\sqrt{3})\,, \quad 0 \leq t \leq 2\sqrt{3}\,.$$

(We can check that we have a natural parametrization by showing that the norm of the derivative equals 1.) The perpendicular trisectors of this segment intersect it at the points given by $t = \frac{1}{3}2\sqrt{3} = \frac{2\sqrt{3}}{3}$ and $t = \frac{2}{3}2\sqrt{3} = \frac{4\sqrt{3}}{3}$, that is, at the points $(-1/3, -1/3, -1/3)$ and $(1/3, 1/3, 1/3)$. We know that the direction vector of a line—in this case, $\frac{1}{\sqrt{3}}(1, 1, 1)$ or, equally good, $(1, 1, 1)$ —gives suitable coefficients for planes (hyperplanes in general) perpendicular to the line. Since the two planes of interest to us pass through $(-1/3, -1/3, -1/3)$ and $(1/3, 1/3, 1/3)$ they are given by the equations

$$x_1 + x_2 + x_3 = -1$$

and

$$x_1 + x_2 + x_3 = 1\,.$$

We consider the plane $x_1 + x_2 + x_3 = -1$, leaving the other plane to the reader. Let us find the intersection of the plane $x_1 + x_2 + x_3 = -1$ with the face

$$\{X : x_3 = 1, |x_1| \leq 1, |x_2| \leq 1\}\,.$$

We must have $x_1 + x_2 = -2$ which happens if and only if $x_1 = x_2 = -1$. Thus the intersection consists of the single point $(-1, -1, 1)$. Similarly, the intersection of the plane $x_1 + x_2 + x_3 = -1$ with the face

$$\{X : x_2 = 1, |x_3| \leq 1, |x_1| \leq 1\}$$

consists of the point $(-1, 1, -1)$ and its intersection with the face

$$\{X : x_1 = 1, |x_2| \leq 1, |x_3| \leq 1\}$$

consists of the point $(1, -1, -1)$.

The first two coordinates of the intersection of the plane $x_1 + x_2 + x_3 = -1$ with the face

$$\{X : x_3 = -1, |x_1| \leq 1, |x_2| \leq 1\}$$

must satisfy $x_1 + x_2 = 0$, $|x_1| \leq 1$, and $|x_2| \leq 1$. Thus the intersection is a line segment with endpoints $(-1, 1, -1)$ and $(1, -1, -1)$. Similarly, the intersection of the plane $x_1 + x_2 + x_3 = -1$ with the face

$$\{X : x_2 = -1, |x_3| \leq 1, |x_1| \leq 1\}$$

is the line segment with endpoints $(1, -1, -1)$ and $(-1, -1, 1)$ and its intersection with the set

$$\{X : x_1 = -1, |x_2| \leq 1, |x_3| \leq 1\}$$

is the line segment with endpoints $(-1, -1, 1)$ and $(-1, 1, -1)$.

We conclude that the intersection of the plane $x_1 + x_2 + x_3 = -1$ with the box $\mathcal{C}$ is the triangle with vertices $(-1, 1, -1)$, $(1, -1, -1)$, and $(-1, -1, 1)$. Here is a natural parametrization:

$$\begin{cases} \left(-1 + \frac{t}{\sqrt{2}}, 1 - \frac{t}{\sqrt{2}}, -1\right) & \text{if} \quad 0 \leq t < 2\sqrt{2} \\ \left(1 - \frac{t - 2\sqrt{2}}{\sqrt{2}}, -1, -1 + \frac{t - 2\sqrt{2}}{\sqrt{2}}\right) & \text{if} \quad 2\sqrt{2} \leq t < 4\sqrt{2} \\ \left(-1, -1 + \frac{t - 4\sqrt{2}}{\sqrt{2}}, 1 - \frac{t - 4\sqrt{2}}{\sqrt{2}}\right) & \text{if} \quad 4\sqrt{2} \leq t < 6\sqrt{2} \end{cases}$$

$$= \begin{cases} \left(-1 + \frac{t}{\sqrt{2}}, 1 - \frac{t}{\sqrt{2}}, -1\right) & \text{if} \quad 0 \leq t < 2\sqrt{2} \\ \left(3 - \frac{t}{\sqrt{2}}, -1, -3 + \frac{t}{\sqrt{2}}\right) & \text{if} \quad 2\sqrt{2} \leq t < 4\sqrt{2} \\ \left(-1, -5 + \frac{t}{\sqrt{2}}, 5 - \frac{t}{\sqrt{2}}\right) & \text{if} \quad 4\sqrt{2} \leq t < 6\sqrt{2}. \end{cases}$$

The perimeter of the triangle equals $6\sqrt{2}$. It is left to the reader to check that the above formula represents a continuous function and that its derivative, where it exists, has a norm of 1.

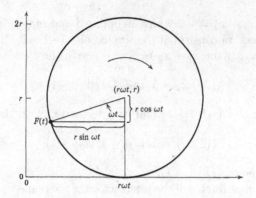

FIGURE 5.1. A rolling wheel

5.2B CYCLOIDS

Let us consider a wheel of radius r rolling in the positive direction along the x-axis. Suppose that at time 0 the center of the wheel is at the point $(0, r)$, so that the edge of the wheel touches the point $(0, 0)$. We imagine that the place on the wheel that initially touches the point $(0, 0)$ has been marked with a blue dot, and we want to parametrically describe the motion of this blue dot, with the parameter representing time elapsed. The path traveled by the blue dot is called a **cycloid**. We suppose that the angular speed of the wheel is constant, say ω radians per unit time, so that after $2\pi/\omega$ units of time the blue dot is at the point $(2\pi r, 0)$. The reader is advised to draw pictures of the wheel after different amounts of elapsed time t, for example, for values of t in the various intervals $(0, \pi/2\omega)$, $(\pi/2\omega, \pi/\omega)$, $(\pi/\omega, 3\pi/2\omega)$, and $(3\pi/2\omega, 2\pi/\omega)$.

Let us first calculate where the center of the wheel is after t units of time have elapsed. The fact that the wheel has rotated through an angle ωt means that the length of that portion of the edge of the wheel that has made contact with the x-axis is equal to $r\omega t$. Since we are assuming rolling without slipping, that is also the length of that portion of the x-axis which has been touched. Therefore the center of the wheel is at the point $(r\omega t, r)$.

To find the (x, y)-coordinates of the blue dot we first find the "coordinates" with respect to the center of the wheel. For this calculation the sine and cosine functions enter in almost the usual way, the difference arising because the motion of the wheel is clockwise rather than counterclockwise and the blue dot begins below rather than to the right of the center of the wheel. With reference to the center of the wheel, the blue dot is located at $(-r\sin\omega t, -r\cos\omega t)$. See Figure 5.1. Thus, with reference to $(0, 0)$, the location of the blue dot is

$$r(\omega t - \sin\omega t, 1 - \cos\omega t) \,.$$

5.2c SIMULTANEOUS MOTIONS

Six spiders sit at the six corners on the floor of a room, the floor having the shape of a regular hexagon with edge length s centimeters. At time 0 each spider begins walking on the floor at the rate of one centimeter per second constantly adjusting its direction so that it is going directly towards the next spider in the counterclockwise direction around the room. Let us describe the motion of each spider.

Let us place the origin of $\mathbb{R}^2$ at the center of the room and denote the position of one particular spider by $F(t)$. By symmetry the matrices of the position vectors of the six spiders are:

$$\begin{bmatrix} \frac{1}{2} & -\frac{\sqrt{3}}{2} \\ \frac{\sqrt{3}}{2} & \frac{1}{2} \end{bmatrix} [F(t)] \, ;$$

$$\begin{bmatrix} \frac{1}{2} & -\frac{\sqrt{3}}{2} \\ \frac{\sqrt{3}}{2} & \frac{1}{2} \end{bmatrix}^2 [F(t)] = \begin{bmatrix} -\frac{1}{2} & -\frac{\sqrt{3}}{2} \\ \frac{\sqrt{3}}{2} & -\frac{1}{2} \end{bmatrix} [F(t)] \, ;$$

$$\begin{bmatrix} \frac{1}{2} & -\frac{\sqrt{3}}{2} \\ \frac{\sqrt{3}}{2} & \frac{1}{2} \end{bmatrix}^3 [F(t)] = \begin{bmatrix} -1 & 0 \\ 0 & -1 \end{bmatrix} [F(t)] = [-F(t)] \, ;$$

$$\begin{bmatrix} \frac{1}{2} & -\frac{\sqrt{3}}{2} \\ \frac{\sqrt{3}}{2} & \frac{1}{2} \end{bmatrix}^4 [F(t)] = \begin{bmatrix} -\frac{1}{2} & \frac{\sqrt{3}}{2} \\ -\frac{\sqrt{3}}{2} & -\frac{1}{2} \end{bmatrix} [F(t)] \, ;$$

$$\begin{bmatrix} \frac{1}{2} & -\frac{\sqrt{3}}{2} \\ \frac{\sqrt{3}}{2} & \frac{1}{2} \end{bmatrix}^5 [F(t)] = \begin{bmatrix} \frac{1}{2} & \frac{\sqrt{3}}{2} \\ -\frac{\sqrt{3}}{2} & \frac{1}{2} \end{bmatrix} [F(t)] \, ;$$

$$\begin{bmatrix} \frac{1}{2} & -\frac{\sqrt{3}}{2} \\ \frac{\sqrt{3}}{2} & \frac{1}{2} \end{bmatrix}^6 [F(t)] = \begin{bmatrix} \frac{1}{2} & -\frac{\sqrt{3}}{2} \\ \frac{\sqrt{3}}{2} & \frac{1}{2} \end{bmatrix}^0 [F(t)]$$

$$= \begin{bmatrix} 1 & 0 \\ 0 & 1 \end{bmatrix} [F(t)] = [F(t)] \, .$$

Alternatively, if $F(t)$ is regarded to be complex rather than an ordered pairs of reals, the positions of the six spiders can be represented in terms of complex multiplication:

$$e^{\pi i/3} F(t), \qquad e^{2\pi i/3} F(t), \qquad e^{\pi i} F(t) = -F(t),$$
$$e^{4\pi i/3} F(t), \qquad e^{5\pi i/3} F(t), \qquad e^{2\pi i} F(t) = F(t) \, .$$

We will not, however, use these complex-arithmetical formulas in the remainder of this discussion about spiders.

The direction of $F'(t)$ is the direction from the position of the spider at $F(t)$ to the position of the next spider, which is the same as that of the vector whose matrix is

$$\begin{bmatrix} \frac{1}{2} & -\frac{\sqrt{3}}{2} \\ \frac{\sqrt{3}}{2} & \frac{1}{2} \end{bmatrix} [F(t)] - \begin{bmatrix} 1 & 0 \\ 0 & 1 \end{bmatrix} [F(t)]$$

$$= \begin{bmatrix} -\frac{1}{2} & -\frac{\sqrt{3}}{2} \\ \frac{\sqrt{3}}{2} & -\frac{1}{2} \end{bmatrix} [F(t)] .$$

For convenience we let

$$B = \begin{bmatrix} -\frac{1}{2} & -\frac{\sqrt{3}}{2} \\ \frac{\sqrt{3}}{2} & -\frac{1}{2} \end{bmatrix} .$$

Thus $F'(t)$ is parallel to the vector whose matrix is $B[F(t)]$. We can calculate the inner product of this vector with itself, using the fact that $B^T B = I$:

$$(B[F(t)])^T (B[F(t)]) = [F(t)]^T (B^T B)[F(t)]$$
$$= [F(t)]^T I [F(t)] = [F(t)]^T [F(t)] = \|F(t)\|^2 .$$

Since we know that $\|F'(t)\| = 1$, we conclude that the matrix of the vector $F'(t)$ is

$$[F'(t)] = \frac{1}{\|F(t)\|} B[F(t)] .$$

We now obtain

$$\frac{d}{dt} \|F(t)\| = \frac{\langle F(t), F'(t) \rangle}{\|F(t)\|} = \frac{[F(t)]^T B[F(t)]}{\|F(t)\|^2} ,$$

which can be explicitly calculated by introducing the coordinate functions of $F(t)$. The result is $-1/2$, which is valid up until the time the spiders meet. It follows that before that time, $\|F(t)\| = \|F(0)\| - \frac{t}{2}$. The spiders meet when $F(t) = 0$, which occurs at time $t = 2\|F(0)\|$. For a regular hexagon it is easy to check that the distance between the center and any vertex is the same as the edge length s. Therefore, $\|F(0)\| = s$, the spiders meet at time $t = 2s$, and

$$\|F(t)\| = s - \frac{t}{2} = \frac{2s - t}{2}$$

for $0 \le t \le 2s$.

In order to give attention to the direction of $F(t)$ we write

$$F(t) = \frac{2s - t}{2} (\cos \theta(t), \sin \theta(t)) .$$

Differentiation gives

$$F'(t) = -\frac{1}{2} (\cos \theta(t), \sin \theta(t)) + \frac{(2s - t)\theta'(t)}{2} (-\sin \theta(t), \cos \theta(t)) .$$

We also get an expression for $F'(t)$ from the previously obtained relation between F' and F:

$$F'(t) = -\frac{1}{2} (\cos \theta(t), \sin \theta(t)) + \frac{\sqrt{3}}{2} (\sin \theta(t), -\cos \theta(t)) .$$

Comparison of the two expressions for $F'(t)$ gives

$$\theta'(t) = \frac{\sqrt{3}}{2s - t},$$

from which we conclude that

$$\theta(t) = \theta(0) - \sqrt{3} \ln \left(\frac{2s - t}{2s} \right).$$

By comparison with Example 11 in Section 4.3d we see that the path of each spider is a spiral. It is, in fact, a naturally parametrized spiral because the speed of each spider equals 1.

Exercises

1. Show that the final formula in Section 5.2a is a natural parametrization of a curve by showing that it represents a continuous function whose derivative exists and has a norm of 1 except at a finite number of points.

2. Complete Secton 5.2a by naturally parametrizing the intersection of the plane $x_1 + x_2 + x_3 = 1$ with the cube.

3. By using perpendicular trisectors of the other three long diagonals of the cube in Section 5.2a, six other triangles are obtained. For each of the eight triangles find a line that passes through $(0, 0, 0)$ and is perpendicular to the plane containing the triangle. Calculate the angles at which the various pairs of these lines meet at $(0, 0, 0)$.

4. Find the intersection of the cube of Section 5.2a with the perpendicular bisector of the diagonal from $(-1, -1, -1)$ to $(1, 1, 1,)$. Parametrize it naturally. Decide if it is a regular polygon. (Remember that for polygons having more than three edges, equiangular is not equivalent to equilateral.)

5. What is the largest speed attained by the blue dot of Section 5.2b? Where is it when it attains that speed?

6. Find a time t_0 and a place (x_0, y_0) at which the blue dot of Section 5.2b attains a speed of 0. For t near t_0 compare dx/dt and dy/dt in order to accurately draw the curve traveled by the blue dot near (x_0, y_0).

7. Calculate the distance traveled by the blue dot in Section 5.2b between successive visits to the x-axis.

8. In the problem in Section 5.2b involving the wheel and the blue dot, replace the x-axis by a fixed circle of radius $2r$ centered at $(0, 2r)$. Calculate the position of the blue dot at time t. Assume that $w = 1$.

9. For the setting of the preceding exercise find the distance traveled by the blue dot between successive visits to the fixed circle.

10. Except for the calculation of $\langle F(t), F'(t) \rangle$, the matrices in Section 5.2c did not play an essential role. Matrices can be avoided altogether because the inner product can be treated by using its connection with angle. Do so.

11. Rework the spider problem of of Section 5.2c for four spiders starting at the corners of a square room having edge length equal to s.

12. Generalize Section 5.2c and the preceding exercise by treating a room in the shape of a regular polygon having m edges. Denote the edge length by s. Can you give a reasonable interpretation of the case $m = 2$?

13. Replace the edge length in the preceding exercise by p/m, where p is a constant. Thus, for each m, the perimeter of the room is p and there are m spiders. Does the distance traveled by the spiders approach ∞ or a finite number as $m \to \infty$? Draw a picture of the limiting path (as $m \to \infty$) traveled by the spiders.

5.3   Gravitational Motion

Sections 5.3b, 5.3c, and 5.3d each depend on Section 5.3a, but can be studied independently of each other.

5.3A   NEWTON'S LAWS

Why is acceleration important? Suppose that we wish to study the motion of a particle, for example, a satellite, a photon of light, an electron, or a baseball. Its motion is described by some unknown function $X(t)$ of time t (in this section we write $X(t)$ instead of $F(t)$ for the position, because we want to reserve the letter F for "force"). Here, the genius of Newton is the pillar of our knowledge. In particular, we invoke *Newton's second law of motion*. It relates acceleration to force: "force equals mass times acceleration",

$$mX'' = F,$$

where m is the mass of the particle involved and F is the net force on the particle.

Implicit in this wonderful formula is the assertion that the force F is a vector. This is not at all a priori obvious; in fact, it is a statement that requires experimental verification—and experiments do confirm that force is a vector. Newton's second law tells us that forces "cause" acceleration and hence "cause" the motion. But in order for it to be useful, we must be able to say what the forces are. Over the years, physics has accumulated theories that describe the forces in various situations.

One of the first and perhaps most widely known force laws is also due to Newton. Say that we have two particles, one of mass m at the point X and another of mass M at the point Y. See Figure 5.2. Newton's *law of gravitational force* states that the force acting on the particle at X due to the particle at Y is

$$F = -\gamma \frac{mM}{\|X - Y\|^2}\mathbf{e},$$

where $\mathbf{e}$ is a unit vector pointing from Y toward X and $\gamma > 0$ is a fixed constant, called the **universal gravitational constant**. The negative sign

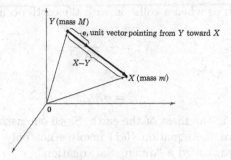

FIGURE 5.2.

means that the force is an attractive force. In view of the relation $\mathbf{e} = (X - Y)/\|X - Y\|$, the formula for the gravitational force on the particle at X due to the particle at Y can be rewritten as

$$F = -\gamma \frac{mM}{\|X - Y\|^3} (X - Y).$$

Combining Newton's second law with this equation for the force, we find that

$$mX'' = -\gamma \frac{mM(X - Y)}{\|X - Y\|^3},$$

an equation that governs motion of all particles moving under the influence of gravity, for example, the moon, spaceships, and baseballs. This is a single simple and profound formula. As an outgrowth of Newton's work, the existence of such universal formulas may well be, in our culture, a basic article of belief. Such is the influence of Newton.

REMARK:

- When we study integration, we will show that if the masses are not point masses, but are distributed symmetrically in a spherical region, such as approximated by the earth or moon, then in applying the equation, it is possible to replace the mass distribution by a *single point* positioned at the center of the body, with mass equalling the total mass of the body.

In the applications in this section we will investigate motion under the gravitational influence of the earth which we will regard as being fixed with its center at the origin. The vector X will denote the location of the object—say, a baseball or artificial satellite—with respect to the center of the earth; thus, Y, the location of the center of the earth, will equal $\mathbf{0}$. The equation governing the motion of the object (until, of course, other forces play a role, such as when a collision with the earth occurs) is

$$X'' = -\beta \frac{X}{\|X\|^3} \, , \tag{5.1}$$

where

$$\beta = \gamma M_e \, ,$$

with M_e denoting the mass of the earth. Since one usually wants to solve for X, and since the equation (5.1) involves not only X but its second derivative, (5.1) is called a "differential equation".

Exercise 13 shows that this differential equation can be simplified because the motion takes place in a plane containing the center of the earth. Accordingly, we will pick our coordinate axes so that, besides the origin being at the center of the earth, the x_3-axis is perpendicular to the plane in which the motion takes place. Since it is then the case that $x_3(t) = 0$ for all t, we can disregard x_3 and regard our gravitational studies as taking place in $\mathbb{R}^2$.

We are now regarding X in (5.1) as an $\mathbb{R}^2$-valued function of time, that is, as a parametrized curve in $\mathbb{R}^2$. In the applications below we solve this "two-dimensional" version of (5.1) under certain simplifying assumptions. These special cases will bring out the beauty and power of Newton's Laws.

5.3B RELATIVELY SMALL MOTIONS

We will assume here that the particle in motion starts near the surface of the earth and does not move very far. There are two consequences of this assumption: (1) that the earth acts as if it were flat, that is, the direction of the gravitational force does not change as the particle moves, and (2) that the norm of the force is essentially constant because, in a relative sense, the distance from the center of the earth hardly changes. These conditions are appropriate, for instance, when the motion of a baseball is studied.

We choose the x_1-axis to be perpendicular to the surface of the earth and to pass through the position of the particle at time 0. As a consequence, the x_2-axis is tangent to the earth there (or lies on the earth if you prefer to think of the earth as flat). Then X is approximately equal to $R_e\mathbf{e}_1$, where R_e denotes the radius of the earth. The right-hand side of the differential equation (5.1) can be approximated as $-(\beta/R_e^2)\mathbf{e}_1$. Thus, we hope that we do not make a significant error in replacing the differential equation (5.1) by the simpler differential equation:

$$X'' = -\frac{\beta}{R_e^2}\mathbf{e}_1 \, .$$

We can rewrite this equation as

$$X'' = -g\mathbf{e}_1 \, , \tag{5.2}$$

where g is related to β by the formula

$$\beta = gR_e^2 . \tag{5.3}$$

The differential equation (5.2) can be written as the two scalar equations

$$x_1''(t) = -g$$
$$x_2''(t) = 0 .$$

We integrate both sides of these equations from 0 to t and obtain

$$x_1'(t) - x_1'(0) = -gt$$
$$x_2'(t) - x_2'(0) = 0 .$$

We will write v_1 for $x_1'(0)$ and v_2 for $x_2'(0)$. Thus, $(v_1, v_2) = X'(0)$, the initial velocity of the particle, and

$$x_1'(t) = -gt + v_1$$
$$x_2'(t) = v_2 .$$

We integrate again from 0 to t, introducing another constant into each equation as follows:

$$x_1(t) = -\frac{g}{2}t^2 + v_1 t + p_1 ;$$
$$x_2(t) = v_2 t + p_2 ,$$

where $P = (p_1, p_2) = (x_1(0), x_2(0))$ denotes the position of the particle at time 0. By the choice of axes made above, $p_2 = 0$. We write the constant p_1 as $R_e + h$ so that h is the height of the particle above the surface of the earth at time 0. We let $w_1(t)$ denote the height of the particle above the surface of the earth at time t; so $x_1(t) = R_e + w_1(t)$. The first of the two equations above can be written as

$$w_1(t) = -\frac{g}{2}t^2 + v_1 t + h .$$

Changing the name of x_2 to w_2 so that we can speak of the two-dimensional vector $W = (w_1, w_2)$, we finally obtain the following simple expression for the location of the particle at time t:

$$W(t) = \left(-\frac{g}{2}, 0\right) t^2 + Vt + (h, 0) .$$

The first coordinate of W describes the height of the particle and the second coordinate describes the horizontal distance from the starting point.

EXAMPLE:

1. Suppose that a particle leaves the ground in a direction that makes an angle $\alpha \in (0, \pi/2]$ with the horizontal, and also suppose that the initial velocity is such that the maximum height the particle reaches is equal to H. Let us find the horizontal distance traveled by the particle before it hits the ground.

   Implicit in the wording of the problem is that an answer is expected that depends on α and H. (Before getting involved in calculations, it is instructive to make considered judgments as to how the answer should depend on these two quantities. For instance, in order to make H larger for a fixed α one would have to make the vertical component of the velocity larger and, therefore, also the horizontal component. Also, the time elapsed before collision with the ground would increase, so the horizontal distance traveled would certainly increase. It is also interesting to contemplate the dependence on α.)

   Our first step will be to make a calculation involving H, so we focus on the function

   $$w_1(t) = -\frac{g}{2}t^2 + v_1 t .$$

   Single-variable calculus calculations tell us that w_1 has an absolute maximum equal to $v_1^2/2g$. Since we are given that the absolute maximum equals H, we conclude that

   $$v_1 = \sqrt{2gH}$$

   and, therefore, that

   $$w_1(t) = -\frac{g}{2}t^2 + \sqrt{2gH}\, t .$$

   Next, by setting $w_1(t) = 0$, we can calculate the duration T of the particle's journey. The positive solution of $w_1(t) = 0$ is

   $$T = \sqrt{8H/g} .$$

   The function $w_2(t) = v_2 t$ describes the horizontal distance traveled by time t. Inserting T for t we obtain

   $$D = v_2\sqrt{8H/g} ,$$

   where D denotes the total horizontal distance traveled up to the time of collision with the ground.

   To finish the solution we need to calculate v_2. Because the initial angle of motion is α, we know that

   $$\begin{aligned} v_1 &= \|V\| \sin\alpha \\ v_2 &= \|V\| \cos\alpha , \end{aligned}$$

and, therefore, that

$$v_2 = v_1 \cot \alpha = \sqrt{2gH} \cot \alpha .$$

Hence,

$$D = 4H \cot \alpha .$$

(One should not regard the example as finished until one has thought deeply about the facts that the answer does not depend on the value of g and that it is a decreasing function of α that approaches ∞ as α decreases to 0.)

There are, of course, problems whose answers do depend on g. Often one would like a numerical approximation of such an answer. Experiments indicate that an approximate value of g is 32 feet per second per second. According to legend Galileo approximated g by dropping stones from the Leaning Tower of Pisa. An approximate value of R_e can be obtained in various ways and from (5.3) we can get an approximation for β. For our study here we will be content to use the approximations

$$g \approx 32 \frac{\text{feet}}{\text{second}^2} , \tag{5.4}$$

$$R_e \approx 4000 \text{ miles} , \tag{5.5}$$

$$\beta \approx 1.26 \times 10^{12} \frac{\text{miles}^3}{\text{hours}^2} . \tag{5.6}$$

Since these approximations do not satisfy the relation (5.3) exactly, different correct approaches using these approximations may yield slightly different numerical answers.

5.3C VERTICAL MOTION

We drop the assumption made in the preceding section that the motion takes place near the surface of the earth, but introduce a condition on the initial velocity that makes the problem one-dimensional. We assume that the initial veocity is vertical, and, thus, if we place the x_1-axis through the center of the earth and the initial position of the particle, it follows that the entire motion takes place on the x_1-axis. In fact, all positions are positive multiples of $\mathbf{e}_1$. Thus we only need consider x_1. We let the initial distance p_1 from the center of the earth be arbitrary (but of course at least as large as the radius of the earth); and we assume that the initial speed is positive, that is, $v_1 > 0$. Initially the object (perhaps a rocket that has no more fuel) is moving away from the earth. We ask whether it continues to move away from the earth for all time or whether it eventually turns around and then subsequently hits the earth.

From the differential equation (5.1) we obtain

$$x_1''(t) = -\frac{\beta}{x_1^2(t)}$$

for $x_1 > 0$.

Looking ahead, we multiply both sides of this differential equation by $2x_1'(t)$ to obtain

$$2x_1'(t)x_1''(t) = -\frac{2\beta x_1'(t)}{x_1^2(t)} .$$

We recognize the left-hand side as the derivative with respect to t of $[x_1'(t)]^2$ and the right-hand side as the derivative with respect to t of $2\beta/x_1(t)$. For two functions to have equal derivatives they must differ by a constant. So,

$$[x_1'(t)]^2 = \frac{2\beta}{x_1(t)} + c , \tag{5.7}$$

for some constant c.

The relation just obtained is itself a differential equation since it involves the derivative of the unknown function x_1. Rather than trying to solve this equation in order to obtain an explicit formula for $x_1(t)$, let us learn something directly from the differential equation. Since x_1' is continuous (a consequence of the existence of x_1''), we know that $x_1'(t)$ will remain positive until (if ever) it equals 0. But it can only equal 0 at a time t for which

$$\frac{2\beta}{x_1(t)} + c = 0 .$$

If $c \geq 0$, then this can never happen so $x_1'(t) > 0$ for all t. Hence, for such c the rocket never approaches the earth. If, on the other hand, $c < 0$, then at a t for which $x_1(t) = 2\beta/|c|$ the velocity will equal 0; and then the velocity becomes negative for larger t since the second derivative is negative. We conclude that if $c \geq 0$, the rocket never returns to earth, but if $c < 0$, the rocket reaches a maximum distance from the center of the earth equal to $2\beta/|c|$ after which it returns to the earth. (There is a small gap in the reasoning in this paragraph. Can you find it? Can you fix it?)

To calculate c we set $t = 0$ in (5.7) to obtain

$$c = v_1^2 - \frac{2\beta}{p_1} .$$

For fixed p_1, the minimal initial speed to ensure non-return is $\sqrt{2\beta/p_1}$.

In case p_1 equals the radius R_e of the earth, the minimum speed is known as the **escape speed from earth** or, more commonly but incorrectly, as the **escape velocity from earth**. This escape speed equals

$$\sqrt{2\beta/R_e} = \sqrt{2gR_e} .$$

The escape speed from earth can be approximated using the approximation (5.5) for R_e together with either the approximation (5.4) for g or the approximation (5.6) for β. We use R_e and g to obtain

$$\sqrt{\frac{2(32 \text{ feet})(4000 \text{ miles})}{\text{seconds}^2}}$$

$$= \sqrt{\frac{(256000 \text{ feet miles })(5280 \text{ feet})}{(\text{seconds}^2)(1 \text{ mile})}}$$

$$= \sqrt{1351680000} \; \frac{\text{feet}}{\text{seconds}} \; ,$$

which is approximately equal to 36765 feet per second. A case can be made for calling 3.7×10^4 feet per second the approximation. It gives an indication of how accurate we believe our approximation to be.

5.3D CIRCULAR MOTION

We assume that an object, for example, a satellite, stays a constant distance R from the center of the earth. The differential equation (5.1) simplifies to

$$X'' = -\frac{\beta}{R^3} X \; .$$

Thus, Exercise 30 of Section 4.2 applies and we conclude that the speed is the constant $\sqrt{\beta/R}$. We are well acquainted with circular motion at constant speed; so, we conclude that

$$X(t) = R \left(\cos \sqrt{\frac{\beta}{R^3}} \, t, \; \sin \sqrt{\frac{\beta}{R^3}} \, t \right) ,$$

provided the axes are oriented so that the object is initially on the x_1-axis. (Notice that the constants have been chosen so that $\|X(t)\| = R$ and $\|X'(t)\| = \sqrt{\beta/R}$.)

The time t_0 for the particle to complete one complete circle is

$$t_0 = 2\pi \sqrt{\frac{R^3}{\beta}} \; .$$

We solve this relation between R and t_0 for R to obtain

$$R = \sqrt[3]{\frac{\beta t_0^2}{4\pi^2}} \; .$$

Let us use the observed value of t_0 for the moon—namely, approximately 27 days—to approximate the distance of the moon from the center of the

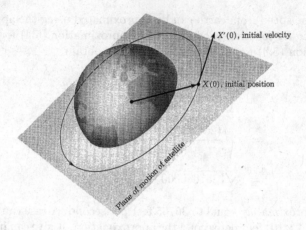

FIGURE 5.3.

earth. Using the approximation for β given in (5.6) we obtain

$$\sqrt[3]{\frac{(1.26)(10^{12})(27^2)\ \text{miles}^3 \text{days}^2}{4\pi^2\ \text{hours}^2}}$$

$$\approx \sqrt[3]{(918)(10^{12})\ \frac{\text{miles}^3 \text{days}^2}{4\pi^2\ \text{hours}^2}\ \frac{24^2\ \text{hours}^2}{1^2\ \text{day}^2}}\ ,$$

which is approximately equal to 2.4×10^5 miles.

Exercises

Motion governed by gravity.

1. This exercise is about relatively small motions; that is, it is to be assumed that the assumptions described in Section 5.3b are appropriate. A projectile is launched with speed s at an angle α with the horizontal from a point on a level plain. Find the maximal height it attains as a function of α and s. Also, find the horizontal distance it travels before hitting the ground, as a function of α and s.

2. This exercise is a continuation of the preceding exercise. For a fixed s find the value of α that maximizes the maximal height attained. Also, for a fixed s, find α so as to make the horizontal distance traveled as large as possible.

3. Suppose that the projectile in Exercise 1 is launched from a point at height $h > 0$ above the surface of the earth. Find the horizontal distance that the projectile travels before hitting the ground, as a function of α, s, and h. Then fix only s and h and find α so as to maximize the horizontal distance traveled.

4. If $X(t)$ is the position of a particle of mass m moving under a gravitational

force due to a mass M at the origin, prove that the **energy**

$$E(t) = \frac{m}{2}\|X'(t)\|^2 - \frac{\gamma m M}{\|X(t)\|}$$

is conserved; that is, prove that $E'(t) = 0$ and conclude that $E(t)$ is a constant function.

5. A meteor falls directly towards the earth. It is sighted $12,000$ miles from the center of the earth, at which time its speed is observed to be 100 miles/hour. Neglecting air resistance, what is its speed when it hits the earth? *Hint:* Use the result in the preceding exercise.

6. A rocket ship is fired vertically from the surface of the earth. At $6,000$ miles from the earth's surface, its rockets turn off. Neglecting the presence of the sun, and so on, what must its speed be at that time in order to escape the gravitational attraction of the earth and "travel to infinity"? Do this exercise by two different methods: (1) by using the result in Exercise 4 and (2) by using a formula from Section 5.3c.

7. Write a short essay relating Exercise 4 to aspects of Section 5.3c.

8. The Syncom satellites are intended to hover over one point of the equator. Thus their orbits are circular with a 24 hour period in order to appear stationary. Find the radius of the orbit and the speed of such a satellite.

9. Suppose that an artificial satellite is in a circular orbit 4000 miles above the surface of the earth. How long, approximately, does it take for it to orbit once about the earth.

10. Is it possible for a satellite, with its rockets turned off, to have a circular orbit that always remains over the Northern Hemisphere of the earth? Explain.

11. The earth's orbit about the sun is roughly a circle of radius $R = 93,000,000$ miles, and its period is roughly $T = 365$ days. If M_s is the mass of the sun, show that

$$\frac{M_s}{M_e} = \frac{4\pi^2 R^3}{gT^2 R_e^2}$$

and evaluate the right-hand side of the equation. Thus you have found the mass of the sun (in units of "masses of earth"). *Hint:* The origin should be at the center of the sun rather than the center of the earth.

12. As indicated in Section 5.3d circular motion resulting from a single gravitational force takes place in a plane. Thus it can be described in polar coordinates. Do so by filling in the blanks.

$$\|X(t)\| = \underline{\hspace{1cm}} \quad \text{and} \quad \theta(t) = \underline{\hspace{1cm}} .$$

13. We say that a particle with position $X(t)$ moves in a **central force field** if $F(X) = \varphi(X)X$, where $F(X)$ is the force on the particle when it is at X and $\varphi(X)$ is a scalar-valued function. Thus, the direction of the force F in a central force field is the same as or opposite that of the position X.

(a) Prove that

$$\frac{d}{dt}(X(t) \times X'(t)) = \mathbf{0} .$$

Conclude from this fact that $Z = X(t) \times X'(t)$ is a constant vector.

(b) Deduce that the position and velocity vectors are both in the plane perpendicular to the constant vector Z. Thus *central force motion always takes place in a plane*. The vector $mZ = X(t) \times mX'(t)$ is the **angular momentum** of a particle of mass m about the origin. The result of part (a) asserts that angular momentum is constant in central force motion.

5.4   Geometry in $\mathbb{R}^n$

We have already seen that notation for vectors, inner products, norms, and linear transformations simplifies work involving lines, planes, line segements, rays, and general curves. We are often able to avoid entirely the introduction of large numbers of scalar variables or at least delay it until some final Gauss-Jordan elimination. Similar economy of notation is possible for many other types of important sets.

5.4A   HALF SPACES

Recall that for every hyperplane in $\mathbb{R}^n$ (that is, for every $(n-1)$-dimensional affine subspace of $\mathbb{R}^n$) there is a vector $Z \in \mathbb{R}^n$ and a scalar k such that the hyperplane equals $\{X : \langle Z, X \rangle = k\}$. We now give names to the sets obtained by changing the equality to an inequality. The sets

$$\{X : \langle Z, X \rangle > k\} \quad \text{and} \quad \{X : \langle Z, X \rangle < k\}$$

are called **open half spaces** and their complements,

$$\{X : \langle Z, X \rangle \leq k\} \quad \text{and} \quad \{X : \langle Z, X \rangle \geq k\}$$

are **closed half spaces**.

The two open half spaces determined by $\langle Z, X \rangle > k$ and $\langle Z, X \rangle < k$ are sometimes called the **two sides** of the hyperplane determined by $\langle Z, X \rangle = k$. We also speak of the half spaces and the hyperplane as **determining** each other.

EXAMPLE:

1. Let us find the intersection of the closed half space

$$x_1 + 2x_2 + 3x_3 \geq 6$$

with the line given parametrically by

$$(6, 3, 2) + t(-7, -5, -5), \quad t \in \mathbb{R}.$$

We merely insert the formula for the line into the inequality describing the half space, obtaining

$$(6 - 7t) + 2(3 - 5t) + 3(2 - 5t) \geq 6,$$

the solution of which is $t \leq 3/8$. Thus the intersection is the ray

$$(6, 3, 2) + t(-7, -5, -5), \quad t \leq 3/8 .$$

When a ray is being described parametrically, some find it pleasing for the parameter value 0 to correspond to the point of emanation of the ray. For this ray we replace t by $t + \frac{3}{8}$ to obtain such a parametrization:

$$(6, 3, 2) + (t + \frac{3}{8})(-7, -5, -5) = (\frac{27}{8}, \frac{9}{8}, \frac{1}{8}) + t(-7, -5, -5) .$$

Since we are claiming that the ray

$$(\frac{27}{8}, \frac{9}{8}, \frac{1}{8}) + t(-7, -5, -5), \quad t \leq 0 ,$$

is the intersection of the line

$$(6, 3, 2) + t(-7, -5, -5), \quad t \in \mathbb{R} ,$$

and the closed half space

$$x_1 + 2x_2 + 3x_3 \geq 6 ,$$

we expect the point of emanation $(\frac{27}{8}, \frac{9}{8}, \frac{1}{8})$ of the ray to lie on the plane $x_1 + 2x_2 + 3x_3 = 6$. A quick insertion in the left-hand side gives $48/8 = 6$, as expected.

As for many concepts, there are special terms for the $\mathbb{R}^2$- and $\mathbb{R}^1$-settings. In $\mathbb{R}^2$ half spaces are called **half planes**, and in $\mathbb{R}^1$ they are called **half lines** or **rays**.

EXAMPLE:

2. Let us describe the intersection of the following four closed half planes in $\mathbb{R}^2$:

$$\{X : x_1 \leq 1\}, \qquad \{X : x_1 \geq -1\},$$
$$\{X : x_2 \leq 1\}, \quad \text{and} \quad \{X : x_2 \geq -1\} .$$

For a point to be in the intersection of these four sets it must satisfy all four of the inequalities. Thus, the intersection equals

$$\{X : |x_1| \leq 1, |x_2| \leq 1\} ,$$

a square region whose boundary is a square with an edge length of 2.

REMARK:

- In the English language, to some geometrical nouns there is a corresponding smoothly flowing adjective, for example, "triangle" and "triangular". Unfortunately, for others the noun also serves as the adjective, for example, "square". So, we speak both of a square and of a square region; and, to make matters worse, in settings where the distinction need not be emphasized, the term "square" might be used for either the boundary of an appropriate region or for the region itself.

EXAMPLES:

3. Let us describe the intersection of the following eight closed half spaces in $\mathbb{R}^3$:

$$\{X : \langle (1,1,1), X \rangle \leq 1\}, \qquad \{X : \langle (1,1,-1), X \rangle \leq 1\},$$
$$\{X : \langle (1,-1,1), X \rangle \leq 1\}, \qquad \{X : \langle (-1,1,1), X \rangle \leq 1\},$$
$$\{X : \langle (1,-1,-1), X \rangle \leq 1\}, \qquad \{X : \langle (-1,1,-1), X \rangle \leq 1\},$$
$$\{X : \langle (-1,-1,1), X \rangle \leq 1\}, \qquad \{X : \langle (-1,-1,-1), X \rangle \leq 1\}.$$

Replacing "≤ 1" by "$= 1$" throughout, we observe, by comparison with the answer to Exercise 12 of Section 5.1, that we get the eight planes requested in that exercise. Therefore, we might guess that the intersection of the eight given closed half spaces is the octahedronal region whose boundary is the octahedron described in Exercise 12 of Section 5.1. To check that this conjecture is true, we need to check whether the inequalities go in the right directions, that is, whether the octahedronal region is on the appropriate side of each of the planes determined by the above half-spaces. Since the origin, which is the center of the octahedron, satisfies all eight inequalities, it must be that the octahedronal region is indeed on the correct side of each of the planes, verifying our conjecture.

4. Let us find the intersection of the following three open half planes in $\mathbb{R}^2$:

$$\{X : x_1 > 0\}, \quad \{X : x_2 > 0\}, \quad \text{and} \quad \{X : x_1 + x_2 > 0\}.$$

A point that lies in both the first two half planes automatically lies in the third half plane, since the sum of two positive numbers is positive. So the answer is the first quadrant (not including any portions of the coordinate axes).

5. Let us find the intersection of the following three open half planes in $\mathbb{R}^2$:

$$\{X : x_1 > 0\}, \quad \{X : x_2 > 0\}, \quad \text{and} \quad \{X : x_1 + x_2 < 0\}.$$

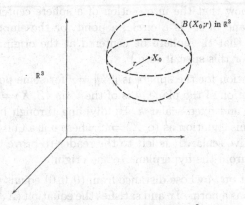

FIGURE 5.4. An open ball in $\mathbb{R}^3$

A point that lies in both the first two half planes cannot lie in the third half plane, since the sum of two positive numbers cannot be negative. So the answer is the empty set.

5.4B BALLS AND SPHERES

We now treat some subsets of $\mathbb{R}^n$ that rival affine subspaces in importance. For $r > 0$, the **open ball** of **radius r centered** at X_0 is the set

$$B(X_0; r) = \{X \in \mathbb{R}^n : \|X - X_0\| < r\} \ .$$

Note that $B(X_0; r)$ does not contain those points whose distance from X_0 is either exactly r or greater than r. See Figure 5.4. A set closely related to the open ball $B(X_0; r)$ is the **closed ball**

$$\bar{B}(X_0; r) = \{X \in \mathbb{R}^n : \|X - X_0\| \leq r\} \ .$$

The open ball $B(X_0; r)$ is a proper subset of $\bar{B}(X_0; r)$.

There are special terms for the $\mathbb{R}$- and $\mathbb{R}^2$-settings. An open ball in $\mathbb{R}$ is a **bounded open interval** and in $\mathbb{R}^2$ it is an **open disk**. A closed ball in $\mathbb{R}$ or $\mathbb{R}^2$ is a **bounded closed interval** or **closed disk**, respectively.

The set of points in $\mathbb{R}^n$ whose distance from a fixed point X_0 equals some positive number r is called the **sphere** of radius r centered at X_0. It is denoted by

$$S(X_0; r) = \{X \in \mathbb{R}^n : \|X - X_0\| = r\}$$

and equals the set of points that belong to $\bar{B}(X_0; r)$ but not to $B(X_0; r)$. Spheres in $\mathbb{R}^3$ are surfaces. Spheres in $\mathbb{R}^2$ are curves, in fact, circles. In $\mathbb{R}^1$ spheres consist of two points. We could call spheres in $\mathbb{R}^n$ "hypersurfaces".

EXAMPLES:

6. Let us show that the intersection of a sphere centered at the origin and a plane in $\mathbb{R}^3$ is a circle, a point, or the empty set. (It is not essential that the sphere be centered at the origin, but notation is simpler for this special case.)

   The equation for the sphere is $\|X\| = r$ for some positive number r. The equation of the plane $\mathcal{A}$ is of the form $\langle Z, X \rangle = k$ for some fixed vector Z and fixed scalar k. By dividing through by $\pm\|Z\|$, we can rewrite this equation as $\langle \mathbf{e}, X \rangle = c$, where $\mathbf{e}$ is a unit vector and c is a nonnegative scalar. It is left to the reader to prove that c equals the distance from the hyperplane to the origin.

   The point on $\mathcal{A}$ whose distance from $(0,0,0)$ equals c is the point $c\mathbf{e}$, since $c\mathbf{e}$ has a norm of c and satisfies the equation $\langle X, \mathbf{e} \rangle = c$. It follows that the intersection of the plane $\mathcal{A}$ and the sphere $S((0,0,0); r)$ is empty if $c > r$ and consists of the single point $c\mathbf{e}$ if $c = r$.

   In case $c < r$ we guess that the intersection of $\mathcal{A}$ and $S((0,0,0); r)$ is a circle centered at $c\mathbf{e}$. To prove this let us write an arbitrary $X \in \mathbb{R}^3$ as $c\mathbf{e} + Y$, that is, we make the substitution $Y = X - c\mathbf{e}$. In order that Y correspond to an X that is a member of both $\mathcal{A}$ and $S((0,0,0); r)$ we must have

$$\langle \mathbf{e}, c\mathbf{e} + Y \rangle = c \quad \text{and} \quad \|c\mathbf{e} + Y\| = r \,,$$

   that is,

$$\langle \mathbf{e}, Y \rangle = 0 \quad \text{and} \quad c^2 + 2c\langle \mathbf{e}, Y \rangle + \|Y\|^2 = r^2 \,.$$

   Using the first condition to simplify the second, we obtain the following equivalent pair of conditions:

$$\langle \mathbf{e}, Y \rangle = 0 \quad \text{and} \quad \|Y\| = \sqrt{r^2 - c^2} \,.$$

   The first condition is the condition that $X \in \mathcal{A}$ and the second is that the distance between $c\mathbf{e}$ and X equal $\sqrt{r^2 - c^2}$. According to the definition in Exercise 21 of Section 4.1, the pair of conditions thus imposed on X are exactly those required for X to belong to a circle of radius $\sqrt{r^2 - c^2}$ centered at $c\mathbf{e}$ and lying in the plane $\mathcal{A}$.

7. In $\mathbb{R}^n$ let us consider a one-dimensional linear subspace $\{tX : t \in \mathbb{R}\}$ for some fixed $X \neq \mathbf{0}$. Let us calculate the points of intersection of this line with the sphere $S(C; r)$, where $C \in \mathbb{R}^n$ and $r \geq 0$. That a point tX be on the sphere is equivalent to

$$\|tX - C\|^2 = r^2 \,.$$

   So, we are interested in solving this equation for t. We rewrite the left-hand side of the equation as follows:

$$\|tX - C\|^2 = \langle tX - C, tX - C \rangle = t^2 \langle X, X \rangle - 2t\langle X, C \rangle + \langle C, C \rangle$$
$$= t^2 \|X\|^2 - 2t\langle X, C \rangle + \|C\|^2 \,.$$

Thus, we should solve the equation

$$t^2 \|X\|^2 - 2t\langle X, C\rangle + \|C\|^2 = r^2 \; ;$$

and for each solution t, the point tX is a point of intersection of the given line and the given sphere. Moreover, every point of intersection can be obtained in this manner. Since the equation is a quadratic in t, we conclude without any further calculation that there are two, one, or zero points of intersection, in agreement with our pictorial understanding in the case where $n \leq 3$.

We solve the quadratic equation for t, and then obtain a second form of the solution in terms of the angle ψ between the vectors X and C by using the relation

$$\cos\psi = \frac{\langle X, C\rangle}{\|X\| \, \|C\|}$$

as follows:

$$t = \frac{\langle X, C\rangle \pm \sqrt{\|X\|^2 r^2 - \|X\|^2 \|C\|^2 + \langle X, C\rangle^2}}{\|X\|^2}$$

$$= \frac{\|C\|}{\|X\|} \left[\cos\psi \pm \sqrt{\left(\frac{r}{\|C\|}\right)^2 - 1 + \cos^2\psi} \, \right].$$

If $r > \|C\|$, then $\mathbf{0} \in B(C; r)$ and so we would expect two points of intersection, one on each of the two rays emanating from $\mathbf{0}$ on the given line. Thus, we expect two solutions of the quadratic equation, one positive and one negative. The quantity under the radical sign is bigger than $\cos^2\psi$, so, indeed there are two solutions of opposite sign.

Suppose $r < \|C\|$. We might ask ourselves: For which X does the line passing through $\mathbf{0}$ and X intersect the sphere in exactly one point and that on the ray emanating from $\mathbf{0}$ and passing through X? In order for this to happen the quantity under the radical sign must equal 0, and, in order that the one solution be positive, $\cos\psi$ must be positive. Therefore, the answer to our question is as follows: those X's for which

$$\frac{\langle X, C\rangle}{\|X\|\|C\|} = \sqrt{1 - \left(\frac{r}{\|C\|}\right)^2} \, .$$

(As will be seen from the forthcoming Section 5.4c and Exercise 9, the set of X's that we have described is a "cone" missing its vertex.)

It is worth noting that the symbolism in this example did not get very messy, even though we were solving a somewhat involved problem for all $\mathbb{R}^n$ (with, of course, most interest being on $\mathbb{R}^2$ and $\mathbb{R}^3$). One reason is that vectors were used throughout and no coordinates were introduced. Some of the exercises will give practice in this technique.

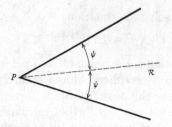

FIGURE 5.5. $y = \|X\| \cot \psi$ in $\mathbb{R}^3$

5.4C   CONES OF REVOLUTION

Let $\mathcal{R}_1$ and $\mathcal{R}_2$ denote rays emanating from a common point P. When we speak of the angle between $\mathcal{R}_1$ and $\mathcal{R}_2$ we refer to an angle in the interval $[0, \pi]$, but not necessarily the interval $[0, \pi/2]$. If X_1 and X_2 are different from P and lie on $\mathcal{R}_1$ and $\mathcal{R}_2$, respectively, then the **angle between the rays** is the angle between the vectors $X_1 - P$ and $X_2 - P$.

Let $n \geq 1$, and let $\mathcal{R}$ denote a ray in $\mathbb{R}^{n+1}$ emanating from a point P. Let $\psi \in (0, \pi/2)$. A **cone of revolution** in $\mathbb{R}^{n+1}$ with **vertex** P, **axis** $\mathcal{R}$, and **angle** ψ is the set of points on all rays emanating from P that make an angle ψ with $\mathcal{R}$. We have chosen to label the dimension of the space in which we are working $n+1$ in order to facilitate the use of notation in the following example. Notice that the angle of a cone is required to be less than $\pi/2$.

EXAMPLES:

8. We label the coordinates of members of $\mathbb{R}^{n+1}$ as follows:

$$ Z = (x_1, x_2, \ldots, x_n, y) \,, $$

which we then abbreviate as (X, y), where $X \in \mathbb{R}^n$ denotes the first n coordinates of the point $Z \in \mathbb{R}^{n+1}$ and y denotes the $(n+1)^{\text{st}}$ coordinate. Let us algebraically describe a cone with an arbitrary angle ψ, vertex $\mathbf{0}$, and axis equal to the set of points of the form $(0, 0, \ldots, 0, y)$, where $y \geq 0$.

We see that $\mathbf{0}$ belongs to the cone, and that for another point (X, y) to belong to the cone it is necessary and sufficient that

$$ \cos \psi = \frac{\langle (X, y), (0, 0, \ldots, 0, 1) \rangle}{\|(X, y)\| \, \|(0, 0, \ldots, 0, 1)\|} \,, $$

which is equivalent to

$$ \cos \psi = \frac{y}{\sqrt{\|X\|^2 + y^2}} \,. $$

We solve for y by squaring both sides (taking care to discard extraneous solutions), obtaining

$$y = \|X\| \cot \psi \, ,$$

which includes the origin (even though initially the origin required special treatment). This last equation is the desired algebraic description of the cone.

9. Suppose $n = 2$ in the preceding example and so we are working in $\mathbb{R}^3$. Let us calculate the intersection of the cone with the plane $x_2 = c$, where c denotes some fixed constant. The set of interest is the set of those (x_1, x_2, y) for which

$$y = (\cot \psi)\sqrt{x_1^2 + c^2} \quad \text{and} \quad x_2 = c \, .$$

In order to rid ourselves of the radical we square both sides of the first equation, but as we do so we must adjoin the condition that $y \geq 0$:

$$y^2 - (\cot^2 \psi)x_1^2 = (\cot^2 \psi)c^2 \quad \text{and} \quad y \geq 0 \quad \text{and} \quad x_2 = c \, ,$$

which we recognize to be one piece of a hyperbola lying in the plane $x_2 = c$ if $c \neq 0$ and to be the union of two rays if $c = 0$.

REMARKS:

- The phrase "cone of revolution" refers to one way in which such cones may be constructed. Start with a single ray that emanates from P and makes an angle ψ with $\mathcal{R}$, and then revolve this ray about $\mathcal{R}$, using P as a pivot. The points swept out by the ray as it revolves about the axis constitute the cone. Note that the intersection of a cone of revolution in $\mathbb{R}^3$ with a plane perpendicular to the axis of the cone is either a circle, a point, or empty. There are more general types of cones, where, for instance, such intersections are ellipses or squares.

- A cone of revolution in $\mathbb{R}^2$ ($n = 1$ in the notation used above) is the union of two rays emanating from a common point.

- A cone of revolution in $\mathbb{R}^3$ is also called a **circular cone**.

- One can define a **double cone of revolution** as the set of all points on *lines* that make some fixed angle $\psi \in (0, \pi/2)$ with a fixed line $\mathcal{S}$ and pass through some fixed point P on $\mathcal{S}$. A double cone is the union of two (single) cones.

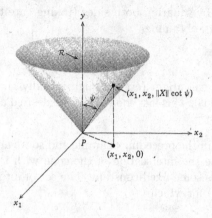

$(x_1, x_2, \|X\| \cot \psi)$

$(x_1, x_2, 0)$

FIGURE 5.6. A cone in $\mathbb{R}^2$

5.4D PARABOLOIDS OF REVOLUTION

Let $n \geq 1$. A **paraboloid of revolution** in $\mathbb{R}^{n+1}$ is the set of points
in $\mathbb{R}^{n+1}$ each of whose distance from some fixed hyperplane $\mathcal{A}$ equals its
distance from some fixed point P not on $\mathcal{A}$. The hyperplane $\mathcal{A}$ is called the
directrix of the paraboloid and P its **focus**. The line through the focus
perpendicular to the directrix is the **axis** of the paraboloid, and the point
where the axis intersects the paraboloid is the **vertex**.

EXAMPLE:

10. The notation introduced in Example 8 of Section 5.4c is used here.
    Thus, a point $Z \in \mathbb{R}^{n+1}$ is written as (X, y), where X has n coordi-
    nates. Let us represent algebraically the parabola whose directrix is
    the hyperplane $y = -c$ and whose focus is the point $(0, 0, \ldots, 0, c)$,
    where c is some fixed positive number. The distance between an ar-
    bitrary point (X, y) and the hyperplane $y = -c$ is $|y + c|$, and the
    distance between (X, y) and the point $(0, 0, \ldots, 0, c)$ is

$$\sqrt{(x_1 - 0)^2 + (x_2 - 0)^2 + \cdots + (x_n - 0)^2 + (y - c)^2}$$
$$= \sqrt{\|X\|^2 + (y - c)^2} \, .$$

Squaring these nonnegative numbers and setting the results equal to
each other, we obtain

$$(y + c)^2 = \|X\|^2 + (y - c)^2 \, ,$$

which is equivalent to

$$4cy = \|X\|^2 \, .$$

REMARKS:

- The above argument is essentially the same one that appears in pre-calculus or second-year high school algebra books for the parabola. And, of course, we will use the term "parabola" rather than "paraboloid" in the $\mathbb{R}^2$-setting.

- The term **circular paraboloid** is sometimes used for a paraboloid of revolution in $\mathbb{R}^3$.

Exercises

1. Let $\mathcal{A}$ denote the hyperplane consisting of those X for which $\langle X, \mathbf{e} \rangle = c$, where $\mathbf{e}$ is a unit vector and $c \geq 0$. Use Exercise 9 of Section 5.1 to prove that c is the distance from the point $\mathbf{0}$ to $\mathcal{A}$.

2. As a check on your understanding of Example 7 in Section 5.4b, use the formula there to calculate all points of intersection of the sphere $S(C; r)$ with the line passing through $\mathbf{0}$ and X, for the following choices of X, C, and r:

    (a) $X = (0, 1)$, $C = (0, 1)$, $r = 2$
    (b) $X = (1, 0)$, $C = (0, 1)$, $r = 1$
    (c) $X = (1, 1)$, $C = (0, 1)$, $r = 1$
    (d) $X = (0, 0, 1)$, $C = (0, 0, 1)$, $r = 2$
    (e) $X = (0, 0, \ldots, 0, 1)$, $C = (0, 0, \ldots, 0, 1)$, $r = 2$
    (f) $X = (2, 1, 1)$, $C = (4, 0, 0)$, $r = \sqrt{6}$
    (g) $X = (2, 1, 1)$, $C = (5, 0, 0)$, $r = \sqrt{6}$.

3. For the case in Example 7 of Section 5.4b where there are two points of intersection, find the midpoint of the line segment connecting those two points. On which of the given quantities does it depend and on which does it not depend?

4. Use Example 7 of Section 5.4b to find all points on the sphere of radius 1 centered at $(1, 1, 0)$ through which there is a tangent line that passes through $\mathbf{0}$. Express the answer in terms of two equations satisfied by the coordinates. *Hint:* Do not work with coordinates until near the end of the solution.

5. Suppose, in Example 7 of Section 5.4b, that $r < \|C\|$ and that the line passing through $\mathbf{0}$ and X intersects the sphere in two distinct points. Do you expect those two points to lie on a common ray emanating from $\mathbf{0}$? Show that your answer to this question is consistent with the formula obtained for t in the example.

6. Suppose, in Example 7 of Section 5.4b, that $r = \|C\|$. Without making any calculations partition $\{X : X \neq \mathbf{0}\}$ into three sets: (1) those X for which the line passing through $\mathbf{0}$ and X intersects the sphere in exactly two points, (2) those X for which the line intersects the sphere in exactly one point, and (3) those X for which the number of intersection points is 0. Then show that these answers are consistent with the formula for t in the example.

7. Suppose the fixed ray in Example 8 of Section 5.4c is replaced by the ray consisting of points of the form $(0, 0, \ldots, 0, y)$, where $y \leq 0$. What does the resulting equation of the cone become?

8. Explain why any cone of revolution in $\mathbb{R}^2$ is the union of two rays.

9. Let $n > 1$ and $r < \|C\|$ in Example 7 of Section 5.4b. Explain why the set of $X \neq \mathbf{0}$, for which the ray emanating from $\mathbf{0}$ and passing through X intersects the sphere in exactly one point, is a cone minus its vertex. Find the vertex, axis, and angle of this cone.

10. Find an equation for each of the two cones of revolution in $\mathbb{R}^3$ whose axes lie on the x_3-coordinate axis and which have a common vertex and angle—namely, vertex equals $(0, 0, 1)$ and angle equals $\pi/4$.

11. Let $r > 0$. Prove that the intersection of all the closed half spaces of the form

$$\{X : \langle \mathbf{e}, X \rangle \leq r\},$$

where $\|\mathbf{e}\| = 1$, equals the closed ball $\bar{B}(\mathbf{0}; r)$. What does the intersection equal if $r = 0$? What does it equal if $r < 0$. *Hint:* Draw some pictures in $\mathbb{R}^2$.

12. Parametrize the intersection of the open disk $B((2, 0); 1)$ and the circle $S(\mathbf{0}; 2)$ in $\mathbb{R}^2$.

13. Parametrize the intersection of $S((2, 0); 2)$ and $\bar{B}(\mathbf{0}; 1)$ in $\mathbb{R}^2$.

14. Parametrize the intersection of the open ball $B(\mathbf{0}; 1)$ and the line passing through the points $(1, 1, 0)$ and $(-1, 0, -1)$ in $\mathbb{R}^3$.

15. Let c and k be positive constants such that $k > 2c$. Let $\mathcal{P}$ denote the set of points in $\mathbb{R}^{n+1}$ each of whose distances from the points $(0, 0, \ldots, c)$ and $(0, 0, \ldots, -c)$ sum to k. Describe the set $\mathcal{P}$ algebraically in as simple a form as possible.

16. The set $\mathcal{P}$ described in the preceding exercise is an "ellipsoid of revolution". Write a general definition. (Of course, this is not a well-posed exercise; you are being asked to make a guess at what the commonly used definition might be.) Then identify the particular situations when the terms "ellipse" and "circular ellipsoid" might be used.

17. Modify Exercise 15 by replacing "$k > 2c$" by "$k = 2c$".

18. Consider the plane

$$\{(1, 1, 2) + r(1, 0, 2) + s(1, 1, 2) : r, s \in \mathbb{R}\}.$$

For each s calculate those values of r which correspond to points that are both on the plane and on the circular paraboloid $x_3 = x_1^2 + x_2^2$. For which values of s are there no corresponding values of r? The intersection of the plane and the paraboloid is a curve. Parametrize the curve. *Hint:* Here, finally, is a problem where it seems best to turn immediately to coordinates. A possible first step in parametrizing the curve is to parametrize half of it.

6

Functions $f : \mathbb{R}^n \to \mathbb{R}$

6.0 Introduction

In this chapter we concentrate on the differential calculus of a real-valued function f with $\mathcal{D}(f) \subseteq \mathbb{R}^n$, that is, a scalar-valued function of a vector variable. Thus, to every vector X in $\mathcal{D}(f)$ (the domain of f) the function f assigns a real number $f(X)$.

Since $X = (x_1, \ldots, x_n)$, one often sees $f(X)$ written as $f(x_1, \ldots, x_n)$. Thus f is a function of n (scalar) variables, hence the name "multivariable calculus". Also in two and three dimensions, custom has established the notation $f(x, y)$ instead of $f(x_1, x_2)$, and $f(x, y, z)$ instead of $f(x_1, x_2, x_3)$. We use these occasionally.

EXAMPLES:

1. *Temperature in a room.* Let $\mathcal{D}(f)$ denote some room, considered as a subset of $\mathbb{R}^3$, and let $f(X)$ be the temperature at the point X. Thus f assigns a real number—temperature—to each point in the room.

2. *The height function of a paraboloid of revolution.* Let $f : \mathbb{R}^2 \to \mathbb{R}$ be given by $f(X) = x_1^2 + x_2^2$. See Figure 6.1. The graph of $z = f(X)$ in $\mathbb{R}^3$ is a circular paraboloid (see Section 5.4d) and the number $f(X)$ tells us the *height* of this surface above the point $(x_1, x_2, 0)$.

3. *Linear functions.* In Section 3.3 we referred to scalar-valued functions $L : \mathbb{R}^n \to \mathbb{R}$ as functionals. There we saw that every linear functional is of the form
$$L(X) = \langle Z, X \rangle ,$$
where Z is a fixed member of $\mathbb{R}^n$ and X is a variable member of $\mathbb{R}^n$.

   The function $L(x_1, x_2) = -x_1 - x_2$ is a linear functional on $\mathbb{R}^2$. Since, $L(1, 0) = -1$, $L(0, 1) = -1$, and $L(0, 0) = 0$, the graph of L is the plane containing the three points $(1, 0, -1)$, $(0, 1, -1)$, and $(0, 0, 0)$. This two-dimensional linear subspace of $\mathbb{R}^3$ is illustrated in Figure 6.2. It is the solution set of the linear equation $z - L(x_1, x_2) = 0$, that is, $z + x_1 + x_2 = 0$.

4. *Affine functions.* An example of an affine function $T : \mathbb{R}^2 \to \mathbb{R}$ is given by $T(X) = 3 - x_1 - x_2$. Since $T(\mathbf{0}) = 3$, its graph is the plane

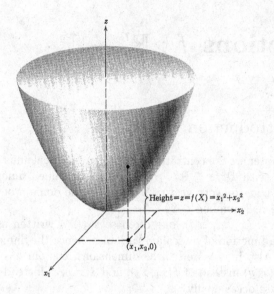

FIGURE 6.1.

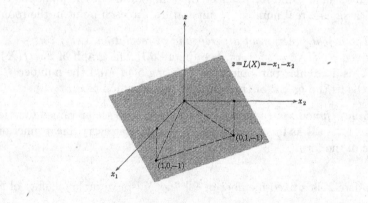

FIGURE 6.2. The graph of a linear map

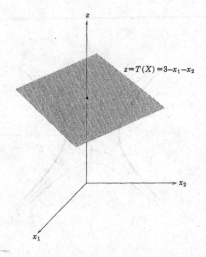

FIGURE 6.3. The graph of an affine map

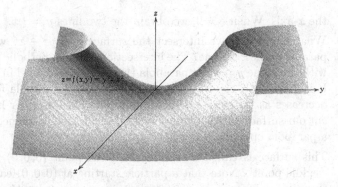

FIGURE 6.4.

of the preceding example translated by three units in the positive z-direction. See Figure 6.3. This plane (affine subspace) is determined by the equation $z - T(x_1, x_2) = 0$, which can be written as $x_1 + x_2 + z = 3$ or $x_1 + x_2 + z - 3 = 0$.

5. *The function $z = f(x, y) = y^2 - x^2$.* The graph of this function is a very fancy surface—a "hyperbolic paraboloid". See Figure 6.4. It is a useful example for illustrating important ideas.

If this surface is cut by a plane parallel to the xy-plane, for example $z = c$, where c is some constant, then the intersection is the curve $c = y^2 - x^2$. See Figure 6.5. If $c > 0$, then this curve is a hyperbola that opens about the y-axis in the plane of all points (x, y, c). If $c < 0$, the curve determined by $c = y^2 - x^2$ is a hyperbola opening about

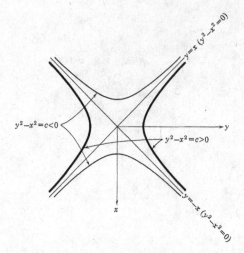

FIGURE 6.5. View of $y^2 - x^2 = z$ from above

the x-axis. When $c = 0$, we obtain the two lines $y = \pm x$.

What happens if we intersect the surface $z = y^2 - x^2$ with planes parallel to the xz-plane? The intersection of the hyperbolic paraboloid with the plane $y = c$ is a parabola that opens downward in this plane. It is given by $z = -x^2 + c^2$, which is clearly a parabola for which z decreases as $|x|$ increases. See Figure 6.6. On the other hand, if we cut our surface with a plane $x = c$ parallel to the yz-plane, we obtain a parabola opening upward.

This surface is called a "saddle" and the point $(0, 0, 0)$ is called a "saddle point". Note that a particle starting at $(0, 0, 0)$ can move on the surface in one direction and go up, or move on the surface in another direction and go down, or even move in a direction so that it neither rises nor falls. We will encounter saddle points again.

We introduce a term useful for the next example. Consider a sphere in $\mathbb{R}^n$ and a hyperplane passing through the center of the sphere. The intersection of the sphere and a half space determined by the hyperplane is called a **hemisphere**, and it is a **closed hemisphere** if the half space is closed. In $\mathbb{R}^2$, the term **semicircle** replaces "hemisphere".

EXAMPLES:

6. *The height function of a closed hemisphere.* Let

$$f(X) = \sqrt{25 - \|X\|^2} \,.$$

Since the domain has not been explicitly specified it is implicitly specified as the set of all X for which the formula for f is meaningful,

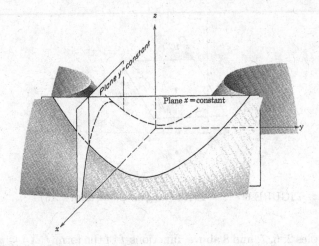

FIGURE 6.6. Parabolas cut out by planes

namely, the closed ball $B(\mathbf{0}; 5)$. The graph of f is a subset of $\mathbb{R}^{n+1}$. The square of the norm of each point on the graph is given by

$$x_1^2 + x_2^2 + \cdots + x_n^2 + f(X)^2$$
$$= \|X\|^2 + f(X)^2 = \|X\|^2 + (25 - \|X\|^2) = 25 .$$

Therefore, the graph of f is a subset of $S(\mathbf{0}; 5)$. We ask: Which points on this sphere actually belong to the graph? As a first step in answering we note that $f(X) \geq 0$ for all $X \in \mathcal{D}(f)$; so, only points on a certain closed hemisphere belong to the graph of f. To show that every point on that closed hemisphere is on the graph, let us consider an arbitrary point $(x_1, x_2, \ldots, x_n, z)$ for which $z \geq 0$ and

$$x_1^2 + x_2^2 + \cdots + x_n^2 + z^2 = 25 .$$

Solving for z we obtain

$$z = \sqrt{25 - \|X\|^2} ,$$

from which it follows that $(x_1, x_2, \ldots, x_n, z)$ belongs to the graph of f.

7. *The height function of a cone.* For $X \in \mathbb{R}^n$, let $f(X) = \|X\|$. The graph of f is a cone of revolution. See Section 5.4c.

8. From Section 5.4d we see that the graph of a function of the form $f(X) = k\|X\|^2$, where k is a nonzero constant, is a paraboloid of revolution.

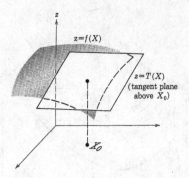

FIGURE 6.7. $z = T(X)$ (tangent plane above X_0)

In Examples 2, 6, 7, and 8 above, functions f of the form $f(X) = g(\|X\|)$, where g is a scalar-valued function of a scalar, were considered. Such functions often arise in applications where there is a considerable amount of symmetry.

Here are the central ideas that will be discussed in this chapter. In the next section limits and continuity will be discussed. After that the "directional derivative" of a function f is introduced. It measures rate of change in a chosen direction. When the chosen direction is that of a coordinate axis, the directional derivative is called a "partial derivative". It can be calculated using the methods of single-variable calculus. Then we turn to the "total derivative" which yields the best affine approximation near some point. Such an approximation is illustrated in Figure 6.7 for the case $n = 2$. It is the total derivative that is the more fundamental concept; the directional derivatives are easily obtained from it. This chapter ends with basic facts about total derivatives.

Exercises

Becoming familiar with the graphs of functions of a *vector* variable.

1. Sketch in $\mathbb{R}^3$ the graphs of the following functions defined on $\mathbb{R}^2$:

    (a) $f(X) = 2 - x_1 - 2x_2$

    (b) $f(x, y) = y - x + 1$

    (c) $f(x, y) = (x - 1)^2 + y^2 + 2$

    (d) $f(X) = 1 + x_1^2 - x_2^2$.

2. Each of the following functions on $\mathbb{R}^2$ describes the temperature at each point of $\mathbb{R}^2$. For each such temperature function draw a sketch of $\mathbb{R}^2$ and on the sketch indicate the places where the temperature is -1, where it is 0, where it is 1, and where it is 4.

    (a) $f(X) = 3 + x_1 - 2x_2$

    (b) $f(X) = x_1^2 + x_2^2$

(c) $f(x, y) = xy - 1$

(d) $f(x, y) = y^2 - x^2$

(e) $f(X) = x_1 - x_2^2$

(f) $f(x, y) = x^2 - 3$.

3. Sketch the graph of $e^{-\|X\|}$ in two cases: $X \in \mathbb{R}^1$ and $X \in \mathbb{R}^2$.

4. Follow the instructions of the preceding exercise for the function $1 \wedge \|X\|$. (Note: $a \wedge b$ denotes the smaller of the real numbers a and b.)

5. Describe or draw the graph of the function $f(x_1, x_2) = |x_1 - x_2|$.

6. Describe the domain of the function

$$f(x_1, x_2, x_3, x_4, x_5, x_6) = \frac{1}{(x_1 - x_4)^2 + (x_2 - x_5)^2 + (x_3 - x_6)^2} .$$

In particular, what kind of object is its complement in $\mathbb{R}^6$ and what is the dimension of its complement? Interpret the function f, possibly using information from Section 5.3a.

6.1   Continuity and Limits

Continuity and limits are discussed in this section. These topics have already been treated in Chapter 4, but for a different setting. In Chapter 4 we generally restricted our attention to functions F for which $\mathcal{D}(F)$ is an interval. For functions of a vector it is also, for some purposes, useful to restrict one's attention to certain types of domains. For variety we treat the two major issues of this section, continuity and limits, in the order opposite from that used in Chapter 4.

6.1A   CONTINUITY

In order to develop a definition of continuity analogous to that given in Chapter 4, we first reformulate the definition of continuity given there for functions of a real variable.

Definition 6.1.1 A function $F : \mathcal{D}(F) \to \mathbb{R}^q$, where $\mathcal{D}(F) \subseteq \mathbb{R}$, is said to be **continuous at a point** $x_0 \in \mathcal{D}(F)$ if and only if for every $\varepsilon > 0$, there is an positive number δ such that

$$\|F(x) - F(x_0)\| < \varepsilon$$

whenever

$$|x - x_0| < \delta \quad \text{and} \quad x \in \mathcal{D}(F) .$$

A function that is continuous at each point in its domain is said to be **continuous**.

REMARK:

- At first, the above definition looks identical to Definition 4.1.2, given in Section 4.1b. The difference between the two definitions is that the domain is assumed to be an interval in Section 4.1b, but no assumption about the domain is made in Definition 6.1.1 above. For this rather technical reason, Definition 6.1.1 above should be regarded as the official definition. We will use it as a model for the definition of continuity for functions of a vector variable.

Definition 6.1.2 A function $f : \mathcal{D}(f) \to \mathbb{R}$, where $\mathcal{D}(f) \subseteq \mathbb{R}^n$ for some n, is said to be **continuous at a point** $X_0 \in \mathcal{D}(f)$ if and only if, for every $\varepsilon > 0$, there is a positive number δ such that

$$|f(X) - f(X_0)| < \varepsilon \,,$$

whenever

$$\|X - X_0\| < \delta \quad \text{and} \quad X \in \mathcal{D}(f) \,.$$

If f is continuous at each point in $\mathcal{D}(f)$, then f is said to be **continuous**.

As usual we say that a function is **discontinuous at a point** in its domain if it is not continuous there. It is **discontinuous** if there is at least one point at which it is discontinuous.

The preceding definition says that a function f is continuous if and only if the values of f at points X near any particular point X_0 are close to the value of f at X_0; of course, only points in the domain of f are considered.

REMARKS:

- Defintion 6.1.2 can be restated using the language of open balls. The function f is continuous at a point $X_0 \in \mathcal{D}(f)$ if and only if for every open interval $B(f(X_0); \varepsilon)$ centered at $f(X_0)$, there is an open ball $B(X_0; \delta)$ centered at X_0 such that

$$f(X) \in B(f(X_0); \varepsilon)$$

whenever

$$X \in B(X_0; \delta) \cap \mathcal{D}(f) \,.$$

The definition of continuity is often used in this form.

- Definition 6.1.1 can be similarly reformulated. The function F is continuous at a point $x_0 \in \mathcal{D}(F)$ if and only if for every open ball $B(F(x_0); \varepsilon)$ centered at $F(x_0)$, there is an open interval $B(x_0; \delta)$ centered at x_0 such that

$$F(x) \in B(F(x_0); \varepsilon)$$

whenever

$$x \in B(x_0; \delta) \cap \mathcal{D}(F) \,.$$

Two questions arise naturally. How can one recognize or create continuous functions? Why is the property of continuity important? We begin with the first of these questions. Our first result says that a function which picks out a particular coordinate of a vector variable is a continuous function of that variable.

Theorem 6.1.1 A function of a vector variable X which assigns to each X its ith coordinate x_i is continuous.

PROOF: Let $f : \mathbb{R}^n \to \mathbb{R}$ be defined by $f(X) = x_i$. Fix an arbitrary point $U \in \mathbb{R}^n$. To show that f is continuous at U let $\varepsilon > 0$ and choose $\delta = \varepsilon$. Suppose $\|X - U\| < \delta$. Then

$$|f(X) - f(U)| = |x_i - u_i| = \sqrt{(x_i - u_i)^2} \leq \|X - U\| < \delta = \varepsilon .$$

Done. <<

Here is a theorem of a type that by now is quite familiar; it tells one how to form new continuous functions from functions already known to be continuous, for example, those described in Theorem 6.1.1.

Theorem 6.1.2 Let f and g be two continuous $\mathbb{R}$-valued functions with $\mathcal{D}(f) = \mathcal{D}(g) \subseteq \mathbb{R}^n$, let h be a continuous $\mathbb{R}$-valued function whose domain contains the image of f, and let $K : \mathcal{D}(K) \to \mathcal{D}(f)$, where $\mathcal{D}(K) \subseteq \mathbb{R}$, also be a continuous function. Then the functions $f + g$, fg, $h \circ f$, and $f \circ K$ are continuous functions as is the function $1/f$ defined on the set $\mathcal{D}(f) \cap \{X : f(X) \neq 0\}$.

PROOF: We will prove that two of the five indicated functions are continuous and leave the other three for the exercises.

To prove that $f + g$ is continuous, let X_0 be an arbitrary member of $\mathcal{D}(f + g)$ (which equals $\mathcal{D}(f) = \mathcal{D}(g)$) and let $\varepsilon > 0$. Since both f and g are given to be continuous, there exist positive numbers α and β such that

$$|f(X) - f(X_0)| < \varepsilon/2 \quad \text{if} \quad X \in B(X_0; \alpha) \cap \mathcal{D}(f)$$

and

$$|g(X) - g(X_0)| < \varepsilon/2 \quad \text{if} \quad X \in B(X_0; \beta) \cap \mathcal{D}(f) .$$

Hence,

$$|f(X) - f(X_0)| < \varepsilon/2 \quad \text{and} \quad |g(X) - g(X_0)| < \varepsilon/2$$
$$\text{if } X \in B(X_0; \alpha) \cap B(X_0; \beta) \cap \mathcal{D}(f) .$$

This assertion can be rewritten as

$$|f(X) - f(X_0)| < \varepsilon/2 \quad \text{and} \quad |g(X) - g(X_0)| < \varepsilon/2$$
$$\text{if } X \in B(X_0; \delta) \cap \mathcal{D}(f) ,$$

where $\delta = \alpha \wedge \beta$ (that is, δ is the smaller of α and β). By using the triangle inequality in $\mathbb{R}^1$ (and it was in looking ahead to this step that we decided to focus our attention on $\varepsilon/2$ once ε had been specified), we conclude that

$$|(f(X) - f(X_0)) + (g(X) - g(X_0))| < \frac{\varepsilon}{2} + \frac{\varepsilon}{2} = \varepsilon$$
$$\text{if } X \in B(X_0; \delta) \cap \mathcal{D}(f) \, .$$

A rearrangement gives the desired conclusion that

$$|(f + g)(X) - (f + g)(X_0)| < \varepsilon$$

whenever $X \in B(X_0; \delta) \cap \mathcal{D}(f)$.

To prove that $f \circ K$ is continuous let x_0 denote an arbitrary member of $\mathcal{D}(f \circ K)$ (which equals $\mathcal{D}(K)$) and let $\varepsilon > 0$. Since f is continuous and $K(x_0) \in \mathcal{D}(f)$, we can, according to Definition 6.1.2, choose $\gamma > 0$ so that

$$|f(Y) - f(K(x_0))| < \varepsilon \quad \text{whenever} \quad Y \in B(K(x_0); \gamma) \cap \mathcal{D}(f) \, .$$

According to Definition 6.1.1 we can choose a positive number δ such that

$$K(x) \in B(K(x_0); \gamma) \quad \text{whenever} \quad x \in B(x_0; \delta) \cap \mathcal{D}(K) \, ,$$

that is, whenever $x \in B(x_0; \delta) \cap \mathcal{D}(f \circ K)$. From the preceding two conclusions and the fact that every $K(x)$ is a member of $\mathcal{D}(f)$, we make the following deduction that completes the proof:

$$|f(K(x)) - f(K(x_0))| < \varepsilon \quad \text{whenever} \quad x \in B(x_0; \delta) \cap \mathcal{D}(f \circ K) \, .$$

As mentioned above, the remaining parts are left for the exercises. $<<$

EXAMPLES:

1. Let g be a continuous scalar-valued function with $\mathcal{D}(g) = [0, \infty)$. Let $f(X) = g(\|X\|)$ for $X \in \mathbb{R}^n$. Let us prove that f is continuous. In view of the preceding theorem we only need prove that the function $\|X\|$ is continuous. By several applications of Theorems 6.1.1 and 6.1.2 above, the function $x_1^2 + x_2^2 + \cdots + x_n^2$ is a continuous function of X. From single-variable calculus, we know that the square root function is continuous. So, another application of Theorem 6.1.2 enables us to conclude that $\|X\|$ is continuous. The adjective **spherically symmetric** is sometimes used for a function of the form of f (whether or not g, and therefore f, happens to be continuous). When the domain is contained in $\mathbb{R}^2$, a spherically symmetric function is sometimes called **radially symmetric**. A spherically symmetric function whose domain is contained in $\mathbb{R}$ is often called an **even** function.

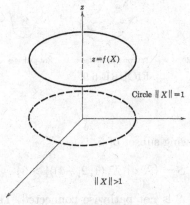

FIGURE 6.8.

2. Figure 6.8 shows, for the case $n = 2$, the graph of the discontinuous spherically (radially) symmetric function

$$f(X) = \begin{cases} 1 & \text{if} \quad \|X\| \leq 1 \\ 0 & \text{if} \quad \|X\| > 1 . \end{cases}$$

There are infinitely many points at which it is discontinuous.

In Exercise 2, the reader is asked to use the preceding Theorems 6.1.1 and 6.1.2 to prove the following theorem, which asserts that scalar-valued affine functions are continuous.

Theorem 6.1.3 Every $\mathbb{R}$-valued affine function defined on $\mathbb{R}^n$ is continuous.

6.1B   PATHWISE CONNECTEDNESS

We know from single-variable calculus that a function with an identically zero derivative is constant. But there is a crucial hypothesis, namely, that the domain be an interval. The same hypothesis is crucial for the Intermediate Value Theorem. For these reasons we are motivated to construct a definition for subsets of $\mathbb{R}^n$ that is analogous to that of being an interval in $\mathbb{R}$.

Definition 6.1.3 A set $S \subseteq \mathbb{R}^n$ is said to be **pathwise connected** if and only if any two members of S can be connected by a curve that is a subset of S.

Two pathwise connected sets and one set that is not pathwise connected are shown in Figure 6.9. The following examples may give a feel for the concept of pathwise connectedness.

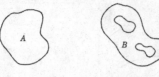

Pathwise connected   Pathwise connected   Not pathwise connected

FIGURE 6.9.

EXAMPLES:

3. Consider the following subset of $\mathbb{R}^3$:

$$S = \{X : |\langle X, (1, 2, -4)\rangle| > 3\} .$$

Let us show that S is not pathwise connected. The points $(2, 1, 0)$ and $(0, 0, 1)$ are both members of S. Consider a curve C that connects these two points. Let F denote a parametrization of C that describes motion from $(2, 1, 0)$ to $(0, 0, 1)$. Thus F is a continuous function with $\mathcal{D}(F)$ equal to a closed interval $[a, b]$, $F(a) = (2, 1, 0)$ and $F(b) = (0, 0, 1)$. Consider the scalar-valued function g defined on the interval $[a, b]$ by the formula

$$g(t) = \langle F(t), (1, 2, -4)\rangle .$$

The function g inherits certain properties from F: it is continuous, $g(a) = \langle (2, 1, 0), (1, 2, -4)\rangle = 4$, and $g(b) = \langle (0, 0, 1), (1, 2, -4)\rangle = -4$. By the Intermediate Value Theorem of single-variable calculus, there is a point $t_0 \in [a, b]$ such that $g(t_0) = 0$. That is,

$$\langle F(t_0), (1, 2, -4)\rangle = 0 .$$

Hence, $F(t_0) \notin S$ and, therefore, $C \not\subseteq S$. Since C was introduced as an arbitrary curve connecting $(2, 1, 0)$ and $(0, 0, 1)$, we conclude that the set S is not pathwise connected.

4. Let

$$S = \{(x_1, x_2) : |x_1| \le |x_2| + 3\} .$$

A picture of S is shown in Figure 6.10, together with a curve within S that connects two particular points of S.

The picture gives us a hint about how we might try to show that S is pathwise connected—namely, by proving that a certain curve connecting two arbitrary points lies in S, that curve being the one that consists of two line segments between the origin and each of the two points. Let P and Q be two members of S. Let

$$F(t) = \begin{cases} -tP & \text{if} \quad -1 \le t \le 0 \\ tQ & \text{if} \quad 0 < t \le 1 \end{cases} .$$

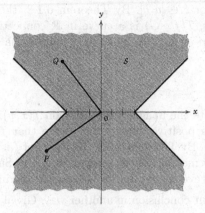

FIGURE 6.10.

It is straightforward to show, directly from the definition of continuiuty, that F is continuous. Also, $F(-1) = P$ and $F(1) = Q$. Thus the image of F is a curve that connects P and Q.

We will complete the proof that S is connected by showing that the image of F is a subset of S. There are two cases. First consider t such that $0 \le -t \le 1$. Then writing $P = (p_1, p_2)$, we have

$$|tp_1| = |t||p_1| \le |t|(|p_2| + 3) = |tp_2| + 3|t| \le |tp_2| + 3 \,.$$

Thus $-tP$, which equals $F(t)$ in this case, is in S. For the second case, in which $0 < t \le 1$, one merely replaces p_1 and p_2 by q_1 and q_2 in this calculation.

Here without proof is an identification of the subsets of $\mathbb{R}^1$ which are pathwise connected.

Theorem 6.1.4 A subset of $\mathbb{R}$ is pathwise connected if and only if it is an interval (bounded or unbounded), a single point, or empty.

We complete this section with a discussion of a result that may be a little more interesting than the preceding results. It is a vector version of the Intermediate Value Theorem.

Theorem 6.1.5 Let f be a continuous scalar-valued function for which $\mathcal{D}(f)$ is pathwise connected. Then the image of f is pathwise connected, that is, $\mathcal{I}(f)$ is an interval.

PROOF: Let p and q be two members of $\mathcal{I}(f)$. Choose P and Q so that $f(P) = p$ and $f(Q) = q$. Since $\mathcal{D}(f)$ is pathwise connected there exists a parametrized curve G with a domain $[a, b]$ such that $G(a) = P$, $G(b) = Q$, and $G(t) \in \mathcal{D}(f)$ for $t \in [a, b]$. By Theorem 6.1.2, the function $f \circ G$ is continuous. That is, $\mathcal{I}(f \circ G)$ is a curve in $\mathbb{R}$ connecting p and q. Since p and q were chosen arbitrarily in $\mathcal{I}(f)$, it follows that the image of f is pathwise connected. $<<$

EXAMPLE:

5. Let us calculate the image of the function $f(x, y) = (x + y) \sin \pi y - y$. Let m be a positive integer and notice that $f(0, m) = -m$ and $f(0, -m) = m$. By Theorem 6.1.5, the image of f is an interval and, so contains every member of the interval $[-m, m]$. Since m is arbitrary, it follows that every real number belongs to the image of f.

Let us state our conclusion in another way. Given any real number z the equation

$$(x + y) \sin \pi y - y = z$$

has at least one solution (x, y). By examining the proof of Theorem 6.1.5 we see that we can conclude more. If $|z| < m$, then on every curve with endpoints $(0, -m)$ and $(0, m)$ there is a point (x, y) which solves the equation

$$(x + y) \sin \pi y - y = z \ .$$

6.1C   LIMIT POINTS

We have defined continuity without explicit reference to limits, but implicit in the definition of continuity is a definition of limit. In preparation for that definition we first define the kind of point at which there can be a limit.

Definition 6.1.4 A point $X_0 \in \mathbb{R}^n$ is called a **limit point** or an **accumulation point** of a subset S of $\mathbb{R}^n$ if and only if every ball $B(X_0; r)$ centered at X_0 contains at least one member of S other than X_0.

REMARK:

• In the preceding definition there is no condition on the membership or lack of membership of X_0 itself in S.

EXAMPLE:

6. Let

$$S = \{(x, y) : y > -1 \text{ and } (x^2 + y^2)(4 - x^2 - y^2) \leq 0\} \ .$$

We give ourselves the task of sketching S and then finding its limit points.

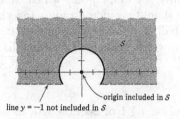

line $y = -1$ not included in S origin included in S

FIGURE 6.11.

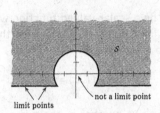

limit points   not a limit point

FIGURE 6.12.

The factor $x^2 + y^2$ is either positive or 0. If it equals 0, then the second inequality is satisifed. If it is positive, then the second inequality is equivalent to $4 - x^2 - y^2 \le 0$. Thus

$$S = \{\mathbf{0}\} \cup \{(x, y) : y > -1 \ \text{ and } \ x^2 + y^2 \ge 4\},$$

which is the union of the one-point set $\{\mathbf{0}\}$ with the intersection of the open half-plane given by $y > -1$ and the complement of the open disc $B(\mathbf{0}; 2)$. The set S is shown in Figure 6.11.

A ball of radius less than 2 centered at $\mathbf{0}$ contains only one member of S, namely $\mathbf{0}$ itself. Thus $\mathbf{0}$ is *not* an limit point of S. A ball centered at any other member of S contains infinitely many members of S, so every member of $S - \{\mathbf{0}\}$ is an limit point of S. There are other limit points as well: each point of the form $(x, -1)$ with $|x| \ge \sqrt{3}$ is a limit point. The set of limit points is illustrated in Figure 6.12. Algebraically it can be described as follows:

$$\{(x, -1) : |x| \ge \sqrt{3}\} \cup (S - \{\mathbf{0}\}).$$

It is worth noting that, except for our claim that $\mathbf{0}$ is not a limit point, we have not actually proved our assertion about the set of limit points. Two things are needed for a complete proof. First, for every claimed limit point and every ball centered at that point, we should prove, say by giving a formula in terms of the radius of the ball, that there is a point in that ball, other than its center, that is also in $\mathcal{S}$. Secondly, for every point which we describe as not being a limit point we should provide a radius together with an argument saying why the ball of that radius centered at the point in question contains no point which belongs to $\mathcal{S}$ other than possibly its center.

6.1D LIMITS

We now modify Definition 4.1.1 of Section 4.1a to fit the present setting. After the preceding discussion of limit points we are prepared to formulate the definition for a function with arbitrary domain.

Definition 6.1.5 Let f be a scalar-valued function with domain $\mathcal{D}(f) \subseteq \mathbb{R}^n$ for some n. Let X_0 be a limit point of $\mathcal{D}(f)$. Then the **limit** of $f(X)$ as X approaches X_0 is equal to t, written as

$$\lim_{X \to X_0} f(X) = t ,$$

if and only if for every positive number ε, there is a positive number δ such that $|f(X) - t| < \varepsilon$ whenever $0 < \|X - X_0\| < \delta$ and $X \in \mathcal{D}(f)$.

REMARK:

- The preceding definition can be reformulated using the language of open balls. We have $f(X) \to t$ as $X \to X_0$ if and only if X_0 is an limit point of $\mathcal{D}(f)$ and for every open interval $B(t; \varepsilon)$ centered at t there is and open ball $B(X_0; \delta)$ such that $f(X) \in B(t; \varepsilon)$ whenever $X \in B(X_0; \delta) \cap \mathcal{D}(f)$ and $X \neq X_0$.

As in Section 4.1a there is a theorem that entitles us to use the adjective "the" before the noun "limit". We leave its proof for the exercises since it is so similar to the proof of Theorem 4.1.1 in Section 4.1a.

Theorem 6.1.6 Let f be a scalar-valued function of a vector variable. If $f(X) \to s$ and $f(X) \to t$ as $X \to X_0$, then $s = t$.

As is to be expected, our first task is to describe how the operation of taking limits works in conjunction with the usual algebraic operations. There are no surprises. We will omit the proof.

Theorem 6.1.7 Suppose that the scalar-valued functions f and g have a common domain. If $f(X) \to s$ and $g(X) \to t$ as $X \to X_0$, then as $X \to X_0$,

1. $(f + g)(X) \to s + t$,

2. $(fg)(X) \to st$,

3. $(1/f)(X) \to 1/s$ provided $s \neq 0$.

We also include, without proof, two theorems giving basic facts involving compositions of functions.

Theorem 6.1.8 Let f be a scalar-valued function with $\mathcal{D}(f) \subseteq \mathbb{R}^n$ for some n, and suppose, for some X_0 and t, that $f(X) \to t$ as $X \to X_0$. Let g be a continuous scalar-valued function whose domain contains both t and the image of f. Then $(g \circ f)(X) \to g(t)$ as $X \to X_0$.

Theorem 6.1.9 Let F be an $\mathbb{R}^n$-valued function with $\mathcal{D}(F) \subseteq \mathbb{R}$, and suppose, for some x_0 and Y, that $F(x) \to Y$ as $x \to x_0$. Let g be a continuous scalar-valued function whose domain contains both Y and the image of F. Then $(g \circ F)(x) \to g(Y)$ as $x \to x_0$.

Some of the preceding discussion involves scalar-valued functions and can be expanded to include the possibility of limits equal to $-\infty$ or ∞. Since the details are so similar to those for single-variable calculus, we will not give them here. Nevertheless, we will, when appropriate, freely speak of limits of scalar-valued functions equaling $-\infty$ or ∞.

6.1E   INTERIOR POINTS

When discussing derivatives in single-variable calculus one usually begins by considering the situation at a point in the interval of definition that is *not* an endpoint. For functions of a vector, we need to identify analogous points in $\mathbb{R}^n$.

Definition 6.1.6 A point X belonging to a subset S of $\mathbb{R}^n$ is called an **interior point** of S if and only if for some positive number r, the open ball $B(X;r)$ is a subset of S. The set of all interior points of the set S is called the interior of S.

Figure 6.13 illustrates a triangular region and three of its members X_0, X_1, and X_2. The first two are interior points, but X_2 is not an interior point.

EXAMPLE:

7. Let us prove that the interior of the closed disk $\bar{B}(0;5)$ is the open disk $B(0;5)$. We must prove the following two things: (1) that every member of $B(0;5)$ is an interior point of $\bar{B}(0;5)$ and (2) that no member of $S(0;5)$ is an interior point of $\bar{B}(0;5)$.

   To prove (1) let X be an arbitrary member of $B(0;5)$. Set $r = 5 - \|X\|$, which is a positive number since $X \in B(0;5)$. We will complete the proof of (1) by showing that $B(X;r) \subseteq \bar{B}(0;5)$. For this purpose we let U be an arbitrary member of $B(X;r)$. By the triangle inequality,

$$\|U\| \leq \|U - X\| + \|X\| < r + \|X\| = 5 .$$

   Thus $U \in \bar{B}(0;5)$ as desired. (In fact, $U \in B(0;5)$.)

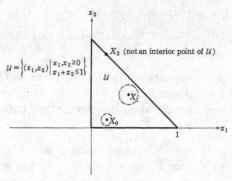

$$\mathcal{U} = \left\{ (x_1, x_2) \,\middle|\, \begin{matrix} x_1, x_2 \geq 0 \\ x_1 + x_2 \leq 1 \end{matrix} \right\}$$

X_2 (not an interior point of $\mathcal{U}$)

X_0, X_1 are interior points of $\mathcal{U}$

FIGURE 6.13.

To prove (2) let X be an arbitrary member of the sphere $S(\mathbf{0}; 5)$ and let r be an arbitrary positive number. We will complete the proof by finding a member of $B(X; r)$ that is not a member of $\bar{B}(\mathbf{0}; 5)$. The reader might draw a picture to see why we are now deciding to let

$$V = X + \frac{r}{10}X .$$

Let us check that $V \in B(X; r)$ and $V \notin \bar{B}(\mathbf{0}; 5)$. That $V \in B(X; r)$ follows from

$$\|V - X\| = \left\| \frac{r}{10}X \right\| = \frac{r}{10}\|X\| = \frac{r}{2} < r .$$

That $V \notin \bar{B}(\mathbf{0}; 5)$ follows from

$$\|V\| = \left\| (1 + \frac{r}{10})X \right\| = (1 + \frac{r}{10})\|X\| = 5 + (r/2) > 5 .$$

Exercises

1. Which theorems of this section permit one to conclude that the function $x_1^2 x_2 - x_2^3$ is a continuous function on $\mathbb{R}^2$?

2. Prove Theorem 6.1.3.

3. Which results in this section permit one to conclude that any polynomial, for example $g(X) = x_1 - 7x_2 + x_1 x_2 + x_1^2 x_2^7$, is continuous? Evaluate

$$\lim_{X \to (2,1)} g(X) .$$

4. Which theorems in this section enable one to deduce the continuity of $f : \mathbb{R}^n \to \mathbb{R}$ given by $f(X) = \langle X, X \rangle$? Compute

$$\lim_{X \to (2,1)} f(X) .$$

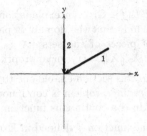

FIGURE 6.14.

5. Which theorems in this section enable one to deduce the continuity of $g : \mathbb{R}^n \to \mathbb{R}$ given by $g(X) = \langle X, L(X) \rangle$, where $L : \mathbb{R}^n \to \mathbb{R}^n$ is linear? Functions of the form of g are known as **quadratic forms** if the matrix of L is symmetric (that is, $[L] = [L]^T$); they are generalizations of functions of the form ax^2.

6. With reference to the definition in the preceding exercise, show that the quadratic form $\langle X, L(X) \rangle$ can be represented as the matrix product

$$[X]^T [L][X] .$$

(Recall that $[X]$ denotes the column matrix corresponding to X and therefore $[X]^T$ denotes the corresponding row matrix.) Strictly speaking, what is being requested is impossible; one cannot show that a matrix equals a number. But in this case the product matrix is a 1-by-1 matrix and we treat it as equal to its one entry.

7. Let

$$f(x,y) = \begin{cases} \frac{xy}{x^2+y^2} & \text{if} \quad (x,y) \neq (0,0) \\ 0 & \text{if} \quad (x,y) = (0,0) . \end{cases}$$

The theorems of this section enable one to easily conclude that f is continuous at every point different from $(0,0)$. This exercise is concerned about what happens at $(0,0)$.

   (a) Show that $f(x,0)$ is a continuous function of x and $f(0,y)$ is a continuous function of y.

   (b) Show that f is not continuous at $(0,0)$ by computing $\lim f(x,y)$ as $(x,y) \to (0,0)$ along the two paths shown in Figure 6.14 and observing that the two limits are different.

   (c) Show that if any one-dimensional linear subspace of $\mathbb{R}^2$ is regarded as the domain of f, then $\lim_{(x,y) \to (0,0)} f(X)$ exists.

8. Which affine subspaces of $\mathbb{R}^n$ are pathwise connected? Explain.

9. Complete the following sentence. The set of limit points of an affine subspace of $\mathbb{R}^n$ is _____ .

10. For appropriate $X \in \mathbb{R}^2$, let

$$f(X) = \ln \left(\frac{2 - \|X\|}{\|X\| - 1} \right) .$$

Sketch a picture of $\mathcal{D}(f)$. Give an explanation, with the aid of pictures, of how one can conclude that the domain is pathwise connected by using curves consisting of pieces of the sets $\{X : \|X\| = r\}$ for appropriate $r \in (0, \infty)$ and $\{X : X/\|X\| = U\}$ for appropriate $U \in S(0; 1)$.

11. What theorems from this section enable one to conclude that the function f defined in the preceding problem is continuous? Of course, one also has to use the fact that ln is a continuous function on its domain.

12. Find the image of the function f defined in Exercise 10.

13. For $x > 0$ and all y, one commonly used definition of x^y is

$$x^y = e^{y \ln x} .$$

(Of course, for such a definition one assumes that the natural logarithmic and natural exponential functions have already been defined—say the natural logarithmic by an integral and the natural exponential as the inverse of the natural logarithmic.) What theorems from this section, combined with the facts that the natural logarithmic and natural exponential functions are continuous, enable one to conclude that x^y is a continuous function on its domain in $\mathbb{R}^2$?

14. Let $g(x, y) = x^y$, as defined in the preceding exercise. Sketch the domain of g. Use your picture to identify the limit points of the domain and also the interior of the domain. For each limit point (u, v) of the domain decide if $\lim_{(x,y) \to (u,v)} g(x, y)$ exists, and if it does exist, evaluate it. *Hint:* Consider different cases, for example, one case is where $u = 0$ and $v > 0$.

15. Explain why 0^v is usually defined to equal 0 in case $v > 0$, defined to equal ∞, if defined at all, in case $v < 0$, and, for many settings, is not defined if $v = 0$.

16. When discussing polynomials, the definition $0^0 = 1$ is often used. Reconcile this fact with your answers to the preceding two exercises.

17. For appropriate $X \in \mathbb{R}^2$ let

$$f(X) = \frac{\sin(x_1 x_2)}{\|X\|} .$$

What is the domain of f? Is f continuous? Why or why not? What are the limit points of $\mathcal{D}(f)$? For each limit point U decide whether $\lim_{X \to U} f(X)$ exists, and, if it does exist, evaluate it.

18. Follow the instructions of the preceding exercise for the function

$$g(X) = \frac{\sin(x_1 x_2)}{x_1 + x_2} .$$

19. Find two different sets in $\mathbb{R}^1$ that have the same interior. Then do the same for $\mathbb{R}^2$ and, more generally, for $\mathbb{R}^n$.

20. Find two sets in $\mathbb{R}^2$ that have the same interior, but whose complements have different interiors.

21. What are the interior points of the interval $(0,1) \subseteq \mathbb{R}$? Also, find all the interior points of

$$\{(x,y) : 0 < x < 1\} \subseteq \mathbb{R}^2 \ .$$

22. Find a subset of $\mathbb{R}^2$ that has infinitely many members, but no interior points. Do the same for $\mathbb{R}^1$.

23. Find a subset of $\mathbb{R}^n$ that equals its own interior.

24. What is the only kind of nonempty subset of $\mathbb{R}$ that has an empty interior and is pathwise connected?

25. Fill in the blanks. Except for _____ whose interior is _____, the interior of any affine subspace of $\mathbb{R}^n$ is _____.

26. Prove Theorem 6.1.6.

27. Complete the proof of Theorem 6.1.2 by showing the following: (a) that $h \circ f$ is continuous, (b) that $1/f$ is continuous, and (c) that fg is continuous. (Part (a) is rather straightforward, but parts (b) and (c) require some ingenuity, in addition to a good understanding of the defintion of a continuous function.)

28. Find all limit points of the set

$$\{(x,y) : |x| < |y| \le 1\} \ .$$

Give a complete proof of your assertions.

29. Let $\mathcal{C}$ be a rectifiable curve in $\mathbb{R}^n$ parametrized by a one-to-one function and having endpoints U and V. For a point $X \in \mathcal{C}$ let $f(X)$ denote the arc length of that portion of $\mathcal{C}$ from U to X. Evaluate $f(U)$. What does $f(V)$ represent? Is f continuous? Explain. What is the image of f?

6.2 Directional Derivatives

6.2A Definition of $\nabla_{\mathbf{e}} f(X_0)$

Think of $f : \mathbb{R}^2 \to \mathbb{R}$ as the temperature function, so that $f(X)$ is the temperature at the point X in the plane. We ask, somewhat vaguely, what is the rate of change of the temperature as one moves from a fixed point X_0? Since the temperature changes at different rates, depending on the direction in which one moves, we refine our question and ask the following:

1. What is the rate of change of the temperature f as one moves from a given point X_0 in the direction of a given unit vector $\mathbf{e}$?

2. Given the point X_0, in what direction $\mathbf{e}$ should one move so that the temperature increases (or decreases) most rapidly?

The first question is answered here, the second later in this chapter (see the "Heat-seeking bug" in Section 6.5).

Say that we are at a fixed point X_0 and intend to move in the direction of the unit vector $\mathbf{e}$ toward $X_0 + \mathbf{e}$. See Figure 6.15. Now all points on the

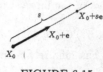

FIGURE 6.15.

line joining X_0 to $X_0 + \mathbf{e}$ are of the form $X_0 + s\mathbf{e}$, where s is a real number. For $s = 0$ we are at X_0 and for $s = 1$ we are at $X_0 + \mathbf{e}$. The difference between the temperature at X_0 and at $X_0 + s\mathbf{e}$ is $f(X_0 + s\mathbf{e}) - f(X_0)$. Thus, just as in elementary calculus, we are led to focus on a limit of a difference quotient.

Definition 6.2.1 Let f be a scalar-valued function for which $\mathcal{D}(f) \subseteq \mathbb{R}^n$. Let X_0 be an interior point of $\mathcal{D}(f)$ and let $\mathbf{e}$ be a unit vector. The **directional derivative** of f at X_0 in the direction of $\mathbf{e}$, denoted by $\nabla_{\mathbf{e}} f(X_0)$, is defined by

$$\nabla_{\mathbf{e}} f(X_0) = \lim_{s \to 0} \frac{f(X_0 + s\mathbf{e}) - f(X_0)}{s}.$$

The symbol ∇ is usually called "del". Note that the directional derivative is a number that depends on both X_0 and the direction $\mathbf{e}$.

REMARKS:

- Note how we are using unit vectors to indicate direction in $\mathbb{R}^n$. Compare this to the way a navigator's compass, for example, uses points on a circle, a "unit circle", as directions. In single-variable calculus, more emphasis is typically given to slope as a direction indicator. In $\mathbb{R}^n$, with $n \geq 3$, there is no natural generalization of slope, but there is a natural generalization of the unit circle—namely, the unit sphere. The points on the unit sphere correspond to unit vectors, and these unit vectors correspond to all possible directions in $\mathbb{R}^n$.

- Often it will be convenient to use vectors other than unit vectors to indicate direction. For instance, the vector $(3, 4)$ indicates the same direction as the unit vector $(\frac{3}{5}, \frac{4}{5})$.

6.2B COMPUTATION OF $\nabla_{\mathbf{e}} f(X_0)$

EXAMPLE:

1. Let $f : \mathbb{R}^2 \to \mathbb{R}$ be the function $f(X) = x_1^2 + x_2^2$. Let us compute the directional derivative of f at $X_0 = (2, 1)$ in the direction of the vector $Z = (3, -4)$.

We first set $\mathbf{e} = Z/\|Z\| = (\frac{3}{5}, -\frac{4}{5})$, the unit vector in the direction of Z. Then $X_0 + s\mathbf{e} = (2 + \frac{3s}{5}, 1 - \frac{4s}{5})$, so that

$$f(X_0 + s\mathbf{e}) = \left(2 + \frac{3s}{5}\right)^2 + \left(1 - \frac{4s}{5}\right)^2$$

$$= 5 + \frac{4s}{5} + s^2.$$

Since $f(X_0) = 5$, we find that $f(X_0 + s\mathbf{e}) - f(X_0) = \frac{4s}{5} + s^2$. Thus

$$\nabla_{\mathbf{e}} f(X_0) = \lim_{s \to 0} \frac{f(X_0 + s\mathbf{e}) - f(X_0)}{s}$$

$$= \lim_{s \to 0} \frac{(4s/5) + s^2}{s}$$

$$= \lim_{s \to 0} \left(\frac{4}{5} + s\right) = \frac{4}{5}.$$

If, for example, $f(X)$ represents temperature in degrees centigrade and distance is measured in feet, then the rate of increase at X_0 is $\frac{4}{5}°$ per foot. In particular, because $\nabla_{\mathbf{e}} f(X_0) = 4/5$ is positive, we conclude that if one starts at X_0 and moves in the direction of $\mathbf{e}$, then the temperature increases.

The "brute force" computation used in the preceding example consists of the following two parts:

1. Compute the difference quotient explicitly as a function of s.

2. Take the limit as s tends to zero.

Both of these parts may be complicated for complicated functions. Later on, we give a much simpler procedure, so simple that the example above becomes a mental computation.

Right now, we present an alternative view of the procedure above. Although this will clean up the computation, we mainly present it for later theoretical purposes. Let $\mathbf{e}$ be as above and define $\varphi(s) = f(X_0 + s\mathbf{e})$, so that, as s varies, $\varphi(s)$ is the temperature at $X_0 + s\mathbf{e}$. Now by the definition of the ordinary derivative,

$$\varphi'(0) = \left.\frac{d\varphi}{ds}\right|_{s=0} = \lim_{s \to 0} \frac{\varphi(s) - \varphi(0)}{s}.$$

But $\varphi(0) = f(X_0)$. Thus $\varphi(s) - \varphi(0) = f(X_0 + s\mathbf{e}) - f(X_0)$. Consequently, we find that

$$\varphi'(0) = \lim_{s \to 0} \frac{f(X_0 + s\mathbf{e}) - f(X_0)}{s},$$

that is,

$$\varphi'(0) = \nabla_{\mathbf{e}} f(X_0).$$

In other words, the directional derivative of f in the direction of the unit vector $\mathbf{e}$ is the ordinary derivative of φ evaluated at $s = 0$.

EXAMPLE:

2. In the preceding example we found that

$$\varphi(s) = f(X_0 + s\mathbf{e}) = 5 + \frac{4}{5}s + s^2 .$$

Thus

$$\varphi'(s) = \frac{4}{5} + 2s .$$

Letting $s = 0$, we have

$$\nabla_{\mathbf{e}} f(X_0) = \varphi'(0) = \frac{4}{5} ,$$

as found earlier. This is easier than the previous method but not as easy as a method to be presented later.

6.2C A GEOMETRIC INTERPRETATION

Let us find a relationship between the directional derivative of f and the graph $z = f(X)$. In order to present a picture we take $\mathcal{D}(F)$ to be a subset of $\mathbb{R}^2$ so that the graph $z = f(X)$ is a surface in $\mathbb{R}^3$. Given a point X_0 and unit vector $\mathbf{e}$ as before, we construct the line of all points $X_0 + s\mathbf{e}$ in the x_1x_2-plane. Having this, let us construct a vertical plane (parallel to the z-axis) that passes through the x_1x_2-plane along the line $X_0 + s\mathbf{e}$. See Figure 6.16. What is the intersection of this vertical plane with the graph $z = f(X)$? Clearly, it is the set of points on the graph that correspond to points on the line $\{X_0 + s\mathbf{e} : s \in \mathbb{R}\} \subseteq \mathcal{D}(f)$. Hence, this intersection is a curve that looks like the graph of the function $f(X_0 + s\mathbf{e})$, the function we called $\varphi(s)$ above.

This gives us an interpretation of the number $\nabla_{\mathbf{e}} f(X_0)$. Since it equals $\varphi'(0)$, it is the *slope* of the curve obtained by intersecting the graph $z = f(X)$ with the vertical plane through the line $X_0 + s\mathbf{e}$, computed at X_0, that is, at $s = 0$.

This interpretation as slope of a curve underscores the fact that the directional derivative is very much a single-variable notion. You should convince yourself—it isn't difficult—that $\nabla_{\mathbf{e}} f(X_0)$ depends on the direction $\mathbf{e}$ and is typically different for different directions $\mathbf{e}$.

EXAMPLE:

3. Again we return to the function treated in Examples 1 and 2: $f(X) = x_1^2 + x_2^2$. The graph of this is the surface $z = x_1^2 + x_2^2$, a paraboloid. At $X_0 = (2, 1)$, the height of this surface is $f(X_0) = 5$. The intersection

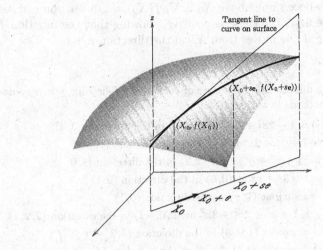

FIGURE 6.16.

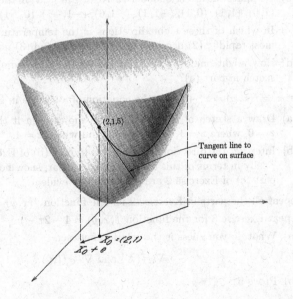

FIGURE 6.17.

of the vertical plane along the line $X_0 + s\mathbf{e}$ is a curve—in fact, it is a parabola—whose slope above X_0 is $\nabla_{\mathbf{e}} f(X_0) = 4/5$. As you can see from Figure 6.17, the slope is positive, showing that the function f is increasing if one moves from X_0 in the direction $\mathbf{e}$.

Exercises

1. Compute the directional derivatives of each of the following; you may use any of the methods in this section.

   (a) $f(x_1, x_2) = 1 - 2x_1 + 3x_2$ at $(2, -1)$ in the direction $(3, 4)$
   (b) $q(x, y) = 7 + x^2 - 2y^2$ at $(1, 1)$ in the direction $(1, -1)$
   (c) $h(x, y) = (1 + x^2 + y^2)^{-1}$ at $(1, 2)$ in the direction $(1, 0)$
   (d) $f(r, s) = 1 - 3r + rs$ at $(0, 6)$ in the direction $(0, 1)$
   (e) $k(x, y) = x \sin y$ at $(1, \pi)$ in the direction $(0, 1)$
   (f) $\varphi(x_1, x_2, x_3) = x_1^2 + 2x_2^2 + 3x_3^2$ at $(0, 0, -1)$ in the direction $(2, 2, 1)$
   (g) $u(x, y, z) = xyz$ at $(1, 0, 3)$ in the direction $(3, 2, -6)$
   (h) $p(x, y, z) = xye^{y+z}$ at $(1, 3, -3)$ in the direction $(4, -2, 4)$.

2. Let $z = f(x, y) = x^2 + y^2$ be the temperature at a point $(x, y) \in \mathbb{R}^2$.

   (a) Draw a sketch indicating the points in $\mathbb{R}^2$ where $z = 0$, where $z = 1$, where $z = 2$, and where $z = 4$.

   (b) Compute the directional derivative of f at $(1, 1)$ in the eight directions $(1, 0)$, $(1, 1)$, $(0, 1)$, $(-1, 1)$, $(-1, 0)$, $(-1, -1)$, $(0, -1)$, and $(1, -1)$.

   (c) In which of these eight directions is the temperature (1) increasing most rapidly, (2) decreasing most rapidly, and (3) not changing?

   (d) How might one have deduced the result in preceding part from the sketch in part (a)?

3. Let $z = f(x, y) = x^2 + y^2$ be the equation of a surface in $\mathbb{R}^3$.

   (a) Draw a sketch of this surface, and indicate on it the places where $z = 0$, where $z = 1$, where $z = 2$, and where $z = 4$.

   (b) Interpret the directional derivatives in part (b) of Exercise 2 geometrically in terms of this surface. In particular, show how the results in part (c) of Exercise 2 are geometrically evident.

4. Repeat all the parts in Exercise 2 for the function $f(x, y) = 4 - 2x - y$.

5. Repeat Exercise 3 for the function $f(x, y) = 4 - 2x - y$.

6. (a) What do you guess is the relation between

$$\nabla_{\mathbf{e}} f(X) \quad \text{and} \quad \nabla_{-\mathbf{e}} f(X) \text{ ?}$$

   (b) Prove it.

7. Let

$$f(X) = \|X\|, \quad X \in \mathbb{R}^q.$$

   For $X \neq 0$ and $\mathbf{e}$ any unit vector, show that

$$\nabla_{\mathbf{e}} f(X) = \frac{\langle X, \mathbf{e} \rangle}{\|X\|}.$$

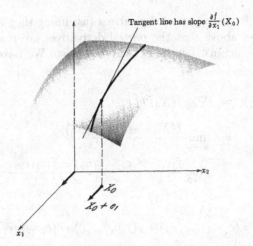

Tangent line has slope $\frac{\partial f}{\partial x_1}(X_0)$

FIGURE 6.18.

6.3 Partial Derivatives

6.3A INTRODUCTORY COMMENT

In this section we define and compute partial derivatives. Partial derivatives are special directional derivatives which are particularly easy to calculate and useful when calculating general directional derivatives.

A picture like that of Figure 6.17 is appropriate, but now the vertical plane must be parallel to one of the coordinate axes of the domain. See Figure 6.18.

6.3B DEFINITION AND COMPUTATION

As in the preceding section, fix an interior point X_0 of $\mathcal{D}(f)$. Recall the standard basis vectors $\mathbf{e}_1, \ldots, \mathbf{e}_n$ of $\mathbb{R}^n$; they are given by

$$\mathbf{e}_1 = (1, 0, \ldots, 0), \quad \mathbf{e}_2 = (0, 1, 0, \ldots, 0), \quad \ldots, \quad \mathbf{e}_n = (0, \ldots, 0, 1).$$

Clearly each of these is a unit vector. Hence, we may use any of them to define a directional derivative.

For instance, let us form the directional derivative $\nabla_{\mathbf{e}_1} f(X_0)$. This is commonly called the **partial derivative** of f with respect to x_1 at the point X_0 and denoted $\frac{\partial f}{\partial x_1}(X_0)$. It is a real number, provided that $\nabla_{\mathbf{e}_1} f(X_0)$ exists. In Figure 6.18, $\nabla_{\mathbf{e}_1} f(X_0) < 0$.

More generally, we define the **partial derivative** of f with respect to x_j at the point X_0 by

$$\frac{\partial f}{\partial x_j}(X_0) = \nabla_{\mathbf{e}_j} f(X_0), \qquad j = 1, \cdots, n.$$

Thus $f : \mathbb{R}^n \to \mathbb{R}$ has n partial derivatives (assuming they all exist).

We mentioned above that the partial derivatives are readily computed using ordinary calculus. Let us see why this is so. We have, for instance, when $X_0 = (c_1, c_2, \ldots, c_n)$,

$$
\begin{aligned}
\frac{\partial f}{\partial x_1}(X_0) &= \nabla_{\mathbf{e}_1} f(X_0) \\
&= \lim_{s \to 0} \frac{f(X_0 + s\mathbf{e}_1) - f(X_0)}{s} \\
&= \lim_{s \to 0} \frac{f(c_1 + s, c_2, \ldots, c_n) - f(c_1, c_2, \ldots, c_n)}{s},
\end{aligned}
$$

since

$$
X_0 + s\mathbf{e}_1 = (c_1, c_2, \ldots, c_n) + s(1, 0, \ldots, 0) = (c_1 + s, c_2, \ldots, c_n).
$$

Let us examine the last limit. Suppose that we form the function $f(x_1, c_2, \ldots, c_n)$ of the single variable x_1; each of the other coordinates slots has been filled by a fixed number $c_2, \ldots, c_n$. Examine the difference quotient appearing in the above limit, namely

$$
\frac{f(c_1 + s, c_2, \ldots, c_n) - f(c_1, c_2, \ldots, c_n)}{s}.
$$

Note that the increment s appears only in the *first* coordinate slot of the function f (in contrast with a general directional derivative). Hence f is being differentiated in its first variable only. Thus, in order to compute $\partial f/\partial x_1(X_0)$, we may proceed as follows (note that $X_0 = (c_1, c_2, \ldots, c_n)$):

1. form the function $f(x_1, c_2, \ldots, c_n)$ of the single variable x_1;

2. differentiate this function with respect to x_1 by the ordinary rules of single-variable calculus;

3. let $x_1 = c_1$ in this derivative to obtain a number $\frac{\partial f}{\partial x_1}(X_0)$.

REMARKS:

- It is clear that the other partial derivatives

$$
\frac{\partial f}{\partial x_2}(X_0), \ldots, \frac{\partial f}{\partial x_n}(X_0)
$$

are obtained by an entirely similar procedure. Observe, however, that we cannot compute arbitrary directional derivatives using this three-step procedure.

- The following alternate notations occur frequently:

$$
\frac{\partial f}{\partial x_j}(X) = f_{x_j}(X) = (D_j f)(X) = D_j f(X) = D_{x_j} f(X) = f_j(X).
$$

Each of these has its virtues and drawbacks. Roughly speaking the various notations can be separated into the following two classes: one consisting of those that focus on the name of the variable with respect to which one is differentiating, especially useful in physical situations where certain letters have conventional meanings (for instance, P for pressure), and a second class consisting of those that focus on the numerical position of the variable with respect to which one is differentiating, thus avoiding ambiguity in situations in which either some particular symbol is being used in different positions or different symbols are being used in similar ways.

- When we use $f(x, y, z)$ instead of $f(x_1, x_2, x_3)$ we write

$$\frac{\partial f}{\partial x}(x, y, z) = f_x(x, y, z),$$

with similar meanings for f_y and f_z.

EXAMPLES:

1. Let $f : \mathbb{R}^2 \to \mathbb{R}$ be given by $f(X) = x_1^2 + 3x_1 x_2 + x_2^3$, and let $X_0 = (c_1, c_2) \in \mathbb{R}^2$. We compute $\frac{\partial f}{\partial x_1}(X_0)$ as follows:

   (a) form $f(x_1, c_2) = x_1^2 + 3x_1 c_2 + c_2^3$, a function of x_1 alone;
   (b) differentiate with respect to x_1, obtaining $2x_1 + 3c_2$;
   (c) let $x_1 = c_1$, obtaining $\frac{\partial f}{\partial x_1}(X_0) = 2c_1 + 3c_2$.

   For instance, if $X_0 = (1, -2)$ then $\frac{\partial f}{\partial x_1}(X_0) = -4$.

   In practice, one does not explicitly introduce $X_0 = (c_1, c_2)$. Instead, in computing $\frac{\partial f}{\partial x_1}(X)$, one thinks of $x_2, \ldots, x_n$ as constants. Thus we would think of x_2 as a constant and compute

$$\frac{\partial f}{\partial x_1}(X) = 2x_1 + 3x_2,$$

   obtaining the same formula as above by essentially the same procedure.

2. Now we compute $(D_2 f)(X)$ for the function of the preceding example. This time we must think of x_1 as a constant. We find

$$(D_2 f)(X) = 3x_1 + 3x_2^2 .$$

   In particular, $(D_2 f)(1, -2) = 15$.

3. Let $g : \mathbb{R}^3 \to \mathbb{R}$ be given by $g(X) = g(x, y, z) = x \sin yz$. We compute g_x by thinking of y and z as constants. Then

$$g_x(x, y, z) = \sin yz.$$

Thus, $g_x(P_0) = 1$ for $P_0 = (5, \pi, \frac{1}{2})$. Similarly we compute g_y by thinking of x and z as constants:

$$g_y(x, y, z) = xz \cos yz,$$

and so $g_y(P_0) = 0$. We leave g_z to you.

4. *Ideal gases.* Recall from physics or chemistry the following formula relating pressure P, volume V, and temperature T of an ideal gas enclosed in a cylinder: $PV = nRT$. (In this formula, T is temperature above absolute zero, and hence is positive.) The numbers n and R are physical constants, fixed throughout our discussion. From the formula, we conclude that changes in any of the variables may cause changes in the other variables. For example, if the temperature T is held constant while the volume V is increased, then the pressure P must decrease, so that PV continues to equal nRT.

Let us examine this situation, using calculus. First, we may write the pressure P as a function $P(V, T)$ of volume V and temperature T as follows:

$$P = P(V, T) = \frac{nRT}{V}.$$

The notation $P(V, T)$ means that pressure depends on volume and temperature. Now we may ask: What is the rate of change of pressure as the volume is increased, assuming that the temperature is held fixed at some $T_0 > 0$? This is a question about partial derivatives. We have

$$\frac{\partial P}{\partial V}(V, T_0) = \frac{\partial}{\partial V}\left(\frac{nRT_0}{V}\right) = \frac{-nRT_0}{V^2},$$

where we have differentiated nRT_0/V with respect to V. We interpret this computation as follows. Since n, R, T_0, and V^2 are positive, $\frac{\partial P}{\partial V}(V, T_0)$ is negative for each V. Hence, the rate of change of pressure with respect to volume (at a fixed temperature) is negative. That is, *pressure decreases as volume increases*, just as we expected.

Likewise, you may show that, for a fixed volume V_0, pressure increases as temperature increases by computing the rate of change $\frac{\partial P}{\partial T}(V_0, T)$ and verifying that it is positive.

6.3C   HIGHER-ORDER PARTIAL DERIVATIVES

If $f(x)$ is a differentiable function of one variable, then its first derivative $f'(x)$ may also be a differentiable function. Hence we are led to consider the second derivative $f''(x)$ and derivatives of higher order $f'''(x)$, $f^{(4)}(x), \ldots, f^{(k)}(x), \ldots$. A similar situation occurs with partial derivatives. Let $f : \mathbb{R}^2 \to \mathbb{R}$. The partial derivatives $\frac{\partial f}{\partial x_1}(X)$ and $\frac{\partial f}{\partial x_2}(X)$, called first partial derivatives for emphasis, are themselves functions of X. As such these partial derivatives may themselves have two partial derivatives each. The partial derivatives of the partial derivatives of f are the second partial derivatives of f. Here is notation for the four second partial derivatives of f:

$$\frac{\partial}{\partial x_1}\left(\frac{\partial f}{\partial x_1}\right), \quad \frac{\partial}{\partial x_2}\left(\frac{\partial f}{\partial x_1}\right), \quad \frac{\partial}{\partial x_1}\left(\frac{\partial f}{\partial x_2}\right), \quad \frac{\partial}{\partial x_2}\left(\frac{\partial f}{\partial x_2}\right).$$

EXAMPLES:

5. We compute the second partial derivatives of the function $f(x_1, x_2) = x_1^2 + x_2^5 e^{-x_1} + 9$, at the same time introducing several useful alternate pieces of notation:

$$\frac{\partial}{\partial x_2}\left(\frac{\partial f}{\partial x_1}\right)(X) = f_{12}(X) = (D_{12}f)(X) = -5x_2^4 e^{-x_1} ;$$

$$\frac{\partial^2 f}{\partial x_1^2}(X) = f_{11}(X) = f_{x_1 x_1}(X) = 2 + x_2^5 e^{-x_1} ;$$

$$\frac{\partial^2 f}{\partial x_2^2}(X) = D_{22}f(X) = 20x_2^3 e^{-x_1} ;$$

$$\frac{\partial}{\partial x_1}\left(\frac{\partial f}{\partial x_2}\right) = f_{x_2 x_1} = \frac{\partial^2 f}{\partial x_1 \partial x_2} = -5x_2^4 e^{-x_1}$$

6. Let $f(x_1, x_2) = 3x_1^2 + x_1 x_2^2$. Then

$$\frac{\partial f}{\partial x_1}(x_1, x_2) = 6x_1 + x_2^2 .$$

Thus

$$\frac{\partial}{\partial x_1}\left(\frac{\partial f}{\partial x_1}\right)(x_1, x_2) = \frac{\partial}{\partial x_1}(6x_1 + x_2^2) = 6 .$$

Also,

$$(D_{12}f)(X) = (D_2(D_1 f))(X) = \frac{\partial}{\partial x_2}(6x_1 + x_2^2) = 2x_2 ,$$

where, as usual, x_1 is regarded as a constant when the differentiation with respect to x_2 is carried out.

7. Let $g(x, y) = x^2 y^3 + x \cos y$. Then $g_x = 2xy^3 + \cos y$ and $g_y = 3x^2 y^2 - x \sin y$. Therefore,

$$(D_{11}g)(x, y) = g_{xx} = \frac{\partial}{\partial x}\left(\frac{\partial g}{\partial x}\right) = 2y^3 ,$$

$$(D_{12}g)(x, y) = g_{xy} = \frac{\partial}{\partial y}\left(\frac{\partial g}{\partial x}\right) = 6xy^2 - \sin y ,$$

$$(D_{21}g)(x, y) = g_{yx} = \frac{\partial}{\partial x}\left(\frac{\partial g}{\partial y}\right) = 6xy^2 - \sin y ,$$

and g_{yy} is left to you.

8. For the function $f(x_1, x_2) = \sin(x_1 x_2)$ we will calculate some third and fourth derivatives. Then

$$f_{x_1} = x_2 \cos(x_1 x_2)$$

and

$$f_{x_1 x_1} = -x_2^2 \sin(x_1 x_2) \, ,$$

so that

$$
\begin{aligned}
f_{112}(X) &= \frac{\partial^3 f}{\partial x_2 \partial x_1^2}(X) = \frac{\partial}{\partial x_2}\left(\frac{\partial^2 f}{\partial x_1^2}\right)(X) \\
&= \frac{\partial}{\partial x_2}(-x_2^2 \sin(x_1 x_2)) \\
&= -2x_2 \sin(x_1 x_2) - x_1 x_2^2 \cos(x_1 x_2) \, ;
\end{aligned}
$$

$$
\begin{aligned}
D_{1121}f(X) &= D_1 D_{112} f(X) \\
&= D_1(-2x_2 \sin(x_1 x_2) - x_1 x_2^2 \cos(x_1 x_2)) \\
&= -3x_2^2 \cos x_1 x_2 + x_1 x_2^3 \sin x_1 x_2 \, ;
\end{aligned}
$$

$$
\begin{aligned}
f_{x_1 x_1 x_1} &= \frac{\partial^3 f}{\partial x_1^3} = \frac{\partial}{\partial x_1}(-x_2^2 \sin(x_1 x_2)) \\
&= -x_2^3 \cos(x_1 x_2) \, ;
\end{aligned}
$$

$$
\begin{aligned}
f_{1112}(X) &= \frac{\partial}{\partial x_2}(-x_2^3 \cos(x_1 x_2)) \\
&= -3x_2^2 \cos(x_1 x_2) + x_1 x_2^3 \sin(x_1 x_2) \, .
\end{aligned}
$$

In this collection of examples, one striking phenomenon may have attracted your attention. Whenever we computed $f_{x_1 x_2}$ and $f_{x_2 x_1}$ (or g_{xy} and g_{yx}) these "mixed partial derivatives" happen to be equal. Is this a coincidence, or is there some theorem lurking in the background? There is a theorem; it says that usually $D_{ij} = D_{ji}$, that is, that partial differentiation with respect to different variables commutes except in rather pathological circumstances.

Theorem 6.3.1 (Equality of mixed partials) Let $f : \mathbb{R}^n \to \mathbb{R}$ and suppose that the second partial derivatives $D_{ij}f$ and $D_{ji}f$ exist and are continuous on some open ball centered at a point X. Then $(D_{ij}f)(X) = (D_{ji}f)(X)$.

REMARKS:

- We do not offer a proof of this theorem. The task for the proof is to show that the order of two limits which define the partial derivatives with respect to the two variables can be interchanged. Thus, the proof is somewhat technical, although not beyond our means. The basic tool is the Mean Value Theorem from single-variable differential calculus.

- In most situations $D_{ij}f$ and $D_{ji}f$ are continuous functions on the set of interior points of $\mathcal{D}(f)$, and thus they are the same function on that set.

- There are examples of functions whose mixed second partial derivatives fail to be continuous at a point and are, in fact, unequal at that point. See Exercise 15.

- Interchanging the order of differentiation enables us to compactify some notation. Thus, if f is suitably differentiable,

$$f_{yxy} = \frac{\partial^3 f}{\partial y \partial x \partial y}$$

may be rewritten as

$$\frac{\partial^3 f}{\partial x \partial y^2}, \qquad \frac{\partial^3 f}{\partial y^2 \partial x}, \qquad f_{xyy}, \qquad \text{or} \quad f_{yyx}.$$

Here we differentiate with respect to y twice and then with respect to x (or x first and then y twice, as suits us).

- In general, interchanging the order of partial differentiation allows us to do all differentiations with respect to one variable before going on to another variable. A typical kth-order derivative would be written

$$\frac{\partial^k f}{\partial x_n^{k_n} \cdots \partial x_1^{k_1}} (x_1, \ldots, x_n),$$

where $k_1 + k_2 + \cdots + k_n = k$. Here we would differentiate k_1 times with respect to x_1, then k_2 times with respect to x_2, and so on. Of course, not all of the k_i need be positive; so, for instance, if $k_2 = 0$, then there are no differentiations with respect to x_2.

6.3D WHERE WE ARE

In discussing the differential calculus of a scalar-valued function $f : \mathbb{R} \to \mathbb{R}$, we have made progress along certain lines.

1. We have defined the directional derivative $\nabla_{\mathbf{e}} f(X_0)$, although we do not yet have a rapid method of computing it.

2. We have defined the partial derivatives $\frac{\partial f}{\partial x_1}(X_0), \ldots, \frac{\partial f}{\partial x_n}(X_0)$, and we can compute them readily.

Still open are the following concerns:

1. an easy way to compute directional derivatives;

2. construction of a single object analogous to the first derivative of a function in single-variable calculus—this is the "total derivative" f', a very handy object that carries *all* the information about the various partial directional derivatives.

3. construction of the best affine approximation $T(X)$ to $f(X)$ at a point.

4. construction of the tangent hyperplane to the graph of f at a point on the graph.

We will see in the following section that all these problems yield to the same attack.

Exercises

1. Compute all the first partial derivatives for the following functions:

   (a) $f(x_1, x_2) = x_1 x_2^2 + 2x_1 - 7$

   (b) $g(x_1, x_2) = x_1 + x_2 \sin 2x_1$

   (c) $h(x, y) = xe^{x+2y} + y^2$

   (d) $f(s, t) = (1 + 2s^2 + t^2)$

   (e) $g(\theta, \varphi) = \cos(\theta - 3\varphi)$

   (f) $f(x, y) = x^y, \quad x > 0$

   (g) $g(x_1, x_2, x_3) = x_1 x_2 x_3$

   (h) $u(x, y, z) = xye^{y+z} + 5$

   (i) $v(x, y, z) = \ln(x^2 + y^2 + z^2 + 3)$

   (j) $p(r, \theta, \varphi) = r \sin\theta \cos\varphi$.

2. Compute all the second partial derivatives of the functions in Exercise 1.

3. Compute all the third partial derivatives for the functions in parts (a), (b), (g), and (h) of Exercise 1.

4. Find an affine function $f : \mathbb{R}^2 \to \mathbb{R}$ such that $f(-1, 2) = 4$ and, for some (x, y), $f_1(x, y) = -1$ and $f_2(x, y) = 3$.

5. Find an affine function $f : \mathbb{R}^2 \to \mathbb{R}$ such that $f(-1, 2) = 4$ and, for all (x, y), $(D_1 f)(x, y) = -1$ and $(D_2 f)(x, y) = 3$.

6. Let $f(X) = \|X\|$ for $X \in \mathbb{R}^2$.

   (a) Show that
   $$\langle X, ((D_1 f)(X), (D_2 f)(X)) \rangle = f(X)$$
   for $X \neq (0, 0)$.

   (b) Show that $((D_1 f)(X), (D_2 f)(X))$ is a (possibly variable) scalar multiple of X for $X \neq (0, 0)$.

7. Let $f(X) = \|X\|$ for $X \in \mathbb{R}^n$. Generalize Exercise 6.

8. Let $\varphi(s, t) = ste^{st}$. Show that $s\varphi_s - t\varphi_t = 0$.

9. Show that the following functions satisfy the equation $u_{xx}+u_{yy}=0$, known as the two-dimensional **Laplace equation**:

   (a) $u(x,y) = x^2 - y^2 - 3xy + 5x - 6$

   (b) $u(x,y) = \ln(x^2 + y^2)$, $(x,y) \neq (0,0)$

   (c) $u(x,y) = e^x \sin y$.

10. Show that the following functions satisfy the equation $u_{tt} = u_{xx} + u_{yy}$, known as the two-space-dimensional **wave equation**:

   (a) $u(x,y,t) = e^{x-t} - 3e^{y+t} + 7xy - t + 13$

   (b) $u(x,y,t) = (3x - 4y + 5t)^\alpha$, α a constant.

11. Pick constants α and β so that $u(x,y) = e^{\alpha x + \beta y}$ satisfies

$$u_{xx} - 5u_{xy} + 6u_{yy} = 0.$$

12. Let A be a square 2-by-2 matrix that equals its own transpose. Let $f: \mathbb{R}^2 \to \mathbb{R}$ be the quadratic form (see Exercises 5 and 6 in Section 6.1) defined by $f(X) = [X]^T A[X]$. Show that

$$(D_1 f)(X) = 2[X]^T A[e_1] \quad \text{and} \quad (D_2 f)(X) = 2[X]^T A[e_2]$$

for all X. Note that these formulas generalize the formula for the derivative of $f(x) = ax^2$.

13. Generalize the preceding exercise to quadratic forms whose domain is $\mathbb{R}^n$.

14. This exercise provides an example of a function $f : \mathbb{R}^2 \to \mathbb{R}$ such that all first partial derivatives exist everywhere, but f is discontinuous. Let

$$f(x,y) = \begin{cases} \frac{2xy^2}{x^2+y^4} & \text{if } (x,y) \neq (0,0) \\ (0,0) & \text{if } (x,y) = (0,0). \end{cases}$$

   (a) Calculate $(D_1 f)(x,y)$ and $(D_2 f)(x,y)$ at all points (x,y) different from $(0,0)$.

   (b) Calculate $(D_1 f)(0,0)$ and $(D_2 f)(0,0)$ directly from the definition of derivative.

   (c) Show that f is discontinuous at $(0,0)$ by showing $\lim_{y\to 0} f(y^2, y) \neq 0$.

15. This exercise provides an example of a function $f : \mathbb{R}^2 \to \mathbb{R}$ such that all first and second partial derivatives exist everywhere, but the function $D_{12}f$ and $D_{21}f$ are different. Let

$$f(x,y) = \begin{cases} \frac{xy(x^2-y^2)}{x^2+y^2} & \text{if } (x,y) \neq (0,0) \\ 0 & \text{if } (x,y) = (0,0). \end{cases}$$

   (a) Calculate $(D_1 f)(x,y)$ and $(D_2 f)(x,y)$ for $(x,y) \neq (0,0)$.

   (b) Use the definition of $(D_1 f)(0,0)$ and $(D_2 f)(0,0)$ as limits to show that they exist and to evaluate them.

   (c) Use the preceding two parts to show that $(D_1 f)(0,y) = -y$ for all y and $(D_2 f)(x,0) = x$ for all x.

(d) Show that $(D_{12}f)(0, y) = -1$ for all y and $(D_{21}f)(x, 0) = 1$ for all x.

(e) How can one now conclude that $(D_{12}f)(0,0) \neq (D_{21}f)(0,0)$?

16. Let f be as defined in the preceding exercise.

(a) Find where each of the two first partial derivatives of f equals 0.

(b) Sketch your answers to the preceding part.

(c) Color code your sketch as follows: one color for the places where both first partial derivatives are positive, a second color for the places where both are negative, a third color for the places where $D_1 f$ is positive and $D_2 f$ is negative, and a fourth color for the places where $D_2 f$ is positive and $D_1 f$ is negative.

17. Let f be as in Exercise 15. Calculate $(D_{12}f)(x, y)$ and $(D_{21}f)(x, y)$ for $(x, y) \neq (0, 0)$.

18. (a) If $u(x, t)$ denotes the displacement, say in centimeters, of a vibrating string at a point x on the string at time t, how would you physically interpret the functions $u_t(x, t)$, $u_{tt}(x, t)$, and $u_x(x, t)$?

(b) Suppose that $u(x, t) = 3 \sin 2x \cos 2\pi t$ is the displacement of a vibrating string of length π stretched between $x = 0$ and $x = \pi$. What can you deduce about the endpoints of the string? What is the initial (at $t = 0$) position of the string? What is the initial velocity of the string? What is the velocity at $x = \pi/4$ for $t = 3$? What is the slope of the string at $x = \pi/4$ for $t = 3$? Draw sketches showing the position of the string at $t = 0$, at $t = 1/4$, at $t = 1/2$, and at $t = 3$.

6.4   Tangency and Affine Approximation

6.4A   INTRODUCTORY EXAMPLE

We have pictured, in Figure 6.19, the graph of the function $f(x_1, x_2) = 1 - x_1^2 - x_2^2$; the graph is a subset of $\mathbb{R}^3$. Above a point $(x_1, x_2, 0)$ we have drawn what looks like the tangent plane to the surface $z = f(X)$. This tangent plane is the graph of an affine function $z = T(X)$. We do not yet know *which* affine function this is. In this section we make a reasonable definition of $T(X)$ and also identify it as the best affine approximation to $f(X)$ at the point X_0. The definition we will give is similar to the corresponding definition in Chapter 4. It will apply to scalar-valued functions of $X = (x_1, \ldots, x_n)$, where n may be larger than two; however, most of our examples and discussion takes place when the domain is a subset of $\mathbb{R}^2$ so that the graph is a subset of $\mathbb{R}^3$.

Even after defining $z = T(X)$ by the appropriate limit requirement, we will *not* yet be able to compute it in a variety of particular cases. The theorems which tell us how to make explicit computations of the best affine approximation are in Section 6.5.

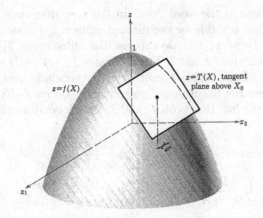

$z = f(X)$

$z = T(X)$, tangent plane above X_0

FIGURE 6.19.

Why do we want the best affine approximation $T(X)$? One reason is this: Since $T(X)$ is affine, its graph $z = T(X)$ is a plane. If we were small enough and lived on the curved surface $z = f(X)$ near the point $(X_0, f(X_0))$, we would not be able to distinguish this surface from the "flat" plane $z = T(X)$. (The earth is flat, as every child knows.) But the function $T(X)$, being affine, is much easier to handle than $f(X)$; we have linear algebra working for us in the affine case.

6.4B   BASIC DEFINITION

Definition 6.4.1 Let f be a scalar-valued function for which $\mathcal{D}(f) \subseteq \mathbb{R}^n$ and let X_0 be an interior point of $\mathcal{D}(f)$. The function f is **differentiable at** X_0 if and only if there exists an affine map $T : \mathbb{R}^n \to \mathbb{R}$ such that:

1. $f(X_0) = T(X_0)$;

2. $\lim_{X \to X_0} \dfrac{f(X) - T(X)}{\|X - X_0\|} = 0$.

If there is such a T, it is the **best affine approximation** to or **tangent map** of f at X_0, and its graph is the **tangent hyperplane** to the graph of f at the point $(X_0, f(X_0))$. The linear map corresponding to T is called the **derivative** or **total derivative** or **first derivative** of f at X_0 and is denoted by $f'(X_0)$.

REMARKS:

- The term "tangent hyperplane" is used in the preceding definition because n is arbitrary. In the important case $n = 2$ the term **tangent plane** suffices, and in case $n = 1$ the familiar term "tangent line" is used.

- Certain uses of the word "the" in the preceding definition indicate that it is not possible for two distinct affine maps to satisfy conditions 1 and 2. To prove this we suppose the affine maps T_1 and T_2 both satisfy conditions 1 and 2. The difference $T_3 = T_1 - T_2$ is affine. Since T_1 and T_2 both satisfy condition 1, we conclude that $T_3(X_0) = 0$. Thus, there is a linear map L_3 such that $T_3(X) = L_3(X - X_0)$ for all X. From the fact that both T_1 and T_2 satisfy condition 2, we conclude that

$$\lim_{X \to X_0} \frac{L_3(X - X_0)}{\|X - X_0\|} = \lim_{X \to X_0} \frac{T_1(X) - T_2(X)}{\|X - X_0\|}$$

$$= \lim_{X \to X_0} \frac{T_1(X) - f(X)}{\|X - X_0\|} + \lim_{X \to X_0} \frac{f(X) - T_2(X)}{\|X - X_0\|} = 0 + 0 = 0 \, .$$

Let V denote some fixed, but arbitrary, nonzero vector, and let $X = X_0 + \varepsilon V$ for $\varepsilon \in \mathbb{R}$. If $\varepsilon \to 0$, then $X \to X_0$. Since $X - X_0 = \varepsilon V$, we conclude from the preceding calculation that

$$\lim_{\varepsilon \to 0} \frac{L_3(\varepsilon V)}{\|\varepsilon V\|} = 0.$$

Using a basic fact about linear maps in the numerator and a basic fact about norms in the denominator, we obtain

$$\lim_{\varepsilon \to 0} \frac{\varepsilon L_3(V)}{|\varepsilon| \|V\|} = 0.$$

Since $\varepsilon / |\varepsilon| = \pm 1$, it follows that $L_3(V) = 0$. Since V was arbitrary, we conclude that L_3 is the zero map and, therefore, that $T_1 = T_2$, as desired.

- Notice that in condition 2 in the definition there is a division by $\|X - X_0\|$, a scalar. Division by the vector $X - X_0$ is not meaningful. We *never* divide by vectors.

- The total derivative $f'(X_0)$ is defined to be the linear map corresponding to the affine map T. Thus

$$T(X) = T(X_0) + f'(X_0)(X - X_0) = f(X_0) + f'(X_0)(X - X_0) \, ,$$

where $f'(X_0)(X - X_0)$ denotes the result of applying the linear map $f'(X_0)$ to the vector $X - X_0$. To see that this is analogous to an important single-variable calculus fact consider the case where $n = 1$ so that X is actually a scalar x. Then the above formula becomes

$$T(x) = f(x_0) + f'(x_0)(x - x_0) \, ,$$

where now $f'(x_0)(x - x_0)$ can be viewed as the product of the two numbers $f'(x_0)$ and $x - x_0$. The graph of T in this case is familiar; it is the line in $\mathbb{R}^2$ that is tangent to the graph of f at the point $(x_0, f(x_0))$.

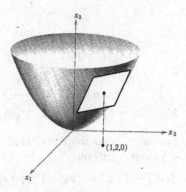

FIGURE 6.20.

- The value $n = 1$ is permitted in Definition 6.4.1. This is a bit troubling since we have long believed that the derivative at a particular point of a function from $\mathbb{R}$ to $\mathbb{R}$ is a number, but now we have a definition that says it is a linear map. However, the difference between the two definitions is one of semantics, not of content. If $f : \mathbb{R} \to \mathbb{R}$, then the matrix $[f'(x_0)]$ of the derivative according to Definition 6.4.1 is a 1-by-1 matrix with one entry. That entry is the derivative as typically defined in single-variable calculus books.

EXAMPLES:

1. Let $f(X) = x_1^2 + x_2^2$. The graph of f is shown in Figure 6.20, along with a tangent plane at the point $(1, 2, f(1, 2)) = (1, 2, 5)$. Let $X_0 = (1, 2) \in \mathbb{R}^2$. Consider the map $T(X) = T(x_1, x_2) = 5 + 2(x_1 - 1) + 4(x_2 - 2)$. We claim that this map is the best affine approximation to f at X_0. (You may well ask where we obtained this $T(X)$. This is another story, to be told in the next section.)

   We must verify the criteria for best affine approximation. We show that T is affine by writing it as

   $$T(X) = L(X) + (T(X_0) - L(X_0)) = T(X_0) + L(X - X_0),$$

   with $T(X_0) = 5$, $X - X_0 = (x_1 - 1, x_2 - 2)$, and $L(X - X_0) = 2(x_1 - 1) + 4(x_2 - 2)$, and observing that L *is linear* with matrix $[2 \ 4]$. Now we observe that $T(X_0) = 5 = f(X_0)$, so that condition 1 of the definition is satisfied.

   To check condition 2, namely, $\lim_{X \to X_0} \frac{f(X) - T(X)}{\|X - X_0\|} = 0$, we first put $f(X)$ into a more useful form with respect to the point $X_0 = (1, 2)$, namely, we rewrite it in powers of $x_1 - 1$ and $x_2 - 2$. We have

   $$x_1 = (x_1 - 1) + 1 \quad \text{and} \quad x_2 = (x_2 - 2) + 2,$$

so that

$$
\begin{aligned}
f(X) &= [(x_1 - 1) + 1]^2 + [(x_2 - 2) + 2]^2 \\
&= [(x_1 - 1)^2 + 2(x_1 - 1) + 1] \\
&\qquad + [(x_2 - 2)^2 + 4(x_2 - 2) + 4] \\
&= 5 + 2(x_1 - 1) + 4(x_2 - 2) + (x_1 - 1)^2 + (x_2 - 2)^2.
\end{aligned}
$$

Although this expression looks more cluttered, it is in a form suitable for simplification in our next step.

Now note that $f(X) - T(X) = (x_1 - 1)^2 + (x_2 - 2)^2$. The limit has become

$$
\lim_{X \to X_0} \frac{[(x_1 - 1)^2 + (x_2 - 2)^2]}{\|X - X_0\|} = \lim_{X \to X_0} \frac{\|X - X_0\|^2}{\|X - X_0\|}
$$
$$
= \lim_{X \to X_0} \|X - X_0\| = 0 \ .
$$

Since conditions 1 and 2 have been verified, we conclude that $T(X) = 5 + 2(x_1 - 1) + 4(x_2 - 2)$ is the best affine approximation to $f(X)$ at the point $X_0 = (1, 2)$. We also conclude that the map L equals the map $f'(X_0)$, the total derivative. Thus $f'(X_0)$ is the linear map—actually a linear functional—from $\mathbb{R}^2$ to $\mathbb{R}$ whose matrix is [2 4], that is,

$$
[f'(X_0)] = [2 \ 4] \ .
$$

2. Let us check some calculations against our geometric intuition for the function $f(X) = 1 - x_1^2 - x_2^2$. From the graph in Figure 6.21, we expect that its best affine approximation at the point $X_0 = (0, 0)$ is the constant map $z = T(X) = 1$ whose graph is the horizontal plane through the point $(0, 0, 1)$ of $\mathbb{R}^3$ with coordinate axes labeled by x_1, x_2, and z.

We note first that $z = T(X) = 1$ is affine. We have $T(0, 0) = 1 = f(0, 0)$, so that condition 1 in Definition 6.4.1 is satisfied. To verify condition 2 of Definition 6.4.1, note that

$$
\frac{f(X) - T(X)}{\|X - X_0\|} = \frac{-x_1^2 - x_2^2}{\|X\|} = -\frac{\|X\|^2}{\|X\|} = \|X\|.
$$

Therefore,

$$
\lim_{X \to 0} \frac{f(X) - T(X)}{\|X\|} = 0.
$$

Thus computation and intuition agree, reassuring us that we have made the "correct" definition of "best affine approximation". (Recall that definitions are made by people and are rejected if they do not prove worthy.)

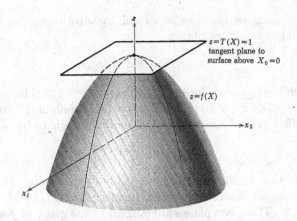

FIGURE 6.21.

Finally, the linear part of $T(X) = 1$ is the linear map $L = O$, since $T(X) = 1 + 0x_1 + 0x_2$. Therefore, the total derivative matrix of f is

$$[f'(\mathbf{0})] = [0 \ \ 0] \ .$$

It is comforting that an important relationship from single-variable calculus carries over, at least in this example as follows: the total derivative is zero at a point corresponding to a horizontal tangent plane.

6.4C TANGENT PLANES

Efficient calculation of total derivatives is discussed in Section 6.5. Here we highlight the fact that one can answer questions about tangent planes (or lines or, more generally, hyperplanes) once one knows the total derivative.

EXAMPLE:

3. Let $f(x_1, x_2) = 1 - x_1^2 - x_2^2$. In Secton 6.5 we will see that the tangent map at a point (c_1, c_2) is

$$T(x_1, x_2) = 1 + c_1^2 + c_2^2 - 2c_1 x_1 - 2c_2 x_2 \ .$$

Let us find all tangent planes to the graph of f that contain the line passing through the points $(1, 0, 2)$ and $(0, 2, 2)$.

A tangent plane will contain this line if and only if it contains two points on the line. So, let us write the conditions for the two given points to lie on the tangent plane:

$$
\begin{aligned}
2 &= 1 + c_1^2 + c_2^2 - 2c_1 \\
2 &= 1 + c_1^2 + c_2^2 - 4c_2 \ .
\end{aligned}
$$

Subtracting we get $c_1 = 2c_2$. Then, by substituting $2c_2$ for c_1 in the second equation we obtain

$$1 = 5c_2^2 - 4c_2 \, ,$$

which has the two solutions $c_2 = 1$ and $c_2 = -1/5$. The corresponding values of c_1 are 2 and $-2/5$, so the two points in $\mathbb{R}^2$ are: $(2,1)$ and $(-2/5, -1/5)$. A quick check, either of our logic or by plugging into

$$2 = 1 + c_1^2 + c_2^2 - 2c_1$$

shows that neither of these two points is extraneous. Thus there are two tangent planes containing the line passing through $(1,0,2)$ and $(0,2,2)$. These two planes are tangent to the graph of f at the points $(2,1,-4)$ and $(-2/5, -1/5, 4/5)$, respectively, and are the graphs of the functions

$$6 - 4x_1 - 2x_2$$

and

$$\frac{6}{5} + \frac{4}{5}x_1 + \frac{2}{5}x_2 \, ,$$

respectively.

6.4D APPROXIMATE VALUES

Since tangent maps are best affine approximations, we can use them to calculate approximate values of particular quantities. The first of the next examples is a single-variable example that we give as preparation for a multivariable example.

EXAMPLES:

4. Let us calculate a close approximation to the number $\sqrt{4.0310}$. We seek a better approximation to the true value of $\sqrt{4.0310}$ than the obvious rough approximate $2.000 (= \sqrt{4.000})$. We will use the tangent line to the graph of $f(x) = \sqrt{x}$ at the point $(4,2)$. Since $f'(4) = 1/4$, the tangent line is the graph of

$$T(x) = 2 + \frac{1}{4}(x - 4) \, .$$

See Figure 6.22. We approximate by evaluating $T(4.0310)$:

$$2 + \frac{0.0310}{4} = 2.00775 \, .$$

This is our approximate value of the square root, that is,

$$\sqrt{4.0310} \approx 2.00775 \, .$$

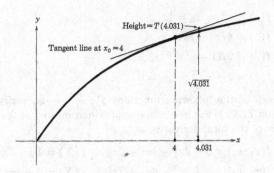

FIGURE 6.22. $4.031^{1/2} \approx T(4.031)$

(The symbol $\approx$ means "is approximately equal to".)

As a check, we square our approximate square root, obtaining

$$(2.00775)^2 \approx 4.0311,$$

which is very close to 4.0310.

5. Let us calculate an approximation to $\sqrt{(3.01)^2 + (3.98)^2}$. It is natural to use the function $f : \mathbb{R}^2 \to \mathbb{R}$ given by

$$f(X) = \|X\|$$

since we can easily calculate that $f(3,4) = 5$ and the point $(3,4)$ is quite close to the point of interest $(3.01, 3.98)$. In the Section 6.5 we will see that, for $X_0 \neq \mathbf{0}$, $f'(X_0)$ is the linear functional given by

$$(f'(X_0))(X) = \frac{\langle X_0, X \rangle}{\|X_0\|}.$$

Thus, the best affine approximation to f at the point $(3,4)$ is the function

$$T(X) = 5 + (f'(3,4))(X - (3,4)) = 5 + \frac{\langle (3,4), (X - (3,4)) \rangle}{5}.$$

We now insert $X = (3.01, 3.98)$ and get (leaving the arithmetic steps to the reader)

$$T(3.01, 3.98) = 4.99.$$

This number is our approximation.

Exercises

1. For each of the functions $f : \mathbb{R} \to \mathbb{R}$, find $f'(x_0)$ and the tangent line to the graph of the curve $y = f(x)$ at x_0. Then verify that condition 2 of Definition 6.4.1 is true. Draw a sketch illustrating the graphs of $f(x)$ and $T(x)$:

   (a) $f(x) = x^2$, $x_0 = 1$

   (b) $f(x) = 4/x$, $x_0 = 2$

   (c) $f(x) = 2/(1 + x^2)$, $x_0 = -1$

   (d) $f(x) = 2 - x^{7/3}$, $x_0 = 0$.

2. For each of the following functions $f : \mathbb{R}^2 \to \mathbb{R}$, verify that the given function $T(X)$ is the best affine approximation to f at X_0, and exhibit the matrix of the total derivative at X_0:

   (a) $f(x_1, x_2) = 3 + 2x_1^2 - 4x_1 + 2x_2^2$, $T(X) = 1$, $X_0 = (1, 0)$

   (b) $f(x_1, x_2) = 1 + 2x_1^2 - 6x_1 + 2x_2^2$, $T(X) = 3 - 6x_1 + 4(x_2 - 1)$,
     $X_0 = (0, 1)$

   (c) $f(X) = x_1^2 - 3x_1 - 7$, $T(X) = -8 - x_1$, $X_0 = (1, -17)$

   (d) $f(X) = x_1^4 + x_2^2$, $T(X) = -2x_2 - 1$, $X_0 = (0, -1)$

   (e) $f(X) = x_1 x_2 + 2x_1 - x_2 - 2$, $T(X) = -2 + 2x_1 - x_2$, $X_0 = (0, 0)$.
     Hint: Prove and use the inequality $2x_1 x_2 \leq x_1^2 + x_2^2$.

3. Let $f : \mathbb{R}^2 \to \mathbb{R}$ have the form $f(x, y) = g(x)$, that is, f does not depend on y. If g, a function of one variable, is differentiable at x_0, prove that f is differentiable at (x_0, y_0) by verifying that the affine function $T(x, y) = g(x_0) + g'(x_0)(x - x_0)$ is the best affine approximation to f at (x_0, y_0). Calculate $[f'(x_0, y_0)]$ in terms of $g'(x_0)$.

4. If a function $f : \mathbb{R}^n \to \mathbb{R}$ is differentiable at X_0, prove that f is continuous there.

5. Let $f : \mathbb{R}^n \to \mathbb{R}$. Assume that there is an affine function $T(X)$ such that $T(X_0) = f(X_0)$ and

$$\lim_{X \to X_0} \frac{f(X) - T(X)}{\|X - X_0\|^2} = 0.$$

Must f be differentiable at X_0? Prove or give a counterexample.

6. For each part, use the formula for the tangent map in Example 3 to find all tangent planes to the graph of $f(x_1, x_2) = 1 - x_1^2 - x_2^2$ with the indicated property:

   (a) tangent at the point $(1, 1, -1)$

   (b) tangent at the point corresponding to the point $(\frac{1}{\sqrt{2}}, \frac{1}{\sqrt{2}})$ in $\mathcal{D}(f)$

   (c) containing the line through the points $(3, 0, 3)$ and $(0, -3, 3)$

   (d) containing the line through the points $(1, 0, 0)$ and $(0, 1, 0)$

   (e) containing the line through the points $(1, 0, 1)$ and $(0, 1, 1)$

   (f) containing the point $(0, 0, 2)$

   (g) containing the point $(0, 0, 0)$

   (h) perpendicular at the point of tangency to a line that passes through the origin.

7. Use the formula for the derivative of $\|X\|$ in Example 5 of Section 6.4d to calculate an approximation to each of the following numbers without first doing the decimal arithmetic underneath the radical symbol:

  (a) $\sqrt{(4.97)^2 + (11.99)^2}$

  (b) $\sqrt{(20.04)^2 + (21.03)^2}$

  (c) $\sqrt{(1.98)^2 + (3.03)^2 + (5.97)^2}$

  (d) $\sqrt{(2.01)^2 + (5.98)^2 + (2.99)^2}$

  (e) $\sqrt{(2.01)^2 + (1.98)^2 + (6.99)^2 + (8.04)^2}$.

6.5   The Main Theorems

6.5A   VARIOUS DERIVATIVES

Let $f : \mathbb{R}^n \to \mathbb{R}$ be a scalar-valued function of the vector $X = (x_1, \ldots, x_n)$. In this chapter we have defined the following kinds of derivatives for f at a point X_0:

1. the directional derivatives $(\nabla_{\mathbf{e}} f)(X_0)$ with $\|\mathbf{e}\| = 1$;

2. the partial derivatives $(D_1 f)(X_0), \ldots, (D_n f)(X_0)$;

3. the total derivative $f'(X_0)$—a linear map from $\mathbb{R}^n$ into $\mathbb{R}$.

In so far as the relations among them are concerned, the only feature we have already described is the fact that partial derivatives are particular directional derivatives. We complete the story in this section. As a by-product we will obtain several calculational tools.

6.5B   RELATIONSHIPS AMONG VARIOUS DERIVATIVES

Consider a function $f : \mathbb{R}^n \to \mathbb{R}$ whose total derivative $f'(X_0)$ exists at a point X_0. We will show, on the one hand, how to calculate directional derivatives and partial derivatives at X_0 from it, and, on the other hand, how to calculate its matrix from the partial derivatives.

In order to describe the full story we introduce one more piece of notation. If each of the partial derivatives of f exists at a point X_0, we call the vector consisting of those partial derivatives the **gradient** of f at X_0. Here are several pieces of notation for the gradient of f:

$$(\text{grad } f)(X_0) = (\nabla f)(X_0) = ((D_1 f)(X_0), \ldots, (D_n f)(X_0)) .$$

(The term "grad" is pronounced to rhyme with "bad".)

A reminder about notation may be appropriate here. A vector Y is not a matrix, but Y does correspond to the following two matrices: the column matrix $[Y]$ and its transpose $[Y]^T$, a row matrix.

Theorem 6.5.1 Let X_0 be an interior point of the domain $\mathcal{D}(f) \subseteq \mathbb{R}^n$ of a scalar-valued function f that is differentiable at X_0. Then the following assertions are true:

1. the directional derivative $(\nabla_{\mathbf{e}} f)(X_0)$ equals $(f'(X_0))(\mathbf{e})$ and, in particular, the partial derivative $(D_j f)(X_0)$ equals $(f'(X_0))(\mathbf{e}_j)$;

2. the 1-by-n matrix of $f'(X_0)$ is given by

$$[f'(X_0)] = [(D_1 f)(X_0) \ \cdots \ (D_n f)(X_0)] ,$$

   which is the same as the row matrix $[(\operatorname{grad} f)(X_0)]^T$;

3. the linear map $f'(X_0)$ is a linear functional whose value at X equals the inner product of X with the vector $(\operatorname{grad} f)(X_0)$, that is,

$$(f'(X_0))(X) = \langle X, (\operatorname{grad} f)(X_0) \rangle .$$

REMARKS:

- We know from our discussion of linear functionals in Section 3.3 that conclusions 2 and 3 are equivalent. Linear functionals are those linear maps whose matrices have only one row, and they can represented by the inner product of an appropriate fixed vector with a vector variable.

- Conclusion 1 tells us that when the total derivative exists, then so do all directional derivatives and, thus, in particular, the partial derivatives and the gradient. The existence of each directional derivative does not have to be assumed; its existence is a consequence of the existence of the total derivative.

- Conclusions 2 and 3 are useful for calculations. They tell us how to use the partial derivatives (the calculations of which are single-variable calculus problems) to calculate the total derivative. There is a catch, however: we can use conclusion 2 or 3 only if we know that the total derivative exists. We will treat this aspect of the issue later in this section.

- Notice that by combining conclusions 1 and 3 we get the following formula for the directional derivative in the direction $\mathbf{e}$:

$$\nabla_{\mathbf{e}} f = \langle \mathbf{e}, \nabla f \rangle .$$

PROOF: (of Theorem 6.5.1) We know that the columns (in the case at hand, just single numbers) of the matrix of a linear transformation are the values of the linear transformation at the basis vectors. Thus (the reader is asked to think about this carefully, since the logic is subtle), we only need prove conclusion 1.

By the definition of the tangent map and total derivative

$$\lim_{s \to 0} \frac{f(X_0 + s\mathbf{e}) - f(X_0) - (f'(X_0))(s\mathbf{e})}{|s|} = 0 .$$

Since the limit is 0, we can remove the absolute value symbol from the denominator, and then, using $(f'(X_0))(s\mathbf{e}) = s(f'(X_0))(\mathbf{e})$, a consequence of the linearity of $f'(X_0)$, we conclude that

$$\lim_{s \to 0} \frac{f(X_0 + s\mathbf{e}) - f(X_0)}{s} = (f'(X_0))(\mathbf{e}) .$$

Since, the left-hand side is the definition of the directional derivative in the direction $\mathbf{e}$, we have completed the proof of conclusion 1, and, therefore, of the theorem. $<<$

The preceding theorem suggests the following calculational approach that combines single-variable calculus with linear algebra: calculate the partial derivatives, form the gradient, identify the total derivative as the linear functional that assigns to a vector variable its inner product with the gradient, and, if desired, calculate any particular directional derivative by inserting the appropriate unit vector for the variable vector. Unfortunately, as mentioned in the above remarks, the applicability of the preceding theorem depends on the existence of the total derivative; it does not assert that the existence of the total derivative is a consequence of the existence of the partial derivatives. Not only does this theorem not assert such an implication, but, as will be seen in Exercise 18, it is possible for the partial derivatives to all exist even if the total derivative fails to exist.

In most cases of interest, however, we are saved by the following theorem, which asserts that the total derivative does exist at a point X_0 if the partial derivatives exist *and are continuous* at X_0. We will not give a proof of the theorem.

Theorem 6.5.2 Let X_0 be an interior point of the domain $\mathcal{D}(f) \subseteq \mathbb{R}^n$ of a scalar-valued function f. Suppose that all n partial derivatives of f exist on some open ball centered at X_0 and are continuous at X_0. Then $f'(X_0)$ exists.

EXAMPLES:

1. Let us use our theorems to compute the total derivative (and show it exists) and best affine approximation of the function $f(X) = 1 - x_1^2 - x_2^2$ at $(1,2)$.

   We compute the partial derivatives: $\frac{\partial f}{\partial x_1} = -2x_1$ and $\frac{\partial f}{\partial x_2} = -2x_2$. These two functions are continuous functions on $\mathbb{R}^2$, so, by Theorem 6.5.2, the total derivative $f'(1,2)$ exists. Thus Theorem 6.5.1 is applicable and gives us

$$[f'(1,2)] = \left[\ \frac{\partial f}{\partial x_1}(1,2) \quad \frac{\partial f}{\partial x_2}(1,2) \ \right] = [\ -2 \quad -4 \ ].$$

It follows that the best affine approximation at X_0 is

$$\begin{aligned} T(X) &= f(1,2) + f'(1,2)(X - (1,2)) \\ &= -4 - 2(x_1 - 1) - 4(x_2 - 2). \end{aligned}$$

2. We compute $\nabla_{\mathbf{e}} f(X_0)$, where $f(X) = x_1^2 + 2x_1 x_2 + 3x_2$, $X_0 = (1,0)$ and $\mathbf{e} = (-1/\sqrt{2}, 1/\sqrt{2})$. We apply the theorems, of course. The partial derivatives of f are the continuous functions $f_{x_1}(X) = 2x_1 + 2x_2$ and $f_{x_2}(X) = 2x_1 + 3$. Thus Theorem 6.5.1 is applicable, $f_{x_1}(X_0) = 2$ and $f_{x_2}(X_0) = 5$. By conclusion 1 of Theorem 6.5.1,

$$\nabla_{\mathbf{e}} f(X_0) = 2 \left(\frac{-1}{\sqrt{2}} \right) + 5 \left(\frac{1}{\sqrt{2}} \right) = \frac{3}{\sqrt{2}}.$$

Actually, we computed similar directional derivatives in Section 6.2 using only the definition. The present method obviates the taking of limits and is much faster.

3. *An economic interpretation.* Let $q(x_1, \ldots, x_n)$ be the steel output at an iron mill as a function of various input variables. For example, x_1 = number of workers, x_2 = amount of iron ore available, x_3 = amount of capital investment (in, say, dollars), and so on. Then the matrix of the total derivative is

$$[q'(X)] = \left[\frac{\partial q}{\partial x_1} \cdots \frac{\partial q}{\partial x_n} \right].$$

The partial derivative $\partial q / \partial x_1$ measures how the productivity changes as the number of workers is changed while all other variables are held constant. Economists refer to q_{x_1} as the "marginal product of labor". Similarly q_{x_2} and q_{x_3} are, respectively, the "marginal product of raw materials" and "marginal product of capital". Thus the matrix $[q'(X)]$ might be called the marginal product matrix. Any time you see the word "marginal" in an economic setting, you can be certain that there is a partial derivative lurking in the background.

Now $q(X)/x_1$ represents the average production of a worker, which we might call the **productivity** per worker. How does the productivity change as the number of workers is changed? To find out, we compute

$$\frac{\partial}{\partial x_1} \left(\frac{q(X)}{x_1} \right) = \frac{x_1 q_{x_1} - q}{x_1^2}.$$

From this we conclude that the productivity per worker increases with the size of the labor force as long as

$$\frac{\partial q}{\partial x_1} > \frac{q}{x_1},$$

while the productivity of each worker decreases if the opposite inequality holds. The factory is operating most efficiently if equality holds, that is, if the marginal productivity of labor equals the average productivity.

The last example makes it seem reasonable that scalar-valued functions of three or more variables can arise in practice. The graph of such a function is a subset of $\mathbb{R}^n$ for some $n \geq 4$. In these situations we speak of the tangent hyperplane at a point rather than the tangent line or tangent plane.

EXAMPLE:

4. Let $f : \mathbb{R}^3 \to \mathbb{R}$ be given by $f(X) = 3x_1^3 + 4x_2^3 + 5x_3^3$. We compute the equation of the tangent hyperplane to $z = f(X)$ in $x_1x_2x_3z$-space "above" the point $(1, 2, -1)$.

To get $z = T(X)$, we take partial derivatives. Thus $f_{x_1}(X) = 9x_1^2$, $f_{x_2}(X) = 12x_2^2$, and $f_{x_3}(X) = 15x_3^2$. It follows that

$$[f'(1, 2, -1)] = [9 \ 48 \ 15] \, .$$

Since $f(1, 2, -1) = 30$, we get

$$
\begin{aligned}
z = T(X) \ &= \ 30 + 9(x_1 - 1) + 48(x_2 - 2) + 15(x_3 + 1) \\
&= \ -60 + 9x_1 + 48x_2 + 15x_3 \, .
\end{aligned}
$$

This equation in x_1, x_2, x_3, and z determines the tangent hyperplane in $\mathbb{R}^4$.

REMARK:

- The last simplification in the preceding example is not necessarily a wise simplification. It effaces the connection with the original point of interest $(1, 2, -1)$. On the other hand, for many calculations, the simplification would probably prove useful.

6.5C DERIVATIVES OF COMBINATIONS OF FUNCTIONS

The following theorem is hardly a surprise. We leave its proof for the exercises, because, except for notational matters, it is very similar to the proof of the corresponding theorem in single-variable calculus.

Theorem 6.5.3 Let f and g be scalar-valued functions that are differentiable at a point $X_0 \in \mathbb{R}^n$. Then the following properties hold:

1. $(f + g)'(X_0) = f'(X_0) + g'(X_0)$;

2. $(fg)'(X_0) = f(X_0)g'(X_0) + f'(X_0)g(X_0)$;

3. $\left(\dfrac{f}{g}\right)'(X_0) = \dfrac{g(X_0)f'(X_0) - f(X_0)g'(X_0)}{(g(X_0))^2}$.

REMARKS:

- The right-hand sides of the equalities in the theorem involve sums of linear maps and products of scalars with linear maps.

- Part of each assertion in the theorem is that the left-hand side exists if the right-hand side does.

We omit the proofs of the forthcoming "chain rules" of which there are two. We suggest that some time be spent on absorbing the meanings of the statements.

Theorem 6.5.4 Let f be a scalar-valued function that is differentiable at a point $X_0 \in \mathbb{R}^n$ and let g be a scalar-valued function that is differentiable at the point $f(X_0) \in \mathbb{R}$. Then

$$(g \circ f)'(X_0) = (g' \circ f)(X_0)f'(X_0) .$$

Theorem 6.5.5 Let K be an $\mathbb{R}^n$-valued function that is differentiable at a point $t_0 \in \mathbb{R}$ and let f be a scalar-valued function that is differentiable at the point $K(t_0) \in \mathbb{R}^n$. Then

$$(f \circ K)'(t_0) = \langle \nabla f(K(t_0)), K'(t_0) \rangle .$$

EXAMPLE:

5. Let $K(t) = (t \cos t, t^2 \sin t)$ for $t \in \mathbb{R}$ and let $f(x_1, x_2) = x_1^2 + x_2^2$. We give ourselves the task of calculating a formula for the derivative of the function $f \circ K : \mathbb{R} \to \mathbb{R}$. The preceding theorem indicates that we should first calculate the following two vectors:

$$K'(t) = (\cos t - t \sin t, \, 2t \sin t + t^2 \cos t) ;$$
$$\nabla f(x_1, x_2) = (2x_1, 2x_2) .$$

Then we take the inner product and substitute $t \cos t$ for x_1 and $t^2 \sin t$ for x_2 as follows:

$$
\begin{aligned}
(f \circ K)'(t) & \\
&= (\cos t - t \sin t)(2t \cos t) + (2t \sin t + t^2 \cos t)(2t^2 \sin t) \\
&= 2t \cos^2 t - 2t^2 \sin t \cos t + 4t^3 \sin^2 t + 2t^4 \sin t \cos t .
\end{aligned}
$$

6.5D THE HEAT-SEEKING BUG

We close this chapter with an amusing problem that also points the way to our study of maxima and minima in the following chapter. Suppose that you are a heat-seeking bug located at the point X_0 on the plane $\mathbb{R}^2$. At each point X on the plane, the temperature is given by some scalar-valued function $f(X)$. You want to travel toward warmth as efficiently as possible. In fact, you *refuse* to visit any point that is cooler than any previous point you have visited. Thus, if the temperature at your present position, namely $f(X_0)$, is not less than the temperature $f(X)$ at any nearby point X —that is, if $f(X_0) \geq f(X)$ —then you will stay where you are. But if not, you will head for warmth the fastest way possible. Which way?

Knowing a bit of vector calculus, you realize that this is a directional derivative problem. You want to find a direction, pointed out by some unit vector $\mathbf{e}$, that has these properties:

1. the rate of change of temperature with respect to distance in the $\mathbf{e}$ direction is positive, $\nabla_{\mathbf{e}} f(X_0) > 0$; in brief, $\mathbf{e}$ points toward relative warmth;

2. the rate of increase in the $\mathbf{e}$ direction is greater than in any other direction; in brief, $\mathbf{e}$ points the quickest way to warmth. In mathematical language, given X_0 and f, you want to find $\mathbf{e}$ such that the value $\nabla_{\mathbf{e}} f(X_0)$ is maximized.

Let us examine the directional derivative $\nabla_{\mathbf{e}} f(X_0)$. We know from Theorem 6.5.1 that

$$\nabla_{\mathbf{e}} f(X_0) = \langle \nabla f(X_0), \mathbf{e} \rangle,$$

where $\nabla f(X_0)$ is the gradient vector $((D_1 f)(X_0), (D_2 f)(X_0))$.

Now we recall from Chapter 2 that

$$\begin{aligned} \langle \nabla f(X_0), \mathbf{e} \rangle &= \|\nabla f(X_0)\| \, \|\mathbf{e}\| \cos\theta \\ &= \|\nabla f(X_0)\| \cos\theta, \end{aligned}$$

where θ is the angle between $\nabla f(X_0)$ and the unit vector $\mathbf{e}$. Note that we have used the fact that $\|\mathbf{e}\| = 1$. See Figure 6.23.

Finally, observe that $\|\nabla f(X_0)\| \cos\theta$ is greatest precisely when $\cos\theta = 1$, that is, when the angle θ is zero. (Recall that X_0 and hence $\nabla f(X_0)$ are fixed and only θ varies.) This means that the directional derivative $\nabla_{\mathbf{e}} f(X_0)$ is greatest when the vector $\mathbf{e}$ is chosen to point in the same direction as the gradient vector $\nabla f(X_0)$. Again see Figure 6.23.

Hence, the vector $\mathbf{e}$ should be defined by $\mathbf{e} = \frac{\nabla f(X_0)}{\|\nabla f(X_0)\|}$. This gives a unit vector pointing in the same direction as $\nabla f(X_0)$. Of course, all this requires that $\nabla f(X_0)$ be different from $\mathbf{0}$. We have proved the following result.

FIGURE 6.23. Choose **e** in same direction as $\nabla f(X_0)$

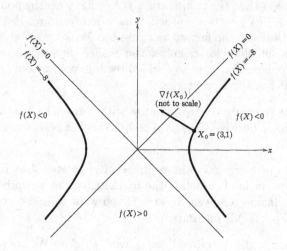

FIGURE 6.24. $\nabla f(X_0)$ points towards warmth

Theorem 6.5.6 Fix $X_0 \in \mathcal{D}(f)$. Then the directional derivative of f at X_0 has its largest value $\|\nabla f(X_0)\|$ in the direction of the gradient vector $\nabla f(X_0)$, provided that $\nabla f(X_0) \neq \mathbf{0}$. All directional derivatives equal 0 if $\nabla f(X_0) = \mathbf{0}$.

You, still imagining yourself to be a heat-seeking bug situated at X_0, should move off in the direction pointed out by $\nabla f(X_0)$. As you travel, at each point X you should recompute $\nabla f(X)$ and move in that direction toward increasing warmth. The next chapter will address issues relevant to the situation in which you reach a point where the gradient equals **0**.

EXAMPLE:

6. Let $f(x,y) = y^2 - x^2$, and suppose that we want to calculate the direction of fastest increase of f at $(3,1)$. See Figure 6.24. We first calculate the gradient:

$$\nabla f(x,y) = (-2x, 2y) .$$

Its value at $(3,1)$ is the vector $(-6,2)$. The unit vector pointing in

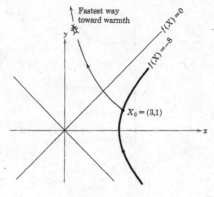

FIGURE 6.25.

the same direction is

$$\frac{(-6,2)}{\|(-6,2)\|} = (-\frac{3}{\sqrt{10}}, \frac{1}{\sqrt{10}}) .$$

Of course, if you are the bug, you should continually revise your direction as you move, so that at each instant you are traveling in the direction of fastest temperature increase. See Figure 6.25.

At the point $X_0 = (0,0)$, the gradient is $\mathbf{0}$, so that no direction is pointed out. See Figure 6.26. Does this mean that there is no nearby warmth to be found? For our present example, the answer is clearly no. If the bug travels along the y-axis in either direction, things get warmer; the temperature at $(0,y)$ is y^2. The gradient method breaks down, because there are two directions of fastest increase. We will need further resources in two-dimensional calculus to prevail in situations like this.

REMARK:

- If we choose θ so as to minimize the expression $\|\nabla f(X_0)\| \cos \theta$, we have the direction of fastest *decrease*. This expression is minimized when $\theta = \pi$, which means that the direction of fastest decrease is given by $-\nabla f(X_0)$. Thus, a "heat-fleeing" bug will head in exactly the opposite direction from a heat-seeking bug.

Exercises

1. Find the matrix of the total derivative at the indicated point of each of the functions in Exercise 1 of Section 6.2.

2. Use Theorem 6.5.1 to redo Exercise 1 of Section 6.2.

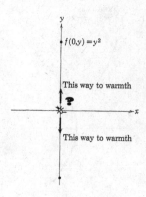

FIGURE 6.26.

3. Find the best affine approximation at the indicated point of each of the functions in Exercise 1 of Section 6.2.

4. (a) Sketch the graph of the surface $z = f(x, y) = 4 - x^2 - y^2$ in $\mathbb{R}^3$.

   (b) Find the best approximation to f near the point $(1, 1)$.

   (c) Draw the graph of this best affine approximation, in the sketch in part (a).

5. Repeat Exercise 4 for $z = f(x, y) = -1 + (x - 1)^2 + (y - 1)^2$.

6. Repeat Exercise 4 for $z = f(x, y) = y^2 - x^2$.

7. Let $z = f(x, y) = x^2 + y^2$. Draw a sketch of $\mathbb{R}^2$ indicating the points where $z = 1$ and $z = 4$. On this same sketch, draw the vector $\nabla f(X)$ at a variety of points $(x, y) \in \mathbb{R}^2$ where $z = 1$.

8. Repeat Exercise 7 for $z = f(x, y) = xy$.

9. Let $f(X) = g(\|X\|)$ for a differentiable function $g : \mathbb{R} \to \mathbb{R}$. For $X_0 \neq \mathbf{0}$, show that
$$[f'(X_0)] = \frac{1}{\|X_0\|} [X_0]^T g'(\|X_0\|) .$$

Hint: First do the following exercise to become comfortable with the notation. Then come back and use Exercise 7 of Section 6.2 and Theorems 6.5.1, 6.5.2, and 6.5.4.

10. Use the preceding exercise to calculate the matrices of the total derivatives of the following functions at places other than the origin:

    (a) $f(X) = \|X\|^2$

    (b) $\varphi(X) = \|X\|$

    (c) $h(X) = e^{-\|X\|}$

    (d) $k(X) = e^{-\|X\|^2/2}$.

11. This is a continuation of Exercise 9. Find necessary and sufficient conditions on g in order that f be differentiable at $\mathbf{0}$. Then find a formula for $f'(\mathbf{0})$ that is valid when f is differentiable at $\mathbf{0}$.

12. Apply the preceding exercise to the functions in Exercise 10.

13. Let $K(t) = e^t(\cos t, \sin t)$ for $t \in \mathbb{R}$ and let $f(X) = \|X\|$ for $X \in \mathbb{R}^2$. Use a chain rule from this chapter to calculate a formula for the derivative of the composition $f \circ K$.

14. Let $K(t) = (\cos t, \sin t, t)$ for $t \in \mathbb{R}$ and let $f(X) = \|X\|$ for $X \in \mathbb{R}^3$. Use a chain rule from this chapter to calculate a formula for the derivative of the composition $f \circ K$.

15. Use a best affine approximation to evaluate the following numbers approximately:

    (a) $\sqrt{(1.91)^3 + (1.09)^3}$

    (b) $2.98e^{.01}$

    (c) $\sqrt{3.9/4.1}$

    (d) $(1.01)^7(3.97)^{1.5}$

    (e) $\frac{1}{2.01} + \frac{1}{2.98} + \frac{1}{6.03}$

    (f) $3.98^{1.49}$

    (g) $\sin(2.01^{4.01} - 4.02^{1.99})$.

16. For each of the following functions, find the direction one must go at the given point in order that the function increase most rapidly, and also the direction for most rapid decrease. Also, find the maximal rate of increase at the given point.

    (a) $f(x_1, x_2) = 3 - 2x_1 + 5x_2$ at $(2, 1)$

    (b) $g(x, y) = e^{2x+y}$ at $(1, -2)$

    (c) $\varphi(x, y, z) = 2x^2 + 3xy + 5z^2 + 4y - y^2 + 7$ at $(1, 0, -1)$

    (d) $h(u, v, w) = uvw$ at $(1, -2, 3)$

    (e) $p(x, y) = xy - x + y - 2$ at $(-1, 1)$.

17. Prove Theorem 6.5.3.

18. Combine Exercise 14 of Section 6.3 and Exercise 4 of Section 6.4 for an interesting example of a non-differentiable function. Show that the values 0 and 0 of the two partial derivatives at $(0, 0)$ are not very useful for "approximating" the function near $(0, 0)$.

19. Let $X(t)$ denote a parametrized curve in $\mathbb{R}^n$. Its image is said to be a **level curve** of a scalar-valued function f on $\mathbb{R}^n$ if $f(X(t))$ is a constant function of t. Show that if the image of X is indeed a level curve for f, then $X'(t)$ is orthogonal to $(\nabla f)(X(t))$ for every t at which both exist. *Hint:* Differentiate and use a chain rule.

6.6 The World of First Derivatives

Function	First (total) derivative at a point	First derivative in matrix form	Best affine approximation or tangent map at a point
$f: \mathbb{R} \to \mathbb{R}$ $y = f(x)$ ordinary calculus	$f'(x_0): \mathbb{R} \to \mathbb{R}$ usually interpreted as slope	$f'(x_0)$, a number	$y = T(X) = f(x_0) + f'(x_0)(x - x_0)$ (tangent line at x_0)
$f: \mathbb{R}^n \to \mathbb{R}$ $z = f(X) = f(x_1, \ldots, x_n)$ scalar-valued (Chapter 6)	Linear map (functional) $f'(X_0): \mathbb{R}^n \to \mathbb{R}$	$[f'(X_0)] = \left[\frac{\partial f}{\partial x_1}(X_0) \cdots \frac{\partial f}{\partial x_n}(X_0)\right]$; also, a gradient vector $\nabla f(X_0) = \left(\frac{\partial f}{\partial x_1}(X_0), \ldots, \frac{\partial f}{\partial x_n}(X_0)\right)$	$\begin{aligned} z &= T(X) = f(X_0) + f'(X_0)(X - X_0) \\ &= f(X_0) + \frac{\partial f}{\partial x_1}(X_0)(x_1 - c_1) \\ &\quad + \cdots + \frac{\partial f}{\partial x_n}(X_0)(x_n - c_n) \end{aligned}$ (here $X_0 = (c_1, \ldots, c_n)$)
$F: \mathbb{R} \to \mathbb{R}^q$ $F(t) = (f_1(t), \ldots, f_q(t))$ parametrized curve, in $\mathbb{R}^q$ (Chapter 4)	Linear map $F'(t_0): \mathbb{R} \to \mathbb{R}^q$ usually drawn as velocity vector $(f'_1(t_0), \ldots, f'_q(t_0))$ at point $F(t_0)$ on curve	$[F'(t_0)] = \begin{bmatrix} f'_1(t_0) \\ \vdots \\ f'_q(t_0) \end{bmatrix}$	$\begin{aligned} T(t) &= F(t_0) + F'(t_0)(t - t_0) \\ &= (f_1(t_0) + f'_1(t_0)(t - t_0), \ldots, \\ &\qquad f_q(t_0) + f'_q(t_0)(t - t_0)) \end{aligned}$
$F: \mathbb{R}^n \to \mathbb{R}^q$ $Y = F(X)$ $= (f_1(x_1, \ldots, x_n), \ldots,$ $f_q(x_1, \ldots, x_n))$ vector-valued function of a vector (Chap. 8)	Linear map $F'(X_0): \mathbb{R}^n \to \mathbb{R}^q$	$[F'(X_0)] = \begin{bmatrix} \frac{\partial f_1}{\partial x_1}(X_0) \cdots \frac{\partial f_1}{\partial x_n}(X_0) \\ \vdots \\ \frac{\partial f_q}{\partial x_1}(X_0) \cdots \frac{\partial f_q}{\partial x_n}(X_0) \end{bmatrix}$	$\begin{aligned} Y &= T(X) = F(X_0) \\ &\quad + F'(X_0)(X - X_0) \\ Y &= (y_1, \ldots, y_q) \\ y_j &= f_j(X_0) + \frac{\partial f_j}{\partial x_1}(X_0)(x_1 - c_1) \\ &\quad + \cdots + \frac{\partial f_j}{\partial x_n}(X_0)(x_n - c_n) \\ &\quad (j = 1, \ldots, q) \end{aligned}$

7

Scalar-Valued Functions and Extrema

7.0   Introduction

Here is a familiar story in single-variable calculus. Let $f: \mathcal{D}(f) \to \mathbb{R}$, where $\mathcal{D}(f)$ an interval. Then f has a local maximum at an interior point $x_0 \in \mathcal{D}(f)$ provided that for all x in some open interval of points on the x-axis containing x_0 we have $f(x) \leq f(x_0)$. The value $f(x_0)$ is a local maximum of f. The story for local minima is similar. And the term "local extremum" encompasses both possibilities.

In the graph pictured in Figure 7.1, the local maxima are at x_0, x_3, and x_5 (the "mountain peaks") and the local minima at x_2 and x_4 (the "valleys"). Note that f may have several local maxima and minima.

One reason we are interested in the locations of the local maxima and minima of a function is that these locations give us a rough picture of its graph.

Recall the following basic result from single-variable calculus.

Theorem 7.0.1 (Local Extrema in One Variable) If $f: \mathbb{R} \to \mathbb{R}$ is a differentiable function with a local extremum at x_0, then the following assertions are true:

1. the tangent line to $y = f(x)$ at $(x_0, f(x_0))$ is horizontal;

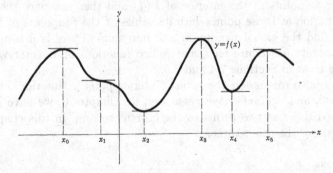

FIGURE 7.1. Local maxima at x_0, x_3, and x_5, local minima at x_2 and x_4, critical point at x_1 at which there is no local extremum

2. $f'(x_0) = 0$.

We note that statements 1 and 2 are equivalent. Also, it is *not* true that there is necessarily a local extremum at a point x if $f'(x) = 0$, as the point $x = x_1$ in the graph in Figure 7.1 illustrates. The slope $f'(x_1) = 0$, but the graph has neither a peak nor a valley there.

Thus we find the local extrema of $y = f(x)$ by the following process:

1. Locate all points x_0 at which $f'(x_0) = 0$. (Such points are the critical points of f.)

2. Decide separately whether each critical point is a local maximum, a local minimum, or neither. One tool for this decision is the second derivative $f''(x_0)$.

Just as in the single-variable case, the problem of locating the local extrema of a scalar-valued function f with $\mathcal{D}(f) \subseteq \mathbb{R}^n$ is of major importance. In Sections 7.1 through 7.3 we examine this problem in some detail. We will see that the statements about the first derivative $f'(x_0)$, quoted above for the single-variable case, have their exact analogs with regard to the total derivative $f'(X_0)$, or, equivalently, the gradient vector $\nabla f(X_0)$. Moreover, the problem of deciding whether a point X_0 is a local maximum or minimum leads us to the notion of the "second derivative" $f''(X_0)$. This, as we shall see, is neither a number, like $F(X_0)$, nor a vector, like $\nabla f(X_0)$, but is represented by an n-by-n matrix.

In single-variable calculus one also is concerned about *global* extrema. The global maximum (or absolute maximum) of a function f is (if it exists) a number $f(x_0)$ such that $f(x_0) \geq f(x)$ for every $x \in \mathcal{D}(f)$. A function can have at most one global maximum, but there may be more than one point in its domain at which the global maximum is achieved. A single-variable calculus theorem asserts that if f is continuous and $\mathcal{D}(f)$ is a bounded closed interval, then f has a global maximum and a global minimum. A standard procedure in such a situation is to search for all local extrema occurring at points in the interior of $\mathcal{D}(f)$ and then compare the values of the function at those points with its values at the endpoints of $\mathcal{D}(f)$ in order to find the global maximum and minimum. There is a similar but more elaborate procedure for scalar-valued functions of a vector variable, which we treat in Sections 7.4 and 7.5.

In the above discussion the terms "interior point", "interior", "open", "bounded", and "closed" have been used. In Chapter 6, we have already generalized the first two terms to the current setting. In this chapter we will generalize the other three.

Exercises

Decide whether each statement is true or false, and why. Except for Exercise 24, assume that all unspecified functions have three or more derivatives and that the domain in each case is $\mathbb{R}$. We suggest using lots of pictures.

1. $f(x) = (x-1)^2$ has a local minimum at $x = 1$.

2. $f(x) = x^3$ has a local minimum at $x = 0$.

3. $f(x) = -x^4$ has a local maximum at $x = 0$.

4. $f(x) = (1+x^2)^{-1}$ has no local minimum.

5. $f(x) = x^5$ has a critical point at $x = 0$.

6. $f(x) = 1 + x + x^3$ has no local extrema.

7. If $f(x)$ has a local minimum at 0, then $f''(0) > 0$.

8. If $f(x)$ has a local maximum at 0, then $f''(0) \leq 0$.

9. If $f'(0) = 0$ and $f''(0) > 0$, then f has a local minimum at 0.

10. If $f'(0) = 0$ and $f''(0) = 0$, then f cannot have a local minimum at 0.

11. If f has local maxima at -2 and 3, then it must have a local minimum at some point c where $-2 < c < 3$.

12. If $f(-1) = 3$, $f(0) = 5$, and $f(2) = -4$, then f has a local maximum at some point c in the interval $(-1, 2)$.

13. If both f and g have local maxima at x_0, then so does $f + g$.

14. If both f and g have local minima at x_0, then so does fg. (Note: Here fg denotes the product of the two functions f and g, *not* the composition.)

15. $f(x,y) = x^2 + (y-1)^2$ has a local minimum at $(0,1)$.

16. $f(x,y) = 1 - (x-y)^2$ has a local maximum at $(1,1)$.

17. $f(x,y) = y^2 - x^2$ has a local maximum at $(0,0)$.

18. $f(x,y) = x^2y^2$ has a local minimum at $(-1,0)$.

19. $f(x,y) = 7 + xy$ has a local maximum at $(0,0)$.

20. $f(x,y) = 3 - x + y$ has no local extrema.

21. $f(x,y) = x + y^2$ has no local extrema.

22. $f(x,y) = x^2 - \cos y$ has a local minimum at $(0,0)$.

23. $f(x,y) = [x^2 + (y-1)^2][x^2 + (y+1)^2]$ has local minima at both $(0,1)$ and $(0,-1)$.

24. If $f(X) \geq 0$ for all X and f has a local minimum at X_0, then $g(X) = f^2(X)$ also has a local minimum at X_0.

25. If f satisfies $f''(x) - \alpha(x)f(x) = 0$, with $\alpha(x) > 0$, then f cannot have a local minimum where f is negative.

26. The function $f(x) = \sqrt{1-x^2}$ has both a global maximum and a global minimum (on its domain).

27. The function $g(x) = (\sin x)/x$ has a global minimum.

28. The function $g(x) = (\sin x)/x$ has a global maximum.

29. For distinct fixed points U and V in $\mathbb{R}^n$, the function $f(X) = \|X - U\| + \|X - V\|$ has a global minimum occurring at infinitely many points in $\mathcal{D}(f) = \mathbb{R}^n$.

7.1   Local Extrema are Critical Points

We begin by formally defining the concepts of global and local extrema for functions of vector variables. Later in this section, we will obtain the analog of the one-dimensional theorem relating local extrema to the first derivative.

Definition 7.1.1 Let f be a scalar-valued function for which $\mathcal{D}(f) \subseteq \mathbb{R}^n$. A number $f(X_0)$ such that $f(X_0) \geq f(X)$ for every $X \in \mathcal{D}(f)$ is a **global maximum** (or **absolute maximum**) of the function f, and f is said to attain its global maximum at X_0.

It should be noted that some functions do not have a global maximum and, on the other hand, many functions that do have a global maximum attain it at more than one point. A **global minimum** is defined similarly using, of course, $f(X_0) \leq f(X)$. A **global extremum** is a global maximum or minimum.

Definition 7.1.2 For f as in Definition 7.1.1, $f(X_0)$ is a **local maximum** if and only if both of the following are true:

1. X_0 is an interior point of $\mathcal{D}(f)$,

2. $f(X_0) \geq f(X)$ for all X in some ball centered at X_0.

When these conditions are satisfied f is said to have a local maximum at X_0.

The first requirement indicates that the term "local maximum" is used only at points in the interior of the domain (and thus at points where the total derivative of f might be defined). The relevant balls in the second condition are those of small radius, since if a ball with large radius works, then so will a smaller one. According to the second condition a function has a local maximum at a point if its value at that point is at least as large as its values at all nearby points, regardless of the possibility of much larger values at far away points. The definition of **local minimum** is similar; the relevant inequality is $f(X_0) \leq f(X)$. A **local extremum** is a local maximum or minimum.

Figure 7.2 illustrates the graph of a function f whose domain is two-dimensional. We see that the local maxima and minima are the heights of the mountain peaks and valleys on the surface that is the graph of f.

The following basic theorem is most useful for locating the local extrema of f. Note the similarity with single-variable calculus.

Theorem 7.1.1 (Local Extrema) Let $f : \mathcal{D}(f) \to \mathbb{R}$ be differentiable. If f attains a local extremum at X_0, then the following assertions are true:

1. the total derivative $f'(X_0)$ is the zero map;

2. the gradient $\nabla f(X_0)$ is the zero vector;

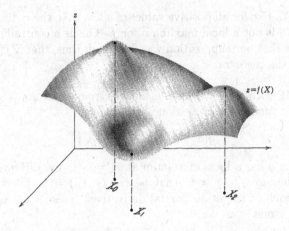

FIGURE 7.2. Local maxima at X_0 and X_2; local minimum at X_1

3. the partial derivatives $(D_1 f)(X_0) = \cdots = D_n f(X_0) = 0$.

REMARKS:

- Assertions 1,2, and 3 are clearly equivalent.

- We will give two proofs, the first of them being "coordinate-free", not involving $x_1, \ldots, x_n$ or the partial derivatives.

- Example 2 below shows that it is possible for assertions 1, 2, and 3 to be true even if f does not have a local extremum at X_0.

FIRST PROOF: Given that f has a local maximum at X_0 (the proof for a local minimum is similar), we show that $\nabla f(X_0) = \mathbf{0}$. For a proof by contradiction suppose that $\nabla f(X_0) \neq \mathbf{0}$. Let $\mathbf{e} = \nabla f(X_0)/\|\nabla f(X_0)\|$ be the unit vector pointing in the same direction as $\nabla f(X_0)$. Define $\varphi(s)$ to be the scalar-valued function of the scalar variable s given by

$$\varphi(s) = f(X_0 + s\mathbf{e}).$$

Since f has a local extremum at X_0, we conclude that X_0 is an interior point of $\mathcal{D}(f)$ and, therefore, $\varphi(s)$ is defined for all s in some open interval containing 0. As we have seen in Section 6.2, $\varphi'(0)$ equals the directional derivative at X_0 in the direction $\mathbf{e}$. Thus

$$\varphi'(0) = \nabla_{\mathbf{e}} f(X_0) = \langle \nabla f(X_0), \mathbf{e} \rangle = \|\nabla f(X_0)\| > 0.$$

Single-variable calculus assures us, therefore, that $\varphi(s)$ is increasing at $s = 0$. Thus, for all sufficiently small positive values of s, $\varphi(s) > \varphi(0)$. But this means that $f(X_0 + s\mathbf{e}) > f(X_0)$. Since any ball centered at X_0 contains the points $X_0 + s\mathbf{e}$ for all positive values of s less than the radius of the ball, the point X_0 is not a local maximum for f. This is a contradiction. Hence we conclude that our supposition was false and, thus, that $\nabla f(X_0) = \mathbf{0}$, as claimed by the theorem. <<

SECOND PROOF: Let f have a local extremum at $X_0 = (c_1, \ldots, c_n)$. Consider the function g of x_1 alone defined by

$$g(x_1) = f(x_1, c_2, \ldots, c_n).$$

The function g has a local extremum at $x_1 = c_1$ (why?). Hence, by single-variable calculus, $g'(c_1) = 0$, that is, $(D_1 f)(X_0) = 0$. Likewise, we can show that each of the other partial derivatives vanishes at X_0, obtaining assertion 3. Done. <<

Just as in single-variable calculus, our search for local extrema has led us to points at which the first derivative vanishes. We define an interior point X_0 of $\mathcal{D}(f)$ to be a **critical point** of f if and only if the total derivative $f'(X_0)$ is the zero map from $\mathbb{R}^n$ to $\mathbb{R}$. This second statement is, of course, equivalent to $\nabla f(X_0) = \mathbf{0}$ or the vanishing of all the partial derivatives of f at X_0. We remark that some people refer to critical points as **stationary** points.

EXAMPLES:

1. Let $f(X) = f(x_1, x_2) = 1 - x_1^2 - x_2^2$. We locate the critical points by taking the partial derivatives

$$\frac{\partial f}{\partial x_1}(X) = -2x_1, \qquad \frac{\partial f}{\partial x_2}(X) = -2x_2$$

and setting both of them equal to zero. See Figure 7.3. We obtain $X_0 = (0,0)$; the origin is the only critical point for f. It is clear from the fact that $f(X) \leq 1 = f(\mathbf{0})$, that f has a local maximum equal to 1 at the origin, and that this local maximum is also the global maximum.

Since there are no critical points besides $(0,0)$, we conclude that f has no local minimum. If a global minimum occurs at an interior point of $\mathcal{D}(f)$, it is also a local minimum; so, no global minimum is attained at an interior point of $\mathcal{D}(f)$. Since $\mathcal{D}(f) = \mathbb{R}^2$ and every point of $\mathbb{R}^2$ is an interior point, f does not have a global minimum.

2. Let $g(x, y) = y^2 - x^2$. In this case $g_x = -2x$ and $g_y = 2y$, so that $X_0 = (0,0)$ is, as in the preceding example, the only critical point. See Figure 7.4. However, there is not a local extremum at $(0,0)$. To see this, note first that $g(0,0) = 0$. Now $g(0,y) = y^2 > 0$ if $y \neq 0$. Hence there are points $(0,y)$ arbitrarily close to $(0,0)$ (just take y

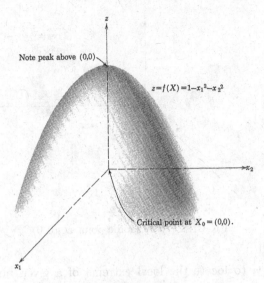

Note peak above $(0,0)$

$z = f(X) = 1 - x_1{}^2 - x_2{}^2$

Critical point at $X_0 = (0,0)$.

FIGURE 7.3.

small) at which the function g assumes larger values than $g(0,0)$. Thus g does *not* have a local maximum at $(0,0)$. Consideration of points $(x,0)$ shows that there is *not* a local minimum at $(0,0)$ either. By an argument similar to that used in the preceding example we conclude, also, that there are no global extrema.

The preceding example shows that not all critical points are points where extrema occur and leads us to make the following definition. The point $(X_0, f(X_0)) \in \mathbb{R}^n$ on the graph of f is a **saddle point** of the scalar-valued function f if and only if the following conditions hold:

1. X_0 is a critical point of f; that is, X_0 is an interior point of $\mathcal{D}(f)$ and $f'(X_0) = O$;

2. there is not a local extremum at X_0; that is, within any ball centered at X_0, there exist points U and V such that $f(U) < f(X_0) < f(V)$.

In Example 2, the origin $(0,0,0)$ was seen to be a saddle point for the function $g(x,y) = y^2 - x^2$.

REMARK:

- Note that a saddle point of a function f is a point $(X_0, f(X_0))$ on the *graph* of f. When we want to speak of the corresponding point X_0 in the domain, we say that f has a saddle point **at** X_0. The definition of saddle point is such that for any critical point X of a function f, either f has a local extremum at X or f has a saddle point at X.

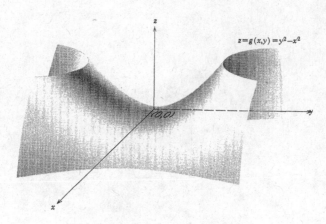

FIGURE 7.4. A saddle point at $(0,0)$

In summary, to locate the local extrema of a given function, do the following:

1. locate its critical points;

2. classify each critical point as the location of a local maximum, local minimum, or saddle point.

To locate the critical points, we know that we must find all points X_0 satisfying the n equations $(D_1 f)(X_0) = 0, \ldots, (D_n f)(X_0) = 0$. Classifying critical points may be more delicate, although in some cases (such as the two examples above) it is quite elementary. We discuss the classification of critical points in Section 7.2. See also Exercise 4.

EXAMPLE:

3. Let us find the distance between $\mathbf{0}$ and the plane $\mathcal{S} + Z$, where $\mathcal{S}$ is a two-dimensional linear subspace of $\mathbb{R}^n$ spanned by the vectors U and V. The points on the plane can be represented by

$$sU + tV + Z$$

for scalars s and t. Let $g(s,t)$ denote the square of its distance from $\mathbf{0}$; that is,

$$
\begin{aligned}
g(s,t) &= \|sU + tV + Z\|^2 \\
&= \langle (sU + tV + Z), (sU + tV + Z) \rangle \\
&= \langle U, U \rangle s^2 + 2\langle U, V \rangle st + \langle V, V \rangle t^2 \\
&\quad + 2\langle U, Z \rangle s + 2\langle V, Z \rangle t + \langle Z, Z \rangle .
\end{aligned}
$$

Our first step in looking for the critical points of g is to find its partial derivatives. Inserting a factor of $1/2$ for convenience we obtain

$$\frac{1}{2}(D_1 g)(s,t) = \langle U, U \rangle s + \langle U, V \rangle t + \langle U, Z \rangle,$$

$$\frac{1}{2}(D_2 g)(s,t) = \langle U, V \rangle s + \langle V, V \rangle t + \langle V, Z \rangle.$$

We set each partial derivative equal to 0 and use Gauss-Jordan elimination to solve for the unknowns s and t. The result is that there is exactly one critical point:

$$s = \frac{\langle U, V \rangle \langle V, Z \rangle - \langle V, V \rangle \langle U, Z \rangle}{\langle U, U \rangle \langle V, V \rangle - \langle U, V \rangle^2},$$

$$t = \frac{\langle U, V \rangle \langle U, Z \rangle - \langle U, U \rangle \langle V, Z \rangle}{\langle U, U \rangle \langle V, V \rangle - \langle U, V \rangle^2}.$$

From the nature of the problem we believe that there is a global minimum. If this conjecture is correct the minimum of the square of the distance must be attained at the critical point. (A technique for verifying such conjectures will be given in Example 6 of Section 7.5.) Insertion of the critical point into the formula for g and then taking square roots gives, after a considerable amount of algebraic manipulation, that the distance between the given plane and the origin equals

$$\sqrt{\|Z\|^2 - \frac{\langle V, V \rangle \langle U, Z \rangle^2 + \langle U, U \rangle \langle V, Z \rangle^2 - 2 \langle U, V \rangle \langle U, Z \rangle \langle V, Z \rangle}{\langle U, U \rangle \langle V, V \rangle - \langle U, V \rangle^2}}.$$

Exercises

1. Find all the critical points of the following functions:

   (a) $f(x_1, x_2) = x_1^2 - 6x_1 + 2x_2^2 + 10$

   (b) $g(x, y) = xy - x + y + 2$

   (c) $h(x, y) = x \sin y$

   (d) $\varphi(x, y) = (x^2 - 1)e^y$

   (e) $p(s, t) = (s^2 - 1)(t^2 - 1)$

   (f) $\psi(x, y) = (x - 2y)^2$

   (g) $u(\xi, \eta) = \xi^2 \eta^2$

   (h) $f(x, y) = \frac{xye^x}{1+y^2}$

   (i) $f(x, y, z) = x^2 + 2x + 3y^2 + z^2 - 4z + 5$

   (j) $g(x, y, z) = xz + 5y^2$

   (k) $f(x, y) = x \sinh y$

(l) $g(x, y) = x \cosh(xy)$.

2. Use the formula in Example 3 to find the distance between the origin and the plane $S + Z$, where S is the plane spanned by U and V. In parts (b), (e), (f) confirm your answers through geometric or other reasoning.

   (a) $Z = (2, 0, -1)$, $U = (0, 1, 7)$, $V = (2, -5, 1)$

   (b) $Z = (0, 0, 0)$, $U = (0, 1, 7)$, $V = (2, -5, 1)$

   (c) $Z = (1, 1, 0, 0)$, $U = (0, -1, 1, 0)$, $V = (2, 1, 0, 1)$

   (d) $Z = (1, 0, 0)$, $U = (1, 1, 0)$, $V = (1, 1, 1)$

   (e) $Z = (1, 100, 100)$, $U = (0, 1, 0)$, $V = (0, 0, 1)$

   (f) $Z = U \times V$, $U \in \mathbb{R}^3$, $V \in \mathbb{R}^3$.

3. Use the formula in Example 3 to find the minimum distance from the origin to the plane $2x + y - 2z = 6$ in $\mathbb{R}^3$. Then find a second method to do this problem.

4. In order to examine the nature of a critical point (a, b) of $f(x, y)$, consider $\varphi(h, k) = f(a + h, b + k) - f(a, b)$ as a function of (h, k) for h and k small, so that $(a + h, b + k)$ is near (a, b). For example, $f(x, y) = x^2 + 2x + 5y^2$ has a critical point at $(-1, 0)$, and so we let $x = -1 + h$ and $y = k$ and examine

$$\varphi(h, k) = (-1 + h)^2 + 2(-1 + h) + 5k^2 + 1 = h^2 + 5k^2 \geq 0 ,$$

that is,

$$f(a + h,\ b + k) - f(a, b) \geq 0 .$$

Hence, f has a local minimum at $(-1, 0)$. If φ were seen to have both positive and negative values for small h and k, we would have concluded instead that f has a saddle point. Use this method to determine if there are local extrema at the critical points in parts (a), (b), (e), (f), (g), (i), and (j) of Exercise 1. For each local extremum decide whether it is a local maximum or a local minimum.

5. Find the points on the surface $z = 2xy + \frac{1}{4}$ nearest the origin.

6. Find the points on the graph of the function $1/xy$ nearest the origin.

7. Let f be a spherically symmetric differentiable function on $\mathbb{R}^n$.

   (a) Show that $\mathbf{0}$ is a critical point.

   (b) Show that if X is a critical point, then every point with norm equal to $\|X\|$ is a critical point.

8. Find all critical points of the function

$$f(X) = \|X\|^2 e^{-\|X\|} .$$

Then find all local and global extrema and the points at which they are attained.

9. Find all critical points of the function
$$f(X) = \|X\|e^{-\|X\|} .$$

Then find all local and global extrema and the points at which they are attained.

10. Let $C \in \mathbb{R}^n$ and $b \in \mathbb{R}$. For $X \in \mathbb{R}^n$, let
$$f(X) = \langle X, C \rangle + b ,$$

and let $g(X)$ denote the distance from the point $(X, f(X))$ on the graph of f in $\mathbb{R}^{n+1}$ to the origin in $\mathbb{R}^{n+1}$. Show that the function g has exactly one critical point, namely
$$\frac{-b}{1 + \|C\|^2} C .$$

11. This exercise concerns three games.

  (a) Frank picks any real number x that he wants. Then, Jerry, after hearing Frank's choice, picks a real number y of his choice and pays Frank $xy - 3x + 2y$ dollars. What number should Frank pick? (A negative value of $xy - 3x + 2y$ indicates a payment from Frank to Jerry.)

  (b) Discuss the modified game in which Jerry chooses first and, then, Frank makes his choice after hearing Jerry's choice.

  (c) Discuss the modification in which each player writes his choice secretly on a piece of paper, before hearing the choice of the other player.

7.2 The Second Derivative

7.2A THE SINGLE-VARIABLE SITUATION

In order to classify critical points as locations of local maxima, local minima, or saddle points we need multivariable versions of the Mean Value Theorem, the second derivative, and Taylor's Theorem. We first review the single-variable situation.

Theorem 7.2.1 (Mean Value Theorem (one variable)) Let g be a differentiable scalar-valued function whose domain is an interval containing the points t_0 and t. Then there exists a point t^* between t_0 and t such that
$$g(t) = g(t_0) + g'(t^*)(t - t_0).$$

This shows that the value $g(t)$ varies from $g(t_0)$ by an amount $g'(t^*)(t - t_0)$, a quantity which depends on the difference $t - t_0$ and the rate of change $g'(t^*)$. Of course, the location of the point t^* depends on t and t_0.

Compare the *equality* in the Mean Value Theorem with the approximate equality in the best affine approximation:
$$g(t) \approx g(t_0) + g'(t_0)(t - t_0).$$

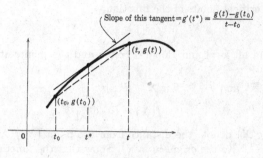

FIGURE 7.5.

Equality is achieved by replacing $g'(t_0)$ by $g'(t^*)$. However, the Mean Value Theorem is only an *existence* theorem; it does not tell us how to find t^* explicitly. This is the price we pay for getting an equality.

Instead of the formula given in the statement of the theorem, you may have seen

$$g'(t^*) = \frac{g(t) - g(t_0)}{t - t_0} .$$

Note that the two formulas are the same. This version may be interpreted as saying that the slope of the tangent line to the graph of g directly above t^* equals the slope of the line through the two points $(t_0,\ g(t_0))$ and $(t,\ g(t))$; the first slope is $g'(t^*)$, and the second is $(g(t) - g(t_0))/(t - t_0)$. See Figure 7.5.

The Mean Value Theorem has many uses. For example, you should now be able to prove that if $g' = 0$ at all points in an interval, then g is constant in that interval.

In the next theorem, we see that if we know a bit more about the derivatives of $g(t)$, we may be able to say even more about the values of the function.

Theorem 7.2.2 (Taylor's Theorem, order two (one variable)) If g is a twice differentiable scalar-valued function whose domain is an interval containing the points t_0 and t, then there exists a number t^* between t_0 and t such that

$$g(t) = g(t_0) + g'(t_0)(t - t_0) + \frac{1}{2}g''(t^*)(t - t_0)^2 .$$

This is an improvement on the Mean Value Theorem. It tells us that the difference between the actual value $g(t)$ and the value $T(t) = g(t_0) + g'(t_0)(t - t_0)$ of the best affine approximation of the function g is measured by the second derivative of g and the quantity $(t - t_0)^2$. (Note that $(t - t_0)^2$ is very small if t is close to t_0.)

The expression $g(t_0) + g'(t_0)(t - t_0) + \frac{1}{2}g''(t^*)(t - t_0)^2$ is "almost" a quadratic polynomial in $t - t_0$; it isn't one because t^* depends on t, so

$g''(t^*)$ is not a constant as t varies and t_0 stays fixed. Nevertheless, the expression is useful because polynomials are more easily treated than are more general functions.

If, as often happens, g'' is a continuous function, we may use Taylor's Theorem to obtain the following well-known result about local extrema in the single-variable case.

Theorem 7.2.3 (Second Derivative Test (one variable)) Let t_0 be a critical point of $g: \mathbb{R} \to \mathbb{R}$. Suppose that g has a second derivative that is continuous at t_0.

1. If $g''(t_0) < 0$, then g has a local maximum at t_0.

2. If $g''(t_0) > 0$, then g has a local minimum at t_0.

REMARKS:

- The second derivative test gives no information in case $g''(t_0) = 0$.

- A technical point that will not play a role in this is book is the fact that the assumption that g'' be continuous at t_0 can be replaced by the weaker assumption that $g''(t_0)$ exist. We have chosen the continuity assumption to parallel the forthcoming multivariable treatment.

7.2B THE MEAN VALUE THEOREM

Theorem 7.2.4 (Mean Value Theorem) Let f be a scalar-valued function that is differentiable at each point on the line segment connecting two points X_0 and X in $\mathbb{R}^n$. Then there exists a point X^* on the line segment between X_0 and X such that

$$f(X) = f(X_0) + f'(X^*)(X - X_0).$$

REMARKS:

- The second term on the right-hand side denotes the operation of the linear map $f'(X^*)$ on the vector $X - X_0$.

- As in the single-variable case, the location of the point X^* depends on X_0 and X. The theorem does not tell us how to find this location.

PROOF: The line segment from X_0 to X consists of all points $X_0 + t(X - X_0)$ with $0 \leq t \leq 1$. See Figure 7.6. Putting $t = 0$ yields X_0, and $t = 1$ yields X.

Now define $g(t) = f(X_0 + t(X - X_0))$, a function of t alone. The single-variable Mean Value Theorem guarantees the existence of a number t^* satisfying $0 < t^* < 1$, such that

$$g(1) = g(0) + g'(t^*)(1 - 0) = g(0) + g'(t^*),$$

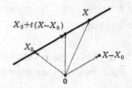

FIGURE 7.6. The line $X_0 + t(X - X_0)$

which can be rewritten as

$$f(X) = f(X_0) + g'(t^*) \, .$$

It remains for us to substitute for the second term on the right-hand side. The function g is the composition of f on the affine function $K(t) = X_0 + t(X - X_0)$, the derivative of which is the vector $X - X_0$. By a chain rule, Theorem 6.5.5 of Section 6.5c, $g'(t^*)$ equals an inner product:

$$g'(t^*) = \langle \nabla f(X_0 + t^*(X - X_0)), (X - X_0) \rangle = \langle \nabla f(X^*), (X - X_0) \rangle \, .$$

This completes the proof since

$$\langle \nabla f(X^*), (X - X_0) \rangle = f'(X^*)(X - X_0) \, .$$

Done. $<<$

We leave it to you to prove the following familiar-looking statement.

Corollary 7.2.1 Let $f : \mathbb{R}^n \to \mathbb{R}$ be a differentiable function. If the total derivative $f'(X) = O$ for all X, then f is a constant function.

7.2C SECOND DERIVATIVES AND THE HESSIAN MATRIX

We will see that the "proper" generalization of the second derivative to scalar-valued functions $f : \mathbb{R}^n \to \mathbb{R}$ is represented by an n-by-n matrix.

A scalar-valued function f is said to be **twice differentiable** at a point X_0 if each of its first partial derivatives is differentiable at X_0. Thus, f is twice differentiable if all its second partial derivatives exist and are continuous, in which case we know that mixed second partial derivatives are equal.

REMARKS:

- The following is a technical point that we will not investigate but which should be remembered: mixed second partial derivatives of a twice differentiable function f are equal (that is, $D_{ij}f = D_{ji}f$), even if the second partial derivatives of f are not continuous.

- Here is a second technical point, one that will not play a significant role in this book: it is possible for the first partial derivatives to be differentiable even if the second partial derivatives are not continuous. Also, it is possible for mixed second partial derivatives to be equal even if they are not continuous.

Definition 7.2.1 Let f be a scalar-valued function that is twice differentiable at a point $X_0 \in \mathbb{R}^n$. The **Hessian** of f at X_0 is the matrix whose entry in position i, j is $(D_{ij}f)(X_0)$.

Here is how the Hessian looks for $n = 3$:

$$[f''(X_0)] = \begin{bmatrix} f_{11}(X_0) & f_{12}(X_0) & f_{13}(X_0) \\ f_{21}(X_0) & f_{22}(X_0) & f_{23}(X_0) \\ f_{31}(X_0) & f_{32}(X_0) & f_{33}(X_0) \end{bmatrix} .$$

A corresponding "once and for all notation" for the Hessian that suppresses mention of a specific function or specific point is

$$\begin{bmatrix} D_{11} & D_{12} & D_{13} \\ D_{21} & D_{22} & D_{23} \\ D_{31} & D_{32} & D_{33} \end{bmatrix} .$$

There are n^2 second partial derivatives of $f : \mathbb{R}^n \to \mathbb{R}$, and we might expect the "total second derivative" to display them all, just as the total first derivative is constructed from all n first partial derivatives. We will in fact use the Hessian matrix to formally define the second derivative. Recall that the first derivative at a point is not a number, but a linear map. The following definition of the second derivative says that it also is not a number, but a function—and that this function is *not* linear, but quadratic.

Definition 7.2.2 Let f denote a scalar-valued function f that is twice differentiable at a point $X_0 \in \mathbb{R}^n$, and let H denote the Hessian of f. The **second derivative** of f (or, for emphasis, its **total second derivative**) at X_0, denoted by $f''(X_0)$, is the function from $\mathbb{R}^n$ into $\mathbb{R}$ given by the formula

$$[X]^T H [X] .$$

REMARKS:

- Recall that $[X]$ denotes the column matrix corresponding to the vector X, and therefore $[X]^T$, the transpose of $[X]$, is a row matrix. The matrix product $[X]^T H [X]$ is a 1-by-1 matrix which, as is customary, we interpret as a scalar.

- Using the notation of the above definition we write

$$f''(X_0)(X) = [X]^T H [X] ,$$

where H is the Hessian of f.

- Because of our earlier remark about the equality of mixed partial derivatives, $D_{ij} = D_{ji}$, so $H = H^T$. A matrix which equals its own transpose is called **symmetric**, so the Hessian is a symmetric matrix. Therefore, to use the terminology introduced in Exercise 5 of Section 6.1, the second derivative is a quadratic form. We will have more to say about quadratic forms in Section 7.3b.

- Even though $f''(X_0)$ is not a linear map, we speak of the Hessian as being its corresponding matrix and write $[f''(X_0)]$ for the Hessian. Of course, since the second derivative is a quadratic form, the relation between the Hessian and the second derivative is not the same as that between a matrix and its corresponding linear map.

EXAMPLE:

1. Let $f(x_1, x_2) = x_1^2 x_2 + \sin x_2$. Then

$$(D_1 f)(X) = 2x_1 x_2 \quad \text{and} \quad (D_2 f)(X) = x_1^2 + \cos x_2 \,.$$

Now we take second partial derivatives:

$$(D_{11} f)(X) = 2x_2 , \qquad (D_{12} f)(X) = 2x_1 ,$$
$$(D_{21} f)(X) = 2x_1 , \qquad (D_{22} f)(X) = -\sin x_2 \,.$$

Since these are all continuous functions, the two first partial derivatives are differentiable (and it is no surprise that the two mixed second partials are equal). Hence f is twice differentiable and the Hessian at an arbitrary point $X_0 = (c_1, c_2)$ equals

$$[f''(X_0)] = \begin{bmatrix} 2c_2 & 2c_1 \\ 2c_1 & -\sin c_2 \end{bmatrix} \,.$$

For instance, setting $X_0 = (-1, \pi/2)$ we obtain

$$[f''(-1, \pi/2)] = \begin{bmatrix} \pi & -2 \\ -2 & -1 \end{bmatrix} \,.$$

To find a formula for the second derivative we calculate the following matrix product:

$$\begin{bmatrix} x_1 & x_2 \end{bmatrix} \begin{bmatrix} \pi & -2 \\ -2 & -1 \end{bmatrix} \begin{bmatrix} x_1 \\ x_2 \end{bmatrix} = [\pi x_1^2 - 4x_1 x_2 - x_2^2] \,.$$

Thus,

$$(f''(-1, \pi/2))(X) = \pi x_1^2 - 4x_1 x_2 - x_2^2 \,.$$

Let us repeat the above calculation for a general $X_0 = (c_1, c_2)$:

$$\begin{bmatrix} x_1 & x_2 \end{bmatrix} \begin{bmatrix} 2c_2 & 2c_1 \\ 2c_1 & -\sin c_2 \end{bmatrix} \begin{bmatrix} x_1 \\ x_2 \end{bmatrix}$$
$$= [(2c_2)x_1^2 + (4c_1)x_1 x_2 + (-\sin c_2)x_2^2] \,.$$

Thus,

$$(f''(X_0))(X) = (2c_2)x_1^2 + (4c_1)x_1x_2 + (-\sin c_2)x_2^2 .$$

7.2D TAYLOR'S THEOREM

Just as in the single-variable case, for which

$$g(t) = g(t_0) + g'(t_0)(t - t_0) + \frac{1}{2}g''(t^*)(t - t_0)^2 ,$$

we will obtain an exact expression for $f(X)$ involving the second derivative.

Theorem 7.2.5 (Taylor's Theorem, order two) Let f denote a scalar-valued function that is twice differentiable at each point on the line segment connecting two points X_0 and X in $\mathbb{R}^n$. Then there exists a point X^* on the line segment between X_0 and X such that

$$f(X) = f(X_0) + (f'(X_0))(X - X_0) + \frac{1}{2}f''(X^*)(X - X_0) .$$

REMARK:

- In analogy with the Mean Value Theorem, this theorem gives an exact expression for $f(X)$ at the expense of including in the expression a point X^* that is not described exactly.

PROOF: We use ideas we have already seen. Given X_0 and X, define, as in the proof of the Mean Value Theorem,

$$g(t) = f(X_0 + t(X - X_0)) ,$$

a scalar-valued function of the scalar t. In the proof of the Mean Value Theorem, we noted that

$$\begin{aligned} g(0) &= f(X_0) , \\ g(1) &= f(X) , \\ g'(0) &= f'(X_0)(X - X_0) . \end{aligned}$$

From the single-variable version of Taylor's Theorem we know that

$$\begin{aligned} g(1) &= g(0) + g'(0)(1 - 0) + \frac{1}{2}g''(t^*)(1 - 0)^2 \\ &= g(0) + g'(0) + \frac{1}{2}g''(t^*) \end{aligned}$$

for some t^* belonging to the interval $[0, 1]$. Thus

$$f(X) = f(X_0) + f'(X_0)(X - X_0) + \frac{1}{2}g''(t^*) .$$

Hence, we must show that

$$g''(t^*) = f''(X^*)(X - X_0)$$

for an appropriate choice of X^*.

Let $K(t) = X_0 + t(X - X_0)$ and, to prepare for work with coordinates, we let $X = (x_1, x_2, \ldots, x_n)$ and $X_0 = (c_1, c_2, \ldots, c_n)$. From the proof of the Mean Value Theorem we have

$$\begin{aligned}
g'(t) &= \langle (X - X_0), \nabla f(X_0 + t(X - X_0)) \rangle \\
&= (x_1 - c_1)((D_1 f) \circ K)(t) + (x_2 - c_2)((D_1 f) \circ K)(t) \\
&\quad + \cdots + (x_n - c_n)((D_n f) \circ K)(t) .
\end{aligned}$$

In order to calculate $g''(t)$ we apply the chain rule, Theorem 6.5.5 of Section 6.5c, term-by-term:

$$\begin{aligned}
g''&(t) \\
&= (x_1 - c_1)\langle ((\nabla D_1 f) \circ K)(t), (X - X_0) \rangle \\
&\quad + \cdots + (x_n - c_n)\langle ((\nabla D_n f) \circ K)(t), (X - X_0) \rangle \\
&= [X - X_0]^T H(K(t))[X - X_0] ,
\end{aligned}$$

where $H(K(t))$ is the Hessian of f at $K(t)$. We have used the fact, which the reader should verify, that the rows of the matrix $H(K(t))$ are the vectors

$$((\nabla D_1 f) \circ K)(t), \ldots, ((\nabla D_n f) \circ K)(t) .$$

We let $t = t^*$ and $X^* = K(t^*)$ to complete the proof. $<<$

REMARKS:

- The proofs of the Mean Value Theorem and Taylor's Theorem given above both depend on the truth of these theorems for functions of one variable.

- It is good practice to work though the proofs, identifying the domain and image of each function, especially whether they consist of vectors or scalars.

- In the following section we use Taylor's Theorem to prove the second derivative test, which is used in classifying critical points.

Exercises

1. Find the Hessian of each of the following functions:

    (a) $f(x_1, x_2) = x_1^2 - 4x_1 x_2 + x_2^2$

    (b) $g(x, y) = 2xy$

(c) $h(x,y) = e^{3x-2y}$

(d) $\varphi(x_1, x_2) = \sin(x_1 x_2)$

(e) $f(x_1, x_2, x_3) = x_1 x_2 x_3$

(f) $u(x,y,z) = xe^{yz}$.

2. For each part of the preceding exercise find the second derivative at

(a) the origin

(b) at a general point X_0.

(Keep in mind that the second derivative at a point is a function, not a vector or a matrix.)

3. Let $f, g: \mathbb{R}^n \to \mathbb{R}$ be given differentiable functions. Prove the following.

(a) If $f'(X) = O$ for all $X \in \mathbb{R}^n$, then f is a constant function.

(b) If $f'(X) = g'(X)$ for all $X \in \mathbb{R}^n$ and $f(X_0) = g(X_0)$ for some fixed X_0, then $f(X) = g(X)$ for all $X \in \mathbb{R}^n$.

4. Find a function $f: \mathbb{R}^2 \to \mathbb{R}$ such that $[f'(X)] = [3 \;\; -2]$ for all $X \in \mathbb{R}^2$ and $f(1, -1) = 4$. Is there more than one such f? Explain.

5. (a) Let $f: \mathbb{R}^2 \to \mathbb{R}$ be a twice differentiable function such that $f''(X) = O$ for all $X \in \mathbb{R}^2$. Prove that f is an affine map.

(b) Generalize to $\mathbb{R}^n$.

6. Find a function $f: \mathbb{R}^2 \to \mathbb{R}$ such that

$$[f''(X)] = \begin{bmatrix} 1 & 2 \\ 2 & -3 \end{bmatrix}$$

for all $X \in \mathbb{R}^2$, $[f'(1,0)] = [\; -2 \;\; 5 \;]$, and $f(1,0) = 7$. Is there more than one such f? Explain.

7. Let $z = T(x,y)$ be the tangent plane to the graph of $f(x,y) = 2x^2 + 3y^2$ in $\mathbb{R}^3$ at some point (x_0, y_0). Show that $f(x,y) \geq T(x,y)$ for every (x,y) by showing that the second derivative term in Taylor's Theorem is never negative.

8. Repeat Exercise 7 for the function $f(x,y) = x^4 + 3x + e^{-y} - 2$.

9. Let $f: \mathcal{D}(f) \to \mathbb{R}$, where $\mathcal{D}(f) \subseteq \mathbb{R}^n$, be a differentiable function, and suppose that $f'(X) = O$ for every $X \in \mathcal{D}(f)$ and that $\mathcal{D}(f)$ is pathwise connected.

(a) Prove that f is a constant function. (Since it is being assumed that f is differentiable at every point in $\mathcal{D}(f)$, then $\mathcal{D}(f)$ must be its own interior.)

(b) Give an example that shows that if we drop the assumption that $\mathcal{D}(f)$ is pathwise connected, then it is possible that f is not a constant function even if its derivative is zero at every point in its domain.

7.3   The Second Derivative Test

7.3A   ONE VARIABLE AND APPROXIMATION BY PARABOLAS

We now begin our discussion of a method for deciding the nature of a critical point. In the single-variable case there are three possibilities for f'' at a critical point x_0: $f''(x_0) > 0$, $f''(x_0) < 0$, and $f''(x_0) = 0$.

Let us examine how these three cases arise from examining Taylor's Theorem. Because x_0 is a critical point the first order term in Taylor's Formula is 0 and, hence

$$f(x) = f(x_0) + \frac{1}{2} f''(x^*)(x - x_0)^2$$

for an appropriate x^* between x_0 and x^*. Since $(x - x_0)^2 \geq 0$, the sign of $f''(x^*)$ tells us how $f(x)$ compares with $f(x_0)$. But calculating x^* seems to be at least as difficult as deciding the nature of the critical point. It would be nice if were we able to replace $f''(x^*)$ by $f''(x_0)$. We can do this if $f''(x_0) \neq 0$ and x is sufficiently close to x_0; because then, $f''(x^*)$ and $f''(x_0)$ have the same sign (provided f'' is continuous). It is only when $f''(x_0) = 0$ that we cannot use this argument to decide whether $f''(x^*)$ is positive, negative, or 0. The result of these considerations is Theorem 7.2.3 stated earlier.

One view of the story just described is that the parabola corresponding to the quadratic function

$$f(x_0) + \frac{1}{2} f''(x_0)(x - x_0)^2$$

approximates the graph of f well enough so that we can decide the nature of the critical point x_0 of f by deciding its nature for the parabola, except in the case that the parabola is degenerate—a horizontal line. Since we are interested in the sign of $f(x) - f(x_0)$ our focus has been on two issues: (1) the behavior of the function $f''(x_0)(x - x_0)^2$ and (2) the question of whether this function and the function $f''(x^*)(x - x_0)^2$ behave similarly. We will confront these same issues in the vector setting.

7.3B   QUADRATIC FORMS

Let $H = [c_{ij}]$ be an n-by-n symmetric matrix, that is, $c_{ij} = c_{ji}$ for all i and j. A function of the form

$$h(X) = [X]^T H [X]$$

from $\mathbb{R}^n$ into $\mathbb{R}$ is called a **quadratic form** (see Exercise 5 in Section 6.1).

REMARK:

- We are often using H rather than A or B to denote a symmetric matrix as a reminder that there is connection to be made with Hessians.

Here are two simple observations. If h is any quadratic form, then $h(\mathbf{0}) = 0$. If $H = I$, the identity, then

$$h(X) = [X]^T X = x_1^2 + \cdots + x_n^2 = \|X\|^2 \,.$$

We divide the class of quadratic forms h into four subclasses as follows:

1. $h(X) > 0$ for all $X \neq \mathbf{0}$, in which case h is said to be **positive definite**;

2. $h(X) < 0$ for all $X \neq \mathbf{0}$, in which case h is said to be **negative definite**;

3. $h(X) > 0$ for some X and $h(X) < 0$ for some other X, in which case h is said to be **indefinite**;

4. h does not satisfy any of the first three conditions, in which case it is called **semidefinite**.

All the above terms used to describe quadratic forms can also be applied to the corresponding symmetric matrices. Thus, a negative definite matrix is a symmetric matrix H such that $C^T H C < 0$ for all nonzero column matrices C having the appropriate number of rows.

REMARK:

- A quadratic form on $\mathbb{R}^1$ is a function of the form $h(x) = ax^2$ for some real number a. Such a function cannot satisfy condition 3 in the above list. Because condition 3 can arise for quadratic forms on $\mathbb{R}^n$ for $n > 1$, we expect a general second derivative test that is more involved than that for the single-variable case.

EXAMPLES:

1. The n-by-n identity I matrix is positive definite, because

$$[X]^T I [X] = x_1^2 + \cdots x_n^2$$

is positive if at least one of the numbers x_j is nonzero.

2. The matrix $-I$ is negative definite.

3. The matrix

$$H = \begin{bmatrix} 2 & 0 \\ 0 & 3 \end{bmatrix}$$

is positive definite, because

$$[x_1 \ x_2] H \begin{bmatrix} x_1 \\ x_2 \end{bmatrix} = 2x_1^2 + 3x_2^2 > 0$$

for $(x_1, x_2) \neq (0, 0)$.

4. The matrix

$$\begin{bmatrix} 2 & 0 \\ 0 & -3 \end{bmatrix}$$

is indefinite, since

$$[x_1 \ x_2] \begin{bmatrix} 2 & 0 \\ 0 & -3 \end{bmatrix} \begin{bmatrix} x_1 \\ x_2 \end{bmatrix} = 2x_1^2 - 3x_2^2$$

is positive if $x_2 = 0 < x_1$ and negative if $x_1 = 0 < x_2$.

5. The quadratic form $h(x_1, x_2) = x_2^2$ corresponding to the matrix

$$\begin{bmatrix} 0 & 0 \\ 0 & 1 \end{bmatrix}$$

is semidefinite since $h(X) \geq 0$ for all X, which rules out indefiniteness and negative definiteness, and $h(1,0) = 0$, which rules out positive definiteness.

An important issue in linear algebra is that of deciding whether a given symmetric matrix is positive definite, negative definite, indefinite, or semidefinite. This issue for an arbitrary symmetric n-by-n matrix may be quite complicated. We content ourselves with fully treating the case where $n = 2$. First, however, we establish one very useful result that applies to symmetric matrices of any size. This is a sufficient condition for indefiniteness.

Theorem 7.3.1 Let $H = [c_{ij}]$ be an n-by-n symmetric matrix. If H has two main diagonal entries of opposite sign, then H is indefinite.

PROOF: You may check that $[\mathbf{e}_k]^T H[\mathbf{e}_k] = c_{kk}$ for each k. The fact that two of these numbers have opposite signs implies that H is indefinite. <<

EXAMPLES:

6. Theorem 7.3.1 enables us to conclude that the following symmetric matrices are indefinite:

$$\begin{bmatrix} 2 & 3 \\ 3 & -1 \end{bmatrix}, \quad \begin{bmatrix} 3 & 0 & 2 \\ 0 & -1 & 1 \\ 2 & 1 & -5 \end{bmatrix}, \quad \begin{bmatrix} 6 & 2 & 3 & 0 & 1 \\ 2 & 0 & -1 & 4 & -1 \\ 3 & -1 & 4 & -4 & 0 \\ 0 & 4 & -4 & 0 & 2 \\ 1 & -1 & 0 & 2 & -4 \end{bmatrix}.$$

7. Even though the matrix

$$\begin{bmatrix} 1 & -5 \\ -5 & 2 \end{bmatrix},$$

has only positive numbers on the main diagonal, the corresponding quadratic form h is indefinite because $h(1,0) = 1 > 0$ and $h(1,1) = -7 < 0$.

For our treatment of quadratic forms on $\mathbb{R}^2$ we use some algebraic facts. Recall that the determinant of any 2-by-2 matrix

$$\begin{bmatrix} a & b \\ c & d \end{bmatrix}$$

equals $ad - bc$. For a symmetric matrix

$$H = \begin{bmatrix} a & b \\ b & d \end{bmatrix}$$

this formula becomes

$$\det H = ad - b^2 .$$

Let h denote the corresponding quadratic form. Straightforward matrix multiplication gives

$$h(x_1, x_2) = ax_1^2 + 2bx_1x_2 + dx_2^2 .$$

Theorem 7.3.2 Let

$$H = \begin{bmatrix} a & b \\ b & d \end{bmatrix}$$

be a 2-by-2 symmetric matrix. Then H is

1. positive definite if $\det H > 0$ and $a > 0$;

2. negative definite if $\det H > 0$ and $a < 0$;

3. indefinite if $\det H < 0$;

4. semidefinite if $\det H = 0$.

REMARK:

- If $a = 0$, then $\det H = -b^2$. It follows that if $\det H > 0$, then $a \neq 0$. Thus the theorem covers all possible cases for the 2-by-2 case.

PROOF: Let h denote the quadratic form corresponding to the matrix H. We consider two different cases: (1) if at least one of a and d is different from zero and (2) if $a = d = 0$.

Case 1: a or d is different from 0. We assume $a \neq 0$; the proof is similar if $d \neq 0$. Completing the square gives

$$h(x_1, x_2) = \frac{1}{a}[(ax_1 + bx_2)^2 + (ad - b^2)x_2^2] ,$$

which you may verify by multiplying out the right-hand side and combining terms to get $ax_1^2 + 2bx_1x_2 + dx_2^2$ back.

Case 1,1: $\det H > 0$ and $a > 0$. In the above expression for h the coefficient $1/a$ is positive. Both terms are nonnegative, and moreover, the second term is positve if $x_2 \neq 0$ and the first term is positive if $x_2 = 0$ and $x_1 \neq 0$. Thus h is positive definite.

Case 1,2: $\det H > 0$ and $a < 0$. The discussion of case 1.1 applies with the modification that the coefficient $1/a$ is negative. Thus H is negative definite.

Case 1,3: $\det H < 0$. Choose $x_2 = 0$ and $x_1 \neq 0$ to make $h(x_1, x_2)$ have the same sign as $1/a$. Choose $x_2 \neq 0$ and x_1 so that $ax_1 + bx_2 = 0$ to make $h(x_1, x_2)$ have the sign opposite that of $1/a$. Thus H is indefinite.

Case 1,4: $\det H = 0$. Then $h(X) = (1/a)(ax_1 + bx_2)^2$. We can choose x_1 and x_2 not both zero, so that $ax_1 + bx_2 = 0$. Hence H is neither positive definite nor negative definite. Also, $h(x_1, x_2)$, when not 0, has the same sign as a. Hence H is not indefinite. Therefore H is semidefinite.

Case 2: $a = d = 0$. Thus, $\det H = -b^2 \leq 0$ and so what we would call cases 2.1 and 2.2 cannot occur. For the other two cases we have

$$h(X) = 2bx_1x_2 \,.$$

Case 2,3: $\det H < 0$. Since $b \neq 0$, $h(x_1, x_2)$ has the same sign as b if x_1 and x_2 have identical signs and it has the sign opposite to that of b if x_1 and x_2 have signs opposite to each other. Thus H is indefinite.

Case 2,4: $\det H = 0$. Thus $b = 0$, so that H is the zero matrix, which is semidefinite. <<

EXAMPLES:

8. To classify the symmetric matrix

$$\begin{bmatrix} 2 & 3 \\ 3 & 8 \end{bmatrix}$$

we first calculate that its determinant equals $7 > 0$ and then observe that the upper left entry of the matrix equals $2 > 0$. Thus the given matrix is positive definite.

9. In order to decide about definiteness and indefiniteness of the symmetric matrix

$$\begin{bmatrix} 1 & 1 \\ 1 & 1 \end{bmatrix}$$

we first calculate its determinant. Getting the result 0 we immediately conclude that it is semidefinite, that is, it is neither indefinite nor positive or negative definite.

10. Let us classify the quadratic form

$$h(x_1, x_2) = [x_1 \ x_2] \begin{bmatrix} 2 & -3 \\ -3 & 1 \end{bmatrix} \begin{bmatrix} x_1 \\ x_2 \end{bmatrix} \,.$$

The determinant of the corresponding matrix is $-7 < 0$, so h is indefinite.

7.3C THE SECOND DERIVATIVE AND CRITICAL POINTS

The first derivative of a scalar-valued function f of a vector variable is **continuous** at a point if the first partial derivatives are continuous there. Thus, there is no distinction between the applicability of the adjective "continuity" to all the first partial derivatives and to the total derivative. On the other hand, for existence we recall that there is a distinction; it is possible for each partial derivative to exist without the total derivative existing. The (total) second derivative of f is **continuous** at a point if all second partial derivatives are continuous there.

Theorem 7.3.3 (Second Derivative Test) Let $X_0 \in \mathbb{R}^n$ be a critical point of a scalar-valued function f with a second derivative that is continuous at X_0. Then the following assertions are true:

1. f has a local maximum at X_0 if $f''(X_0)$ is negative definite;

2. f has a local minimum at X_0 if $f''(X_0)$ is positive definite;

3. f has a saddle point at X_0 if $f''(X_0)$ is indefinite;

4. the second derivative $f''(X_0)$ does not give sufficient information to decide among local maximum, local minimum, or saddle point if it is semidefinite.

PROOF: Taylor's Theorem gives

$$f(X) = f(X_0) + \frac{1}{2}f''(X^*)(X - X_0) \, ,$$

for some X^* on the line segment between X and X_0.

Suppose that $f''(X_0)$ is negative definite. Thus, $f''(X_0)(X - X_0) < 0$ for all $X \neq X_0$. Since f'' is continuous at X_0 we get the same conclusion if X_0 is replaced by Y that is sufficiently close to X_0:

$$f''(Y)(X - X_0) < 0 \, .$$

If X is sufficiently close to X_0, then X^* is such a Y, and so, by the formula above from Taylor's Theorem, $f(X) < f(X_0)$.

Similar arguments hold in case $f''(X_0)$ is positive definite or indefinite.

We will omit the proof for the case where $f''(X_0)$ is semidefinite, but the idea of the proof will become clear in the examples. <<

REMARKS:

- The second paragraph of the preceding proof has a defect. An exercise requests that you find it.

- The preceding theorem does not completely solve the problem of classifying critical points. Rather, it replaces that problem with another, that of classifying a Hessian. Section 7.3b contains a solution of this latter problem in case $n = 2$, but not for $n > 2$. The full solution of the problem for general n involves certain quantities called the "eigenvalues" of the matrix H, and is beyond the scope of this book.

We look at several examples in which the goal is to find and classify all critical points. We will set the first derivative equal to 0 to find the critical points and then, at least as a first try, use the second derivative test to classify them.

EXAMPLES:

11. Let $f(x_1, x_2) = 1 - x_1^2 - x_2^2$. We found in Example 1 of Section 7.1 that $(0,0)$ is the only critical point of f. We readily compute

$$[f''(x_1, x_2)] = \begin{bmatrix} -2 & 0 \\ 0 & -2 \end{bmatrix}$$

for all (x_1, x_2), not just $(0,0)$. This is negative definite by Theorem 7.3.2, and so, there is a local maximum at $(0,0)$ by Theorem 7.3.3.

12. Let $g(x, y) = 3 - 2x + x^2 + 4y^2$. To find the critical points we set the first partial derivatives equal to zero:

$$g_x(x, y) = -2 + 2x = 0$$
$$g_y(x, y) = 8y = 0 ,$$

obtaining one critical point $(1, 0)$. The Hessian at $(1, 0)$ (and at every other point) equals

$$\begin{bmatrix} 2 & 0 \\ 0 & 8 \end{bmatrix} .$$

This matrix is positive definite (why?), and so there is a local minimum at the critical point $(1, 0)$.

13. Let $f(x, y) = [x^2 + (y+1)^2][x^2 + (y-1)^2]$. Before computing blindly, we note that $f(x, y) \geq 0$ and $f(x, y) = 0$ only if $(x, y) = (0, -1)$ or $= (0, 1)$. Hence, we deduce that there is a local minimum (in fact global minimum) at $(0, -1)$ and at $(0, 1)$ (why?). But there may be others; so, let us engage in systematic calculation.

As our first step we compute the partial derivative with respect to x:

$$f_x(x, y) = [x^2 + (y-1)^2]2x + [x^2 + (y+1)^2]2x$$
$$= 2x[2x^2 + (y-1)^2 + (y+1)^2],$$

which equals 0 if and only if $x = 0$. Likewise, we compute

$$f_y(x, y) = 2(y+1)[x^2 + (y-1)^2] + 2(y-1)[x^2 + (y+1)^2].$$

Since we have already seen that x must equal 0 at any critical point, we substitute 0 for x in the expression for $f_y(x, y)$:

$$f_y(0, y) = 2(y+1)(y-1)^2 + 2(y-1)(y+1)^2$$
$$= 2(y+1)(y-1)[(y-1) + (y+1)] = 4y(y+1)(y-1),$$

which equals 0 if and only if $y = -1$, $y = 1$ or $y = 0$.

Hence the critical points of f are $(0, -1)$, $(0, 1)$, and $(0, 0)$, the first two of which we found before we began systematic calculation. The third critical point $(0, 0)$ is a revelation. At $(0, 0)$ is there a local maximum, a local minimum, or a saddle point?

We examine the Hessians. The partial derivatives are given by

$$f_{xx}(x, y) = 12x^2 + 4y^2 + 4,$$
$$f_{xy}(x, y) = 8xy,$$
$$f_{yy}(x, y) = 4x^2 + 12y^2 - 4.$$

Thus the Hessians for $(0, -1)$, $(0, 1)$, and $(0, 0)$, which we obtain by substitution, equal

$$\begin{bmatrix} 8 & 0 \\ 0 & 8 \end{bmatrix}, \quad \begin{bmatrix} 8 & 0 \\ 0 & 8 \end{bmatrix}, \quad \text{and} \quad \begin{bmatrix} 4 & 0 \\ 0 & -4 \end{bmatrix},$$

respectively. We see that the first two (identical) matrices are positive definite, and so there are local minima at $(0, -1)$ and $(0, 1)$ as expected. On the other hand, the third matrix is indefinite, which implies that $(0, 0)$ is a saddle point. The two minima and the saddle point are illustrated in Figure 7.7.

14. Let $f(x, y, z) = 1 - 2x + 3x^2 - xy + xz - z^2 + 4z + y^2 + 2yz$. To find the critical points we first set the three partial derivatives equal to 0 as follows:

$$-2 + 6x - y + z = 0$$
$$-x + 2y + 2z = 0$$
$$4 + x + 2y - 2z = 0.$$

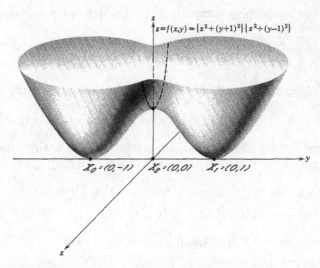

FIGURE 7.7. Global minima at X_0 and X_1; saddle point at X_2

This is a system of three linear equations in the unknowns x, y, and z. By eliminating variables as usual, we find that the unique solution to this system is $(x, y, z) = (0, -1, 1)$. Hence $(0, -1, 1)$ is the unique critical point of f.

The second partial derivatives are given by

$$f_{xx} = 6, \qquad f_{xy} = -1, \qquad f_{xz} = 1,$$
$$f_{yy} = 2, \qquad f_{yz} = 2,$$
$$f_{zz} = -2.$$

(Note that $f_{xy} = f_{yx}$, and so on.) We see that the Hessian is the constant matrix

$$\begin{bmatrix} 6 & -1 & 1 \\ -1 & 2 & 2 \\ 1 & 2 & -2 \end{bmatrix}.$$

Now we observe that the main diagonal includes both positive and negative numbers. By Theorem 7.3.1, the Hessian is indefinite, and, therefore, by Theorem 7.3.3, the critical point $X_0 = (0, -1, 1)$ is a saddle point for function f.

The given function in this example is a polynomial of degree two, a quadratic. As you may easily check,

$$f(X) = f(0) + f'(0)(X) + \frac{1}{2} f''(0)(X).$$

Thus it was to be expected that the Hessian would be constant. What we have calculated is the matrix that represents the second degree term in the polynomial, that term being a quadratic form, and we have assessed that the quadratic form is indefinite.

15. Now we consider a situation where the second derivative test fails to apply (see Exercise 11 for others). Consider the two functions $f(x, y) = x^4 + y^4$ and $g(x, y) = -f(x, y) = -x^4 - y^4$. Both functions have the same critical point $(0, 0)$.

At $(0, 0)$, each function has the same Hessian, namely, the zero matrix. Since $f(0, 0) = 0$ and $f(X) \geq 0$ for all X, f has a global and, therefore, local minimum, at $(0, 0)$. Similarly g has a local (and global) maximum there. The *same* Hessian

$$\begin{bmatrix} 0 & 0 \\ 0 & 0 \end{bmatrix}$$

cannot be expected to inform us of the *different* natures of f and g. Not only does the second derivative test fail for these two functions, but the arguments show that there is no "improved second derivative test" that would work for these two functions.

REMARK:

- The examples above illustrate a method for locating and classifying extrema. Roughly speaking, the search for extrema has been reduced to solving equations to find the critical points and then using linear algebra (studying a certain symmetric matrix) to classify the critical points. To get to this stage, however, we have made use of many monuments of calculus: the directional derivative, a chain rule, the Mean Value Theorem, the second derivative, the Hessian, and Taylor's Theorem.

Exercises

1. Determine if the following matrices are positive definite, negative definite, indefinite, or semidefinite.

(a) $\begin{bmatrix} 1 & 0 \\ 0 & 2 \end{bmatrix}$

(b) $\begin{bmatrix} 1 & 1 \\ 1 & -2 \end{bmatrix}$

(c) $\begin{bmatrix} 3 & 5 \\ 5 & 4 \end{bmatrix}$

(d) $\begin{bmatrix} -3 & 2 \\ 2 & -4 \end{bmatrix}$

(e) $\begin{bmatrix} 1 & 2 \\ 2 & 4 \end{bmatrix}$

(f) $\begin{bmatrix} -1 & 2 \\ 2 & -4 \end{bmatrix}$

(g) $\begin{bmatrix} 0 & 0 \\ 0 & 3 \end{bmatrix}$

(h) $\begin{bmatrix} 1 & 1 \\ 1 & 1 \end{bmatrix}$

(i) $\begin{bmatrix} 1 & 2 & 3 \\ 2 & -4 & 5 \\ 3 & 5 & 6 \end{bmatrix}$

(j) $\begin{bmatrix} 1 & 0 & 0 \\ 0 & 2 & 0 \\ 0 & 0 & 4 \end{bmatrix}$

(k) $\begin{bmatrix} 3 & 2 & 0 \\ 2 & 4 & 0 \\ 0 & 0 & 1 \end{bmatrix}$.

2. Find and classify the critical points of the following functions, even if the second derivative test fails.

(a) $f(x_1, x_2) = x_1^2 + 4x_1 + 2x_2^2 + 10$

(b) $\varphi(x, y) = 3 - 2x + 2y + x^2y^2$

(c) $f(x, y) = xy - x + y - 2$

(d) $u(x_1, x_2) = x_1^2 - x_2^2$

(e) $g(x, y) = y^3 - 2x^2 - 2y^2 + y$

(f) $v(x, y) = (x^2 + y^2)^2 - 8y^2$

(g) $w(x, y) = x^2 - 2xy + \frac{1}{3}y^3 - 3y$

(h) $\psi(x_1, x_2) = \dfrac{\cos x_2}{1 + x_1^2}$

(i) $h(x, y) = \dfrac{\cosh(2x)}{1 + 2y^2}$

(j) $f(x, y) = \dfrac{3x^4 + 4x^3 - 12x^2 + 6}{1 + y^2}$

(k) $\psi(y, z) = \dfrac{z^3 - 3z}{1 + y^2}$

(l) $k(x_1, x_2) = (x_1^2 - 1)e^{x_2}$

(m) $g(x, y, z) = \dfrac{xye^y}{1 + x^2} + z^2 + 2z$

(n) $h(x, y, z) = \dfrac{3x^4 - 6x^2 + 1}{1 + z^2} - y^2$

(o) $\varphi(x, y, z) = x^3 - 3x + y^2 + z^2$

(p) $r(x, y, z) = (1 + 2x + 3y - z)^2$.

3. Prove that if $f(x, y)$ is twice differentiable and has a local minimum at (a, b), then $f_{xx}(a, b) \geq 0$ and $f_{yy}(a, b) \geq 0$. *Hint:* Consider the behavior of f on the lines $x = a$ and $y = b$.

4. Prove that if $f(x, y)$ satisfies $3f_{xx} + 4f_{yy} = -1$ everywhere in $\mathbb{R}^2$, then f cannot have a local minimum anywhere.

5. Let $f \colon \mathbb{R} \to \mathbb{R}$ be twice differentiable. A continuous function f is called **concave upwards** if

$$f\left(\frac{a+b}{2}\right) \leq \frac{1}{2}[f(a) + f(b)] \; \bullet$$

for every a and b.

(a) Prove that f is concave upwards if $f''(x) \geq 0$ for all x.

(b) If $f''(x) \geq 0$ for all x, prove that the graph of the curve $y = f(x)$ lies above its tangent line at any point.

6. A function $f \colon \mathbb{R}^n \to \mathbb{R}$ is called **concave upwards** if

$$f\left(\frac{X+Y}{2}\right) \leq \frac{1}{2}[f(X) + f(Y)]$$

for all X and Y.

(a) Prove that a twice differentiable function f is concave upwards if $f''(X)$ is positive definite for all X.

(b) If $f''(X)$ is positive definit for all X, prove that the graph of $z = f(X)$ lies above the graph of the tangent map $z = T(X)$ at any point X_0.

7. Let H be a symmetric n-by-n matrix and let h denote the corresponding quadratic form.

(a) Show that $(\nabla h)(X)$ equals twice $L(X)$ where L is the linear map whose matrix is H.

(b) Show the Hessian of h is $2H$ (at any point).

8. Let H be a symmetric n-by-n matrix, Z a fixed vector in $\mathbb{R}^n$, and define f by

$$f(X) = [X]^T H[X] - 2\langle X, Z \rangle \,.$$

(a) Prove that X is a critical point of f if and only if $H[X] = [Z]$.

(b) Show that the Hessian of f is $2H$ (at any point).

9. Let g be the quadratic polynomial

$$g(x_1, x_2) = ax_1^2 + bx_1x_2 + cx_2^2 + dx_1 + ex_2 + f \,,$$

where $a, b, c, d, e,$ and f are fixed real numbers.

(a) Compute the (total) first and second derivatives of g.

(b) Show that g can be written in the form

$$g(X) = [X]^T A[X] + \langle X, Z \rangle + k ,$$

where A is a fixed symmetric 2-by-2 matrix, Z is a fixed vector in $\mathbb{R}^2$, and k is a fixed real number. Relate A, Z, and k to g and its derivatives.

10. Let

$$A = \begin{bmatrix} a & b \\ b & c \end{bmatrix}$$

be positive definite. Prove that

$$B = \begin{bmatrix} a & b & 0 \\ b & c & 0 \\ 0 & 0 & \alpha \end{bmatrix}$$

is positive definite $\Leftrightarrow \alpha > 0$.

11. This exercise shows that the semidefinite matrix

$$\begin{bmatrix} 2 & 0 \\ 0 & 0 \end{bmatrix}$$

can be the Hessian at a saddle point and that it can also be the Hessian at a point where there is a local minimum.

(a) Classify the critical points of $f(x, y) = x^2 + y^3$.

(b) Classify the critical points of $f(x, y) = x^2 + y^4$.

12. The aim of this exericse is to show that the nature of a critical point cannot be determined by merely approaching it along straight lines. Let $f(x, y) = (y - 4x^2)(y - x^2)$.

(a) Make a sketch of $\mathbb{R}^2$, indicating the points where $f(x, y) = 0$, where $f(x, y) > 0$, and where $f(x, y) < 0$.

(b) Show that the origin is a critical point of f.

(c) Show that on any straight line through the origin, the function f has a local minimum at the origin.

(d) Use some other path to show that $(0, 0, 0)$ is actually a saddle point of f.

13. For each of the following statements give either a proof or counterexample.

(a) If a matrix is positive definite, then all its entries are positive.

(b) If all the elements of a symmetric matrix are positive, then the matrix is positive definite.

(c) A diagonal matrix is positive definite $\Leftrightarrow$ all of the diagonal elements are positive. (A matrix $[c_{ij}]$ is a **diagonal** matrix if and only if $c_{ij} = 0$ whenever $i \neq j$.)

14. Find the defect in the second paragraph of the proof of the second derivative test. Then fix it.

7.4   Global Extrema

7.4A   OPEN SETS, CLOSED SETS, AND BOUNDED SETS

In the single-variable setting we know that a continuous function whose domain is a bounded closed interval has a global maximum and a global minimum. In this section we identify the vector version of a bounded closed interval. One feature to notice about a closed interval is that every member of its complement is an interior point of its complement. When we generalize to $\mathbb{R}^n$, it is this feature of the complement that is our starting point.

A subset of $\mathbb{R}^n$ is said to be **open** if it equals its own interior, that is, if every one of its members is an interior point. According to this definition open balls are open sets. A subset of $\mathbb{R}^n$ is said to be **closed** if its complement is open. Closed balls and affine subspaces are examples of closed sets. Any member of a closed subset of $\mathbb{R}^n$ which is not an interior point of that subset is called a **boundary point** of the subset. The **boundary** of a subset of $\mathbb{R}^n$ is equal to the set consisting of its boundary points. The boundary of the closed ball $\bar{B}(X;r)$ is the sphere $S(X;r)$. A subset of $\mathbb{R}^n$ is said to be **bounded** if it is a subset of some closed (or open) ball.

EXAMPLE:

1. Let us reason informally to classify the following eight subsets of $\mathbb{R}^2$ according to openness, closedness, and boundedness:

$$
\begin{aligned}
\mathcal{Q}_1 &= \{X : x_1 + x_2 \le 1\}\,; \\
\mathcal{Q}_2 &= \{X : x_1 + x_2 < 1\}\,; \\
\mathcal{Q}_3 &= \{X : |x_1| + |x_2| \le 1\}\,; \\
\mathcal{Q}_4 &= \{X : |x_1| + |x_2| < 1\}\,; \\
\mathcal{Q}_5 &= \{X : x_1 < 1 \text{ and } x_2 \le 1\}\,; \\
\mathcal{Q}_6 &= \{X : |x_1| < 1 \text{ and } |x_2| \le 1\}\,; \\
\mathcal{Q}_7 &= \text{the empty set}\,; \\
\mathcal{Q}_8 &= \mathbb{R}^2\,.
\end{aligned}
$$

For fixed x_2, the other coordinate x_1 can be chosen arbitrarily far in the negative direction without the point X falling outside $\mathcal{Q}_1$. So $\mathcal{Q}_1$ is unbounded. Any open ball centered at the point $(1, 0)$ contains a point of the form $(1 + \varepsilon, 0)$ for some $\varepsilon > 0$, a point that does not belong to $\mathcal{Q}_1$. So $\mathcal{Q}_1$ is not open. If $x_1 + x_2 > 1$, then the inequality $w_1 + w_2 > 1$ will hold for all points W in a sufficiently small ball centered at X. So, the complement of $\mathcal{Q}_1$ is open and, hence, $\mathcal{Q}_1$ is closed.

The set Q_2 is unbounded for the same reason that Q_1 is. Arguments similar to those of the preceding paragraph show that Q_2 is open and not closed.

In so far as openness and closedness are concerned, Q_3 is like Q_1 and Q_4 is like Q_2. But Q_3 and Q_4 are both subsets of $\bar{B}(0;1)$ and thus are bounded.

The ball $\bar{B}(0;\sqrt{2})$ contains Q_6, so Q_6 is bounded. The set Q_5 is not bounded because it contains all points both of whose coordinates are negative. The point $(0,1)$ belongs to both Q_5 and Q_6. In view of the fact that any ball centered at $(0,1)$ contains points of the form $(0,1+\varepsilon)$ for some $\varepsilon > 0$, we see that neither Q_5 nor Q_6 is open. The point $(1,0)$ belongs to the complement of both Q_5 and Q_6. Any ball centered there contains points of the form $(1-\delta,0)$ for $\delta > 0$, but less than 2. Such a point belongs to both Q_5 and Q_6. So the complements of Q_5 and Q_6 are not open. Thus neither Q_5 nor Q_6 is closed.

Clearly $\mathbb{R}^2$ is unbounded and the empty set is bounded. The set $\mathbb{R}^2$ is open because every ball (no matter where it is centered and what its radius equals) is a subset of $\mathbb{R}^2$. The empty set is open because there are no points in the empty set that are not interior points (there are no points at all in the empty set, interior or otherwise). Since $\mathbb{R}^2$ and the empty set are complements of each other and both are open, it follows that both are also closed.

7.4B   EXISTENCE OF GLOBAL EXTREMA

For a proof of the following theorem see an advanced calculus book.

Theorem 7.4.1 Let f be a scalar-valued continuous function whose domain is a closed and bounded subset of $\mathbb{R}^n$. Then f has a global maximum and a global minimum.

EXAMPLE:

2. Let us find the global maximum and the global minimum of the function f defined by

$$f(x,y) = x\sin(\pi xy)\,, \quad -1 \le x \le 1,\ 0 \le y \le 1\,.$$

Note that the domain of f is a closed rectangular region. The preceding theorem assures us that f has global extrema of both types.

If a global extremum occurs at an interior point of $\mathcal{D}(f)$, then there must also be a local extremum there, and, thus, it must be a critical point. So we begin by looking for critical points in the interior of $\mathcal{D}(f)$. The equations obtained from setting the two first partial derivatives equal to 0 are

$$\sin(\pi xy) + \pi xy \cos(\pi xy) = 0$$
$$\pi x^2 \cos(\pi xy) = 0 .$$

The second equation can be satisfied by making either $\cos(\pi xy)$ or x equal to 0. If $\cos(\pi xy) = 0$, then the left-hand side of the first equation equals ± 1, and thus the first equation is not satisfied. So, we take $x = 0$, and, when we do, both equations are satisfied for whatever y equals. Let us calculate the values of the given function at all these critical points:

$$f(0, y) = 0 ,$$

whatever the value of y.

Now we examine f at those points of its domain that are not interior points. These points consist of four line segments, the four edges of the rectangular region that makes up the domain of f. One of these four edges consists of points of the form $(x, 0)$ for $|x| \leq 1$. For such points

$$f(x, 0) = 0 .$$

There are three more edges to consider. For one of those edges we have

$$f(1, y) = \sin(\pi y) , \ 0 \leq y \leq 1,$$

the maximum and minimum values of which we can obtain by one-variable calculus techniques, not forgetting to consider the endpoints for y, namely 0 and 1. But we know enough about the sine function to conclude immediately that the maximum value of $f(1, y)$ is 1, achieved by setting $y = 1/2$, and the minimum value is 0, obtained by setting $y = 0$ or $y = 1$.

We note that $f(-1, y) = f(1, y)$, so it remains only to consider the fourth edge:

$$f(x, 1) = x \sin(\pi x) , \ |x| \leq 1 .$$

Since $|x| \leq 1$, both x and $\sin(\pi x)$ have the same sign, or one (or both) is 0. So $f(x, 1) \geq 0$ for $|x| \leq 1$. Also

$$|f(x, 1)| = |x \sin(\pi x)| = |x||\sin(\pi x)| \leq 1 \cdot 1 = 1$$

for $|x| \leq 1$.

Let us summarize. We have found points on the edges of the rectangular region at which f takes the values 0 and 1. We have also discovered that $0 \leq f(x, y) \leq 1$ along these edges. In the interior of the rectangular region, we have found infinitely many critical points, and at all of these critical points, f takes the value 0. Since a differentiable function can only attain its global extremum at a critical point in the interior of the domain or at a non-interior point of the domain, we may conclude without further investigation that the global maximum of f is 1 and the global minimum is 0. There is no need to use the second derivative test or any other means to classify the critical points of f.

Exercises

1. Decide whether the given subset of $\mathbb{R}^2$ is bounded or not, open or not, closed or not, and pathwise connected or not.

   (a) $\{X : x_1 + x_2^2 \leq -4\}$
   (b) $\{X : 1 \leq \|X\| < 5\}$
   (c) $\{X : |x_2| > 3\}$
   (d) $\{X : |x_2| > x_1\}$
   (e) $\{X : \|X\| \text{ is an integer}\}$
   (f) $\{X : x_2 = \sin(1/x_1)\}$
   (g) $\{X : x_2 = \sin(1/x_1) \text{ or } X = \mathbf{0}\}$
   (h) $\{X : x_2 = \sin(1/x_1)\} \bigcup \{X : x_1 = 0 \text{ and } \|x_2\| \leq 1\}$
   (i) $\{X : x_2 = x_1 \sin(1/x_1)\}$
   (j) $\{X : x_2 = x_1 \sin(1/x_1) \text{ or } X = \mathbf{0}\}$
   (k) $\{X : |x_2| < 1/(x_1^2 + 1)\}$
   (l) the spiral in $\mathbb{R}^2$ represented parametrically by $e^t(\cos t, \sin t)$, $t \in \mathbb{R}$.

2. Decide whether the given subset of $\mathbb{R}^3$ is bounded or not, open or not, closed or not, and pathwise connected or not.

   (a) $\{X : \|X\| = 3 \text{ and } x_3 > 0\}$
   (b) $\{X : 0 < x_1 < 1\}$
   (c) $\{X : 0 < x_1 < 1 \text{ and } 0 < x_2 < 1\}$.

3. Find the global maximum and global minimum of the function $f(x, y) = x \sin(\pi x y)$ with the indicated domain. Also, find the points in the domain where the value of the function equals a global extremum.

   (a) $\{(x, y) : 0 \leq x \leq 1, \ 0 \leq y \leq 1/2\}$
   (b) $\{(x, y) : 1/2 \leq x \leq 1, \ 0 \leq y \leq 1/2\}$
   (c) $\{(x, y) : 1/2 \leq x \leq 1, \ 1/2 \leq y \leq 1\}$
   (d) $\{(x, y) : 1/2 \leq x \leq 1, \ 1/2 \leq y \leq 3/4\}$
   (e) $\{(x, y) : -1 \leq x \leq 1, \ 1/2 \leq y \leq 3/4\}$.

4. Does the function $x \sin(\pi x y)$, defined on all of $\mathbb{R}^2$, have a global maximum? If so, find it. Does it have a global minimum? If so, find it.

5. Does the function $x \sin(\pi x y)$, defined on all of $\{(x, y) : y > 0\}$ have a global maximum? If so, find it. Does it have a global minimum? If so, find it.

7.5 Constrained Extrema

7.5A THE ISSUE

Suppose we are interested in the global extrema of a scalar-valued function f which is defined on some set $\mathcal{D}(f)$ contained in $\mathbb{R}^n$. In many practical problems, we will not want to consider the values $f(X)$ for all $X \in \mathcal{D}(f)$. For example, there are many problems for which it is unnatural to allow the coordinates of X to be negative, even if f happens to be defined at such points X. Thus we are often led to global extremum problems in which the domain of f is restricted in some way so that it is smaller than the set of points at which the function f is usually defined. If, as is often the case, the restricted domain of f is a closed bounded set, then we know that our search for global extrema will not be in vain, since Theorem 7.4.1 tells us in this case that f does have at least one global extremum of each type on the restricted domain.

EXAMPLES:

1. *Extremes of temperature.* Let $\mathcal{K}$ be the unit disk in $\mathbb{R}^2$,

$$\mathcal{K} = \{(x,y) \in \mathbb{R}^2 : x^2 + y^2 \leq 1\},$$

   representing a heated metal plate of radius 1. Suppose that the temperature at (x,y) in the plate is given by $f(x,y) = 60(y^2 - x^2)$ for $(x,y) \in \mathcal{K}$. Which are the hottest and coldest points on the plate?

2. *Maximum economic utility.* The cost per unit of commodities A and B is a and b, respectively. The total amount you are to spend is c. You will purchase x units of A and y units of B. Your purchase is, therefore, constrained by the "budget equation", which says that the total amount spent must be c, that is,

$$ax + by = c.$$

   Suppose that you are given a function $f(x,y)$ whose numerical value is the "utility index". This is a pure number that economists use to indicate the satisfaction or utility one receives from having x units of A and y units of B. Constructing such a "utility index function" $f(x,y)$ requires economic judgments, of course. Given such a function, however, it is natural to ask for the values of x and y satisfying the budget equation that maximize this function. That is, how much of A and how much of B should be acquired to produce the greatest satisfaction $f(x,y)$, while adhering to the limited budget? Thus $\mathcal{D}(f)$ is the line in $\mathbb{R}^2$ given by $ax + by = c$.

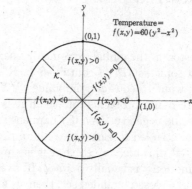

FIGURE 7.8.

REMARKS:

- As these examples indicate, it is natural to be concerned with only a subset of the set of variables (x, y) at which the function of interest is defined. A physicist or engineer is concerned with temperature only in the plate he or she is studying; an economist must deal with a situation in which resources are limited by a budget.

- These two examples are representative of the two kinds of constrained extremal problems. In the temperature example, the points (x, y) of interest were constrained to be on the unit disk $x^2 + y^2 \leq 1$, thus, by an inequality. On the other hand, the problem of maximum utility involved an constraint given by an equality, namely, $ax + by = c$. The remaining material in this chapter is concerned with problems involving domains that are restricted either by "inequality constraints" or by "equality constraints".

7.5B CONSTRAINTS BY INEQUALITIES

In Example 2 of Section 7.4b, we have treated one situation where the constraints are inequalities. A pleasant feature of that example is that the boundaries of the domain are parallel to the axes. In the next example, the boundary is a circle.

EXAMPLE:

3. Consider the temperature function $f(x, y) = 60(y^2 - x^2)$ given earlier, with domain equal to the closed disk $\bar{B}((0,0); 1)$. See Figure 7.8. Thus, we wish to locate the point (or points) (x_0, y_0) in $\bar{B}((0,0); 1)$ at which the global maximum and global minimum of the function f are achieved.

To do this, we first look for critical points. Thus we solve the system

$$f_x(x,y) = -120x = 0$$
$$f_y(x,y) = 120y = 0.$$

There is only one critical point, namely $(0,0)$. Since the second derivative test is so easy to apply, we do so and find out that the critical point is a saddle point. Thus, both the global maximum and the global minimum are attained on the boundary of the disk, which is the unit circle $S((0,0);1)$, and nowhere else. (Even if the second derivative test were to show the critical point to be the location of a relative extremum, we would still have to look at the boundary points to decide the locations of global extrema. For this reason the second derivative test is sometimes not used when global extrema are the issue.)

We now ask: Precisely where on the unit circle is it the hottest and where is it the coldest?

To answer this, note that the unit circle is given by $x^2 + y^2 = 1$, so that $y^2 = 1 - x^2$. If we restrict $f(x,y) = 60(y^2 - x^2)$ to the unit circle, we may eliminate the y^2 by substitution and obtain a new function, defined for $-1 \leq x \leq 1$ as follows:

$$\varphi(x) = 60(1 - 2x^2).$$

At each point (x,y) on the unit circle, the temperature is $\varphi(x)$. Now we have a single-variable calculus problem. The global maximum of the function φ equals 60 and occurs at the interior point 0 of the interval $[-1,1]$; and the global minimum is -60 and occurs at the two endpoints -1 and 1.

The value $x = 0$, which is where the global maximum of φ occurs, corresponds to the points $(0,1)$ and $(0,-1)$ on the unit circle, since $y^2 = 1 - x^2$. Thus the two hottest points are $(0,-1)$ and $(0,1)$, at which the maximum temperature 60 is attained. The minimum, equal to -60, occurs at the points $(1,0)$ and $(-1,0)$. See Figure 7.9.

The ideas that appeared here are typical of extremal problems with inequality constraints. There are often two steps. The first is to check for critical points in the interior of the constraint region and to evaluate the function f at those points. The second is to check the values of f on the boundary of the constrained region. Such a boundary is usually given by an *equality* constraint (or possibly by several equality constraints, if the boundary consists of several pieces, such as the edges of a rectangular region). Thus, in a global extremum problem with an inequality constraint, we often need to solve a global extremum problem with an equality constraint. We have illustrated one method for solving equality constraint problems in

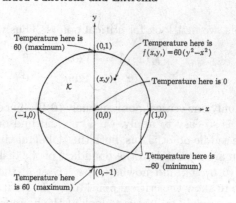

FIGURE 7.9.

Example 3. We will discuss this method further in Section 7.5c. A second method for solving problems involving equality constraints will be discussed in Section 7.5d.

7.5C EQUALITY CONSTRAINTS TREATED BY ELIMINATION

Here is a general description of the problem. We are interested in the global extrema of a function $f : \mathcal{D}(f) \to \mathbb{R}$, where $\mathcal{D}(f)$ is a subset of $\mathbb{R}^n$, namely the set of points in $\mathbb{R}^n$ where f is defined. For one reason or other, we wish to restrict the domain of f by imposing an equality constraint. This means that we only wish to consider points X in $\mathcal{D}(f)$ such that $g(X) = c$, where g is a scalar-valued function whose domain contains $\mathcal{D}(f)$, and c is a real number. In Example 3 in Section 7.5b, we had $f(x, y) = 60(y^2 - x^2)$, and, once we restricted our attention to the boundary of the circle, we had the equality constraint $x^2 + y^2 = 1$, so $g(x, y) = x^2 + y^2$ and $c = 1$.

If we can solve the equality constraint for one of the coordinates, then we may substitute for that coordinate in the function f, obtaining a new function with one fewer variable. The global extrema of this new function, *unrestricted* by any equality (or inequality) constraints, are the global extrema of the original function f, subject to the constraint $g = c$. We have, in effect, eliminated the equality constraint by substituting it into the function f. This method has already been used in Example 3 in Section 7.5b.

EXAMPLES:

4. We will use the elimination technique to solve the maximum utility problem raised in the introduction. Let the budget equation be

$$g(x, y) = 2x + 3y = 10.$$

This constraint restricts our attention to the points on a line in $\mathbb{R}^2$.

Suppose that the utility index function is

$$f(x, y) = 2xy + 3 .$$

(This is a reasonable index of satisfaction for the following two reasons: (1) increasing either x or y increases the satisfaction $f(x, y)$ and (2) if $f(x, y)$ is held constant and one variable increases, then the other variable must decrease. Property 2 says that if satisfaction stays constant while we obtain more of one commodity, it must be because we are losing the other commodity.)

To find the maximum utility subject to our budgetary constraint, use the fact that we may solve $g(x, y) = 10$ for y,

$$y = \frac{1}{3}(10 - 2x) .$$

Now substitute for y in $f(x, y)$, obtaining

$$f(x, y) \;=\; \frac{2}{3}x(10 - 2x) + 3$$
$$=\; -\frac{4}{3}x^2 + \frac{20}{3}x + 3 .$$

Call this new function $h(x)$. It is crucial to note that if $h(x_0)$ is an extremum of $h(x)$ in the sense of single-variable calculus, then for f restricted to the line $2x + 3y = 10$ there is an extremum of the same kind at (x_0, y_0), where $y_0 = \frac{1}{3}(10 - 2x_0)$.

Setting $h'(x) = -\frac{8}{3}x + \frac{20}{3} = 0$, we see that $x_0 = \frac{5}{2}$ is the only location of an extremum of h and that h has a global maximum equal to $h(5/2) = 34/3$ there.

Thus f, constrained to the line $2x + 3y = 10$, has a global maximum equal to $11\frac{1}{3}$ at the point $(x_0, y_0) = (\frac{5}{2}, \frac{5}{3})$. The greatest utility or satisfaction accrues when one devotes resources to obtain $\frac{5}{2}$ units of commodity A and $\frac{5}{3}$ of B. The utility index is then $f(x_0, y_0) = 11\frac{1}{3}$.

Notice that the constrained domain of f in this problem is a line, and, thus, an unbounded set. Nevertheless, we were able to find a global maximum. There is no global minimum.

5. Let us minimize the function

$$f(x, y, z) = 5x + 4y + 9z$$

subject to the constraints that

$$g(x, y, z) = x^2yz = 225$$

and all variables be positive.

It is convenient to solve the equality constraint for y and substitute into the formula for $f(x, y, z)$, obtaining a function of two variables that is to be minimized, subject to no constraint other than that of the two remaining variables being positive:

$$h(x, z) = 5x + \frac{900}{x^2 z} + 9z .$$

The next step is to minimize this function of two variables; so we look for its critical points by solving

$$5 - \frac{1,800}{x^3 z} = 0$$

$$-\frac{900}{x^2 z^2} + 9 = 0 .$$

Multiply the first equation by x and the second by z. Combining the results, we obtain $5x/2 = 900/x^2 z = 9z$. Solve the equation $5x/2 = 9z$ for z, and substitute for z in the equation $5x/2 = 900/x^2 z$ to determine that $x^4 = 2^4 \cdot 3^4$, whence $x = 6$. Therefore $z = 5x/18 = 5/3$ and $y = 225/x^2 z = 15/4$, and so, if there is a global minimum,

$$(5)(6) + (4)(\frac{15}{4}) + (9)(\frac{5}{3}) = 60$$

is it. The argument that there is a global minimum is left for the exercises.

7.5D EQUALITY CONSTRAINTS AND LAGRANGE MULTIPLIERS

We continue to treat the problem of maximizing a function f with a domain equal to the set of points satisfying some equality. Note that by moving all terms to the left-hand side of the equality, we can always write it in the form $g(X) = 0$. For example, the equality $x^2 + y^2 = 1$ can be written as $x^2 + y^2 - 1 = 0$, or $g(x, y) = 0$, where $g(x, y) = x^2 + y^2 - 1$. Thus, the problem we treat is that of maximizing (or minimizing) the function f subject to the constraint $g(X) = 0$. The method—known as the "method of Lagrange multipliers"—that we now give has the virtue of treating all variables on an equal basis.

Let $f : \mathcal{D}(f) \to \mathbb{R}$ and $g : \mathcal{D}(g) \to \mathbb{R}$ be given functions, where $\mathcal{D}(f)$ and $\mathcal{D}(g)$ are subsets of $\mathbb{R}^n$. For simplicity, we assume that $\mathcal{D}(g)$ contains $\mathcal{D}(f)$, so that the equality constraint $g(X) = 0$ is a condition that makes sense for all $X \in \mathcal{D}(f)$. To maximize $z = f(X)$ subject to the constraint $g(X) = 0$, we introduce a new function of the $n+1$ variables $(X, \lambda) = (x_1, \ldots, x_n, \lambda)$,

$$h(X, \lambda) = f(X) - \lambda g(X) .$$

We state the following thereom without proof.

Theorem 7.5.1 Let f, g, and h be as described above and assume that f constrained by $g = 0$ has a global extremum at a point X_0, and that f and g are both differentiable at X_0. Then either $g'(X_0) = 0$ or there exists a real number λ_0 such that (X_0, λ_0) is a critical point of h.

Thus, to find candidates for the global extrema of f subject to the constraint $g = 0$, look at the first n coordinates of the critical points of $h(X, \lambda) = f(X) - \lambda g(X)$. The only other candidates will be points X in $\mathcal{D}(f)$ such that $g(X) = 0$ and either f or g is not differentiable at X. In practice it often turns out that, because of some symmetry in the problem, finding the critical points of h is an easier task than solving the constrained extremum problem by some other method, such as the elimination technique.

REMARKS:

- Notice that the values of f and h agree at places where $g = 0$ regardless of the value of λ.

- For (X_0, λ_0) to be a critical point of h it is necessary, in particular, that its partial derivative with respect to λ equals 0. That partial derivative is $g(X)$, so for any critical point (X_0, λ_0) of h, the constraint $g(X) = 0$ at $X = X_0$.

EXAMPLES:

6. A cylindrical metal can is to contain 2 quarts of liquid. How should it be designed to minimize the quantity of metal used?

    If the height of the can is y and its radius is x, we want to minimize the area, which is the sum of the areas of the top, bottom, and cylindrical surface:
    $$f(x, y) = 2\pi x^2 + 2\pi xy$$
    subject to the constraint $2 = \pi x^2 y$, which we write as $g(x, y) = 0$, where
    $$g(x, y) = \pi x^2 y - 2 .$$
    Let
    $$\begin{aligned} h(x, y, \lambda) &= f(x, y) - \lambda g(x, y) \\ &= 2\pi x^2 + 2\pi xy - \lambda(\pi x^2 y - 2) . \end{aligned}$$

    To find the critical points of h we set its three partial derivatives equal to 0, and, for the purposes of simplification, divide the resulting equations by various constants to obtain
    $$\begin{aligned} 2x + y - \lambda xy &= 0 \\ 2x - \lambda x^2 &= 0 \\ \pi x^2 y - 2 &= 0 . \end{aligned}$$

From the second equation, we have $x = 0$ or $\lambda x = 2$. But $x = 0$ violates the third equation. Hence $\lambda x = 2$. Substituting this into the first equation, we find that $2x + y - 2y = 0$, that is, $y = 2x$. This means that the height y equals the diameter $2x$. The third equation thus yields $\pi 2x^3 = 2$. Thus $x = \pi^{-1/3}$, $y = 2(\pi)^{-1/3}$, and the minimum area equals $6\pi^{1/3}$. We could solve for λ quite easily, but there is no need.

There is a need, however, to convince ourselves that we have found a global minimum (rather than, say, a global maximum). We first note that, because of the nature of the problem, both x and y are restricted to be positive. Let us futher restrict them to satisfy $x \leq 100$ and $y \leq 100$. By drawing a picture it can be seen that the set of points (x, y) which satisfy these inequalities and the condition $g(x, y) = 0$ is closed and bounded. Thus, subject to all these constraints, f has both a global maximum and a global minimum. Our goal is to show that this global minimum is also a global minimum without the constraints $x \leq 100$ and $y \leq 100$.

To perform this task we estimate the values of f when $x > 100$ or $y > 100$, and hope to show, when $g = 0$ is also satisfied, that they are all larger than $6\pi^{1/3}$. If $x > 100$, then $2\pi x^2 > 60000$, so every corresponding value of f is greater that 60000, which is certainly larger than $6\pi^{1/3}$. Now suppose, instead, that $y > 100$. Then the square of f is larger than

$$4\pi(\pi x^2 y)y = 4\pi(2)y > 800\pi .$$

Its square root is larger than $28\pi^{1/2}$ which is certainly larger than $6\pi^{1/3}$, as desired.

7. A steel factory makes a profit of 1, 2, and 2 in dollars per ton on three different kinds of steel. It costs $x^2 + 2y^2 + 4z^2$ to make x, y, and z tons, respectively, of these three kinds of steel. If the firm has \$1600 capital (per hour), how much should they make of each kind of steel (per hour) to maximize profits?

The problem is to maximize

$$p = x + 2y + 2z \quad \text{subject to} \quad x^2 + 2y^2 + 4z^2 = 1600 .$$

Notice that the set of points satisfying the constraints is closed and bounded, so that there definitely is a global maximum (and a global minimum) for p when subjected to the constraint.

Let

$$h(x, y, z, \lambda) = x + 2y + 2z - \lambda(x^2 + 2y^2 + 4z^2 - 1600) .$$

The conditions that the four partial derivatives of h equal 0 are:

$$1 = 2\lambda x$$
$$2 = 4\lambda y$$
$$2 = 8\lambda z$$
$$x^2 + 2y^2 + 4z^2 = 1600 .$$

With the goal of continuing to treat the original variables in an even-handed manner (as is often wise in order to take advantage of a certain amount of "symmetry"), we solve the first three equations for x, y, and z, respectively, and substitute into the fourth equation, obtaining

$$\frac{1}{4\lambda^2} + \frac{2}{4\lambda^2} + \frac{4}{16\lambda^2} = 1600 .$$

This equation has two solutions for λ: $-1/40$ and $1/40$. From the first of these two values we obtain negative values for x, y, and z, giving a negative profit, the global minimum. So, to get the global maximum we must use $\lambda = 1/40$, which gives

$$x = 20, \quad y = 20, \quad z = 10, \quad \text{and} \quad p = \$80 .$$

Whether this profit is adequate, depends on "turn-around" time. If the steel company must wait a year, on average, to recoup the 1600 together with the profit of 80, then a price rise is warranted. If the turn-around time is 3 months, then the "same" 1600 can be invested four times during the year for a nice 20% profit of 320.

In this example it would have been more natural to have adjoined the constraints $x \geq 0$, $y \geq 0$, and $z \geq 0$ at the beginning. Then the possibility of a global minimum at negative values of the variables would not have arisen. Another problem would have arisen, however; Theorem 7.5.1 would not have been applicable at points where one or more of x, y, or z were equal to 0.

Exercises
Finding global extrema.

1. Locate the points in the given set where the given function attains its global extrema.

   (a) $f(x,y) = 3x + 4y$, $\quad x^2 + y^2 \leq 1$
   (b) $g(x,y) = x^2 + y^2$, $\quad x^2 + y^2 \leq 4$
   (c) $h(x,y) = x^2 + y^2 - 1$, $\quad |x| \leq 1$, $\quad |y| \leq 1$
   (d) $k(x,y) = 5x^2 - 6x + 10y^2$, $\quad x^2 + y^2 \leq 1$
   (e) $\varphi(x,y) = 4 + x - y$, $\quad -1 \leq x \leq 2$, $\quad 3 \leq y \leq 5$
   (f) $\psi(x,y) = (3 + 2\cos y)\sin x$, $\quad 0 \leq x \leq 2\pi$, $\quad 0 \leq y \leq 2\pi$.

2. Locate, using both elimination and Lagrange multipliers, the point(s) where the given function attains its global minimum if it has one and its global maximum if it has one.

   (a) $f(x, y) = x^2 + y^2$ on $x - y = 3$

   (b) $g(x, y) = x + y$ on $xy = 1$

   (c) $g(x, y) = x + y$ on $xy = 1$, $x > 0$

   (d) $f(xy) = xy$ on $x + y = 1$, $x \geq 0$, $y \geq 0$

   (e) $h(x, y, z) = xyz$ on $x^2 + y^2 + z^2 = 12$

   (f) $\varphi(x, y, z) = (x - y)^2 + z^2$ on $x^2 + y^2 + z^2 = 18$.

3. Use both elimination and Lagrange multipliers.

   (a) Let x, y, and z be three positive numbers such that $x + y + z = a$, where a is a fixed positive number. How large can their product xyz be?

   (b) If x, y, and z are nonnegative, prove that

   $$(xyz)^{1/3} \leq \frac{x + y + z}{3}.$$

4. Use the method illustrated in Example 6 to show that there is a global minimum in Example 5.

5. A post office regulation states that parcels can have a maximum size of 6 feet in combined length and girth. What dimensions maximize the volume? Use both elimination and Lagrange multipliers.

6. A rectangular box without top is to hold 4 cubic feet. How should the box be designed to use the least amount of material? Use both elimination and Lagrange multipliers.

7. Repeat Exercise 6, but this time assume that the box has a top and that the volume is to be 8 cubic feet.

8. Design a cylindrical can with a circular base but without a top so that its volume is 1 quart and the minimum amount of material is used. Use both elimination and Lagrange multipliers.

9. If your utility function is $g(x, y) = 4xy - x^2 - 3y^2$ and the budget equation is $2x + 3y = 45$, find the values of x and y that maximize the utility. Use both elimination and Lagrange multipliers.

10. Ten years from now you are on a local playground committee with the task of designing an oval-shaped playground consisting of a rectangle with two half-circles at the opposite ends of the playground. The rectangular part is to have an area of 5,000 square yards, and the entire oval is to be enclosed by a fence. How should you design the playground to minimize the amount of fencing needed? Use both elimination and Lagrange multipliers.

11. A wine dealer has space for 100 cases of California, French, and German wines in her cellar. Customers have already ordered 10 cases of California wine, but she knows that she cannot sell more than 20 cases of it. Although she wants no more than 60 cases of French wine, she wants at least as much

French as California wine. If her profit is $10 per case for California, $15 per case for French, and $12 per case for German wines, how much of each should she order to maximize her profit, assuming that she ends up selling the entire stock?

8

Vector Functions $F: \mathbb{R}^n \to \mathbb{R}^q$

8.0 Introduction

Now we discuss the general case of $F: \mathbb{R}^n \to \mathbb{R}^q$, with n and q arbitrary positive integers. We have dealt with special cases before: $n = q = 1$ yields ordinary single-variable calculus; $n = 1$ with q arbitrary yields curves in $\mathbb{R}^q$; and $q = 1$ with n arbitrary yields scalar-valued functions of a vector variable.

If $Y = F(X)$, then each coordinate of $Y = (y_1, \ldots, y_q)$ depends on the coordinates $x_1, \ldots, x_n$ of X, that is,

$$y_j = f_j(x_1, \ldots, x_n) = f_j(X),$$

with $f_j: \mathbb{R}^n \to \mathbb{R}$ for $j = 1, \ldots, q$. The scalar-valued functions $f_1, \ldots, f_q$ are the **coordinate functions** of F. Vector symbols enable us to use the following uncluttered notation:

$$F(X) = (f_1(X), \ldots, f_q(X)),$$

or even

$$F = (f_1, \ldots, f_q).$$

How do we interpret these maps? One standard way is to think of F as a geometric transformation which takes $\mathcal{D}(F)$, a subset of $\mathbb{R}^n$, and changes it into $\mathcal{I}(F)$, a subset of $\mathbb{R}^q$. Such a transformation might involve a lot of stretching, twisting, collapsing, and folding, as well as other, less familiar, types of changes.

EXAMPLES:

1. Let $F: \mathbb{R}^2 \to \mathbb{R}^2$ be the affine map given by

$$\begin{aligned} y_1 &= f_1(x_1, x_2) = 2 + x_1 - 2x_2 \\ y_2 &= f_2(x_1, x_2) = 1 + x_1 + x_2. \end{aligned}$$

The origin $(0,0)$ is mapped to the point $(2,1)$ and the x_1-axis, given by $x_2 = 0$, is mapped to the set of points $(y_1, y_2) = (2 + x_1, \, 1 + x_1)$, that is, the line $y_1 - y_2 - 1 = 0$. The x_2-axis is mapped onto the line $y_1 + 2y_2 - 4 = 0$, as you should verify.

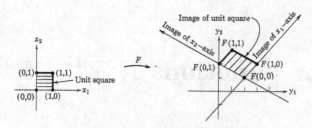

FIGURE 8.1. An affine map

The shaded region in Figure 8.1 indicates the image of the unit square

$$\{X : 0 \le x_1 \le 1, 0 \le x_2 \le 1\}.$$

2. Let a and b denote fixed scalars for which $0 < b < a$ and let

$$G(\theta, \varphi) = ((a + b\cos\varphi)\cos\theta, \, (a + b\cos\varphi)\sin\theta, \, b\sin\varphi)$$

for $0 \le \theta \le 2\pi$ and $0 \le \varphi \le 2\pi$. Thus, $\mathcal{D}(G)$ is a square region in $\mathbb{R}^2$ and the target of g is $\mathbb{R}^3$. The coordinates of $Y = G(\theta, \varphi)$ are given by

$$
\begin{aligned}
y_1 &= g_1(\theta, \varphi) &= (a + b\cos\varphi)\cos\theta \\
y_2 &= g_2(\theta, \varphi) &= (a + b\cos\varphi)\sin\theta \\
y_3 &= g_3(\theta, \varphi) &= b\sin\varphi.
\end{aligned}
$$

We now show that $\mathcal{I}(G)$ is a "torus" in $\mathbb{R}^3$, that is, the shell of a doughnut. To do this, we examine where G maps the vertical segments in its domain. These vertical segments are of the form

$$\{(\theta, \varphi) : 0 \le \varphi \le 2\pi, \theta = \theta_0\},$$

where θ_0 is a some constant between 0 and 2π. See Figure 8.2.

First let $\theta_0 = 0$. The coordinate formulas given above become

$$G(0, \varphi) = (a + b\cos\varphi, 0, b\sin\varphi) = (a, 0, 0) + b(\cos\varphi, 0, \sin\varphi).$$

We recognize this as a parametrized circle in $\mathbb{R}^3$ with radius b and center $(a, 0, 0)$. Since the second coordinate of $G(0, \varphi)$ is 0, this circle lies in the plane $y_2 = 0$. Now replace 0 by the general constant θ_0 to obtain

$$G(\theta_0, \varphi) = a(\cos\theta_0, \sin\theta_0, 0) + b(\cos\theta_0\cos\varphi, \sin\theta_0\cos\varphi, \sin\varphi).$$

It is easy to check directly by matrix multiplication that

$$[G(\theta_0, \varphi)] = \begin{bmatrix} \cos\theta_0 & -\sin\theta_0 & 0 \\ \sin\theta_0 & \cos\theta_0 & 0 \\ 0 & 0 & 1 \end{bmatrix} [G(0, \varphi)].$$

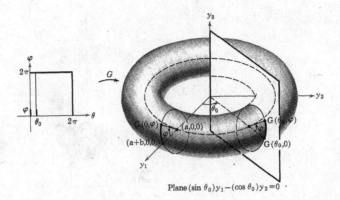

FIGURE 8.2. A torus

We recognize the 3-by-3 matrix as the matrix of an isometry (Section 3.8c). This isometry rotates $\mathbb{R}^3$ through the angle θ_0 counterclockwise about the y_3-axis (when view from above). Therefore, if we rotate the circle parametrized by $G(0, \varphi)$ in this fashion about the y_3-axis, we obtain the curve parametrized by $G(\theta_0, \varphi)$. It follows that this curve is also a circle of radius b. Its center is $a(\cos\theta_0, \sin\theta_0, 0)$, and it lies in the vertical plane given by

$$(\sin\theta_0)y_1 - (\cos\theta_0)y_2 = 0 \ .$$

Thus for each choice θ_0, as φ varies from 0 to 2π, we get a small vertical circle of radius b traced out by $G(\theta_0, \varphi)$. The centers of these circles are points of the form $a(\cos\theta_0, \sin\theta_0, 0)$, so as θ_0 varies between 0 and 2π, these centers trace out the large horizontal circle of radius a centered at the origin. The corresponding small circles trace out a torus (hollow doughnut) in $\mathbb{R}^3$.

The function G is not one-to-one since the values obtained when $\theta = 2\pi$ are the same as those obtained when $\theta = 0$. Similarly for φ. We could have arranged for G to be one-to-one by specifying its domain to be

$$\{(\theta, \varphi) : 0 \leq \theta < 2\pi \,, 0 \leq \varphi < 2\pi\} \ .$$

On the other hand, we could have obtained a four-to-one function by using

$$\{(\theta, \varphi) : 0 \leq \theta < 4\pi \,, 0 \leq \varphi < 4\pi\}$$

for the domain.

We urge you to work through this example with care. You will see the torus again in the following sections. We dwell on it because it is geometrically interesting, easily visualized, and amenable to explicit

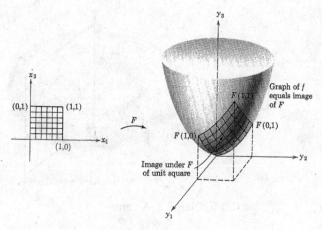

FIGURE 8.3.

computation. Notice that we have given no attention to the graph of G, which is a subset of $\mathbb{R}^5$. Instead, it is the image of G, a subset of $\mathbb{R}^3$, that has received our attention.

3. As discussed in Sections 5.4d and 6.0, the graph of $z = f(x_1, x_2) = x_1^2 + x_2^2$ is a circular paraboloid in $\mathbb{R}^3$. Now we claim that this graph is also the *image* of a vector-valued map $F : \mathbb{R}^2 \to \mathbb{R}^3$ that is related to the scalar-valued map f. The appropriate function $F = (f_1, f_2, f_3)$ is given by

$$
\begin{aligned}
y_1 &= f_1(X) = x_1 \\
y_2 &= f_2(X) = x_2 \\
y_3 &= f_3(X) = f(X) = x_1^2 + x_2^2 \,.
\end{aligned}
$$

Hence F maps the $x_1 x_2$-plane into $y_1 y_2 y_3$-space as a "surface", which we know to be a circular paraboloid and which we are now viewing as $\mathcal{I}(F)$ for F as just defined. See Figure 8.3.

The map F may be described more succinctly by

$$
F(x_1, x_2) = (x_1, x_2, f(x_1, x_2))
$$

or

$$
F(x_1, x_2) = (x_1, x_2, x_1^2 + x_2^2) \,.
$$

Thus, we see that functions f from $\mathbb{R}^2$ into $\mathbb{R}$ provide us with functions from $\mathbb{R}^2$ into $\mathbb{R}^3$ in a natural manner; the function f is attached as the third coordinate with the identity map on $\mathbb{R}^2$ providing the first two coordinates.

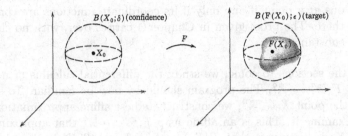

FIGURE 8.4. F maps a "confidence ball" in the domain to a "target ball"

Since limits and continuity have been treated rather completely in Chapter 4 for vector-valued functions of a scalar variable and again in Chapter 6 for scalar-valued functions of a vector variable, we will be brief in our discussion of these concepts for vector-valued functions of a vector variable, leaving some tasks for the exercises.

Definition 8.0.1 A function $F : \mathcal{D}(F) \to \mathbb{R}^q$, where $\mathcal{D}(F) \subseteq \mathbb{R}^n$, is said to be **continuous at a point** $X_0 \in \mathcal{D}(F)$ if and only if for every $\varepsilon > 0$, there is an positive number δ such that

$$\|F(X) - F(X_0)\| < \varepsilon$$

whenever

$$\|X - X_0\| < \delta \quad \text{and} \quad X \in \mathcal{D}(F) .$$

A function that is continuous at each point in its domain is said to be **continuous**.

REMARKS:

- Just as in Chapter 6, this definition can be restated using the language of open balls. The function F is continuous at a point $X_0 \in \mathcal{D}(F)$ if and only if for every open ball $B(F(X_0); \varepsilon)$ centered at $F(X_0)$, there is an open ball $B(X_0; \delta)$ centered at X_0 such that

$$F(X) \in B(F(X_0); \varepsilon)$$

  whenever

$$X \in B(X_0; \delta) \cap \mathcal{D}(f) .$$

  See Figure 8.4.

- As stated in Theorem 4.1.8 of Section 4.1b for functions of a scalar variable, it turns out that a function of a vector variable is continuous at a point if and only if its coordinate functions are continuous there. The proof given in Chapter 4 carries over, with no change in substance.

In the sections to come, we study the differential calculus of nonlinear maps $F: \mathbb{R}^n \to \mathbb{R}^q$. The program should by now be familiar. To study F near the point $X_0 \in \mathbb{R}^n$, we construct its best affine approximation at X_0 and examine it. This is an affine map $T: \mathbb{R}^n \to \mathbb{R}^q$ that approximates F near X_0 in the sense that $T(X_0) = F(X_0)$ and $T(X)$ is close to $F(X)$, provided that X is close to X_0. Having defined the map T, we must learn how to compute it (partial derivatives again) and interpret it.

Exercises

1. Consider the maps $F: \mathbb{R}^2 \to \mathbb{R}^2$, defined as follows:

   (a) $y_1 = 1 + x_1 + x_2$
   $y_2 = -2 + x_1 - x_2$;

   (b) $y_1 = x_1 - x_2$
   $y_2 = x_1^2 + x_2^2$;

   (c) $y_1 = x_1 \cos \pi x_2$
   $y_2 = x_1 \sin \pi x_2$.

   For each of these draw sketches, as in Example 1, showing the square region with vertices at $(1,0)$, $(2,0)$, $(2,1)$, and $(1,1)$ together with the image of this square region. *Hint:* Find the images of the edges of the square.

2. Consider the maps $F: \mathcal{D}(F) \to \mathbb{R}^3$, $\mathcal{D}(F) \subseteq \mathbb{R}^2$, defined as follows:

   (a) $\mathcal{D}(F) = \{(x_1, x_2): x_1^2 + x_2^2 \le 1\}$,
   $y_1 = x_1$
   $y_2 = x_2$
   $y_3 = \sqrt{4 - x_1^2 - x_2^2}$;

   (b) $\mathcal{D}(F) = \{(x_1, x_2): 0 \le x_1 \le \frac{\pi}{2}, \ 0 \le x_2 \le \frac{\pi}{2}\}$,
   $y_1 = \sin x_1 \cos x_2$
   $y_2 = \sin x_1 \sin x_2$
   $y_3 = \cos x_1$;

   (c) $\mathcal{D}(F) = \{(x_1, x_2): 1 \le x_1 \le 2, \ 0 \le x_2 \le 2\}$,
   $y_1 = x_1 \cos \pi x_2$
   $y_2 = x_1 \sin \pi x_2$
   $y_3 = 4x_1^{-2}$.

   For each of these draw sketches of $\mathcal{I}(F)$.

3. Let G be defined by the formula in Example 2, but for various domains as described below. In each case discuss the differences and similarities with Example 2.

   (a) $\{(\theta, \varphi) : |\theta| \le \pi, \ |\varphi| \le \pi\}$

(b) $\{(\theta, \varphi) : |\theta| \leq 2\pi, |\varphi| \leq 3\pi\}$

(c) $\{(\theta, \varphi) : |\theta| \leq \pi/2, |\varphi| \leq \pi/2\}$

(d) $\mathbb{R}^2$

(e) $\{(\theta, \varphi) : |\theta| \leq \pi\}$

(f) $\{(\theta, \varphi) : |\varphi| \leq \pi\}$.

4. Define the concept of "limit" for vector-valued functions of a vector variable.

5. State some theorems about limits of combinations of functions for vector-valued functions of a vector variable.

6. State some theorems about continuity of combinations of continuous vector-valued functions of vector variables. Make sure to include a theorem about compositions of functions.

8.1 Affine Approximation and Tangency

8.1A BASIC DEFINITIONS

We begin by mimicking Definition 6.4.1 in Section 6.4b.

Definition 8.1.1 Let F be an $\mathbb{R}^q$-valued function for which $\mathcal{D}(f) \subseteq \mathbb{R}^n$ and let X_0 be an interior point of $\mathcal{D}(F)$. The function F is **differentiable** at X_0 if and only if there exists an affine map $T : \mathbb{R}^n \to \mathbb{R}^q$ such that the following two conditions hold:

1. $F(X_0) = T(X_0)$;

2. $\lim_{X \to X_0} \dfrac{F(X) - T(X)}{\|X - X_0\|} = 0$.

If there is such a T, it is the **best affine approximation** or **tangent map** to F at X_0, its graph is the **tangent affine subspace to the graph** of F at the point $(X_0, F(X_0))$, and, provided T is one-to-one, $\mathcal{I}(T)$ is the **tangent affine subspace to the image** at the point $F(X_0)$. The linear map corresponding to T is called the **total derivative** or **first derivative** of F at X_0 and is denoted by $F'(X_0)$.

Note how this definition agrees with the those in Chapters 4 and 6. Since the best affine approximation T is affine, it may be put in two standard forms

$$T(X) = L(X) + (T(X_0) - L(X_0)) = T(X_0) + L(X - X_0) ,$$

the first form giving the sum of a linear map and a constant vector and the second incorporating part of the constant vector into the linear map in order to focus attention on the behavior near X_0. Here L is the total derivative at X_0, so we may also write

$$T(X) = F(X_0) + F'(X_0)(X - X_0),$$

an expression that is familiar from earlier chapters.

8.1B TOTAL DERIVATIVES AND PARTIAL DERIVATIVES

In Section 6.5b there are two results of importance. One result tells us how the total derivative can be calculated in terms of the partial derivatives when the total derivative exists. The other result tells us one way of deciding if the total derivative exists by looking at the partial derivatives. These theorems apply, of course, to coordinate functions of vector-valued functions of a vector variable. On the other hand, limits of coordinate functions of a function G exist, as in Theorem 4.1.3 of Section 4.1a, if and only if the corresponding limit of G itself exists, and the coordinate functions of the limit equal the limits of the respective coordinate functions. These two facts can be pieced together to give proofs of theorems for the current setting on the basis of the corresponding theorems for the earlier specialized settings.

Accordingly we give here without proofs the analogs of the two theorems in Section 6.5b. Since the first derivative in the current setting is a linear map from $\mathbb{R}^n$ to $\mathbb{R}^q$ it is represented by a matrix. We advise the reader to think carefully about why the matrix given in the following theorem is the appropriate matrix rather than its transpose.

Theorem 8.1.1 Let X_0 be an interior point of the domain $\mathcal{D}(F) \subseteq \mathbb{R}^n$ of an $\mathbb{R}^q$-valued function F that is differentiable at X_0. Then the following assertions are true:

1. each coordinate function f_i of F is differentiable at X_0 and, in particular, all partial derivatives $D_j f_i$ exist at (X_0);

2. the first derivative $F'(X_0)$ has the matrix

$$[F'(X_0)] = \begin{bmatrix} (D_1 f_1)(X_0) & \cdots & (D_n f_1)(X_0) \\ \vdots & & \vdots \\ (D_1 f_q)(X_0) & \cdots & (D_n f_q)(X_0) \end{bmatrix}.$$

REMARKS:

- This matrix is called the **Jacobian** of F, after the nineteenth-century mathematician C.J.G. Jacobi.

- The rows of F' are the gradients of the coordinate functions of F.

Theorem 8.1.2 Let X_0 be an interior point of the domain $\mathcal{D}(F) \subseteq \mathbb{R}^n$ of an $\mathbb{R}^q$-valued function F. Suppose that all n partial derivatives of each of the q coordinate functions of F exist on some open ball centered at X_0 and are continuous at X_0. Then $F'(X_0)$ exists.

EXAMPLES:

1. Consider the affine map F of Section 8.0, given by

$$f_1(x_1, x_2) = 2 + x_1 - 2x_2$$
$$f_2(x_1, x_2) = 1 + x_1 + x_2 .$$

We will compute its best affine approximation $T(X)$ at $X_0 = 0 = (0,0)$. To do this, we must compute the vector $F(0)$, compute the matrix of $F'(0)$, and use these to form $T(X)$.

Clearly $F(0,0) = (f_1(0,0), \ f_2(0,0)) = (2,1)$. According to the theorem, we compute the partial derivatives

$$\frac{\partial f_1}{\partial x_1}(X) = 1, \qquad \frac{\partial f_1}{\partial x_2}(X) = -2,$$
$$\frac{\partial f_2}{\partial x_1}(X) = 1, \qquad \frac{\partial f_2}{\partial x_2}(X) = 1.$$

Hence the matrix $[F'(0)]$ is

$$[F'(0)] = \begin{bmatrix} 1 & -2 \\ 1 & 1 \end{bmatrix}.$$

Thus,

$$[T(x_1, x_2)] = \begin{bmatrix} 2 \\ 1 \end{bmatrix} + \begin{bmatrix} 1 & -2 \\ 1 & 1 \end{bmatrix} \begin{bmatrix} x_1 \\ x_2 \end{bmatrix} = \begin{bmatrix} 2 + x_1 - 2x_2 \\ 1 + x_1 + x_2 \end{bmatrix}.$$

We see that T is the original function F —hardly surprising since an affine map should be its own best affine approximation.

2. We will compute the best affine approximation at $(a + b, 0, 0)$ to the torus that is the image of $G: \mathbb{R}^2 \to \mathbb{R}^3$ as treated in Example 2 of Section 8.0. We recall that this map is given by $G(\theta, \varphi) = (g_1(\theta, \varphi), g_2(\theta, \varphi), g_3(\theta, \varphi))$, where

$$g_1(\theta, \varphi) = (a + b\cos\varphi)\cos\theta,$$
$$g_2(\theta, \varphi) = (a + b\cos\varphi)\sin\theta,$$
$$g_3(\theta, \varphi) = b\sin\varphi.$$

Since $G(0,0) = (a + b, 0, 0)$ is the point of interest on the torus, we will focus on $(0,0) \in \mathcal{D}(G)$.

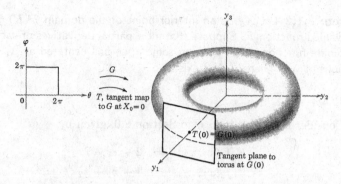

FIGURE 8.5.

We compute partial derivatives, writing $g_{1\theta}$ for $\partial g_1/\partial \theta$, and so on,

$$g_{1\theta}(\theta, \varphi) = -(a + b \cos \varphi) \sin \theta \, , \qquad g_{1\varphi}(\theta, \varphi) = -b \sin \varphi \cos \theta \, ,$$
$$g_{2\theta}(\theta, \varphi) = (a + b \cos \varphi) \cos \theta \, , \qquad g_{2\varphi}(\theta, \varphi) = -b \sin \varphi \sin \theta \, ,$$
$$g_{3\theta}(\theta, \varphi) = 0 \, , \qquad\qquad\qquad\quad g_{3\varphi}(\theta, \varphi) = b \cos \varphi \, .$$

At $(0,0)$, we obtain the Jacobian:

$$[G'(0,0)] = \begin{bmatrix} 0 & 0 \\ a+b & 0 \\ 0 & b \end{bmatrix} .$$

Therefore, the tangent map $T(X)$ is given by

$$
\begin{aligned}
[T(X)] &= [G(0)] + [G'(0)][X] \\
&= \begin{bmatrix} a+b \\ 0 \\ 0 \end{bmatrix} + \begin{bmatrix} 0 & 0 \\ a+b & 0 \\ 0 & b \end{bmatrix} \begin{bmatrix} \theta \\ \varphi \end{bmatrix} = \begin{bmatrix} a+b \\ (a+b)\theta \\ b\varphi \end{bmatrix} ,
\end{aligned}
$$

that is, $T(X) = Y = (y_1, y_2, y_3)$, where

$$y_1 = a + b, \qquad y_2 = (a+b)\theta, \qquad y_3 = b\varphi.$$

It is the image of T that is the tangent affine subspace to the torus at the point

$$G(0,0) = T(0,0) = (a+b, 0, 0) \, .$$

Since the y_1-coordinate of $T(X)$ is always $a + b$, independent of θ and φ, we see that the image of T must be a vertical plane that is perpendicular to the y_1-axis and intersects it at the coordinate $a + b$. See Figure 8.5.

3. This time we approximate the map G of the preceding example at a different point $(\pi/4, \ \pi/4)$. (Note that this point is a point in the

domain of G, not a point on the torus. The phrasing needs to be different to refer to a point in the image.) Thus we must construct a new affine map $T_1(X)$ (different from the map $T(X)$ obtained in the preceding example) of the form

$$T_1(\theta, \varphi) = G(\frac{\pi}{4}, \frac{\pi}{4}) + G'(\frac{\pi}{4}, \frac{\pi}{4})((\theta, \varphi) - (\frac{\pi}{4}, \frac{\pi}{4})) .$$

Using $\cos \pi/4 = \sin \pi/4 = \sqrt{2}/2$, we compute

$$G(\frac{\pi}{4}, \frac{\pi}{4}) = \frac{1}{2}(a\sqrt{2} + b, \ a\sqrt{2} + b, \ b\sqrt{2}) .$$

In the preceding example, we obtained formulas for the relevant partial derivatives. Inserting $(\frac{\pi}{4}, \frac{\pi}{4})$, we obtain the 3-by-2 Jacobian

$$[G'(\frac{\pi}{4}, \frac{\pi}{4})] = \begin{bmatrix} -\frac{1}{2}(a\sqrt{2} + b) & -\frac{b}{2} \\ \frac{1}{2}(a\sqrt{2} + b) & -\frac{b}{2} \\ 0 & \frac{b\sqrt{2}}{2} \end{bmatrix} .$$

We conclude that $T_1(\theta, \varphi) = Y$ is given by

$$y_1 = \frac{1}{2}(a\sqrt{2} + b) - \frac{1}{2}(a\sqrt{2} + b)\left(\theta - \frac{\pi}{4}\right) - \frac{b}{2}\left(\varphi - \frac{\pi}{4}\right),$$

$$y_2 = \frac{1}{2}(a\sqrt{2} + b) + \frac{1}{2}(a\sqrt{2} + b)\left(\theta - \frac{\pi}{4}\right) - \frac{b}{2}\left(\varphi - \frac{\pi}{4}\right),$$

$$y_3 = \frac{b\sqrt{2}}{2} \qquad\qquad\qquad\qquad + \frac{b\sqrt{2}}{2}\left(\varphi - \frac{\pi}{4}\right).$$

Note the presence of $\theta - \pi/4$ and $\varphi - \pi/4$, the coordinates of the vector $X - X_1$.

The image $\mathcal{I}(T_1)$ of T_1 is the plane tangent to the torus at the point

$$\frac{1}{2}(a\sqrt{2} + b, \ a\sqrt{2} + b, \ b\sqrt{2}) .$$

See Figure 8.6.

4. Let $f: \mathbb{R}^2 \to \mathbb{R}$ be a differentiable function. We have two ways to calculate the tangent plane to the graph of f at a particular point $(c_1, c_2, f(c_1, c_2))$. The first is to find the *graph* of the tangent map at (c_1, c_2). The second is to find the *image* of the tangent map of the function

$$F(x_1, x_2, f(x_1, x_2))$$

at the point (c_1, c_2). Let us call the two tangent maps t and T, respectively. Thus t is a function from $\mathbb{R}^2$ into $\mathbb{R}$ and its *graph* is a

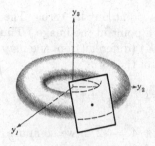

FIGURE 8.6.

two-dimensional affine subspace of $\mathbb{R}^3$, and T is a function from $\mathbb{R}^2$ into $\mathbb{R}^3$ and so its *image* is also an affine subspace of $\mathbb{R}^3$. We hope that the image T is two-dimensional, in fact, that it equals the graph of t.

The matrix of $f'(c_1, c_2)$ is

$$[\ D_1 f(c_1, c_2) \quad D_2 f(c_1, c_2)\ ]$$

and so

$$t(x_1, x_2) = f(c_1, c_2) + (D_1 f(c_1, c_2)(x_1 - c_1) + D_2 f(c_1, c_2)(x_2 - c_2).$$

We turn to F and the corresponding tangent map T. You should be able to show that the Jacobian at (c_1, c_2) equals

$$[F'(X)] = \begin{bmatrix} 1 & 0 \\ 0 & 1 \\ D_1 f(c_1, c_2) & D_2 f(c_1, c_2) \end{bmatrix}.$$

Thus

$$\begin{aligned}
[T(X)] &= [F(c_1, c_2)] + [F'(c_1, c_2)] \begin{bmatrix} x_1 - c_1 \\ x_2 - c_2 \end{bmatrix} \\
&= \begin{bmatrix} c_1 \\ c_2 \\ f(c_1, c_2) \end{bmatrix} \\
&\quad + \begin{bmatrix} x_1 - c_2 \\ x_2 - c_2 \\ D_1 f(c_1, c_2)(x_1 - c_1) + D_2 f(c_1, c_2)(x_2 - c_2) \end{bmatrix} \\
&= \begin{bmatrix} x_1 \\ x_2 \\ t(x_1, x_2) \end{bmatrix}.
\end{aligned}$$

In coordinate form,

$$T(x_1, x_2) = (x_1, x_2, t(x_1, x_2)).$$

Therefore the image of T and the graph of t are obviously identical.

5. Let $f: \mathbb{R}^n \to \mathbb{R}$ be scalar-valued, as in Chapter 6. Then we may think of f' as a map from $\mathbb{R}^n$ to $\mathbb{R}^n$; to each $X \in \mathbb{R}^n$, f' assigns the 1-by-n matrix given by $[(D_1 f)(X) \cdots (D_n f)(X)]$, which corresponds to the vector $((D_1 f)(X) \cdots (D_n f)(X))$. Since $f': \mathbb{R}^n \to \mathbb{R}^n$, it is a function of the type being discussed in this chapter. The total derivative of f' is represented by an n-by-n Jacobian whose entries are the various second partial derivatives of the original function f. This matrix is, therefore, the Hessian of Chapter 7.

Exercises

1. Find the total derivative matrix of each of the following maps at the indicated points:

   (a) $y_1 = 1 + x_1 + x_2$
       $y_2 = -2 + x_1 - x_2$
       at $(3, -4)$;

   (b) $y_1 = x_1 - x_2$
       $y_2 = x_1^2 + x_2^2$
       at $(-1, 2)$;

   (c) $z_1 = y_1 \cos \pi y_2$
       $z_2 = y_1 \sin \pi y_2$
       at $(2, 0)$;

   (d) $y_1 = 3x_1 - 7$
       $y_2 = x_1^2 + 4x_2^2$
       $y_3 = x_2$
       at $(1, -2)$;

   (e) $x = \sin \varphi \cos \theta$
       $y = \sin \varphi \sin \theta$
       $z = \cos \varphi$
       at $(\theta, \varphi) = (\pi/4, \pi/4)$;

   (f) $u = e^x \cos y$
       $v = e^x \sin y$
       $w = 3z - 5$
       at $(0, \pi, 2)$.

2. Find the best affine approximations to the maps in the preceding exercise at the given points.

3. Find a map $F: \mathbb{R}^2 \to \mathbb{R}^3$ such that $F(0, 1) = (1, 0, 2)$ and

$$[F'(X)] = \begin{bmatrix} 1 & -2 \\ 2 & 3 \\ -1 & 0 \end{bmatrix}$$

for all $X \in \mathbb{R}^2$.

4. (a) Let $F: \mathbb{R}^2 \to \mathbb{R}^3$ have the property that $F'(X) = 0$ for all $X \in \mathbb{R}^2$. What can you conclude about F? Proof?

(b) Let F and G map $\mathbb{R}^2$ to $\mathbb{R}^3$. If $F'(X) = G'(X)$ for all X, and $F(X_0) = G(X_0)$ for some $X_0 \in \mathbb{R}^2$, what can you conclude? Why?

(c) Let $F: \mathbb{R}^2 \to \mathbb{R}^3$ have the property that $F'(X) = A$ for all $X \in \mathbb{R}^2$, where A is a constant matrix. If in addition $F(X_0) = Y_0$, where X_0 and Y_0 are fixed vectors, what can you conclude about F?

8.2   Rules for Calculating Derivatives

8.2A   SUMS AND PRODUCTS

The first theorem is about sums and the second about products.

Theorem 8.2.1 Suppose that F and G are $\mathbb{R}^q$-valued functions that are differentiable at a point $X_0 \in \mathbb{R}^n$. Then $F + G$ is differentiable at X_0 and

$$(F + G)'(X_0) = F'(X_0) + G'(X_0) .$$

We omit the proof because it is similar to that of other versions of this theorem.

Theorem 8.2.2 Let F be an $\mathbb{R}^q$-valued function and h an $\mathbb{R}$-valued function, both of which are differentiable at a point $X_0 \in \mathbb{R}^n$. Then the product hF is differentiable at X_0 and

$$[(hF)'(X_0)] = [F(X_0)] [h'(X_0)] + h(X_0)[F'(X_0)] .$$

The proof of this theorem is more difficult than that for sums, but most of the difficulty comes in sorting out meanings. Accordingly, we will omit the proof, but discuss meanings in the following remark.

REMARK:

- In the preceding theorem hF is a function from $\mathbb{R}^n$ to $\mathbb{R}^q$. Therefore, the matrix $[(hF)'(X_0)]$ has q rows and n columns. The symbol $[F(X_0)]$ represents the column vector corresponding to the vector $F(X_0)$. Thus it has q rows and 1 column. The matrix $[h'(X_0)]$ has 1 row and n columns, since it is the matrix of a linear functional on $\mathbb{R}^n$. Thus, the product in the first term on the right-hand side of the equation has q rows and n columns, in agreement with the left-hand side. (Notice that the order of the two factors in the first term on the right-hand side is important, but it is not important in the other two terms.) The second term on the right-hand side, being the product of a scalar with a matrix, is the same shape as the matrix $[F'(X_0)]$, namely q-by-n, in agreement with the other two terms.

EXAMPLE:

1. For $X \in \mathbb{R}^n$ and different from $\mathbf{0}$, let

$$G(X) = \|X\| X .$$

To calculate the derivative of G we use what should be a memorized formula for the derivative of $\|X\|$; it is the linear tranformation with matrix

$$\frac{1}{\|X\|} [X]^T .$$

The derivative of X is the linear map whose matrix is the n-by-n identity matrix. Thus the preceding theorem gives a formula for the linear map $G'(X)$. Its matrix is

$$[X] \frac{1}{\|X\|} [X]^T + \|X\| I ,$$

which is equal to $1/\|X\|$ multiplied by a matrix whose entry in row i and column j is $x_i x_j$ if $i \neq j$ and $x_i^2 + \|X\|^2$ if $i = j$.

8.2B   THE CHAIN RULE

We have already seen three chain rules: Theorem 4.2.5 in Section 4.2c and Theorems 6.5.4 and 6.5.5 in Section 6.5c. These theorems are all special cases of Theorem 8.2.3 given below. Another special case is the single-variable chain rule: $(f \circ g)' = (f' \circ g)g'$. One notices in this simplest case that the derivative of a composition is a product. One is thus tempted to hope that in cases involving the composition of vector-valued functions of a vector variable, there is a chain rule in terms of products of matrices. This is the case.

Let $\mathbb{R}^n \xrightarrow{F} \mathbb{R}^q \xrightarrow{G} \mathbb{R}^r$ be mappings. We suppose that G is defined for all vectors in $\mathcal{I}(F)$, so that if $X \in \mathbb{R}^n$, then $G(F(X))$ is defined. Thus we have a composite map $G \circ F : \mathbb{R}^n \to \mathbb{R}^r$ given by

$$(G \circ F)(X) = G(F(X)).$$

We have discussed the composition of *linear* maps in Chapter 3.

From knowledge of the derivatives F' of F and G' of G it is possible to obtain the derivative of $G \circ F$ in a straightforward manner. This is the content of the "chain rule". Before stating it, we recall that for each point $X_0 \in \mathbb{R}^n$ the first derivative $F'(X_0)$ may be given as a q-by-n matrix, and for each $Y_0 \in \mathbb{R}^q$ the first derivative $G'(Y_0)$ may be given as an r-by-q matrix. Finally, the first derivative $(G \circ F)'(X_0)$ is given as an r-by-n matrix.

Theorem 8.2.3 (Chain Rule) Let F be an $\mathbb{R}^q$-valued function that is differentiable at a point $X_0 \in \mathbb{R}^n$ and let G be an $\mathbb{R}^r$-valued function that is differentiable at the point $f(X_0) \in \mathbb{R}^q$. Then

$$[(G \circ F)'(X_0)] = [G'(F(X_0))][F'(X_0)].$$

REMARK:

- The derivative matrix of the composite map is the product of the derivative matrices of the functions being composed. Thus the chain rule is actually an assertion about the product of certain matrices. Imagine the formulas that we would face if instead of writing everything in terms of linear maps, vectors, and matrices, we wrote out formulas in terms of individual coordinates and entries.

Before discussing the proof of the chain rule, we illustrate its meaning.

EXAMPLE:

2. Let $\mathbb{R}^2 \overset{F}{\to} \mathbb{R}^2 \overset{G}{\to} \mathbb{R}^3$ be mappings given by

$$F(X) = (f_1(X), \ f_2(X)) = (x_1^2 - x_2^2, \ 2x_1 x_2) \,,$$

$$G(Y) = (g_1(Y), \ g_2(Y), \ g_3(Y)) = (3 + y_1 + y_2, \ y_1 y_2, \ y_1 - 2y_2) \,.$$

We will compute the matrix of $(G \circ F)'(X_0)$ at $X_0 = (1, 1)$.

We have $F(X_0) = (0, 2)$. We will first calculate general formulas for $[F'(X)]$ and $G'(Y)$, after which we will insert $(1, 1)$ for X and $(0, 2)$ for Y. We have

$$[F'(X)] = \begin{bmatrix} D_1 f_1(X) & D_2 f_1(X) \\ D_1 f_2(X) & D_2 f_2(X) \end{bmatrix} = \begin{bmatrix} 2x_1 & -2x_2 \\ 2x_2 & 2x_1 \end{bmatrix}$$

and

$$[G'(Y)] = \begin{bmatrix} D_1 g_1(Y) & D_2 g_1(Y) \\ D_1 g_2(Y) & D_2 g_2(Y) \\ D_1 g_3(Y) & D_2 g_3(Y) \end{bmatrix} = \begin{bmatrix} 1 & 1 \\ y_2 & y_1 \\ 1 & -2 \end{bmatrix}.$$

We immediately obtain

$$[F'(1, 1)] = \begin{bmatrix} 2 & -2 \\ 2 & 2 \end{bmatrix}, \quad [G'(0, 2)] = \begin{bmatrix} 1 & 1 \\ 2 & 0 \\ 1 & -2 \end{bmatrix}.$$

According to the chain rule, the Jacobian of the map $G \circ F$ at $(1, 1)$ is given by the product

$$[(G \circ F)'(X_0)] = \begin{bmatrix} 1 & 1 \\ 2 & 0 \\ 1 & -2 \end{bmatrix} \begin{bmatrix} 2 & -2 \\ 2 & 2 \end{bmatrix} = \begin{bmatrix} 4 & 0 \\ 4 & -4 \\ -2 & -6 \end{bmatrix}.$$

It is now easy to obtain the best affine approximation $T_{G \circ F}$ at X_0. We have $G(F(X_0)) = G(0,2) = (5,0,-4) \in \mathbb{R}^3$. Thus

$$[T_{G \circ F}(X)] = \begin{bmatrix} 5 \\ 0 \\ -4 \end{bmatrix} + \begin{bmatrix} 4 & 0 \\ 4 & -4 \\ -2 & -6 \end{bmatrix} \begin{bmatrix} x_1 - 1 \\ x_2 - 1 \end{bmatrix}$$

$$= \begin{bmatrix} 5 + 4(x_1 - 1) \\ 4(x_1 - 1) - 4(x_2 - 1) \\ -4 - 2(x_1 - 1) - 6(x_2 - 1) \end{bmatrix}.$$

Writing this in terms of coordinates, we have

$$T_{G \circ F}(X) = (5 + 4(x_1 - 1), \ 4(x_1 - 1) - 4(x_2 - 1),$$
$$-4 - 2(x_1 - 1) - 6(x_2 - 1)).$$

PROOF (OF THE CHAIN RULE): We begin by observing that if F is differentiable at X_0, then the function $E(X, X_0)$ (E for error) defined by

$$E(X, X_0) = \frac{F(X) - T_F(X)}{\|X - X_0\|}$$

has the property that $E(X, X_0) \to 0$ as $X \to X_0$. Now we can write the equation above as

$$F(X) = T_F(X) + E(X, X_0)\|X - X_0\|,$$

which we think of as

$$F(X) = T_F(X) + \text{small},$$

where "small" means that $E(X, X_0)\|X - X_0\|$ approaches 0 rapidly (faster than $\|X - X_0\|$) as $X \to X_0$. Similarly,

$$G(Y) = T_G(Y) + \text{small}.$$

Thus we have

$$\begin{aligned} G(F(X)) &= T_G(T_F(X) + \text{small}) + \text{small} \\ &= T_G(T_F(X)) + G'(F(X_0))(\text{small}) + \text{small} \\ &= T_G(T_F(X)) + \text{small} \\ &= G(F(X_0)) + G'(F(X_0))F'(X_0)(X - X_0) + \text{small}. \end{aligned}$$

The second equality comes from the formula for T_G and the fourth equality involves a short calculation that the reader is advised to do on paper. The third equality says that

$$G'(F(X_0))(\text{small}) = \text{small}.$$

The proof is now complete since we have shown that $G'(F(X_0)) \circ F'(X_0)$ satisfies the defining property of $(G \circ F)'(X_0)$.

EXAMPLES:

3. Let $\mathbb{R}^2 \xrightarrow{F} \mathbb{R}^2 \xrightarrow{g} \mathbb{R}$, be given by

$$F(X) = (x_1^2 - x_2^2, \ 2x_1 x_2), \qquad g(Y) = y_1 y_2.$$

We compute the partial derivative of $h(X) = g(F(X))$ with respect to x_1. It is equal to the first entry in the row vector that represents the total derivative $h'(X)$. According to the chain rule, this is the same as the entry in the first column, first row of the the matrix $g'(F(X)) F'(X)$, which equals

$$[(\nabla g)(Y)]^T [(\frac{\partial y_1}{\partial x_1}(X), \frac{\partial y_2}{\partial x_1}(X))]$$

$$= \ \frac{\partial g}{\partial y_1}(Y) \frac{\partial y_1}{\partial x_1}(X) + \frac{\partial g}{\partial y_2}(Y) \frac{\partial y_2}{\partial x_1}(X)$$

$$= \ y_2(2x_1) + y_1(2x_2)$$

$$= \ (2x_1 x_2)2x_1 + (x_1^2 - x_2^2)2x_2$$

$$= \ 6x_1^2 x_2 - 2x_2^3 \ .$$

For instance, if $X_0 = (1,1)$, then $(\partial h / \partial x_1)(X_0) = 4$.

Another way to obtain the expression for $(\partial h / \partial x_1)$ is to note that $h(X) = g(x_1^2 - x_2^2, \ 2x_1 x_2) = (x_1^2 - x_2^2)2x_1 x_2$ and then take the partial derivative in the usual way:

$$\frac{\partial h}{\partial x_1}(X) = \frac{\partial}{\partial x_1} 2(x_1^3 x_2 - x_1 x_2^3) = 6x_1^2 x_2 - 2x_2^3 \ .$$

If the derivatives of F and g are already known, however, this second method may be wasteful; it may be simpler to apply the chain rule.

4. Let u be a differentiable scalar-valued function defined on $\mathbb{R}^n$, and for $X \in \mathbb{R}^n$, define $\varphi(X, t) = u(tX)$. We think of φ as the composition of the function u on the function $v(X, t) = tX$. Thus,

$$[\varphi'(X, t)] = [u'(tX)][v'(X, t)] \ .$$

Since u is a scalar-valued function, $u' = \nabla u$. The Jacobian matrix of $v(X, t) = (tx_1, tx_2, \ldots, tx_n)$ is

$$[v'(X, t)] = \begin{bmatrix} t & 0 & \cdots & 0 & x_1 \\ 0 & t & \cdots & 0 & x_2 \\ & & \vdots & & \\ 0 & 0 & \cdots & t & x_n \end{bmatrix} \ .$$

Thus the partial derivatives of φ with respect to $x_1, \ldots, x_n$ are

$$\varphi_{x_k}(X, t) = t u_k(tX)$$

for $1 \le k \le n$, and the partial derivative of φ with respect to t is

$$\varphi_t(X, t) = x_1 u_1(tX) + \cdots + x_n u_n(tX) .$$

Exercises

1. Suppose that F and G are both functions from $\mathbb{R}^3$ into $\mathbb{R}^2$ and that

$$[F'(3, 5, 2)] = \begin{bmatrix} 2 & 0 & -7 \\ 1 & 3 & 0 \end{bmatrix}$$

and

$$[G'(3, 5, 2)] = \begin{bmatrix} -2 & 3 & 8 \\ 0 & -1 & 2 \end{bmatrix} .$$

Calculate $[(F + G)'(3, 5, 2)]$.

2. Can the function in Example 1 be defined at $\mathbf{0}$ to be continuous there? If so, is it differentiable there? If so, what matrix represents its derivative there?

3. Calculate the derivatives of the following functions defined for all $X \in \mathbb{R}^n$ different from $\mathbf{0}$.

   (a) $\dfrac{1}{\|X\|} X$

   (b) $\|X\|^2 X$

   (c) $\ln(\|X\|) X$.

4. For each function in the preceding exercise, decide if it can be defined at $\mathbf{0}$ to be continuous there? If so, is it differentiable there? If so, what matrix represents its derivative there?

5. Given the following maps F and G, decide if $F \circ G$, or $G \circ F$, or both, or neither makes sense:

   (a) $F: \mathbb{R}^2 \to \mathbb{R}^2$, $G: \mathbb{R}^2 \to \mathbb{R}^3$

   (b) $F: \mathbb{R}^2 \to \mathbb{R}^2$, $G: \mathbb{R}^1 \to \mathbb{R}^2$

   (c) $F: \mathbb{R}^1 \to \mathbb{R}^1$, $G: \mathbb{R}^2 \to \mathbb{R}^1$

   (d) $F: \mathbb{R}^2 \to \mathbb{R}^3$, $G: \mathbb{R}^2 \to \mathbb{R}^1$

   (e) $F: \mathbb{R}^2 \to \mathbb{R}^3$, $G: \mathbb{R}^3 \to \mathbb{R}^2$

   (f) $F: \mathbb{R}^3 \to \mathbb{R}^1$, $G: \mathbb{R}^1 \to \mathbb{R}^3$.

6. Let $F: \mathbb{R}^2 \to \mathbb{R}^2$ and $G: \mathbb{R}^3 \to \mathbb{R}^2$ be defined by

$$\begin{aligned} F(x, y) &= (e^{x+y^2}, e^{y+x^2}), \\ G(r, s, t) &= (r + s^2 + t^2, s + t^2 + r^3) . \end{aligned}$$

(a) Compute F' and G'.

(b) If $G \circ F$ makes sense, compute $(G \circ F)'$ at $(-1, -1)$; if $F \circ G$ makes sense, compute $(F \circ G)'$ at $(-1, 0, 0)$.

7. Compute the partial derivatives of f given by

   (a) $f(x, y) = g(u, v) = uv^2$, where $u = \sin x$, $v = y \cos x$ at $(\pi, 1)$

   (b) $f(x, y) = g(r, s) = e^{rs}$, where $r = x^2$, $s = xy$ at $(1, -1)$.

8. Let $F(X) = (x_2 - e^{x_1 - 2x_2}, x_1, x_2)$, $G(Y) = (y_2 + y_3 \sin y_1, (y_1 + y_2)^2)$.

   (a) Compute F' at $X_0 = (-2, 1)$ and G' at $Y_0 = F(X_0)$.

   (b) Let $H = G \circ F$. Compute H' at $X_0 = (-2, 1)$.

9. (a) Let $f(x, y) = xy - e^x$. If $x(t) = 1 - t^2$, $y(t) = 2t - 3$, and $\varphi(t) = f(x(t), y(t))$, find $\varphi'(t)$.

   (b) Repeat this for $f(x, y) = y \cos(x + y + 1)$.

10. Let $\varphi(x, y, t) = xy^2 - t \cos y$. If $x(t) = \sin \pi t$ and $y(t) = t^3 - 1$, let $h(t) = \varphi(x(t), y(t), t)$. Compute $h'(1)$.

11. Let $\varphi(x, s, t) = xs + xt + st$. If $x(t) = t^3 - 7$, compute the total derivative of $f(s, t) = \varphi(x(t), s, t)$ at $(3, 2)$.

12. Let φ be a differentiable function from $\mathbb{R}$ into $\mathbb{R}$.

    (a) If $u(x, y) = \varphi(2x - 3y)$, show that $3u_x + 2u_y = 0$.

    (b) If $u(x, y) = \varphi(ax + by)$, show that $bu_x - au_y = 0$.

    (c) If $u(x, y) = \varphi(xy)$, show that $xD_1u(x, y) - yD_2u(x, y) = 0$.

    (d) If $u(x, y) = \varphi(x/y)$, show that $xD_xu(x, y) + yD_yu(x, y) = 0$ for $y \neq 0$.

    (e) If $u(x_1, x_2) = \varphi(\|X\|^2)$ show that $x_2 D_1 u(X) - x_1 D_2 u(X) = 0$.

13. Let φ be a twice differentiable function from $\mathbb{R}$ into $\mathbb{R}$.

    (a) If $u(x, t) = \varphi(x + ct)$, show that $u_{tt} - c^2 u_{xx} = 0$.

    (b) If $u(x, y) = \varphi(ax + by)$, show that $u_{xx}u_{yy} - u_{xy}^2 = 0$.

14. The temperature at a point $X \in \mathbb{R}^3$ is $f(X) = \|X\|^2$. A particle travels on the curve $X(t) = (\sin \pi t, \cos \pi t, t^2 - 2t + 2)$. What is the coolest point on the trajectory of the particle?

15. Let $X(t) \in \mathbb{R}^3$ be the position of a particle at time t. Compute the derivatives of $\|X(t)\|$ and $1/\|X(t)\|$, assuming that $X(t) \neq 0$.

16. Suppose $f: \mathbb{R}^2 \to \mathbb{R}$ is differentiable at the point $(1, 1)$, with $(D_1 f)(1, 1) = 2$ and $(D_2 f)(1, 1) = -3$. Let $g(s, t) = f(2s - t, s^2)$. Compute the total derivative $g'(1, 1)$.

17. This exercises involves two function u and v and four constants a, b, c, and d satisfying the relation

$$v(s, t) = u(as + bt, cs + dt).$$

    (a) show that $D_1 v = aD_1u + cD_2u$ and $D_2 v = bD_1u + dD_2u$. What single matrix equality is equivalent to these two equalities?

(b) If $D_1 u - D_2 u = 0$, pick constants a, b, c, d so that $D_1 v = 0$ and the matrix

$$\begin{bmatrix} a & b \\ c & d \end{bmatrix}$$

is invertible.

(c) Use the preceding parts to show that the most general function satisfying $D_1 u - D_2 u = 0$ is of the form $u(x, y) = \varphi(x + y)$, where φ is any differentiable function.

(d) If, in addition, you know that $u(x, 0) = x^2$, what must the functions u and φ in part (c) be?

18. Apply the chain rule to $u(x, y) = f(x)g(y)$, with $x = t$, $y = t$, to deduce the standard formula for the derivative of $f(t)g(t)$.

19. Let $f : \mathbb{R} \to \mathbb{R}$ be a differentiable function with the properties $f(1) = 7$ and $f'(1) > 0$. Consider the set of points $X = (x, y) \in \mathbb{R}^2$ such that $f(xy) = 7$. This gives a curve in $\mathbb{R}^2$. Show that the line $x + y = 2$ is tangent to the curve at $(1, 1)$ independent of the particular form of f. What is the equation of the tangent line to this curve at (a, b)?

20. (Euler) An $\mathbb{R}$-valued function f with $\mathcal{D}(f)$ consisting of all points in $\mathbb{R}^n$ except $\mathbf{0}$ is **homogeneous of degree** λ if $f(tX) = t^\lambda f(X)$ for all $t > 0$ and all $X \neq \mathbf{0} \in \mathbb{R}^n$.

(a) Show that $f(x_1, x_2) = x_1^2 x_2 - 5x_2^3$ is homogeneous of degree 3 but that $f(x_1, x_2) = x_1^2 x_2 - 5x_2^4$ is not homogeneous of any order.

(b) Show that $\langle X, \operatorname{grad} f(X) \rangle = \lambda f(X)$ if f is differentiable and homogeneous of degree λ.

(c) Conversely, show that if f satisfies the equation in the preceding part, then f is homogeneous of degree λ. *Hint:* Show that

$$\partial(t^{-\lambda} f(tX)) / \partial t = 0 .$$

See Example 4.

21. Let u be an $\mathbb{R}$-valued twice-differentiable function whose domain is the complement of $\{\mathbf{0}\}$ in $\mathbb{R}^2$ and which satisfies the **Laplace equation**

$$D_{11} u + D_{22} u = 0$$

on its domain. Suppose that $u(X) = \varphi(\|X\|)$.

(a) Show that $r\varphi''(r) + \varphi'(r) = 0$.

(b) Deduce that $u(X) = a \log \|X\| + b$ for constants a and b.

22. Make the appropriate changes in both the statement and the solution to the preceding exercise in order to replace $\mathbb{R}^2$ by $\mathbb{R}^3$.

8.3   Surfaces in $\mathbb{R}^q$

8.3A   PARAMETRIZED SURFACES

Recall that in Chapter 4, we defined a parametrized curve in $\mathbb{R}^q$ to be a continuous function from an interval in $\mathbb{R}$ to $\mathbb{R}^q$. We now wish to define a "parametrized surface" in $\mathbb{R}^q$. This will be a continuous function from an appropriate subset of $\mathbb{R}^2$ to $\mathbb{R}^q$.

Definition 8.3.1 Let F be a continuous function from a subset $\mathcal{D}(F)$ of $\mathbb{R}^2$ into $\mathbb{R}^q$. Suppose that $\mathcal{D}(F)$ is pathwise connected, and that every member of $\mathcal{D}(F)$ is either an interior point of $\mathcal{D}(F)$, or a limit point of its interior. Then F is called a **parametrized surface** in $\mathbb{R}^q$.

REMARK:

- The condition required of the domain of F in the definition has an unpleasantly technical character. The purpose of this condition is to ensure that the domain has (in a sense that we will leave imprecise) a truly 2-dimensional nature, thus avoiding certain degenerate cases. See Example 2 below. Note that pathwise connected domains which are open sets automatically satisfy the condition in the definition.

EXAMPLES:

1. This is a continuation of Example 2 of Section 8.0 in which G was defined by

$$G(\theta, \varphi) = ((a + b\cos\varphi)\cos\theta, \, (a + b\cos\varphi)\sin\theta, \, b\sin\varphi)$$

   with

$$\mathcal{D}(G) = \{(\theta, \varphi) : 0 \leq \theta \leq 2\pi, \, 0 \leq \varphi \leq 2\pi\}.$$

   Each coordinate function of G is continuous, so G is continuous. The domain of G is a pathwise connected square region. Every point in that domain is either in the interior of the square region, or a boundary point of that interior. Thus G is a parametrized surface.

2. Let $F(x_1, x_2) = (x_1, x_2, x_1 + x_2)$ with

$$\mathcal{D}(F) = \{(x_1, x_2) : 0 < x_1 = x_2 < 1\}.$$

   The function F is continuous. However, since the interior of $\mathcal{D}(F)$ is empty, and since the boundary of the empty set is also empty, *none* of the points in $\mathcal{D}(F)$ satisfy the condition of Definition 8.3.1, even though $\mathcal{D}(F)$ is pathwise connected. The story here is that $\mathcal{D}(F)$ is a "one-dimensional" object, so F "should not" be a parametrized surface, which we expect to be a "two-dimensional" object.

3. Let U, V, and W be fixed vectors in $\mathbb{R}^q$ and define

$$G(s, t) = sU + tV + W .$$

The domain of G is $\mathbb{R}^2$ (which is open and pathwise connected), and G is continuous (since it is an affine map), so G is a parametrized surface. In fact, G is a parametrization of the plane $\mathcal{A} = \mathcal{P} + W$, where $\mathcal{P}$ is the plane through the origin spanned by the vectors U and V.

Recall that in Chapter 4 we defined a parametrized curve as a function, and then defined its image to be a curve. In analogy, we say that the image of a parametrized surface is a **surface**. Thus the torus and plane in the preceding examples are surfaces.

EXAMPLES:

4. This is a continuation of Example 3 of Section 8.0 in which the function

$$F(x_1, x_2) = (x_1,\ x_2,\ x_1^2 + x_2^2)$$

was discussed. We let $\mathcal{D}(F)$ be the natural domain, namely $\mathbb{R}^2$, which is pathwise connnected and open. The function F is continuous since each of its coordinate functions is continuous. Thus the circular paraboloid in $\mathbb{R}^3$ is a surface. (If the circular paraboloid had not turned out to be a surface, mathematicians would not be satisfied with our definition of surface.)

5. Let G be a parametrized surface in $\mathbb{R}^q$. The graph of G is a subset of $\mathbb{R}^{2+q}$. Here is an argument similar to that in the preceding example that shows that the graph of G is a surface in $\mathbb{R}^{2+q}$.

The graph of G is the image of the function F defined by the formula

$$F(X) = (X, G(X)) ,$$

with $\mathcal{D}(F) = \mathcal{D}(G)$. We have used the notation $(X, G(X))$ to denote the member of $\mathbb{R}^{2+q}$ whose first 2 coordinates are the coordinates of the vector X and whose last q coordinates are those of the vector $G(X)$. The function F is continuous since each of its coordinate functions is continuous (remember, since G is a parametrized surface, each of its coordinate functions is continuous). Thus, F gives a parametrization of the graph of G.

6. Let $G : \mathbb{R}^2 \to \mathbb{R}^q$ be a constant function. Clearly G is continuous, and the domain of G is open and pathwise connected, so G is a parametrized surface, and its image is a surface. However, the image of G is just a single point in $\mathbb{R}^q$, which is not what we expect a

surface to look like. A similarly unfortunate example can be given for parametrized curves. We will see in the next section that if a surface has a tangent plane, then it "looks" the way we expect a surface to look.

8.3B Tangent Planes of Surfaces

Let G be a parametrized surface in $\mathbb{R}^q$ and let S be the corresponding surface; that is, $S = \mathcal{I}(G)$. Let X_0 be a point in the interior of the domain of G, and assume that G is differentiable at X_0. The tangent map

$$T(X) = G'(X_0)(X - X_0) + G(X_0)$$

is an $\mathbb{R}^q$-valued function whose domain is $\mathbb{R}^2$, so T is also a parametrized surface in $\mathbb{R}^q$. Since T is an affine map with a 2-dimensional domain, the image of T is either a point, a line, or a plane in $\mathbb{R}^q$. If the image of T is a plane, we call it the **tangent plane** at $G(X_0)$ of the surface S. In this case, we also say that G has a tangent plane at $G(X_0)$.

REMARK:

- The image of an affine map $T : \mathbb{R}^2 \to \mathbb{R}^q$ is a plane if T is one-to-one, or equivalently, if the linear map L associated with T is one-to-one. Thus, a parametrized surface G has a tangent plane at $G(X_0)$ if $G'(X_0)$ is one-to-one.

We have already computed the tangent plane of the graph of a function $g : \mathbb{R}^2 \to \mathbb{R}$. Let us look at the connection between the tangent plane of a graph and the tangent plane of a surface. Recall from Section 8.3a that the graph of g is the image of the parametrized surface given by

$$G(x_1, x_2) = (x_1, x_2, g(x_1, x_2)) .$$

The matrix of $G'(X)$ is

$$[G'(X_0)] = \begin{bmatrix} 1 & 0 \\ 0 & 1 \\ g_{x_1}(X_0) & g_{x_2}(X_0) \end{bmatrix} .$$

Using the methods of Chapter 3, you may check that the null space of the linear transformation $G'(X)$ is trivial, so that $G'(X)$ is one-to-one. Therefore, the image of the affine approximation

$$T(X) = G'(X_0)(X - X_0) + G(X_0)$$

is the tangent plane of the graph of g at the point $(X_0, g(X_0))$. We may write the formula for T in coordinate form as follows:

$$T(x_1, x_2) = (x_1, x_2, \langle X - X_0, \nabla g(X_0) \rangle + g(X_0)) .$$

The image of T is clearly the graph of the function

$$t(X) = \langle X - X_0, \nabla g(X_0) \rangle + g(X_0) \, ,$$

which is an affine linear function from $\mathbb{R}^2$ to $\mathbb{R}$. This graph is a plane in $\mathbb{R}^3$, as expected. The discussion just given implies the following result:

Theorem 8.3.1 Let g be a scalar-valued function whose domain is a subset of $\mathbb{R}^2$. If g is differentiable at a point X_0, then the surface which is the graph of g has a tangent plane at the point $(X_0, g(X_0))$. This tangent plane is the graph of the function

$$t(X) = g(X_0) + \langle X - X_0, \nabla g(X_0) \rangle \, .$$

EXAMPLE:

7. Consider the saddle surface $z = g(x, y) = x^2 - y^2$. The gradient of g is $(2x, -2y)$, so the tangent plane at the point $(2, -3, g(2, -3)) = (2, -3, -5)$ is the graph of

$$t(x, y) = -5 + \langle (4, 6), (x - 2, y + 3) \rangle = 4x + 6y + 5 \, ,$$

which has the equation $z = 4x + 6y + 5$.

If G is an $\mathbb{R}^q$-valued function defined on a domain in $\mathbb{R}^2$, and if G is differentiable at X_0, the matrix $[G'(X_0)]$ is an n-by-2 matrix. The first column of this matrix has the same entries as the vector

$$(\frac{\partial g_1}{\partial x_1}(X_0) \, , \ \frac{\partial g_2}{\partial x_1}(X_0) \, , \ \ldots \, , \ \frac{\partial g_n}{\partial x_1}(X_0)) \, ,$$

and the second column has the same entries as the vector

$$(\frac{\partial g_1}{\partial x_2}(X_0) \, , \ \frac{\partial g_2}{\partial x_2}(X_0) \, , \ \ldots \, , \ \frac{\partial g_n}{\partial x_2}(X_0)) \, ,$$

where $g_1, g_2, \ldots, g_n$ are the coordinate functions of G. It is natural to write

$$\frac{\partial G}{\partial x_1}(X_0) \quad \text{and} \quad \frac{\partial G}{\partial x_2}(X_0)$$

for these two vectors. The tangent map of G at the point $X_0 = (c_1, c_2)$ may be written in terms of these vectors as

$$T(x_1, x_2) = (x_1 - c_1)\frac{\partial G}{\partial x_1}(c_1, c_2) + (x_2 - c_2)\frac{\partial G}{\partial x_2}(c_1, c_2) + G(c_1, c_2) \, .$$

The image of T is clearly a plane if the two vectors $\partial G / \partial x_1$ and $\partial G / \partial x_2$ are linearly independent. We have proved the following:

Theorem 8.3.2 The parametrized surface G has a tangent plane at $G(X_0)$ if the vectors

$$U = \frac{\partial G}{\partial x_1}(X_0) \quad \text{and} \quad V = \frac{\partial G}{\partial x_2}(X_0)$$

are linearly independent. In this case, the tangent plane is the affine subspace $\mathcal{P} + G(X_0)$, where $\mathcal{P}$ is the plane through the origin spanned by the vectors U and V, and a parametrization for the tangent plane is

$$sU + tV + G(X_0),$$

where $(s, t) \in \mathbb{R}^2$.

EXAMPLES:

8. Once again, we consider the torus of Example 2 in Section 8.0. We compute

$$\frac{\partial G}{\partial \theta}(\theta, \varphi) = (-(a + b \cos \varphi) \sin \theta, (a + b \cos \varphi) \cos \theta, 0)$$

and

$$\frac{\partial G}{\partial \varphi}(\theta, \varphi) = (-b \sin \varphi \cos \theta, -b \sin \varphi \sin \theta, b \cos \varphi).$$

At the point $G(\pi/4, \pi/4)$, the tangent plane is given by

$$s\left(-\frac{a\sqrt{2} + b}{2}, \frac{a\sqrt{2} + b}{2}, 0\right) + t\left(-\frac{b}{2}, -\frac{b}{2}, \frac{b\sqrt{2}}{2}\right)$$
$$+ \left(\frac{a\sqrt{2} + b}{2}, \frac{a\sqrt{2} + b}{2}, \frac{b\sqrt{2}}{2}\right),$$

for $(s, t) \in \mathbb{R}^2$. Compare this parametrization of the tangent plane of G with the expression for the tangent plane found in Example 3 in Section 8.1.

9. Let G be a parametrized surface in $\mathbb{R}^q$ that is differentiable at the point $X_0 = (c_1, c_2)$. Consider the function

$$F(r) = G(r, c_2),$$

with domain equal to some open interval containing c_1. Since G is continuous, the function F is also continuous, so F is a parametrized curve in $\mathbb{R}^q$. The image of F is a curve on the surface parametrized by G, a curve that goes through the point $G(c_1, c_2)$. We note that

$$F'(c_1) = \frac{\partial G}{\partial x_1}(c_1, c_2),$$

so the tangent line to F at the point $F(c_1) = G(c_1, c_2)$ is given parametrically by

$$r \frac{\partial G}{\partial x_1}(c_1, c_2) + G(c_1, c_2)$$

for r real. Similarly, the line

$$s \frac{\partial G}{\partial x_2}(c_1, c_2) + G(c_1, c_2)$$

is tangent to the curve parametrized by the function $H(s) = G(c_1, s)$ for s in some interval containing c_2. As you may easily check, the tangent plane of the surface parametrized by G at the point X_0 is the plane that contains these two lines.

8.3C LEVEL SETS

Let g be a scalar-valued function whose domain is a subset of $\mathbb{R}^q$. For c a real number, the set

$$\{X \in \mathcal{D}(g) : g(X) = c\}$$

is called a **level set** of g, or also the level set of g **at level** c. Many curves and surfaces are naturally expressed as level sets.

EXAMPLES:

10. Let $f(x, y) = x^2 + y^2$ for $(x, y) \in \mathbb{R}^2$. The level set of f at level c is the circle of radius $\sqrt{c}$ centered at the origin if $c > 0$. The level set of f at level 0 is just a single point, namely the origin, while the level sets of f at levels $c < 0$ are empty.

11. The level set of the function $g(x, y, z) = x^2 + y^2 + z^2$ at level c is $S((0,0,0); \sqrt{c})$ if $c > 0$.

12. Let F be the parametrized curve

$$F(\theta) = (a \cos \theta, a \sin \theta, 0) ,$$

for $0 \leq \theta \leq 2\pi$ and some fixed real number $a > 0$. The image of F is the horizontal circle of radius a centered at the origin in $\mathbb{R}^3$. For any point (x, y, z) in $\mathbb{R}^3$, let $g(x, y, z)$ equal the distance between that point and the circle parametrized by F. In the exercises, the reader is asked to show that for $0 < b < a$, the level set of g at level b is the torus that has already been discussed several times in this chapter.

13. Let $f(x, y) = xy$. For $c \neq 0$, the level set of f at level c is the hyperbola $y = c/x$, with asymptotes equal to the two coordinate axes. If $c > 0$, the two branches of the hyperbola lie in the first and third quadrants, while if $c < 0$, they lie in the second and fourth quandrants. The level set at level 0 is the set $\{(x, y) : xy = 0\}$, which is just the union of the two coordinate axes.

It appears from the examples just given that the level sets of functions defined on $\mathbb{R}^2$ are often curves, and the level sets of functions defined on $\mathbb{R}^3$ are often surfaces. There are some exceptions: some of the level sets of the function $x^2 + y^2$ are empty, and the hyperbolas of the last example are not curves, but instead unions of two separate curves. The following theorem, whose proof we omit, says that in many circumstances, the level sets of functions defined on $\mathbb{R}^2$ or $\mathbb{R}^3$ are "locally" curves or surfaces.

Theorem 8.3.3   1. Let f be a scalar-valued function whose domain is a subset of $\mathbb{R}^2$. Let (x_0, y_0) be a point in the level set of f at level c for some fixed real number c. Assume that the gradient of f exists and is continuous in an open ball centered at (x_0, y_0). If $\nabla f(x_0, y_0) \neq \mathbf{0}$, then there exists a real number $\delta > 0$ such that the set

$$\{(x, y) \in \mathcal{D}(f) : f(x, y) = c\} \cap B((x_0, y_0); \delta)$$

is a curve. Furthermore, this curve has a tangent line at the point (x_0, y_0), and the equation for this tangent line is

$$\langle \nabla f(x_0, y_0), (x - x_0, y - y_0) \rangle = 0 .$$

2. Let g be a scalar-valued function whose domain is a subset of $\mathbb{R}^3$. Let (x_0, y_0, z_0) be a point in the level set of g at level c for some fixed real number c. Assume that the gradient of g exists and is continuous in an open ball centered at (x_0, y_0, z_0). If $\nabla g(x_0, y_0, z_0) \neq \mathbf{0}$, then there exists a real number $\delta > 0$ such that the set

$$\{(x, y, z) \in \mathcal{D}(g) : g(x, y, z) = c\} \cap B((x_0, y_0, z_0); \delta)$$

is a surface. Furthermore, this surface has a tangent plane at the point (x_0, y_0, z_0), and the equation for this tangent plane is

$$\langle \nabla g(x_0, y_0, z_0), (x - x_0, y - y_0, z - z_0) \rangle = 0 .$$

REMARKS:

- The theorem is part of an important result from advanced calculus called the "Implicit Function Theorem". There are generalizations to level sets of functions whose domains lie in $\mathbb{R}^n$ for $n > 3$.

- Although the proof of the complete theorem is beyond the scope of this book, we can give a partial justification of the formula for the tangent plane in the second part of the theorem. Let $F(t)$ be a parametrized curve whose image lies on the level set $g = c$. Assume that $F(t_0) = (x_0, y_0, z_0)$ for some real number t_0 in the domain of F. Further assume that F is differentiable at t_0. Since the image of F lies on the level set, $g(F(t)) = c$ for all t in the domain of F. Thus,

$$\frac{dg(F(t))}{dt} = 0 .$$

By the chain rule,

$$\frac{dg(F(t))}{dt}(t_0) = [\nabla g(x_0, y_0, z_0)]^T [F'(t_0)] \ .$$

It follows that $\nabla g(x_0, y_0, z_0)$ is orthogonal to $F'(t_0)$. This implies that the line $\mathcal{L}$ given by

$$t\nabla g(x_0, y_0, z_0) + (x_0, y_0, z_0)$$

is perpendicular to the tangent line of any differentiable parametrized curve F that lies on the level set $g = c$ and passes through the point (x_0, y_0, z_0). The line $\mathcal{L}$ is also the normal line of the tangent plane given in the theorem (can you see why?). It seems reasonable, then, that this plane is indeed tangent to the surface given by $g = c$. A similar but simpler argument applies to the formula for the tangent line in the first part of the theorem.

EXAMPLES:

14. The saddle surface given by $z = x^2 - y^2$ is the graph of the function $f(x,y) = x^2 - y^2$. Earlier, in Example 7 of Section 8.3b, we found that the tangent plane of this graph at the point $(2, -3, -5)$ has the equation $z = 4x + 6y + 5$. The saddle surface is also the level set at level 0 of $g(x, y, z) = x^2 - y^2 - z$. The gradient of g at the point $(2, -3, -5)$ is $(4, 6, -1)$. Thus, according to the theorem, the tangent plane at $(2, -3, -5)$ of this level set has the equation

$$\langle (4, 6, -1), (x - 2, y + 3, z + 5) \rangle = 0 \ ,$$

which simplifies to $z = 4x + 6y + 5$, in agreement with our earlier calculation.

15. The function $f(x, y) = 2x^2 + 6y^2$ has level sets at level $c > 0$ that are ellipses. Consider the level set at level 8, given by $2x^2 + 6y^2 = 8$. The point $(-1, 1)$ is on this ellipse. The gradient of f at this point is $(-4, 12)$, so theorem says that the corresponding tangent line is

$$\langle (-4, 12), (x + 1, y - 1) \rangle = 0 \ ,$$

or $-4x + 12y = 16$. You may want to verify this by using methods that you learned in single-variable calculus. The point $(-2, 0)$ is also on the ellipse, at its extreme left end. The tangent line at $(-2, 0)$ is given by

$$\langle (-8, 0), (x + 2, y) \rangle = 0 \ ,$$

which is the same as $x = -2$. This is a vertical tangent line, as expected. Single-variable calculus does not handle vertical tangent lines so naturally.

Exercises

1. For each of the following curves, find the tangent line at the indicated point, if it exists. If the tangent line does not exist, explain with pictures why it does not.

   (a) The level set at level 1 of the function $f(x,y) = ye^x + e^y$ at $(0,0)$;

   (b) The level set at level 1 of the function $f(x,y) = |x| + |y|$ at $(0,1)$;

   (c) The level set at level 2 of the function $f(x,y) = |x| + |y|$ at $(1,-1)$;

   (d) The graph of the function $f(x) = \sqrt{x^2 + 1}$ at $(-2, \sqrt{5})$;

   (e) The parametrized curve

   $$F(t) = ((a + b \cos t) \cos t, (a + b \cos t) \sin t, b \sin t)$$

   at $F(\pi/4)$.

2. For parts (b),(c),(d), and (e) of the preceding exercise, draw the curve. Also show the tangent line if it exists.

3. For each of the following surfaces, find the tangent plane at the indicated point, if it exists. If the tangent plane does not exist, explain with pictures why it does not.

   (a) The graph of the function $f(x,y) = \sqrt{x^2 + y^2 + 1}$ at $(2,1,\sqrt{6})$;

   (b) The graph of the function $f(x,y) = \cos(x+y)$ at $(\frac{\pi}{2}, \pi, 0)$;

   (c) The torus of Example 2 in Section 8.0 at $G(\frac{\pi}{3}, \frac{\pi}{6})$;

   (d) The parametrized surface $G(s,t) = (s - 2t, 2s - 4t, 2t - s)$ at $G(1,2)$;

   (e) The parametrized surface $G(\theta,t) = (a\cos\theta, b\sin\theta, t)$, where a and b are fixed positive real numbers, at $G(\frac{\pi}{3}, 2)$;

   (f) The level set at level 8 of the function $f(x,y,z) = xyz$ at $(4,-2,-1)$;

   (g) The level set at level 1 of the function $f(x,y,z) = x^{\frac{1}{3}} + y^{\frac{1}{3}} + z^{\frac{1}{3}}$ at $(1,0,0)$;

   (h) The level set at level 2 of the function $f(x,y,z) = x^{\frac{1}{3}} + y^{\frac{1}{3}} + z^{\frac{1}{3}}$ at $(8,-8,8)$.

4. Draw the surface in part (e) of the preceding exercise.

5. Show that for $0 < b < a$ the level set at level b of the function g of Example 12 is the torus, as claimed in that example. *Hint:* First verify that the intersection of the level surface with the plane $y = 0$ is the circle of radius b about the point $(a,0,0)$ in that plane. Then use rotational symmetry.

6. Draw the level set of the function g in Example 12 at level b for $b = \frac{3}{2}a$. Indicate points on your drawing where the tangent plane does not exist. *Hint.* First, restrict your attention to the plane $y = 0$.

7. (a) Find a parametrization of the cone of revolution $y = \|X\| \cot\psi$ in $\mathbb{R}^3$.

   (b) Write the cone of part (a) as the level set of some function.

8. Let
$$G(\theta, \varphi) = (a\cos\theta\sin\varphi, b\sin\theta\sin\varphi, c\cos\varphi),$$
where a, b, and c are fixed positive constants.

(a) Draw the image of G for $0 \le \theta \le 2\pi$ and $0 \le \varphi \le \pi$.

(b) Prove that the surface of part (a) is the level set at level 1 of the function
$$f(x, y, z) = \frac{x^2}{a^2} + \frac{y^2}{b^2} + \frac{z^2}{c^2}.$$

(c) Find the tangent plane of the surface at the point $(\frac{a}{\sqrt{3}}, \frac{b}{\sqrt{3}}, \frac{c}{\sqrt{3}})$.

9. Let $F : \mathbb{R}^2 \to \mathbb{R}^3$ be defined by
$$F(X) = \begin{cases} (X, \sqrt{1 - \|X\|^2}) & \text{if} \quad \|X\| \le 1 \\ \frac{1}{\|X\|}(\frac{X}{\|X\|}, -\sqrt{\|X\|^2 - 1}) & \text{if} \quad 1 < \|X\| < \infty \end{cases}$$

Show that the image of F is the sphere $S((0, 0, 0); 1)$, except that the "south pole" $(0, 0, -1)$ is missing.

10. In this exercise we study a parametrization of the unit sphere in $\mathbb{R}^3$. Let F denote the map
$$F(\theta, \varphi) = (\cos\theta\sin\varphi, \sin\theta\sin\varphi, \cos\varphi)$$
for $0 \le \theta \le 2\pi$ and $0 \le \varphi \le \pi$.

(a) Show that $\mathcal{I}(F)$ is a subset of the sphere $S((0, 0, 0); 1)$.

(b) Show that $S((0, 0, 0); 1)$ is a subset of $\mathcal{I}(F)$.

(c) Repeat the preceding part with the modification that
$$\mathcal{D}(F) = \{(\theta, \varphi) : 0 \le \theta < 2\pi, 0 \le \varphi \le \pi\}.$$

See Figure 8.7.

(d) Find the best affine approximation to this map at $(0, \frac{\pi}{2})$.

(e) Find the best affine approximation to this map at $(\frac{\pi}{2}, \frac{\pi}{4})$.

(f) Find the best affine approximation to this map at $(\frac{\pi}{3}, \pi)$. Why can't you use this affine approximation to find a tangent plane of the surface at $F(\frac{\pi}{3}, \pi)$? Does this mean that there is no tangent plane at that point? Explain.

(g) Show that the perpendicular at the point of tangency to any tangent plane of the sphere passes through the origin. Does your argument work for *all* points on the sphere? If not, can you fix it so that it does?

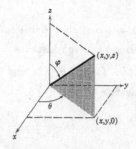

FIGURE 8.7.

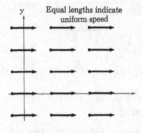

FIGURE 8.8.

8.4   Vector Fields

Now we discuss a new interpretation of functions $F: \mathbb{R}^n \to \mathbb{R}^n$. Consider a thin layer of fluid (such as water or a gas) flowing across a flat plane surface. The fluid is shallow; we are not concerned with its depth. We may suppose that the plane surface is actually the xy-plane $\mathbb{R}^2$. At each point (x, y) of the plane, we attach the tail of an arrow with the following properties:

1. The arrow points in the direction of the flow.

2. The length of the arrow is proportional to the speed of the particle at X_0, so that the longer the arrow, the faster the particle at X_0 is moving.

EXAMPLES:

1. The arrows in Figure 8.8 indicate that the flow is from left to right. Since all appear to have the same length, the speed at each point is the same.

2. In Figure 8.9 the flow is again from left to right, but now the arrows farther to the right are longer. As a particle flows to the right, it speeds up.

3. In Figure 8.10, all particles flow with uniform speed toward a single point (called a **sink**, for obvious reasons; if the arrows all were

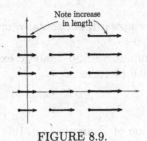

FIGURE 8.9.

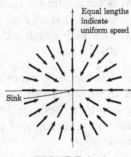

FIGURE 8.10.

reversed, that point would be a **source**).

It is customary to think of these arrows as vectors. If we denote by $F(x, y)$ the vector whose tail is attached to the point (x, y), we are led to the realization that all these vectors are given by a function $F: \mathbb{R}^2 \to \mathbb{R}^2$. If F is such a function and (x, y) a point in its domain, then $F(x, y)$ is some vector in $\mathbb{R}^2$. We think of this vector as an arrow and attach its tail to the point (x, y).

We say that a **vector field** in an open subset S of $\mathbb{R}^n$ is a rule that assigns to each point $X \in \mathbb{R}^n$ a vector $F(X) \in \mathbb{R}^n$. It is general practice to use the term "vector field" only when one is envisaging $F(X)$ as describing an arrow emanating from X. In applications vector fields have other names. For instance, **velocity field** is used when the arrows indicate velocity. If $F(X)$ represents the force acting at X, then F is a **force field**.

For a vector field F in $\mathbb{R}^2$ we often use p and q to denote its first and second coordinate functions, respectively, as in Figure 8.11. If, for example, F is a velocity field and $p(x, y) < 0$ at (x, y), then the particle is moving

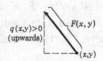

FIGURE 8.11.

toward the left and either upward, horizontally, or downward, accordingly as $q(x,y) > 0, = 0$, or < 0.

In the next three examples we find analytic expressions for the vector fields described earlier with pictures.

EXAMPLES:

4. This is a continuation of Example 1. In this case the same vector is attached to each point (x,y). If we agree that its length is 1, then it is the vector $(1,0)$ since it is horizontal (so that $q = 0$) and points from left to right. Thus we conclude that $F(x,y) = (1,0)$, independent of the point (x,y). If these all had length $c > 0$, then, of course, we would have $F(x,y) = (c,0)$.

5. This is a continuation of Example 2. Since each of these vectors is horizontal, $q(x,y) = 0$ as in the preceding example. Now $p(x,y)$ is positive and increasing as x increases, but does not vary with y. One candidate for such a function is $p(x,y) = e^x$ (there are many others). Using this choice, we have

$$F(x,y) = (p(x,y), q(x,y)) = (e^x, 0).$$

6. This is a continuation of Example 3. Now $F(X)$ has constant length, which we suppose to be 1, and from X towards the origin. Such a vector is given by $F(X) = -X/\|X\|$, as you should check (why the minus sign?). In coordinate notation (less elegant and less clear in this case) we have $F(x,y) = (p(x,y), q(x,y))$, where

$$p(x,y) = \frac{-x}{\sqrt{x^2+y^2}} \quad \text{and} \quad q(x,y) = \frac{-y}{\sqrt{x^2+y^2}}.$$

7. The vector field pictured in Figure 8.12 can be regarded as the force field resulting from the charge felt by a charged particle at X due to a particle of like charge at the origin. Note that the repellent force dies off (the arrows become shorter) as the distance from the particle at the origin increases.

REMARK:

- The vector fields we gave above were "time-independent" in the sense that we assigned the same vector $F(X)$ to a fixed point X for all time. In practice, however, one often encounters "time-dependent" vector fields $F(X,t)$, where $t =$ time. A simple example is $F(X,t) = (e^{xy}, x^2 + \sin t)$.

A time-dependent vector field is illustrated in Figure 8.13, which depicts a moving vector assigned to the point $\mathbf{0}$, one that points toward

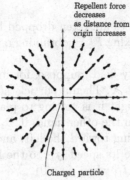

Repellent force
decreases
as distance from
origin increases

Charged particle

FIGURE 8.12.

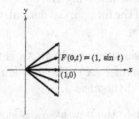

FIGURE 8.13.

the right but "wags" up and down as time passes. You might think of a weather vane, always fixed at a point, but changing direction with the wind.

We now introduce an important scalar-valued function related to a vector field.

Definition 8.4.1 Let F denote a differentiable vector field in an open subset $\mathcal{D}(F)$ of $\mathbb{R}^n$. The function

$$\text{div } F = D_1 f_1 + \cdots + D_n f_n$$

is called the **divergence** of F.

Another notation for $\text{div} F$ is $\langle \nabla, F \rangle$. This is not an actual inner product but suggests a sum with corresponding "coordinates" of ∇ and F being combined to make the various summands.

EXAMPLE:

8. Let $F(x_1, x_2) = (x_1^2 x_2^2, -7x_2 e^{x_1})$. Then $(\text{div } F)(X) = 2x_1 x_2^2 - 7e^{x_1}$.

In the preceding example the parentheses around $\text{div } F$ indicates that this is a symbol for a single function; it is this that is being applied to X.

However, these parentheses are often dropped because it would be meaningless to think of first taking $F(X)$ obtaining a vector and then calculating its divergence.

Let F denote a velocity field. Suppose, for instance, that $D_1 f_1(X) > 0$ and $f_1(X) > 0$ at a point X. Then a particle at X is moving to the right, and as it moves to the right it is getting to places where its speed to the right is getting larger. On the other hand, if $D_1 f_1(X) > 0$ and $f_1(X) < 0$, the particle at X is moving toward the left, and as it moves to the left it is getting to places where its speed $|f_1|$ to the left is getting larger (since smaller f_1 means larger $|f_1|$ when $f_1 < 0$).

By using a more elaborate version of this argument, it is possible to arrive at similar conclusions involving the sign of the full divergence of f, rather than just the sign of the first partial derivative of f_1. These conclusions are summarized as follows:

1. If div $F(X) > 0$, then the particle at X_0 is tending to speed up as it flows in the direction of $F(X)$. The flow is thus "diverging" from X_0 (whence the name "divergence").

2. If div $F(X_0) < 0$, then the flow at X_0 is slowing or faltering.

3. If div $F(X_0) = 0$, then the flow at X_0 is "steady", neither speeding nor slowing.

For instance, let $F(x, y) = (e^x, 0)$, as in Example 5. At any (x, y), we have div $F(x, y) = e^x > 0$. This shows that as the fluid flows from left to right at each point (x, y) it is tending to speed up.

In contrast, consider $G(x, y) = (e^{-x}, 0)$. This also produces a horizontal flow from left to right. But now div $G(x, y) = -e^{-x} < 0$, reflecting the fact that the velocity in the direction of motion is tending to decrease at each point (x, y). This phenomenon could not persist in the case of flowing water, which is an "incompressible" fluid, since the "early" water would not be moving fast enough to make room for the unrushing "later" water.

We discuss vector fields and divergence further in Chapter 10.

Exercises

1. Draw a sketch illustrating the given vector fields.

    (a) $F(x, y) = (e^{-y}, 0)$

    (b) $F(x, y) = (1, 2)$

    (c) $F(x, y) = (x, y)$

    (d) $F(x, y) = (y, 1)$

    (e) $F(x, y) = (y, -x)$

    (f) $F(x, y) = \left(\frac{y}{x^2 + y^2}, \frac{-x}{x^2 + y^2} \right)$, $(x, y) \neq 0$

    (g) $F(x, y) = (x, -y)$

(h) $F(x, y) = (2x, y)$

(i) $F(x, y) = (x^2, 0)$.

2. Compute div F for each of the vector fields in the preceding exercise. Briefly interpret your results in terms of the sketch of the vector field.

3. A vector field $F = (p, q)$ is called **irrotational**, or **conservative** if there is a scalar-valued function φ such that $F = \operatorname{grad} \varphi$. Then φ is called a **potential function** for the field.

   (a) Show that $F(x, y) = (2x, 2y)$ is irrotational.

   (b) If a vector field $F = (p, q)$ is irrotational and has continuous first partial derivatives, show that $p_y = q_x$. *Hint:* If $\varphi(x, y)$ has two continuous derivatives then $\varphi_{xy} = \varphi_{yx}$.

   (c) Show that the vector fields in parts (a), (d), and (e) of Exercise 1 are not irrotational, but the remaining vector fields in that exercises are irrotational (for part (f), consider $\arctan(x/y)$ for $y \neq 0$).

4. Let $F: \mathbb{R}^3 \to \mathbb{R}^3$ be $F(X) = -X/\|X\|^3$. Except for a constant scalar factor, this is the gravitational force field due to a single mass. Draw a sketch of it. Show that div $F = 0$ and that it is conservative for $X \neq 0$. *Hint:* Consider $\varphi(X) = 1/\|X\|$.

5. If $f, g: \mathbb{R}^2 \to \mathbb{R}$ and $F: \mathbb{R}^2 \to \mathbb{R}^2$ are sufficiently differentiable, show that:

   (a) div $(\operatorname{grad} f) = f_{xx} + f_{yy}$ (we write $f_{xx} + f_{yy}$ as $\nabla^2 f$ or as Δf and call $\nabla^2 = \Delta = \partial^2/\partial x^2 + \partial^2/\partial y^2$ the **Laplace operator**);

   (b) div $(fF) = \langle \operatorname{grad} f, F \rangle + f \operatorname{div} F$;

   (c) div $(f \operatorname{grad} g - g \operatorname{grad} f) = f\Delta g - g\Delta f$;

   (d) $\Delta(fg) = f\Delta g + 2\langle \operatorname{grad} f, \operatorname{grad} g \rangle + g\Delta f$.

6. Let $F(X): \mathbb{R}^3 \to \mathbb{R}^3$ denote the force at a point X, and assume that the force is conservative, so that $F = -\operatorname{grad} \varphi$ for some scalar-valued function (we have adopted the usage in physics here and added the minus sign, which could have been absorbed into φ). If $X(t)$ denotes the position of a particle of mass m moving in the force field, we define its energy by

$$E(t) = \frac{m}{2}\|X'(t)\|^2 + \varphi(X(t)).$$

Use Newton's second law, $mX'' = F$, to show that energy is conserved; that is, $E'(t) = 0$ (this justifies the name "conservative force" for $F = -\operatorname{grad} \varphi$).

9

Integration in $\mathbb{R}^n$

9.0 Introduction

Two elementary problems arise naturally and lead to the concept of integration. The first is that of finding the area or volume of some set. The second, to be discussed later, is finding the mass of some region, given its density. Let us discuss volume now. Rather than construct a theoretical edifice, we work informally and rely on geometric intuition. The theory of integration involves some subtle and technical issues that would be out of place at the beginning.

Say that we have a solid with vertical sides in $\mathbb{R}^3$ whose base is a set $\mathcal{B}$ in the xy-plane and whose top is the surface given by $z = f(x, y)$, where f is a nonnegative continuous function. See Figure 9.1. Our problem is to find the volume V of this solid.

We note that if the height $z = f(x, y)$ is a constant, independent of (x, y), then the volume is given by

$$V = (\text{height})(\text{area of } \mathcal{B}),$$

and so if we can compute the area of $\mathcal{B}$, we can compute V.

To deal with the general case where $f(x, y)$ varies with (x, y), we proceed as follows. First we partition the base $\mathcal{B}$ into smaller sets $\mathcal{B}_1, \mathcal{B}_2, \ldots, \mathcal{B}_r$. We denote the volume above each $\mathcal{B}_i$ (and below the surface $z = f(x, y)$,

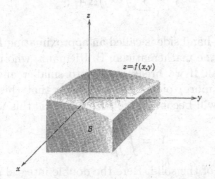

FIGURE 9.1. The solid between $\mathcal{B}$ and the surface $z = f(x, y)$

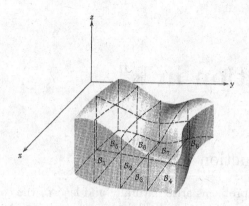

FIGURE 9.2. Partitioning the solid

of course) by ΔV_i (the ith little piece of V). See Figure 9.2. Thus

$$V = \sum_i \Delta V_i = \Delta V_1 + \Delta V_2 + \cdots + \Delta V_r .$$

Here we use the notation $\sum_i$ to mean "sum over all the indices i that occur in the present discussion". Since $i = 1, 2, \ldots, r$ in the present situation, we could be more formal and write $V = \sum_{i=1}^{r} \Delta V_i$. But we continue to be informal.

Second, we approximate V by approximating each ΔV_i. To do this, pick a point X_i in $\mathcal{B}_i$, and compute $f(X_i)$. The volume ΔV_i is approximately equal to the height $f(X_i)$ times the area of $\mathcal{B}_i$, which we denote ΔA_i; that is,

$$\Delta V_i \approx f(X_i)\Delta A_i ,$$

so that we have the approximation

$$V \approx \sum_i f(X_i)\Delta A_i .$$

The sum on the right-hand side is called an approximating **Riemann sum** to the integral, after the mathematician B. Riemann, who investigated the theory of the integral. If we break up $\mathcal{B}$ into smaller and smaller (consequently, more and more) pieces $\mathcal{B}_i$, we expect that this Riemann sum approximation to V will become more accurate. It is this which motivates the notation

$$V = \iint_{\mathcal{B}} f \, dA$$

for the *exact* volume of the solid. Here the double integral signs should be thought of as an elongated SS, to stand for (double) sum. We refer to f as the **integrand**. The sum is "double" because the base $\mathcal{B}$, the so-called

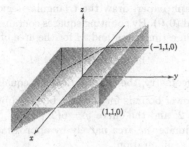

FIGURE 9.3.

region of integration, is two-dimensional. We also say that V is given by a **double integral**.

Even when f has negative values, sums of the form

$$\sum_i f(X_i)\Delta A_i$$

are of interest. These sums approximate the double integral

$$\iint_B f\, dA,$$

but the integral no longer equals volume. Rather it equals the difference between two volumes: the volume of the set that lies between the xy-plane and that portion of the graph of f where f is positive, *minus* the volume of the set that lies between the xy-plane and that portion of the graph of f where f is negative. For instance, if B is the square region in the xy-plane with vertices at $(1,1,0)$, $(1,-1,0)$, $(-1,-1,0)$, and $(-1,1,0)$, and the surface is given by the plane $z = y$ (that is, $f(x,y) = y$), then the double integral is zero, because the volume where $y \leq 0$ equals the volume where $y \geq 0$. See Figure 9.3.

So far we have only raised the issue "find the volume" and introduced some notation. In Section 9.1, we approximate some volumes in the elementary way outlined above. Then, in Section 9.2, we learn how to calculate some volumes exactly. Some of the theoretical aspects of integration are discussed in Section 9.3. In Section 9.4, we learn how to change variables in double (and triple) integrals. Surface area is considered in Section 9.5.

Exercises

1. Find the volume of the set of points in $\mathbb{R}^3$ that are above the xy-plane and below the graph of the function

$$f(x,y) = \begin{cases} 2 & \text{if } 0 \leq x < 1 \text{ and } 0 \leq y < 1 \\ 3 & \text{if } -1 \leq x < 0 \text{ and } |y| \leq 1 \\ 0 & \text{otherwise}. \end{cases}$$

2. On a sheet of graph paper, draw the triangular region A with vertices at $(-2, 2)$, $(1, 1)$, and $(0, 0)$. By counting squares contained in A and also those touching A, obtain estimates m and M for the area of A, such that

$$m \leq (\text{ area of } A) \leq M .$$

(By elementary geometry, the actual area of A equals 2.)

3. Find upper and lower bounds for the area bounded by the x-axis, the lines $x = -1$ and $x = 2$, and the curve $y = 6(2 + x^2)^{-1}$. Do not evaluate any integrals, but estimate the area naïvely by approximating the region by a simple geometric configuration.

4. Use a simple geometric argument to show that the volume V of a ball of radius 1 in $\mathbb{R}^3$ satisfies $\frac{8}{3\sqrt{3}} < V < 8$.

9.1   Estimating the Value of Integrals

9.1A   SOME NUMERICAL EXAMPLES

Now we leave the descriptive introduction and begin the real work of obtaining numerical answers. Remember, the point is to rely on intuition gained from past experience. Before estimating a double integral, we recall a simpler example from elementary calculus.

EXAMPLE:

1. Let us estimate the area (not volume now) between the curve $y = f(x) = 6/(2 + x^2)$, the segment $0 \leq x \leq 2$ on the x-axis, and the vertical lines $x = 0$ and $x = 2$. See Figure 9.4. Thus we want to estimate the (single) integral

$$J = \int_0^2 \frac{6}{2 + x^2} \, dx .$$

If we denote the interval $0 \leq x \leq 2$ by $\mathcal{B}$, then this integral could be written

$$J = \int_{\mathcal{B}} f(x) \, dx ,$$

which is closer to our notation for the double integral. Since the maximum height is 3 $(= f(0))$ and the minimum is 1 $(= f(2))$, it is clear that

$$2 \times 1 < J < 2 \times 3 .$$

Here $2 \times 3 = 6$ is the area of the large rectangular region that contains the area we are approximating.

This first approximation $2 < J < 6$ can be improved, that is, we can calculate numbers k_1 and k_2 such that

$$2 < k_1 < J < k_2 < 6 .$$

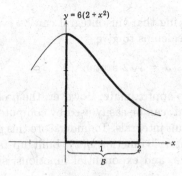

FIGURE 9.4. The area between $\mathcal{B}$ and a curve

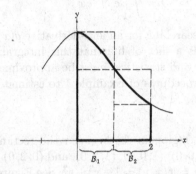

FIGURE 9.5. Partitioning the region

To do this, we break the area into two areas by breaking the base interval $\mathcal{B}$ into two intervals $0 \le x \le 1$ and $1 \le x \le 2$, denoted by $\mathcal{B}_1$ and $\mathcal{B}_2$, respectively. Then we estimate the areas above each interval separately, using inscribed and circumscribed rectangular regions as before. Let J_1 and J_2 denote the exact areas above $\mathcal{B}_1$ and $\mathcal{B}_2$, respectively. See Figure 9.5. Then we see that

$$1 \times f(1) = 2 < J_1 < 1 \times f(0) = 3 \, ,$$

and

$$1 \times f(2) = 1 < J_2 < 1 \times f(1) = 2 \, .$$

Since $J = J_1 + J_2$, we conclude that $2 + 1 < J < 3 + 2$, that is,

$$3 < J < 5 \, .$$

This new approximation sharpens the original estimate $2 < J < 6$. We now have a somewhat more refined idea of the size of the area J. Even better accuracy can be had if we again subdivide $\mathcal{B}_1$ and $\mathcal{B}_2$. This process can be repeated until the area is estimated as closely as we desire.

Some might object, saying that this integral can be evaluated explicitly by the methods of basic calculus to give

$$J = 3\sqrt{2} \arctan \sqrt{2} \,,$$

and so it is unnecessary to approximate. However, the procedure used here, or slight modifications of it, can be easily used by computers to numerically evaluate much more difficult integrals. To underscore this point, we mention that there is no "elementary" function $g(x)$, built up from polynomials, trigonometric, logarithmic, and exponential functions, such that $g'(x) = e^{-x^2}$. Hence

$$J = \int_{-1}^{1} e^{-x^2}\, dx$$

cannot be computed by searching for an anti-derivative $g(x)$ in the familiar manner of calculus. It is a fact of life that this integral, which is very important in probability and statistics, *must* be approximated.

Now we imitate the procedures of Example 1 to estimate volumes.

EXAMPLE:

2. Consider the solid region in $\mathbb{R}^3$ between the rectangular region $\mathcal{B}$ with vertices at $(0,0,0)$, $(1,0,0)$, $(1,2,0)$, and $(0,2,0)$ and the surface whose equation is $z = f(x,y) = 1 + x^2 + y^2$. See Figure 9.6. As above, we denote the volume of this solid by

$$J = \iint_{\mathcal{B}} (1 + x^2 + y^2)\, dA \,.$$

It is clear from Figure 9.6 that

$$
\begin{aligned}
1 \times 2 \;=\; & \text{(minimum height)(area of } \mathcal{B}) < J \\
< \;& \text{(maximum height)(area of } \mathcal{B}) = 6 \times 2 \,,
\end{aligned}
$$

that is, $2 < J < 12$. This gives a first, crude estimation of J.

A better estimate can be found by regarding the solid region as built up from two regions obtained by breaking the base $\mathcal{B}$ into two rectangular regions, $\mathcal{B}_1$ and $\mathcal{B}_2$. See Figure 9.7. Then

$$J = \iint_{\mathcal{B}} f(x,y)\, dA = \iint_{\mathcal{B}_1} f(x,y)\, dA + \iint_{\mathcal{B}_2} f(x,y)\, dA = J_1 + J_2 \,,$$

and we can estimate the integrals J_1 and J_2 separately, just as we did for J above.

For (x,y) in $\mathcal{B}_1$ we see by inspection that

$$1 \leq 1 + x^2 + y^2 \leq 1 + 1 + 1 = 3 \,.$$

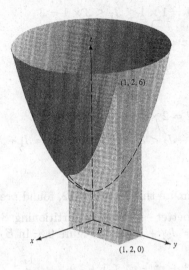

FIGURE 9.6. The first approximation

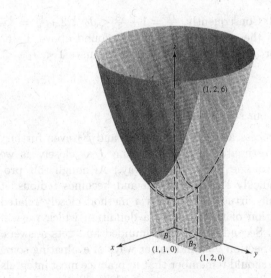

FIGURE 9.7. The second approximation

Since the area of $\mathcal{B}_1$ is 1, we know that

$$1 = 1 \times 1 \leq J_1 \leq 3 \times 1 = 3 .$$

Similarly, in $\mathcal{B}_2$ we have

$$2 \leq 1 + x^2 + y^2 \leq 6 ,$$

so that

$$2 = 2 \times 1 < J_2 < 6 \times 1 = 6 .$$

Therefore, $1 + 2 < J_1 + J_2 < 3 + 6$, and since $J = J_1 + J_2$, we conclude that

$$3 < J < 9 ,$$

which is a better estimate than $2 < J < 12$, found previously.

We can get an even better estimate by partitioning $\mathcal{B}_2$ into $\mathcal{B}_{21}$ and $\mathcal{B}_{22}$, and writing $J_2 = J_{21} + J_{22}$. See Figure 9.8. In $\mathcal{B}_{21}$ we find that

$$2 \leq 1 + x^2 + y^2 \leq 1 + 1 + \frac{9}{4} = \frac{17}{4} .$$

Since $\mathcal{B}_{21}$ has area $\frac{1}{2}$, we see that $1 < J_{21} < \frac{17}{8}$. Similarly, in $\mathcal{B}_{22}$,

$$1 + 0 + \frac{9}{4} = \frac{13}{4} \leq 1 + x^2 + y^2 \leq 1 + 4 + 1 = 6 ,$$

so that $\frac{13}{8} < J_{22} < 3$. Consequently, $\frac{21}{8} = 1 + \frac{13}{8} < J_2 < 3 + \frac{17}{8} = \frac{41}{8}$, which is better than the estimate $2 < J_2 < 6$, found above. Using this new estimate for J_2 together with the old estimate $1 < J_1 < 3$, we have

$$\frac{29}{8} < J < \frac{65}{8} ,$$

which is better than our second estimate $3 < J < 9$.

By repeating this process and partitioning $\mathcal{B}_1$ and $\mathcal{B}_2$ even further, we presumably can estimate the exact volume J as closely as we desire. (This exact volume is $\frac{16}{3}$, by the way.) Although this process of estimation quickly loses its charm and becomes tedious, it has several important virtues. First, it is a method closely related to the precise definition of the integral, a definition which we will study in Section 9.3. Second, it is simple-minded and gets answers. Although we will, in Section 9.2, find easier ways of evaluating some integrals exactly, one should remember that in practice most integrals are actually evaluated by computers, using methods like the one just outlined.

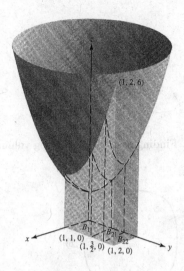

$(1, 2, 6)$

x

B_{11} $(1, 1, 0)$ B_{21} B_{22}

$(1, \frac{3}{2}, 0)$ $(1, 2, 0)$

y

FIGURE 9.8. A still better approximation

The simplest integrand is the function 1. According to the following calculation, a double integral with this integrand equals the area of its region of integration:

$$J = \int \int_{\mathcal{B}} 1 \, dA = \text{ height} \times (\text{area of } \mathcal{B}) = 1 \times (\text{area of } \mathcal{B}) \, .$$

See Figure 9.9. (Do not worry about units of measure here. The point is that the number of cubic units in the three-dimensional solid equals the number of square units in $\mathcal{B}$. Were one to change units the integrand would no longer equal 1, the above calcuation would not apply, and, of course, the number of square units in the region of integration would no longer numerically equal the number of cubic units in the solid.) For a region of integration that is a set whose area we know, the observations in this paragraph are of little interest. More interesting is the case where the region of integration $\mathcal{B}$ is more complicated. Then the double integral

$$\int \int_{\mathcal{B}} 1 \, dA$$

is a formula for the area of $\mathcal{B}$. At this point this formula is not of much use since the work we would do to approximate the integral is the same as the work we would do to approximate the area of $\mathcal{B}$. Such an approximation is illustrated in the following example.

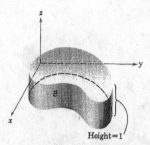

FIGURE 9.9. Finding area by calculating volume

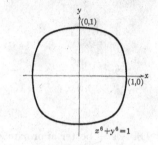

FIGURE 9.10.

EXAMPLE:

3. Estimate the area of the region $\mathcal{B}$ inside the curve $x^6 + y^6 = 1$. See Figure 9.10. From the discussion above, the area of $\mathcal{B}$ is given by the double integral

$$J = \int\int_{\mathcal{B}} 1 \, dA \, .$$

This number is estimated from below by inscribing a square inside $\mathcal{B}$. Note that points (c, c), $(-c, c)$, $(-c, -c)$, and $(c, -c)$, where $c = (\frac{1}{2})^{1/6}$, are on the oval $x^6 + y^6 = 1$ and are the vertices of a square in the xy-plane. Now each side of this square has length $2 \times c = 2 \times (\frac{1}{2})^{1/6}$. Thus the area of the square region is $(2 \times (\frac{1}{2})^{1/6})^2 = 4(\frac{1}{2})^{1/3} < J$.

By circumscribing another square (outside), we easily find that $J < 4$. These estimates of J, the area of $\mathcal{B}$, can be refined by using many smaller such regions. In fact, one might estimate this area by drawing $\mathcal{B}$ on graph paper and counting the number of squares inside $\mathcal{B}$ and also those touching $\mathcal{B}$.

9.1B   DENSITY

Double integrals arise naturally when finding the total mass (or weight) of a thin plate $\mathcal{B}$ in the xy-plane given its "area density function" f, in pounds per square foot for example.

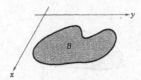

FIGURE 9.11. A thin plate

First, what do we mean by "area density function"? We suppose that some parts of the plate are heavier than others of the same size situated at different locations in the plate. This might occur because the plate is cast from some inhomogeneous alloys. See Figure 9.11. At the point X in question, we center a small square region. Let ΔM be the mass and ΔA the area of this small square region. Then the ratio $\Delta M/\Delta A$ of mass per unit area (pounds per square foot) is roughly the area density at X. It would be exact if the mass were constant around X. The exact **area density** $f(X)$ at X is defined as the limit of $\Delta M/\Delta A$ as the square shrinks down around X, and so

$$f(X) = \lim_{\Delta A \to 0} \frac{\Delta M}{\Delta A} \ .$$

Since the function f describes the area density at each point it is called the **area density function** or, more briefly, the **density function** when there can be no confusion. (In all this, of course, we ignore the thickness of the plate. For our purposes, the plate is two-dimensional.)

To find the total mass of the plate $\mathcal{B}$ with a given density function $f(X)$, we subdivide $\mathcal{B}$ into smaller pieces $\mathcal{B}_i$, each with mass ΔM_i and area ΔA_i. See Figure 9.12. If X_i is a point in the piece $\mathcal{B}_i$, then we approximate

$$\Delta M_i \approx f(X_i)\Delta A_i \ ,$$

since, by the discussion above, $f(X_i) \approx \Delta M_i/\Delta A_i$. Thus the total mass M is approximated by

$$M = \sum_i \Delta M_i \approx \sum_i f(X_i)\Delta A_i \ .$$

In the limit, using smaller and smaller subdivisions, we conclude that the total mass, like volume, is a double integral,

$$M = \int\int_{\mathcal{B}} f \, dA \ .$$

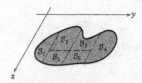

FIGURE 9.12. Partitioning the plate

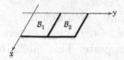

FIGURE 9.13. Two plates whose union is $\mathcal{B}$

EXAMPLE:

4. Consider a rectangular plate $\mathcal{B}$ in $\mathbb{R}^2$ with corners at the points $(0,0)$, $(1,0)$, $(1,2)$, and $(0,2)$ and whose density is $f(X) = f(x,y) = 1 + x^2 + y^2$. Hence the plate is "light" near $(0,0)$ and "heavy" near $(1,2)$. The total mass of this solid is

$$M = \int\int_{\mathcal{B}} (1 + x^2 + y^2)\, dA\,.$$

True to form, we estimate the mass of this plate. Since the minimum and maximum densities are 1 and 6, respectively, that is, $1 \le 1 + x^2 + y^2 \le 6$ for all (x,y) in $\mathcal{B}$, it is intuitively clear on physical grounds that

$$
\begin{aligned}
1 \times 2 \;&=\; \text{(minimum density)(area)} \;<\; M \\
&<\; \text{(maximum density)(area)} \;=\; 6 \times 2\,,
\end{aligned}
$$

that is,

$$2 < M < 12\,.$$

This is a crude estimate of M. By regarding the plate as the union of two plates obtained by partitioning $\mathcal{B}$ into two rectangular regions (see Figure 9.13), we can use the same method to estimate the mass of the two parts separately and then add the result. Of course, all of this is a repetition of Example 2, since both examples involve the same double integral.

From the point of view of computation, this density presents nothing new. We remark, however, that contrary to volume there is a very natural situation in which the density of $f(X)$ may be negative as well as positive, namely electric charge density, for charge may be $+$ or $-$. In this case $\int\int_{\mathcal{B}} f\, dA$ is the total (net) electric charge on $\mathcal{B}$. Many other kinds of problems lead to integrals with negative integrands.

We have seen, in Examples 2 and 4, how to use approximations for double integrals in which the integrand is not a constant function. In Example 3 we saw how to approximate the integral of a constant function when the region of integration is not a rectangular region. The following example illustrates the finding of approximations of a double integral when the integrand is not constant *and* the region of integration is not a rectangular region.

EXAMPLE:

5. Let $\mathcal{B}$ be the region illustrated in Figure 9.10, and let $f(X) = 3 + \|X\|$ for $X \in \mathbb{R}^2$. We will approximate $\int \int_{\mathcal{B}} f \, dA$. Recall from Example 3 that the square region with corners at $(c, c), (c, -c), (-c, c)$, and $(-c, -c)$ is contained in $\mathcal{B}$, where $c = \left(\frac{1}{2}\right)^{1/6}$. Also, $\mathcal{B}$ is contained in the square region with corners at $(1, 1), (1, -1), (-1, 1)$, and $(-1, -1)$. Call the smaller square region $\mathcal{Q}_1$ and the larger square region $\mathcal{Q}_2$. We have

$$\int \int_{\mathcal{Q}_1} f \, dA \leq \int \int_{\mathcal{B}} f \, dA \leq \int \int_{\mathcal{Q}_2} f \, dA .$$

We can underestimate the double integral of f over $\mathcal{Q}_1$ by using the minimum value of f on $\mathcal{Q}_1$, which is 3:

$$\int \int_{\mathcal{Q}_1} f \, dA \geq 3 \times (\text{area of } \mathcal{Q}_1) = 3 \times 4\left(\frac{1}{2^{1/3}}\right) = 12\left(\frac{1}{2^{1/3}}\right) .$$

We overestimate the double integral of f over $\mathcal{Q}_2$ by using the maximum value of f on $\mathcal{Q}_2$, namely $3 + \sqrt{2}$:

$$\int \int_{\mathcal{Q}_2} f \, dA \geq (3 + \sqrt{2}) \times (\text{area of } \mathcal{Q}_2) = (3 + \sqrt{2}) \times 4 = 12 + 4\sqrt{2} .$$

If we let J be the integral of f over $\mathcal{B}$, we have

$$12\left(\frac{1}{2^{1/3}}\right) < J < 12 + 4\sqrt{2} ,$$

which gives us a crude idea of the value of J. We could improve the upper bound by noting that the values of f outside of $\mathcal{B}$ are irrelevant to the double integral of f over $\mathcal{B}$. Thus, we could set f equal to 0 outside of $\mathcal{B}$. Then the maximum value of f on the region $\mathcal{Q}_2$ is taken at the point (c, c). This value is

$$f(c, c) = 3 + \sqrt{\frac{2}{2^{1/3}}} = 3 + 2^{1/3} ,$$

so

$$J < 12 + 4 \cdot 2^{1/3} ,$$

giving us a slightly better upper bound.

To obtain better estimates for J, we would need to use smaller squares $\mathcal{B}_i$, just as we need to do when estimating the area of $\mathcal{B}$. If ΔA_i stands for the area of the small box $\mathcal{B}_i$, a lower estimate of J is

$$J > \sum f(X_i)\Delta A_i \,,$$

with the sum being taken over those values of i such that $\mathcal{B}_i$ is contained in $\mathcal{B}$, and with each point X_i being chosen in $\mathcal{B}_i$ so as to minimize the value of f on $\mathcal{B}_i$. Similarly, an upper estimate of J is

$$J < \sum f(Y_i)\Delta A_i \,,$$

where now we include enough squares to cover the region $\mathcal{B}$, and pick each point Y_i in $\mathcal{B}_i$ so as the maximize f on $\mathcal{B}_i$.

9.1C   TRIPLE INTEGRALS

All we have said concerning double integrals extends readily to triple integrals. One encounters triple integrals when finding the total mass of a solid $\mathcal{B}$. See Figure 9.14. We describe this now. If X is a point of the solid and ΔV is the volume of a small cube of mass ΔM centered at X, then the **volume density** at X is defined by

$$f(X) = \lim_{\Delta V \to 0} \frac{\Delta M}{\Delta V} \,.$$

Of course, the function f is called the **volume density function** or, more briefly, the **density function**. We are often given a solid $\mathcal{B}$ with a volume density function f and asked to find its total mass M. Subdividing $\mathcal{B}$ into smaller parts $\mathcal{B}_i$, each with mass ΔM_i and volume ΔV_i, we have the now familiar type of approximation to the mass:

$$M = \sum_i \Delta M_i \approx \sum_i f(X_i)\Delta V_i \,,$$

where X_i is some point of $\mathcal{B}_i$. In the limit as each $\mathcal{B}_i$ becomes small, we obtain the exact mass as the triple integral

$$M = \iiint_\mathcal{B} f \, dV \,.$$

If the density f is identically 1 ($f \equiv 1$), then the mass of $\mathcal{B}$ is numerically equal to the volume V:

$$V = \iiint_\mathcal{B} 1 \, dV \,.$$

We may regard the three-dimensional region $\mathcal{B}$ as carrying an electrical charge throughout. This leads to a charge density f that may be positive

FIGURE 9.14. Volume density

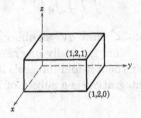

FIGURE 9.15. A box in $\mathbb{R}^3$

or negative. Of course, if this physical interpretation is alien to your mode of thought, then you are free to think of the triple integral $\iiint_{\mathcal{B}} f \, dV$ in some other way.

Let us estimate the value of a triple integral. No new ideas are involved.

EXAMPLE:

6. Let $\mathcal{B} = \{(x, y, z) \in \mathbb{R}^3 : 0 \le x \le 1, \ 0 \le y \le 2, \ 0 \le z \le 1\}$, which is a box in $\mathbb{R}^3$. See Figure 9.15. Let the density $f(x, y, z) = xz + y$. We are to estimate

$$ J = \iiint_{\mathcal{B}} f \, dV = \iiint_{\mathcal{B}} (xz + y) \, dV . $$

A naïve upper bound for J results from observing that the maximum density in $\mathcal{B}$ occurs at the corner $(1, 2, 1)$, so that $xz + y \le 3$. Since the volume of $\mathcal{B}$ is 2, we conclude that $J < 3 \times 2 = 6$. It should be clear that $0 < J$. Hence we have $0 < J < 6$.

A better estimate can be found by partitioning $\mathcal{B}$ into two parts, $\mathcal{B}_1$, where $0 \le y \le 1$, and $\mathcal{B}_2$, where $1 \le y \le 2$. Then $J = J_1 + J_2$, where J_1 and J_2 are the obvious triple integrals taken over $\mathcal{B}_1$ and $\mathcal{B}_2$. Now in $\mathcal{B}_1$ we have $0 \le xz + y \le 2$, so that $0 < J_1 < 2$. Similarly, in $\mathcal{B}_2$ we have $1 \le xz + y \le 3$, so that $1 < J_2 < 3$. Thus we estimate $1 < J < 5$, which is better than $0 < J < 6$.

REMARKS:

- In the preceding discussion of triples integrals with an integrand f, no mention was made of the *graph* of f. The set on which we focused was the region of integration. The graph of f is an set in $\mathbb{R}^4$.

- A subset of $\mathbb{R}^4$ might also arise as a region of integration. Then one is led to **quadruple integrals** written as

$$ \iiiint_{B} f \, dV \, , $$

where B is a set in $\mathbb{R}^4$ and dV signifies a tiny element of four-dimensional volume. More generally one can work with general multiple integrals, calling them n-fold integrals if one wants to emphasize that the region of integration is a subset of $\mathbb{R}^n$.

Exercises

1. Use a compass to draw a disk of radius 10 on a sheet of graph paper. Estimate the area of the disk from above and below by counting squares. Then use the result and the formula for the area of a disk to estimate the value of π.

2. Estimate the areas of the following sets in $\mathbb{R}^2$ by the methods in this section (do not use any techniques of antidifferentiation from single-variable calculus):

    (a) $2 \le x \le 4$, $-1 \le y \le x^2$
    (b) $x^2 + y^2 \le 4$
    (c) $0 \le x \le 1$, $x^2 \le y \le 1$
    (d) $-1 \le x \le 2$, $-1 \le y \le (4+x)^{-1}$
    (e) $x^4 + y^4 \le 16$
    (f) $x^2 \le y \le 2 - x^2$.

3. Let B be the rectangular region in $\mathbb{R}^2$ with corners at $(-1,0)$, $(1,0)$, $(1,2)$, and $(-1,2)$. Estimate each of the following integrals. You should first obtain a crude estimation, then a second, more refined estimation.

    (a) $J = \iint_B (x^2 + y^2) \, dA$
    (b) $J = \iint_B (1 + x^2 y) \, dA$
    (c) $J = \iint_B (1 - 2x^2 - y^2) \, dA$
    (d) $J = \iint_B x(1 - 2y) \, dA$
    (e) $J = \iint_B \frac{1}{5 + x + y} \, dA$
    (f) $J = \iint_B \frac{1}{1 + x^2 + y} \, dA$
    (g) $J = \iint_B x(1 + y) \, dA$

(h) $J = \int\int_B xy^2 \, dA$

(i) $J = \int\int_B e^{-xy-y} \, dA$

(j) $J = \int\int_B e^{-x^2-y^2} \, dA$.

4. Crudely estimate the integrals in parts (b) and (d) of Exercise 3, this time letting B be the set

$$\{(x,y) : -2 \leq x \leq 2, 0 \leq y \leq 1+x^2\} \, .$$

5. Let $B \subseteq \mathbb{R}^3$ be the set $0 \leq x \leq 2, 0 \leq y \leq 3, 1 \leq z \leq 2$. Estimate the following integrals as in Exercise 3:

(a) $J = \int\int\int_B x \, dV$

(b) $J = \int\int\int_B (x+3y+z^2) \, dV$

(c) $J = \int\int\int_B xyz \, dV$

(d) $J = \int\int\int_B (x^2y + y^2z) \, dV$.

6. Estimate the volume of the bounded region between the paraboloid $z = x^2 + y^2$ and the plane $z = 4$.

7. Estimate the volume of the bounded region between the paraboloids $z = x^2 + y^2$ and $z = 2 - x^2 - y^2$.

8. Estimate the volume of the region bounded from above by $z = 1+x^2+y^2$, from below by $z = -2x^2-2y^2$, and on the sides by the cylinder $x^2+y^2 = 1$.

9. Let $f: \mathbb{R} \to \mathbb{R}$ be a continuous function. Find the exact area of the region bounded from below by $y = f(x)$ and from above by $y = f(x) + 7$ for $2 \leq x \leq 4$. *Hint:* This needs a simple mental geometric argument.

10. Let $f: \mathbb{R}^2 \to \mathbb{R}$ be a continuous function. Find the exact volume of the solid region bounded from below by $z = f(x,y) - 1$ and bounded from above by $z = f(x,y) + 3$ for $1 \leq x \leq 5$ and $-2 \leq y \leq 2$.

11. Let $B \subseteq \mathbb{R}^2$ be the set

$$\{(x,y) : y \geq 0, x^4 + y^4 \leq 16\} \, .$$

Estimate:

(a) $J = \int\int_B \frac{1}{1+(x^4+y^4)^{1/4}} \, dA$

(b) $J = \int\int_B \frac{3}{8+\sqrt{x^4+y^4}} \, dA$.

12. Let $B \subseteq \mathbb{R}^3$ be the set

$$\{(x,y,z) : x^4 + y^4 + z^4 \leq 16\} \, .$$

Estimate:

(a) $J = \int\int\int_B (x^2+y^2)z^2 \, dV$

(b) $J = \int\int\int_B (x^2+y^2) z \, dV$.

13. (a) Briefly explain why the integral in Exercise 3(h) must be zero.

(b) Do the same for the integrals in part (d) of Exercise 3 and parts (d), (g), and (h) of Exercise 4 and part (b) of Exercise 12.

(c) Let $\mathcal{B} \subseteq \mathbb{R}^2$ be the set in Exercise 3. By inspection, evaluate

$$J = \int\int_{\mathcal{B}} (1 + 2x - xy^2) \, dA .$$

14. Let A be the area under the graph of $f(x) = 1/x$ for $1 \leq x \leq 2$ and B the area under the same graph for $2 \leq x \leq 4$. Give a geometric argument to show that $A = B$.

15. Let $\mathcal{B} \subseteq \mathbb{R}^2$ have the property that

$$a \leq \text{(area of } \mathcal{B}) \leq \alpha ,$$

and let f be a function with the property $m \leq f(X) \leq M$ for all $X \in \mathcal{B}$. Briefly explain why

$$ma \leq \int\int_{\mathcal{B}} f \, dA \leq M\alpha .$$

16. Let

$$\mathcal{B}(t) = \{(x, y) : \alpha \leq x \leq \beta, a \leq y \leq t\} ,$$

whose height depends on t, and let $g(y)$ be a continuous increasing function depending on y only. Define $M(t)$ by

$$M(t) = \int\int_{\mathcal{B}(t)} g(y) \, dA .$$

(a) Explain why $M(a) = 0$ and, for $h > 0$,

$$g(t)(\beta - \alpha) \leq \frac{M(t+h) - M(t)}{h} \leq g(t+h)(\beta - \alpha) .$$

(b) Deduce from this that $dM/dt = g(t)(\beta - \alpha)$ and conclude that

$$\int\int_{\mathcal{B}(t)} g(y) \, dA = (\beta - \alpha) \int_a^t g(y) \, dy .$$

17. Let $\mathcal{B} = \{(x, y) : 2 \leq x \leq 6, 1 \leq y \leq 3\}$. Use the preceding exercise to evaluate

(a) $\int\int_{\mathcal{B}} (2x + x^3) \, dA$

(b) $\int\int_{\mathcal{B}} \sqrt{1 + 4x} \, dA$

(c) $\int\int_{\mathcal{B}} y^2 \, dA$

(d) $\int\int_{\mathcal{B}} \log(xy^2) \, dA$.

18. Let

$$\mathcal{B}(\tau) = \{(x, y) : \alpha \leq x \leq \tau, a \leq y \leq b\} ,$$

and let $f(x)$ and $g(y)$ be continuous increasing functions, with $g(y) \geq 0$. Define $S(\tau)$ by

$$S(\tau) = \int\int_{\mathcal{B}(\tau)} f(x)g(y) \, dA .$$

Use part (b) of Exercise 16 to show that for $h > 0$,

$$f(\tau) \int_a^b g(y)\, dy \le \frac{S(\tau + h) - S(\tau)}{h} \le f(\tau + h) \int_a^b g(y)\, dy \,,$$

and deduce as in Exercise 16 that

$$\iint_{\mathcal{B}(\tau)} f(x)g(y)\, dA = \left(\int_\alpha^\tau f(x)\, dx \right) \left(\int_a^b g(y)\, dy \right) \,.$$

19. Let $\mathcal{B} = \{(x, y) : 2 \le x \le 6, 1 \le y \le 3\}$. Use the preceding exercise to evaluate

   (a) $\iint_{\mathcal{B}} xy\, dA$

   (b) $\iint_{\mathcal{B}} (x + xy)\, dA$

   (c) $\iint_{\mathcal{B}} (1 - 2xy)\, dA$

   (d) $\iint_{\mathcal{B}} e^{2x + 3y}\, dA$.

9.2 Computing Integrals Exactly

9.2A ITERATED INTEGRALS

We now have some practice in estimating the number $\iint_{\mathcal{B}} f\, dA$; we would like to compute it exactly, if possible. Although the previous section might have been called "Evaluation of Integrals for Computers", this one is "Evaluation of Integrals for Humans". We present a method of computation that works, provided the integrand f and the domain $\mathcal{B}$ of integration are not too complicated or pathological. The method reduces this computation of a double integral to the computation of two ordinary integrals from single-variable calculus.

For our first example we suppose that $\mathcal{B}$ is the rectangular region in the xy-plane determined by $a \le x \le b$ and $c \le y \le d$, and that a nonnegative function $z = f(x, y)$ is given. We wish to evaluate $\iint_{\mathcal{B}} f\, dA$, that is, the volume V of the solid in Figure 9.16. We proceed as follows. Slice the solid into thin slabs parallel to the xz-plane. The jth slab has width Δy_j and volume ΔV_j. It lies between the parallel planes $y = y_j$ and $y = y_{j+1} = y_j + \Delta y_j$. Its volume ΔV_j is approximately equal to $A(y_j)\, \Delta y_i$,

$$\Delta V_j \approx A(y_j)\, \Delta y_j,$$

where $A(y_j)$ is the exact cross-sectional area of the slice through y_j.

Now it is crucial to note that we can compute $A(y_j)$ exactly. By ordinary calculus, we have

$$A(y_j) = \int_a^b f(x, y_j)\, dx \,,$$

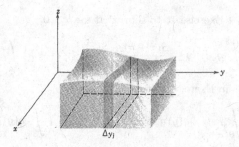

FIGURE 9.16. Slicing a solid

where y_j is kept fixed during this integration.

Now we have the desired volume V approximated as

$$V = \sum_j \Delta V_j \approx \sum_j \left(\int_a^b f(x, y_j) \, dx \right) \Delta y_j \ .$$

If we allow the number of slabs to increase indefinitely and their widths Δy_j to tend to zero, this approximation tends to the exact volume V, and, moreover, the limit equals $\int_c^d (\int_a^b f(x, y) \, dx) \, dy$, that is,

$$V = \int\int_{\mathcal{B}} f(x, y) \, dA = \int_c^d \left(\int_a^b f(x, y) \, dx \right) dy \ .$$

We call the integral on the right-hand side an **iterated integral**. Its virtue is that its evaluation involves two single integrals, each of which may be evaluated by the techniques of ordinary calculus.

Note that we could have equally well taken slices parallel to the yz-plane. This would have given

$$V = \int\int_{\mathcal{B}} f(x, y) \, dA = \int_a^b \left(\int_c^d f(x, y) \, dy \right) dx \ ,$$

in which we integrate first with respect to the variable y.

EXAMPLE:

1. Find the volume of the solid between the rectangular region

$$\mathcal{B} = \{(x, y) : 0 \le x \le 1, \, 0 \le y \le 2\}$$

in the xy-plane and the surface $z = f(x, y) = 1 + x^2 + y^2$. This is Example 2 in Section 9.1a. The volume is given by the integral

$$V = J = \int\int_{\mathcal{B}} (1 + x^2 + y^2) \, dA \ .$$

If we take slices parallel to the xz-plane, we have the iterated integral

$$J = \int_0^2 \left(\int_0^1 (1 + x^2 + y^2)\, dx \right) dy \,.$$

The inner integral is evaluated by treating y as a constant, and so

$$\int_0^1 (1 + x^2 + y^2)\, dx = x + \frac{x^3}{3} + y^2 x \Big]_{x=0}^{x=1} = 1 + \frac{1}{3} + y^2 = \frac{4}{3} + y^2 \,.$$

This provides us with an integrand for the second integration.

$$J = \int_0^2 \left(\frac{4}{3} + y^2 \right) dy = \frac{4}{3} y + \frac{y^3}{3} \Big]_{y=0}^{y=2} = \frac{8}{3} + \frac{8}{3} = \frac{16}{3} \,.$$

Thus the desired volume is $\frac{16}{3}$. How good were our estimates in Section 9.1? How might they have been improved?

Had we taken slices parallel to the yz-plane, we would have an integral with y as the first variable of integration:

$$\int_0^1 \left(\int_0^2 (1 + x^2 + y^2)\, dy \right) dx.$$

This should of course give the same volume $\frac{16}{3}$. We encourage you to verify that it does. It is easy.

9.2B DOMAINS WITH CURVED BOUNDARIES

This same procedure can be applied to domains $\mathcal{B}$ that are more general than rectangular regions. For convenience in picturing, we draw only the set $\mathcal{B}$ in the xy-plane (a top view) and ask you to imagine the surface $z = f(x, y)$ above this plane. Suppose that $\mathcal{B}$ has the form

$$\alpha(y) \le x \le \beta(y), \quad c \le y \le d\,,$$

where $x = \alpha(y)$ and $x = \beta(y)$ are functions of y. Thus two sides of $\mathcal{B}$ are *curves*: the graphs of the functions α and β. In this case it is natural to consider slabs in three dimensions parallel to the xz-plane. The base of such a slab rests in the xy-plane. See Figure 9.17. A typical slab is determined by slicing $\mathcal{B}$ parallel to the x-axis through a typical y. This slice extends from $x = \alpha(y)$ to $x = \beta(y)$, so that the cross-sectional area of the three-dimensional solid at y is

$$A(y) = \int_{\alpha(y)}^{\beta(y)} f(x, y)\, dx \,,$$

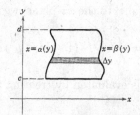

FIGURE 9.17. The base of a slab

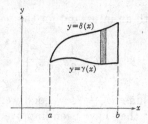

FIGURE 9.18. Slices parallel to the y-axis

which rightly depends on y. Since y varies from $y = c$ to $y = d$, we have

$$\int\int_{\mathcal{B}} f \, dA = \int_c^d \left(\int_{\alpha(y)}^{\beta(y)} f(x, y) \, dx \right) dy .$$

This yields a number, because the second integral is taken from $y = c$ to $y = d$, and these limits are numbers. We will compute an example in a moment.

If, on the other hand, the base $\mathcal{B}$ has the form

$$a \le x \le b, \quad \gamma(x) \le y \le \delta(x) ,$$

then it is natural to take slices parallel to the y-axis. See Figure 9.18. This leads to

$$\int\int_{\mathcal{B}} f \, dA = \int_a^b \left(\int_{\gamma(x)}^{\delta(x)} f(x, y) \, dy \right) dx .$$

EXAMPLES:

2. Let us evaluate $J = \int\int_{\mathcal{B}} x \, dA$, where $\mathcal{B}$ is the region given by

$$y^2 \le x \le \sqrt{y}, \quad 0 \le y \le \frac{1}{2}.$$

We slice parallel to the x-axis. See Figure 9.19. Such a slice extends from $x = \alpha(y) = y^2$ to $x = \beta(y) = \sqrt{y}$. The values of y range from

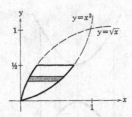

FIGURE 9.19. Slices parallel to the x-axis

$y = 0$ up to $y = \frac{1}{2}$. Hence we must compute

$$J = \int_0^{1/2} \left(\int_{y^2}^{\sqrt{y}} x \, dx \right) dy \, .$$

The inner integral is

$$\int_{y^2}^{\sqrt{y}} x \, dx = \left. \frac{x^2}{2} \right]_{x=y^2}^{x=\sqrt{y}} = \frac{y}{2} - \frac{y^4}{2} \, .$$

Note that there is no longer any trace of the variable x; it has been "integrated out".

Now we obtain the value of J by integrating with respect to y,

$$J = \int_0^{1/2} \left(\frac{y}{2} - \frac{y^4}{2} \right) dy = \left. \frac{y^2}{4} - \frac{y^5}{10} \right]_0^{1/2} = \frac{19}{320} \, .$$

We have computed $\iint_B x \, dA$ exactly.

3. Let us compute $J = \iint_B xy \, dA$, where B is determined by

$$-1 \le x \le 1, \quad 1 - x^2 \le y \le 2 + x^2.$$

We take slices parallel to the y-axis. See Figure 9.20. The slice through a typical x extends in B from $y = 1 - x^2$ to $y = 2 + x^2$. Then x extends from -1 to 1. This leads to the iterated integral

$$J = \int_{-1}^1 \left(\int_{1-x^2}^{2+x^2} xy \, dy \right) dx.$$

The first integral is

$$\int_{1-x^2}^{2+x^2} xy \, dy = \left. \frac{xy^2}{2} \right]_{y=1-x^2}^{y=2+x^2} = \frac{x}{2}(2 + x^2)^2 - \frac{x}{2}(1 - x^2)^2.$$

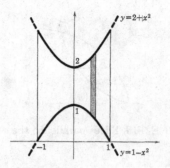

FIGURE 9.20.

Thus the second integral becomes

$$\int_{-1}^{1}\left[\frac{x}{2}(2+x^2)^2 - \frac{x}{2}(1-x^2)^2\right]dx = \left[\frac{(2+x^2)^3}{12} + \frac{(1-x^2)^3}{12}\right]_{-1}^{1}.$$

This is 0 by inspection, so that $J = 0$ (in fact, we could have deduced this by considering the behavior of the integrand xy in the domain $\mathcal{B}$). Note that a certain amount of judgment was made in deciding to take slices parallel to the y-axis. If we had taken slices parallel to the x-axis, the computation would have been considerably more complicated, since for $y > 2$ and $y < 1$, each slice itself consists of two pieces.

4. Sometimes it is equally reasonable to integrate with respect to either variable first, because of the nature of $\mathcal{B}$. Consider the problem of evaluating $J = \iint_{\mathcal{B}}(x - 2y)^2\,dA$ where $\mathcal{B}$ is the triangular region bounded by the lines $x = 1$, $y = -2$, and $y + 2x = 4$.

For our first solution we slice parallel to the x-axis through a typical y. See Figure 9.21. This slice extends from $x = 1$ to $x = 2 - \frac{1}{2}y$. Then y extends from $y = -2$ up to $y \doteq 2$. Thus we have

$$J = \int_{-2}^{2}\left(\int_{1}^{2-(1/2)y}(x - 2y)^2\,dx\right)dy\,.$$

The first integration is

$$\int_{1}^{2-(1/2)y}(x - 2y)^2\,dx = \left.\frac{1}{3}(x - 2y)^3\right]_{x=1}^{x=2-(1/2)y}$$

$$= \frac{1}{3}\left(2 - \frac{5}{2}y\right)^3 - \frac{1}{3}(1 - 2y)^3.$$

Note that x has been integrated out.

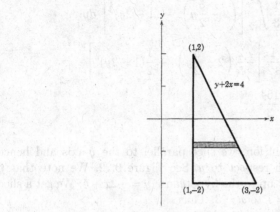

FIGURE 9.21. Integrating over a triangular region (first solution)

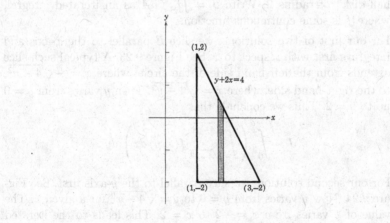

FIGURE 9.22. Integrating over a triangular region (second solution)

Now we obtain

$$J = \frac{1}{3} \int_{-2}^{2} \left[\left(2 - \frac{5}{2}y \right)^3 - (1 - 2y)^3 \right] dy$$

$$= \frac{1}{12} \left[-\frac{2}{5} \left(2 - \frac{5}{2}y \right)^4 + \frac{1}{2}(1 - 2y)^4 \right]_{y=-2}^{y=2}$$

$$= \frac{164}{3} .$$

Done.

For our second solution, we slice parallel to the y-axis and hence integrate first with respect to y. See Figure 9.22. We note that y varies from $y = -2$ up to the hypotenuse $y = -2x + 4$. We get a slice at each x from $x = 1$ to $x = 3$. Thus

$$J = \int_{1}^{3} \left(\int_{-2}^{-2x+4} (x - 2y)^2 \, dy \right) dx .$$

Note that after performing the interior integration and evaluating at the designated limits, we are left with a function of x alone. Then integration with respect to x completes the calculation. You should verify that this procedure gives $J = \frac{164}{3}$ as computed earlier.

5. The emphasis of the next example is on determining the limits of integration. Let $\mathcal{B} = \{(x, y) \in \mathbb{R}^2 : x^2 + y^2 \le 4, \ y \ge 0\}$ be the "upper half-disk" of radius 2. Write $J = \iint_{\mathcal{B}} f \, dA$ as an iterated integral, where f is some continuous function.

For our first of two solutions, we slice $\mathcal{B}$ parallel to the x-axis and integrate first with respect to x. See Figure 9.23. A typical such slice extends from the left-hand side of the circle, where $x = -\sqrt{4 - y^2}$, to the right-hand side, where $x = \sqrt{4 - y^2}$. Then y varies from $y = 0$ up to $y = 2$. Thus we conclude that

$$J = \int_{0}^{2} \left(\int_{-\sqrt{4-y^2}}^{\sqrt{4-y^2}} f(x, y) \, dx \right) dy .$$

For our second solution, we slice parallel to the y-axis first. See Figure 9.24. Now y varies from $y = 0$ to $y = \sqrt{4 - x^2}$ for a given x. The value of x varies from $x = -2$ to $x = 2$. This leads to the iterated integral

$$J = \int_{-2}^{2} \left(\int_{0}^{\sqrt{4-x^2}} f(x, y) \, dy \right) dx .$$

This integral and the one obtained above yield the same number for J, of course. However, depending on the nature of $f(x, y)$, one or the other might be easier to compute.

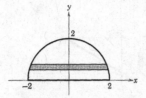

FIGURE 9.23. Integrating over a half-disk (first solution)

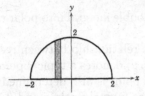

FIGURE 9.24. Integrating over a half-disk (second solution)

Domains somewhat more complicated than the ones just treated may often be decomposed into unions of simpler kinds and handled by the methods above. See Figure 9.25.

9.2C   INTEGRATION ON DISKS; POLAR COORDINATES

There is a particular kind of double integral that is worthy of special mention. The problem is this. Compute $\iint_{B(0;a)} f \, dA$, where $B(0;a)$ is the disk of radius $a > 0$ centered at the origin in the xy-plane. Very often in problems involving such a region, the integrand f is given in terms of coordinates x, y, but it can be seen to have some kind of symmetry with respect to the origin. For example, consider $\iint_{B(0;a)} (x^2 + y^2) \, dA$. Here the integrand $f(x,y)$ attains the same value at two different points (x_1, y_1) and (x_2, y_2) if and only if they are equidistant from the origin. To capitalize on these circular symmetries in the disk $B(0;a)$ and in the integrand

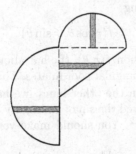

FIGURE 9.25. A more complicated region

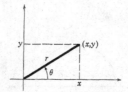

FIGURE 9.26. Polar coordinates

$f(x, y)$, we transform the double integral into polar coordinates, which were introduced in Section 5.1e.

Let us briefly review the relationship between rectangular and polar coordinates. Suppose that (x, y) denotes a typical point in the plane different from the origin. Then this point is also determined by the two numbers r and θ, were r is the distance from the origin and θ is the counterclockwise angle, in radians, from the x-axis to the segment, or arrow, from the origin to the point. Thus $r > 0$, and we may take $0 \le \theta < 2\pi$, since θ and $\theta + 2\pi$ refer to the same angle in the plane. The pair (r, θ) gives the polar coordinates of the point. The polar and rectangular coordinates are related by trigonometry (see Figure 9.26) as follows:

$$x = r \cos \theta, \qquad y = r \sin \theta,$$

whence

$$r = \sqrt{x^2 + y^2}, \ \tan \theta = \frac{y}{x}.$$

We note that the function given by $f(x, y) = x^2 + y^2$ in rectangular coordinates has the very simple form r^2 in polar coordinates. The fact that the coordinate θ does not appear in this polar representation of f shows vividly that f depends only on distance from the origin.

Here is an example of some notation we use. Given $f(x, y) = x^2 + y^2$ again, we use the formulas for x and y above to write

$$f(x, y) = f(r \cos \theta, \ r \sin \theta) = (r \cos \theta)^2 + (r \sin \theta)^2 = r^2,$$

since $\cos^2 \theta + \sin^2 \theta = 1$. The moral is this: Given f in terms of x, y we may rewrite it in terms of r, θ, using

$$f(x, y) = f(r \cos \theta, \ r \sin \theta).$$

We might call the last expression $\hat{f}(r, \theta)$, the hat above the f reminding us that r, θ are polar and not rectangular coordinates. Thus, if $f(x, y) = x^2 + y^2$ again, then $\hat{f}(r, \theta) = r^2$; if, on the other hand, we had written $f(r, \theta)$ for this f, we would have committed the sin of ambiguity becuase $f(r, \theta)$ might stand for $r^2 + \theta^2$ as well as r^2. You should make your own private peace with this issue.

Now we state a result that enables us to integrate functions over disks by iterating ordinary integrals.

Theorem 9.2.1 Let $B(0; a)$ denote the disk of radius a centered at the origin. Then for any continuous function f,

$$\iint_{B(0;a)} f(x,y)\, dA = \int_0^{2\pi} \int_0^a f(r\cos\theta,\, r\sin\theta) r\, dr\, d\theta.$$

REMARK:

- On the right-hand side, r varies from 0 to a and θ varies from 0 to 2π.

Before attempting to justify this theorem, let us apply it in a case which is already familiar. We know that

$$\iint_{B(0;a)} 1\, dA = \text{(area of disk of radius } a) = \pi a^2.$$

Here the integrand $f(x,y) = 1$ identically. The theorem tells us that

$$\iint_{B(0;a)} 1\, dA = \int_0^{2\pi} \left(\int_0^a 1 \cdot r\, dr \right) d\theta = \int_0^{2\pi} d\theta \cdot \int_0^a r\, dr$$

$$= \theta \big]_0^{2\pi} \cdot \frac{r^2}{2} \Big]_0^a = 2\pi \cdot \frac{a^2}{2} = \pi a^2,$$

as desired. This is reassuring. Note that we were able to break the iterated double integral into a product of two single integrals, one in θ and one in r. This is because the integrand was independent of θ (see Exercise 18 of Section 9.1).

ROUGH IDEA OF PROOF: Recall that $\iint_{B(0;a)} f(x,y)\, dA$ is obtained by breaking up the domain of integration—in this case $B(0; a)$—into small pieces of area ΔA_i and considering sums of the form $\sum_i f(X_i)\, \Delta A_i$, where X_i is a point in the small piece corresponding to ΔA_i. Since $B(0; a)$ is a disk, it is natural to partition it into small "curved rectangular regions" as pictured in Figure 9.27. The straight sides of this curved rectangular region have length Δr, and the curved side closest to the origin has length $r\, \Delta\theta$. This is because an entire circle of radius r has length $2\pi r$ and the curved side is only a fraction of the entire circle, $\Delta\theta/2\pi$, to be exact. It follows that the area of the "curved rectangular region" is approximately $r\, \Delta r\, \Delta\theta$.

Also, since $x = r\cos\theta$ and $y = r\sin\theta$,

$$f(x,y) = f(r\cos\theta,\, r\sin\theta) \, .$$

Hence, the integral $\iint_{B(0;a)} f(x,y)\, dA$ may be obtained by considering sums of the form

$$\sum f(r\cos\theta,\, r\sin\theta) r\, \Delta r\, \Delta\theta.$$

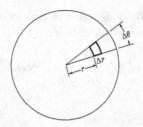

FIGURE 9.27. A polar coordinates rectangular region

If we let the "curved rectangular regions" that pave the disk shrink in size and increase in number and consider the limit of these sums as r varies from 0 to a and θ varies from 0 to 2π, that is, r and θ vary over the disk $B(\mathbf{0}; a)$, then we obtain the iterated integral

$$\int_0^{2\pi} \int_0^a f(r\cos\theta, r\sin\theta)r\,dr\,d\theta,$$

as claimed in the theorem. $<<$

We mention once again that in many cases the formidable expression $f(r\cos\theta, r\sin\theta)$ readily reduces to a function of r alone.

EXAMPLES:

6. Let us compute $\iint_{B(\mathbf{0};a)}(1 + x^2 + y^2)^{1/3}\,dA$.

   By the theorem this is equal to

   $$\int_0^{2\pi} \left(\int_0^a (1 + r^2)^{1/3}\,r\,dr \right) d\theta = \int_0^{2\pi} d\theta \cdot \int_0^a (1 + r^2)^{1/3}r\,dr$$

   $$= \ \theta \Big]_0^{2\pi} \cdot \frac{3}{8}(1 + r^2)^{4/3} \Big]_0^a$$

   $$= \ \frac{3\pi}{4} \cdot ((1 + a^2)^{4/3} - 1).$$

   We remark that this would have been much more difficult if we had tried to use rectangular coordinates x, y.

7. Now we observe that polar coordinates may be used on appropriate subsections of the disk as well. Let us compute $\iint_B 2xy\,dA$, where B is the "piece of pie" shown in Figure 9.28.

   Here the radius $a = 1$, and the angle θ varies from $\theta = 0$ to $\theta = \pi/3$ (*not* to $\theta = 2\pi$, as in the full disk). Hence we have

   $$\iint_B 2xy\,dA \ = \ \int_0^{\pi/3} \left(\int_0^1 (2r^2\cos\theta\sin\theta)r\,dr \right) d\theta$$

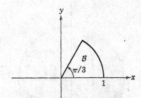

FIGURE 9.28. The area of a piece of pie

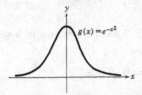

FIGURE 9.29. The bell-shaped curve

$$= \int_0^{\pi/3} 2\sin\theta\,\cos\theta\,d\theta \cdot \int_0^1 r^3\,dr$$

$$= \left.\sin^2\theta\,\right]_0^{\pi/3} \cdot \left.\frac{r^4}{4}\right]_0^1 = \frac{3}{4}\cdot\frac{1}{4} = \frac{3}{16}\,.$$

8. Finally, we apply polar coordinates to a famous computation, that of computing the area under the bell-shaped curve, the graph of $f(x) = e^{-x^2}$. See Figure 9.29. This function is very important in probability and statistics.

We will compute $J = \int_{-\infty}^{\infty} e^{-x^2}\,dx$ by using a clever trick. We note that

$$J^2 = \int_{-\infty}^{\infty} e^{-x^2}\,dx \cdot \int_{-\infty}^{\infty} e^{-y^2}\,dy\,.$$

By Exercise 18 in Section 9.1, we convert the product of integrals into the iterated integral

$$J^2 = \int_{-\infty}^{\infty}\int_{-\infty}^{\infty} e^{-(x^2+y^2)}\,dx\,dy.$$

This is an integral over the entire xy-plane. Hence we may replace x, y by polar coordinates r, θ, with $0 \le r < \infty$, $0 \le \theta < 2\pi$. Carrying this out, we obtain

$$J^2 = \int_0^{2\pi} d\theta \cdot \int_0^{\infty} e^{-r^2} r\,dr = 2\pi \lim_{b\to\infty} \int_0^b e^{-r^2} r\,dr$$

$$= -\pi \lim_{b\to\infty} e^{-r^2}\Big]_0^b = \pi,$$

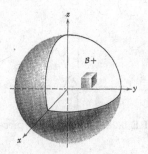

FIGURE 9.30. The volume of a solid ball

since $\lim_{b\to\infty} e^{-b^2} = 0$. Thus

$$J = \int_{-\infty}^{\infty} e^{-x^2} dx = \sqrt{\pi}\,.$$

Done.

9.2D   TRIPLE INTEGRALS

How do we compute $\iiint_{\mathcal{B}} f\, dV$? It is not surprising, but nevertheless reassuring, that the approach used in Section 9.2b may be extended to triple integrals as well. We now compute some of these integrals as iterated integrals. The most difficult aspect of these problems is very likely the determination of the limits of integration. Unfortunately, there is no recipe. Only practice and experience, along with geometric visualization, can guide you.

We have already given an informal definition of $\iiint_{\mathcal{B}} f\, dV$, in terms of limits of sums $\sum_i f(X_i)\,\Delta V_i$, and an interpretation as well: We spoke of the triple integral as the total mass of the solid $\mathcal{B}$, provided that $f = f(x,y,z)$ was the density at the point (x,y,z). In the case where the function f is the constant function equal to 1, the triple integral yields the volume of $\mathcal{B}$. Let us discuss volume first.

EXAMPLES:

9. Let us compute the volume of the three-dimensional ball $B(\mathbf{0};a)$ as an iterated integral. See Figure 9.30. We know that $B(\mathbf{0};a)$ is the set of all points (x,y,z) satisfying

$$x^2 + y^2 + z^2 \le a^2\,.$$

It is clear that the volume of $B(\mathbf{0};a)$ is eight times the volume of that part $B_{+++}(\mathbf{0};a)$ in the first octant; $B_{+++}(\mathbf{0};a)$ is determined by the conditions

$$x^2 + y^2 + z^2 \le a^2 \text{ and } x \ge 0,\, y \ge 0,\, z \ge 0.$$

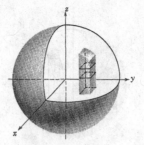

FIGURE 9.31. Stuffing a ball with boxes

It is easier to draw and deal with $B_{+++}(\mathbf{0}; a)$.

Let us compute the volume V_{+++} of $B_{+++}(\mathbf{0}; a)$. You will recall that this volume may be approximated by stuffing $B_{+++}(\mathbf{0}; a)$ with "bricks" or "boxes" whose edges are parallel to the coordinate axes and then adding up the volumes of all these boxes. Now we apply ordinary calculus to this approach.

Let (x, y, z) be one corner of a small box inside $B_{+++}(\mathbf{0}; a)$ whose length, breadth, and height are Δx, Δy, and Δz. We may construct a vertical column or stack of boxes, each having a base of the same dimensions, Δx by Δy. The height of such a column is, of course, obtained by adding the heights of the various boxes. See Figure 9.31. The column extends from the xy-plane $z = 0$ up to the surface of the ball, given by $z = \sqrt{a^2 - x^2 - y^2}$ at one corner. It follows that

$$\text{height of column in one corner} = \int_0^{\sqrt{a^2-x^2-y^2}} dz \ ,$$

and so

$$\text{volume of column} \approx \left(\int_0^{\sqrt{a^2-x^2-y^2}} dz \right) \Delta x \, \Delta y \ .$$

Now we consider a slab parallel to the xz-plane and constructed from vertical columns. The volume of such a slab is obtained by adding the volumes of the various columns. See Figure 9.32. A typical slab has a base in the xy-plane extending from $x = 0$ to $x = \sqrt{a^2 - y^2}$ on one side. We suppose that the thickness of each column, and hence of the slab, is Δy. Adding the volumes of the columns in the manner of ordinary calculus gives

$$\text{volume of slab} \approx \left(\int_0^{\sqrt{a^2-y^2}} \left(\int_0^{\sqrt{a^2-x^2-y^2}} dz \right) dx \right) \Delta y \ .$$

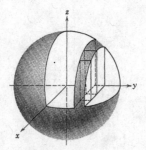

FIGURE 9.32. Cutting a ball into slabs

It should be clear what we do now. We obtain the volume V_{+++} of $B_{+++}(\mathbf{0}; a)$ by adding up the volume of the various slabs. These slabs extend from $y = 0$ to $y = a$, so that

$$V_{+++} = \int_0^a \left(\int_0^{\sqrt{a^2-y^2}} \left(\int_0^{\sqrt{a^2-x^2-y^2}} dz \right) dx \right) dy . \qquad (9.1)$$

Thus we have expressed $V_{+++} = \iiint_{B_{+++}(0;a)} dV$ as an iterated triple integral. Note that we integrate first with respect to z, then x, then y. Note, further, that we could have carried things out in a different order. For example, we also have

$$V_{+++} = \int_0^a \left(\int_0^{\sqrt{a^2-z^2}} \left(\int_0^{\sqrt{a^2-y^2-z^2}} dx \right) dy \right) dz .$$

We are not finished yet. Now let us use ordinary calculus to get a formula for the volume of the ball. We discuss the integral (9.1), integrating in the order z, x, y. The first integration yields

$$V_{+++} = \int_0^a \left(\int_0^{\sqrt{a^2-y^2}} \sqrt{a^2 - x^2 - y^2}\, dx \right) dy .$$

This inner integral in x is best attacked by making the trigonometric substitution

$$x = \sqrt{a^2 - y^2} \sin\theta, \quad dx = \sqrt{a^2 - y^2} \cos\theta\, d\theta ,$$

where we regard y as being fixed. You can compute that this inner integral in x gives $(\pi/4)(a^2 - y^2)$, so that

$$V_{+++} = \frac{\pi}{4} \int_0^a (a^2 - y^2)\, dy.$$

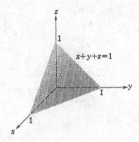

FIGURE 9.33. The volume of a pyramid

This yields $V_{+++} = (\pi/6)a^3$. Since the volume of the ball is eight times V_{+++}, we have the famous formula

$$\text{Volume of ball} = \frac{4}{3}\pi a^3 \; .$$

10. Now we integrate a nonconstant function over a pyramid in $\mathbb{R}^3$. We compute $J = \iiint_P (x - yz)\, dV$, where P is the pyramid bounded by the planes $x = 0$, $y = 0$, $z = 0$ and $x + y + z = 1$. See Figure 9.33.

As an iterated triple integral

$$J = \int_{z_0}^{z_1} \int_{y_0}^{y_1} \int_{x_0}^{x_1} (x - yz)\, dx\, dy\, dz.$$

Here the order of integration we have chosen is clearly first x, then y, then z, and the limits of integration x_0, x_1, y_0, and so on, must be found. To obtain x_0 and x_1, observe that a horizontal column built of small bricks of the Δx by Δy by Δz variety extends from $x = x_0 = 0$ to the slant face $x = x_1 = 1 - y - z$; that is,

$$x_0 = 0, \quad x_1 = 1 - y - z.$$

Since we chose to integrate next with respect to y, we now build a horizontal slab from these columns as Figure 9.34. A typical slab extends from $y = y_0 = 0$ to $y = y_1 = 1 - z$. Finally, these horizontal slabs are piled from $z = z_0 = 0$ to $z = z_1 = 1$. Thus we conclude that

$$y_0 = 0, \quad y_1 = 1 - z, \quad z_0 = 0, \quad z_1 = 1.$$

Now we compute

$$
\begin{aligned}
J &= \int_0^1 \int_0^{1-z} \int_0^{1-y-z} (x - yz)\, dx\, dy\, dz \\
&= \int_0^1 \int_0^{1-z} \left[\frac{x^2}{2} - xyz \right]_{x=0}^{x=1-y-z} dy\, dz
\end{aligned}
$$

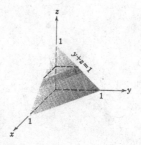

FIGURE 9.34. A slab parallel to the xy-plane

$$= \frac{1}{2} \int_0^1 \int_0^{1-z} [(1-y-z)^2 - 2(1-y-z)yz]\, dy\, dz$$

$$= \frac{1}{6} \int_0^1 \left[-(1-y-z)^3 - 6\left(\frac{1}{2}(1-z)y^2z - \frac{1}{3}y^3z\right)\right]_{y=0}^{y=1-z} dz$$

$$= \frac{1}{6} \int_0^1 (1-z)^4 dz = \frac{1}{30}.$$

11. We conclude our discussion of triple integrals with a very useful observation. Let us compute $\iiint_{B(0;a)} x^2 y^2 z \, dV$.

The integral has the value 0. We see this without any computation since the domain $B(\mathbf{0}; a)$ is symmetric about the plane $z = 0$. Moreover, $f(x, y, z) = x^2 y^2 z$ is *odd* with respect to the variable z:

$$f(x, y, -z) = -f(x, y, z).$$

If we denote the northern hemisphere of the ball (points with $z \geq 0$) by $B_{..+}(\mathbf{0}; a)$ and the southern hemisphere by $B_{..-}(\mathbf{0}; a)$, then

$$\iiint_{B(0;a)} x^2 y^2 z \, dV$$

$$= \iiint_{B_{..+}(0;a)} x^2 y^2 z \, dV + \iiint_{B_{..-}(0;a)} x^2 y^2 z \, dV$$

$$= \iiint_{B_{..+}(0;a)} x^2 y^2 z \, dV - \iiint_{B_{..+}(0;a)} x^2 y^2 z \, dV = 0.$$

No need to perform any integrations. The moral here is: Look for symmetries and odd functions, and don't compute blindly.

Is it clear to you now that

$$\iiint_{B(0;1)} (3 + (x^2 + y^2 + z^2)\sin z)\, dV = 4\pi \, ?$$

Exercises

1. For each of the following, sketch the domain of integration and evaluate the integral:

   (a) $\int_3^4 \left(\int_{-1}^2 (x + x^2 y - 1) \, dx \right) dy$

   (b) $\int_0^1 \left(\int_y^{2y} e^{x-y} \, dx \right) dy$

   (c) $\int_0^2 \left(\int_0^{\sqrt{4-x^2}} 6xy^2 \, dy \right) dx$

   (d) $\int_0^{2\pi} \left(\int_{-\sin x}^x y^2 \, dy \right) dx$.

2. For each of the following regions $\mathcal{B}$, use the methods in Section 9.1 to estimate, roughly,

   $$ J = \iint_{\mathcal{B}} xy \, dA . $$

   Then write it as an iterated integral in two different ways, as

   $$ \int \left(\int xy \, dy \right) dx \quad \text{and as} \quad \int \left(\int xy \, dx \right) dy $$

   and evaluate *both* iterated integrals for the following:

   (a) $\mathcal{B}$ is the rectangular region with vertices at $(1,1)$, $(1,5)$, $(3,1)$, and $(3,5)$

   (b) $\mathcal{B}$ is the triangular region with vertices at $(1,1)$, $(3,1)$, and $(3,5)$

   (c) $\mathcal{B}$ is the curvilinear triangular region enclosed on the left by the line $x = 1$, above by the line $y = 8$, and on the right by the curve $y = x^3$

   (d) $\mathcal{B}$ is the region enclosed by the curves $y = x^2$ and $y = \sqrt{x}$

   (e) $\mathcal{B}$ is the triangular region with vertices at $(-1,1)$, $(0,0)$, and $(2,1)$

   (f) $\mathcal{B}$ is the region enclosed by the curves $y = x^2$ and $y = 2 - x^2$

   (g) $\mathcal{B}$ is the parallelogram region with vertices at $(0,0)$, $(2,1)$, $(1,2)$, and $(3,3)$. Don't evaluate this one. It's too messy to calculate it using the methods we have learned so far. In Section 9.4a, we will learn a cleaner way to compute integrals over parallelogram regions.

3. Evaluate

   $$ \iint_{\mathcal{B}} \sin \pi (2x + y) \, dA , $$

   where $\mathcal{B}$ is the triangular region bounded by the lines $x = 1$, $y = 2$, and $x - y = 5$.

4. Determine the limits of integration but do *not* evaluate

   $$ \iint_{\mathcal{B}} \sqrt{2x + y} \, dA , $$

   where $\mathcal{B}$ is the region defined by

(a) $x + y \geq 1$, $x + y \leq 2$, $x \geq 0$, and $x \leq 1$

(b) $x + y \geq 1$, $x + y \leq 2$, $x \geq 0$, and $y \geq 0$.

5. Evaluate:

(a) $\int_1^2 \left(\int_{-1}^1 \left(\int_{-2}^1 (2xy + z^2) \, dz \right) dx \right) dy$

(b) $\int_1^2 \left(\int_x^2 \left(\int_{x-y}^{x+y} (4x + 3) \, dz \right) dy \right) dx$

(c) $\int_0^1 \left(\int_0^{\sqrt{1-y^2}} \left(\int_{x^2+y^2}^{2-x^2-y^2} y \, dz \right) dx \right) dy$

(d) $\int_1^2 \left(\int_1^z \left(\int_{z+y}^{z+2y} (2x - 6) \, dx \right) dy \right) dz$.

6. For each of the following regions $\mathcal{B}$, use the methods in Section 9.1 to estimate, roughly,

$$ J = \iiint_{\mathcal{B}} 6z \, dV \ . $$

Then write this as some iterated integral and evaluate.

(a) $\mathcal{B}$ is the rectangular box region defined by $0 \leq x \leq 1$, $-1 \leq y \leq 3$, and $1 \leq z \leq 2$

(b) $\mathcal{B}$ is the pyramid bounded by the planes $x = 0$, $y = 0$, $z = 0$, and $2x + y + z = 2$

(c) $\mathcal{B}$ is the solid enclosed by the paraboloid $y = x^2 + z^2$ and the plane $y = 4$

(d) $\mathcal{B} = B((0,0,0); a)$

(e) $\mathcal{B}$ is the solid between the paraboloids $z = x^2 + y^2$ and $z = 2 - x^2 - y^2$

(f) $\mathcal{B}$ is the region bounded by the cone $x^2 = y^2 + z^2$ and between the planes $x = 1$ and $x = 3$.

7. Use polar coordinates to evaluate the following integrals:

(a) $\iint_{B(0;1)} 8(1 - x^2 - y^2)^7 \, dA$

(b) $\iint_{\mathcal{B}} \frac{x}{\sqrt{x^2+y^2}} \, dA$, $\mathcal{B}$ is the ring $1 \leq x^2 + y^2 \leq 4$

(c) $\iint_{B_+(0;3)} xy(x^2 + y^2)^{7/2} \, dA$, $B_+(0;3)$ is the half-disk $x^2 + y^2 \leq 9$, $y \geq 0$

(d) $\iint_{\mathcal{B}} \frac{y}{x^2+y^2} \, dA$, $\mathcal{B}$ is the portion of the ring $1 \leq x^2 + y^2 \leq 4$ in the second quadrant (where $y \geq 0$ and $x \leq 0$)

(e) $\iint_{\mathcal{B}} \exp(2x^2 + 2y^2) \, dA$, $\mathcal{B}$ is the region bounded by $x^2 + y^2 = 4$, $x^2 + y^2 = 25$, $y = x$, and $x = 0$, with $x > 0$

(f) $\iint_{B(0;2)} \sin(x^2 + y^2) \, dA$.

8. Evaluate $\iint_{\mathcal{B}} x^y \, dA$, where

$$ \mathcal{B} = \{ (x, y) : 0 \leq x \leq 1, \, 1 \leq y \leq 2 \} \ . $$

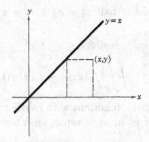

FIGURE 9.35.

9. Evaluate $\iint_{B(0;1)} x^2 y^2 \, dA$.

10. Evaluate $\iint_{B_+(0;5)} \sqrt{25 - x^2 - y^2} \, dA$, where $B_+(0;5)$ is the portion of the disk $x^2 + y^2 \le 25$ in the right half-plane $x \ge 0$.

11. Evaluate by first interchanging the order of integration.

   (a) $\int_0^1 \left(\int_y^1 y e^{x^3} \, dx \right) dy$

   (b) $\int_0^{\pi/2} \left(\int_x^{\pi/2} \frac{\sin y}{y} \, dy \right) dx$.

12. In computing the mass M of a plate of density f, it was found that

$$ M = \int_{-3}^0 \left(\int_0^{3-2x-x^2} f(x,y) \, dy \right) dx + \int_{-1}^3 \left(\int_0^{3-y} f(x,y) \, dx \right) dy. $$

   Sketch the domain of integration.

13. Find the volume of the ellipsoid $(x/a)^2 + (y/b)^2 + (z/c)^2 \le 1$.

14. (a) By interchanging the order of integration, show that if $x \ge 0$,

$$ \int_0^x \left(\int_0^s f(t) \, dt \right) ds = \int_0^x (x - t) f(t) \, dt. $$

   (b) $\int_0^x \left(\int_0^r \left(\int_0^s f(t) \, dt \right) ds \right) dr = ?$, where $x \ge 0$.

15. (a) Let $\mathcal{B}$ be a rectangular region with vertices at $P = (a, c)$, $Q = (b, c)$, $R = (b, d)$, and $S = (a, d)$, where $a < b$ and $c < d$. If all second partial derivatives of f exist and are continuous, show that

$$ \iint_{\mathcal{B}} f_{xy}(x, y) \, dA = f(P) + f(R) - f(Q) - f(S). $$

   (b) Use this to again evaluate the result in part (a) of Exercise 2.

   (c) If $u(x, y)$ satisfies $u_{xy}(x, y) = 0$ for $0 < y < x$, $u(x, x) = 0$ and $u(x, 0) = x \sin x$, find $u(x, y)$ for all points (x, y) in the wedge $0 < y < x$. See Figure 9.35.

16. Find the volume of the ball $x_1^2 + x_2^2 + x_3^2 + x_4^2 \leq a^2$ of radius a in $\mathbb{R}^4$. Generalize.

17. Let $\mathcal{B}$ be the rectangular region $a \leq x \leq b, c \leq y \leq d$. Simplify

$$\iint_{\mathcal{B}} (f_x(x,y) - g_y(x,y)) \, dA,$$

where f and g are given functions with continuous first partial derivatives. Note that the result is, in some sense, an integral around the boundary of $\mathcal{B}$.

18. Let f be a given continuous function (perhaps the height or temperature) defined on a region $\mathcal{B}$ in $\mathbb{R}^2$. Then its **average value** is defined to be

$$f_{av} = \frac{1}{\text{area of } \mathcal{B}} \iint_{\mathcal{B}} f \, dA.$$

For $\mathcal{B}$ equal to the rectangular region defined by $-2 \leq x \leq 2$ and $0 \leq y \leq 3$, compute the average value of f

(a) $f(x,y) = 3x^2 y$

(b) $f(x,y) = y \sin x$.

19. Jill waits on a certain corner for a bus every day. There are two different bus lines that stop at this corner, each going to her destination (the dentist). Each line runs buses every 10 minutes but they begin at varying times every day. What is the average time Jill must wait before a bus comes?

Suppose that service on the route is improved so that three bus lines stop there, each on 10 minute schedules. What is Jill's average waiting time?

9.3   Theory of the Integral

9.3A   INTRODUCTION

It has come to this: We have an intuitive understanding of integrals that is sufficient to estimate them (Section 9.1), and we can use repeated integration to evaluate many simple integrals exactly (Section 9.2). But we do not have a precise definition of the integral, nor do we have precise proofs of the assertions made so far in this chapter. In fact, filling these gaps is an ambitious task that is reserved to a more advanced course.

Our intention here is more humble. We just want to mention what the problem is, why it is difficult, and to state some results. As usual, we restrict our attention mainly to double integrals.

9.3B   PROPERTIES OF THE INTEGRAL

The following properties of the integral are basic:

1. Additivity: $\iint_{\mathcal{B}} (f + g) \, dA = \iint_{\mathcal{B}} f \, dA + \iint_{\mathcal{B}} g \, dA$;

2. Homogeneity: $\iint_{\mathcal{B}} cf \, dA = c \iint_{\mathcal{B}} f \, dA$, where c is a constant;

3. Positivity: $f \geq 0$ in $\mathcal{B} \Rightarrow \iint_{\mathcal{B}} f \, dA \geq 0$;

4. Normalization: $\iint_{\mathcal{R}} dA = $ (area of $\mathcal{R}$), where $\mathcal{R}$ is a rectangular region in the plane.

In all these formulas, we naturally assume that the given integrals exist. They should be very plausible in view of the interpretation of the integral

$$\iint_{\mathcal{B}} f \, dA$$

as the volume of a solid with base $\mathcal{B}$ and height $z = f(x, y)$.

We cannot prove these, since any such proof must be founded on a precise definition of the integral, which we do not give here. But there is another attitude one can take. One can feel that any concept of integration in $\mathbb{R}^2$ must have the four properties given above. From this viewpoint, these four properties become the axioms for the integral. It turns out that these few axioms are all that are needed for the integral; no additional axioms are required. In other words, if one uses *only* Properties 1 to 4, then one can deduce *all* other properties of the integral.

It is instructive to deduce some other properties; they are both useful and intuitive. In order not to repeat ourselves, we assume once and for all that all the integrals we write do exist (as finite numbers).

EXAMPLES:

1. Let us prove that

$$f \geq g \text{ in } \mathcal{B} \Rightarrow \iint_{\mathcal{B}} f \, dA \geq \iint_{\mathcal{B}} g \, dA \,.$$

Note that $f \geq g$ implies that $f - g \geq 0$. Now use Property 3 to deduce that

$$\iint_{\mathcal{B}} (f - g) \, dA \geq 0 \,,$$

and then Properties 1 and 2 to get

$$\iint_{\mathcal{B}} f \, dA - \iint_{\mathcal{B}} g \, dA \geq 0 \,.$$

2. Let us show that

$$\left| \iint_{\mathcal{B}} f \, dA \right| \leq \iint_{\mathcal{B}} |f| \, dA \,.$$

Note that $-|f| \le f \le |f|$. Therefore, by the preceding example,

$$- \iint_B |f| \, dA \le \iint_B f \, dA \le \iint_B |f| \, dA \, .$$

This is just another way of writing the desired conclusion.

3. We will prove that if $|f| \le M$ in a rectangular region R, then

$$\iint_R |f| \, dA \le MA \, ,$$

where A is the area of R.

By Example 1, since $|f| \le M$ in R, we have

$$\iint_R |f| \, dA \le \iint_R M \, dA = M \iint_R dA = MA \, .$$

Note that Property 4 was needed to prove this.

There is one further result from advanced calculus that we quote, since it is often needed. It is a theorem about single integrals.

Theorem 9.3.1 Let $f(x, y)$ be a differentiable function of x, with $f(x, y)$ and $f_x(x, y)$ continuous for all x, y. If

$$\varphi(x) = \int_a^b f(x, y) \, dy,$$

then we can compute the derivative of φ as

$$\varphi'(x) = \int_a^b f_x(x, y) \, dy.$$

More generally, if a and b also depend on x, and are differentiable, then

$$\varphi'(x) = f(x, b(x))b'(x) - f(x, a(x))a'(x) + \int_{a(x)}^{b(x)} f_x(x, y) \, dy \, .$$

EXAMPLE:

4. Let

$$\varphi(x) = \int_0^{1+x^2} \frac{e^{xy}}{1+y^2} \, dy.$$

Then

$$\varphi'(x) = \frac{e^{x+x^3}}{1+(1+x^2)^2} 2x + \int_0^{1+x^2} \frac{ye^{xy}}{1+y^2} \, dy \, .$$

9.3C DEFINITION AND EXISTENCE OF THE INTEGRAL: THE ISSUES

On Section 9.1 we discussed estimation of integrals without ever actually defining that which we were estimating. Then in Section 9.2 we learned how to exactly calculate these undefined objects. We leave it to the reader to pick her or his favorite cliche about this way of doing business.

We can turn the estimation procedures described in Section 9.1 into a definition. To be specific let us discuss a double integral

$$\int\int_{B} f \, dA \, ,$$

where we assume that f is a nonnegative function. Recall that in Section 9.1, particularly in Example 5, two kinds of sums were used for approximation. In both, areas of non-overlapping (except possibly on edges) rectangular regions were muliplied by values of f and the resulting products were summed. In one case we chose one or more non-overlapping rectangles B_i contained in B, and for each such rectangle, we chose a point X_i in B_i such that $f(X_i)$ was the minimum value of f on B_i. Letting ΔA_i be the area of B_i, we used the sum

$$\sum_i f(X_i)\Delta A_i$$

as a lower estimate for the double integral of f over the region B. Let us call the value of a sum obtained in this manner a **lower sum**. For an upper estimate, we might use a different set of rectangles, say R_j. These rectangles are required to completely cover B, and for each such rectangle, chose $Y_j \in R_j$ to maximize f on R_j. The corresponding sum

$$\sum_i f(X_j)\Delta A_j$$

is an upper estimate for the double integral, where ΔA_j stands for the area of R_j. We call the value of a sum obtained in the manner an **upper sum**.

Without defining what we were approximating in Section 9.1, we acted as if it were at least as large as every lower sum and no bigger than any upper sum. Is there such a number? If so, is there only one such number? If the answer to both questions is "yes" then we define $\int\int_{B} f \, dA$ to be the unique number that is no smaller than any lower sum and no larger than any upper sum. Otherwise, we say that the double integral does not exist.

There are two things that can cause a double integral to fail to exist: a bad region of integration or a bad integrand. A region of integration might be so bad that the double integral of the constant function 1 does not exist. The area of such a region would not even be defined.

Let us investigate how it might happen that the area of a region may not exist. One possibility is that it is too large (infinite). But there are also other possibilities. For example, let $\mathcal{Q}$ denote the unit square region defined by $0 \leq x \leq 1$ and $0 \leq y \leq 1$. Let $\mathcal{B}$ denote the set of all points (x, y) in $\mathcal{Q}$, where x and y are both rational numbers. See Figure 9.36. Thus we "throw away" an infinite number of points from $\mathcal{Q}$. What is the area of $\mathcal{B}$? There are no lower sums, since we can't fit any rectangles into $\mathcal{B}$. So our lower estimate for the area of $\mathcal{B}$ is 0, and this estimate cannot be improved. On the other hand, all of the upper sums for the area of $\mathcal{B}$ are at least as large as 1. Thus the area of $\mathcal{B}$ is not defined.

We now describe one type of region $\mathcal{B}$ in $\mathbb{R}^2$ that is nice enough so that its area is defined. First we look at closed sets $\mathcal{B}$. Suppose that the boundary of $\mathcal{B}$ is the union of a finite number of curves, each of which is the image of a differentiable function defined on a bounded closed interval in $\mathbb{R}$. For example, the boundary of the square region

$$\{(x, y) : 0 \leq x, y \leq 1\}$$

consists of four line segments, each of which may clearly be parametrized by a differentiable function defined on a closed bounded interval. Then we will call $\mathcal{B}$ a **regular closed set**. More generally, if a region $\mathcal{B}$ can be obtained by starting with a regular closed set $\bar{\mathcal{B}}$ and then possibly removing part or all of its boundary, then we call $\mathcal{B}$ a **regular set**. See Figure 9.37 for an illustration of a regular region in $\mathbb{R}^2$.

A similar definition can be made for regions in $\mathbb{R}^3$. The appropriate condition on a closed set $\mathcal{B}$ is that its boundary be the union of a finite number of surfaces, each of which may be parametrized by a differentiable function whose domain is closed and bounded. General regular regions in $\mathbb{R}^3$ are formed from closed regular regions by possibly removing part or all of the boundary points. An example of a regular closed region in $\mathbb{R}^3$ is

$$\{(x, y, z) : 4 \leq x^2 + y^2 + z^2 \leq 9\}$$

The boundary of consists of the two spheres $S((0, 0, 0); 2)$ and $S((0, 0, 0); 3)$, each of which may be parametrized by a differentiable function defined on a bounded closed subset of $\mathbb{R}^2$. If we "strip off" the outer boundary piece, we obtain the regular set

$$\{(x, y, z) : 4 \leq x^2 + y^2 + z^2 < 9\} \, .$$

The definition of regular may also be generalized to regions in $\mathbb{R}^n$.

Most bounded sets of practical interest are regular. Furthermore, the double or triple integrals of bounded continuous functions over regular regions always exist, as do any iterated integrals that might arise in the computation of such double integrals. All of the bounded regions considered in Sections 9.0, 9.1, and 9.2 are regular.

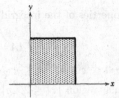

FIGURE 9.36. A wild set in $\mathbb{R}^2$

FIGURE 9.37. Area exists for this set

REMARKS:

- The most important feature in the preceding discussion is the fact that the rather abstract definition of integrals is closely tied to the simplest type of approximation schemes for their values. It is this connection that enables workers in various sciences to set up integrals that represent quantities of interest.

- For nonnegative integrands it is quite easy to introduce improper integrals in case either the region of integration or the integrand is unbounded. In particular, it is possible to speak of the "area" (or "volume") of unbounded sets whose intersections with each ball are regular, but of course, the area or volume might equal ∞.

Exercises

1. Use only the four properties of the integral stated in Section 9.3b to evaluate

$$ J = \iint_{\mathcal{B}} 3 \, dA \,, $$

where $\mathcal{B}$ is the rectangular region $-1 \le x \le 5$, $2 \le y \le 4$.

2. Let

$$ f(x,y) = \begin{cases} 2 & \text{if} & -1 \le x \le 1, \ 0 \le y \le 3 \\ -1 & \text{if} & 2 \le x \le 3, \ -4 \le y \le -2 \\ 0 & \text{otherwise} \,. \end{cases} $$

Use only the four properties of the integral in Section 9.3b to evaluate

$$\iint f \, dA$$

in the region $\{(x,y): -2 \leq x \leq 4, \; -4 \leq y \leq 4\}$.

3. Let $\mathcal{B}$ be the rectangular region $|x| \leq 2$, $|y| \leq 1$ $\mathbb{R}^2$. Show that

$$\iint_{\mathcal{B}} \frac{x^2 + y^2}{2 + \sin(xy)} \, dA \leq 40 \,.$$

4. If $f(t) = \int_1^4 e^{tx^2} \, dx$, find $f'(t)$.

5. If $g(t) = \int_t^{t^2} e^{tx^2} \, dx$, find $g'(t)$.

6. If $u(x) = \int_0^x (x - t)^2 \sin t \, dt$, find u' and u''.

7. If $u(x) = \int_0^x (x - t)e^{2(x-t)} f(t) \, dt$, where f is continuous, show that $u'' - 4u' + 4u = f$.

8. If $h(s) = \int_0^1 \int_0^s f(x, y) \, dx \, dy$, where f is continuous, find h'.

9. If $v(x, t) = \int_{2x-t}^{2x+t} g(s) \, ds$, where g is a given continuous function, find v_x and v_t. Further, assuming that g is differentiable, show that $v_{xx} = 4v_{tt}$.

10. Let $w(x, y) = \int_0^y f(x + s - y, s) \, ds$, where $f(r, s)$ is a given differentiable function. Show that $w_x + w_y = f$.

11. If $u(x, y) = \int_0^1 (x + y^2 + t)^{15} \sin t \, dt$, show that $u_y = 2yu_x$.

9.4   Change of Variables

We have already seen that, when computing double integrals exactly, it can be useful to "change variables " to polar coordinates. And in single-variable calculus, change of variables, or "substitution" is a standard technique of integration. In this section, we will systematically develop the ideas needed to change variables in double and triple integrals.

9.4A   Affine Changes of Variables

Let $\mathcal{B}$ be the region inside the parallelogram in $\mathbb{R}^2$ with vertices at $(1,1)$, $(4,2)$, $(5,4)$, and $(2,3)$, as shown in Figure 9.38. This region is the image of the unit square region $\mathcal{Q}$ (with vertices at $(0,0)$, $(1,0)$, $(0,1)$, and $(1,1)$) under the affine map $T : \mathbb{R}^2 \to \mathbb{R}^2$ defined by

$$[T(u,v)] = \begin{bmatrix} 1 & 3 \\ 2 & 1 \end{bmatrix} \begin{bmatrix} u \\ v \end{bmatrix} + \begin{bmatrix} 1 \\ 1 \end{bmatrix}.$$

In coordinate form, $T(u, v) = u(1, 2) + v(3, 1) + (1, 1)$.

Consider the problem of integrating a function $f(x, y)$ over $\mathcal{B}$. Since none of the sides of $\mathcal{B}$ are parallel to the coordinate axes, a direct approach would

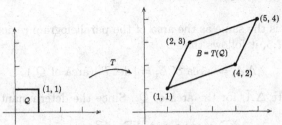

FIGURE 9.38. The parallelogram region $\mathcal{B}$

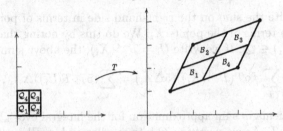

FIGURE 9.39. Partitioning $\mathcal{B}$ by partitioning $\mathcal{Q}$

require us to break $\mathcal{B}$ into several simpler regions and then to integrate separately over each of these regions. In other words, we could follow the suggestion made at the end of Section 9.2b for problems in which the region of integration is "complicated".

There is another, somewhat cleaner approach. We first partition the unit square region $\mathcal{Q}$ into smaller square regions $\mathcal{Q}_i$. The images under T of these smaller square regions are parallelogram regions $\mathcal{B}_i$ contained in $\mathcal{B}$. Thus, partitioning $\mathcal{Q}$ into square regions corresponds to a natural partition of $\mathcal{B}$ into parallelogram regions. See Figure 9.39. The integral of f over $\mathcal{B}$ may be approximated, in the spirit of the Riemann sum approximation, by

$$\sum_i f(X_i)\Delta A_i ,$$

where X_i is a point in the parallelogram region $\mathcal{B}_i$, and ΔA_i is the area of the parallelogram region $\mathcal{B}_i$.

We may write this sum in another form by using the affine map T, after which we will recognize it as an approximation for an integral of a function over $\mathcal{Q}$. Since T is affine, we may write $T(u,v) = T'(u,v) + (1,1)$, where T' is the derivative of T, or in other words, the linear map corresponding to T. The matrix of T' is given by

$$[T'] = \begin{bmatrix} 1 & 3 \\ 2 & 1 \end{bmatrix} .$$

The only difference between the affine map T and the linear map T' is in adding the vector $(1,1)$. Therefore, the area of each parallelogram region

$\mathcal{B}_i = T(\mathcal{Q}_i)$ is the same as the area of the parallelogram region $T'(\mathcal{Q}_i)$. By Theorem 3.7.5, it follows that

$$\Delta A_i = \text{ area of } \mathcal{B}_i = |\det T'|(\text{area of } \mathcal{Q}_i) \ .$$

We will write $\Delta \bar{A}_i$ for the area of $\mathcal{Q}_i$. Since the determinant of T' is -5, we have

$$\sum_i f(X_i) \Delta A_i = \sum_i f(X_i)(5\Delta \bar{A}_i) \ .$$

Let us write the sum on the right-hand side in terms of points U_i in $\mathcal{Q}_i$ instead of in terms of the points X_i. We do this by noting that if $X_i \in \mathcal{B}_i$, then $T^{-1}(X_i) \in \mathcal{Q}_i$. If we write $U_i = T^{-1}(X_i)$, the above sum becomes

$$\sum_i f \circ T(T^{-1}(X_i))(5\Delta \bar{A}_i) = \sum_i (5f \circ T(U_i))\Delta \bar{A}_i \ ,$$

which is a Riemann sum approximation for the integral over $\mathcal{Q}$ of the function $g = 5f \circ T$. As we partition $\mathcal{Q}$ into smaller and smaller square regions $\mathcal{Q}_i$, this Riemann sum approximation converges to the double integral

$$\int_{\mathcal{Q}} 5f \circ T \, dA \ .$$

Here we use a single integral sign as a shorthand for a double or triple integral. It can be shown (and this should not surprise you) that the corresponding approximations of the integral of f over $\mathcal{B}$, using smaller and smaller parallelogram regions $\mathcal{B}_i$ to partition $\mathcal{B}$, converges to the integral of f over $\mathcal{B}$. Thus

$$\int_{\mathcal{B}} f \, dA = \int_{\mathcal{Q}} 5f \circ T \, dA \ .$$

It is helpful to rewrite this equation using variables. To emphasize that the two integrals are taken over different regions, we use (x, y) to label the points in $\mathcal{B}$, and (u, v) to label the points in $\mathcal{Q}$. We think of a point (x, y) in $\mathcal{B}$ as the image under T of a point (u, v) in $\mathcal{Q}$, that is,

$$(x, y) = T(u, v) \ .$$

The following equation thus expresses a change of variables from (x, y) to (u, v):

$$\int_{\mathcal{B}} f(x, y) \, dx \, dy = \int_{\mathcal{Q}} 5f(T(u, v)) \, du \, dv \ .$$

As noted above, writing the integral on the left-hand side as an iterated integral with the appropriate limits of integration would require breaking $\mathcal{B}$ into several pieces. But there is no problem in doing so with the integral on the right-hand side, since $\mathcal{Q}$ is such a simple region:

$$\int_{\mathcal{Q}} 5f(T(u, v)) \, du \, dv = \int_0^1 \int_0^1 5f(T(u, v)) \, du \, dv \ .$$

EXAMPLE:

1. Let $\mathcal{B}$, $\mathcal{Q}$, and T be as in the preceding discussion. Let $f(x, y) = \cos(2x - 4y)$. Using the expression for T, we have

$$(x, y) = u(1, 2) + v(3, 1) + (1, 1),$$

or

$$x = u + 3v + 1 \text{ and } y = 2u + v + 1.$$

Thus

$$(f \circ T)(u, v) = \cos(2(u + 3v + 1) - 4(2u + v + 1)) = \cos(-6u + 2v - 2).$$

We may now compute the integral of f over $\mathcal{B}$ as follows:

$$\int_{\mathcal{B}} \cos(2x - 4y)\, dx\, dy = \int_0^1 \int_0^1 5 \cos(-6u + 2v - 2)\, du\, dv$$

$$= -\frac{5}{6} \int_0^1 (\sin(-6 + 2v - 2) - \sin(2v - 2))\, dv$$

$$= \frac{5}{12}(\cos(-6) - \cos(-8) - \cos(0) + \cos(-2)).$$

REMARKS:

- Note that if $\mathcal{B}$ is shifted by subtracting the vector $(1, 1)$, so that one of its vertices is moved to the origin, then the resulting figure is the parallelogram region generated by the columns of T'. Thus, the matrix of T' is easy to calculate from the vertices of $\mathcal{B}$.

- For any parallelogram region $\mathcal{B}$ in $\mathbb{R}^2$, there are several affine maps T such that $T(\mathcal{Q}) = \mathcal{B}$. For example, if a different vertex of $\mathcal{B}$ is used in the previous remark, a different matrix would be the result.

- By Theorem 3.7.5,

$$|\det T'| = \text{ area of } \mathcal{B}.$$

Thus, even though T is not uniquely determined by $\mathcal{B}$, any choice for T in the example would lead to the same factor 5.

- In single-variable calculus, the main reason to change variables is to simplify the integrand in some way. In Example 1, the change of variables served to simplify the *region* of integration, *not* the integrand. There are also multiple integrals for which a change of variables makes a convenient simplification of the integrand, just as in single-variable calculus. The point is that in multiple integration, a change of variables alters both the region and the integrand, often in significant ways.

- The most important changes of variables are not affine, as we will see in subsequent sections. Nevertheless, it is important to understand affine changes of variables, because, in keeping with one of the main themes in this book, we will handle the general nonlinear case by using affine approximations.

Example 1 can be easily generalized. We may let T be any affine map from $\mathbb{R}^2$ to $\mathbb{R}^2$. There is no reason to insist on using the unit square region $\mathcal{Q}$. Any rectangular region $\mathcal{R}$ with sides parallel to the coordinate axes will do. We can generalize to three dimensions by looking at parallelepiped regions $\mathcal{B}$ and rectangular boxes $\mathcal{R}$. In three dimensions, Theorem 3.7.6 would be used instead of Theorem 3.7.5. Generalizations to higher dimensions are also possible. Here are the appropriate formulas for dimensions 1,2 and 3:

1. Let $\mathcal{R}$ be the closed interval $[a, b]$, let T be an affine map from $\mathbb{R}$ to $\mathbb{R}$, and let $\mathcal{B} = T(\mathcal{R})$. (Note that in this case, $\mathcal{B}$ is an interval, since it is the image of an interval under an affine map.) Then for any continuous function $f : \mathcal{B} \to \mathbb{R}$,

$$\int_{\mathcal{B}} f(x)\, dx = \int_a^b f(T(u))|\det T'|\, du\,.$$

In this formula, if $T(u) = mu + c$, then T' is the 1-by-1 matrix $[m]$ whose determinant is the single entry m.

2. Let $\mathcal{R}$ be the rectangular region $[a_1, b_1] \times [a_2, b_2]$, let T be an affine map from $\mathbb{R}^2$ to $\mathbb{R}^2$, and let $\mathcal{B} = T(\mathcal{R})$. Then for any continuous function $f : \mathcal{B} \to \mathbb{R}$,

$$\int_{\mathcal{B}} f(x, y)\, dA = \int_{a_1}^{b_1} \int_{a_2}^{b_2} f(T(u, v))|\det T'|\, dv\, du\,.$$

3. Let $\mathcal{R}$ be the rectangular box $[a_1, b_1] \times [a_2, b_2] \times [a_3, b_3]$, let T be an affine map from $\mathbb{R}^3$ to $\mathbb{R}^3$, and let $\mathcal{B} = T(\mathcal{R})$. (Note that in this case, $\mathcal{B}$ is a parallelepiped region.) Then for any continuous function $f : \mathcal{B} \to \mathbb{R}$,

$$\int_{\mathcal{B}} f(x, y, z)\, dV = \int_{a_1}^{b_1} \int_{a_2}^{b_2} \int_{a_3}^{b_3} f(T(u, v, w))|\det T'|\, dw\, dv\, du\,.$$

We do not prove this result any more than we proved the special case that appeared in the Example 1. However, the justification used in that example carries over completely to the general case. The only difference is in the terminology used for the geometric objects $\mathcal{B}$ and $\mathcal{R}$ in different dimensions.

EXAMPLES:

2. Let us see what the change of variables formula says about the case of $\mathbb{R}^1$, already familiar to you from single-variable calculus. Let $\mathcal{R} = [0,3]$ and $T(u) = -2u+1$. Then $\mathcal{B} = [-5,-1]$ and $\det T'$ is -2. Thus, the change of variables formula says that for any continuous function f defined on the interval $[-5,-1]$,

$$\int_{-5}^{-1} f(x)\,dx = \int_0^3 f(-2u+1)|-2|\,du = \int_0^3 2f(-2u+1)\,du \ .$$

In single-variable calculus, this example would have appeared in a different but equivalent form:

$$\int_0^3 f(-2u+1)\,du = -\frac{1}{2}\int_{-1}^{-5} f(x)\,dx = \frac{1}{2}\int_{-5}^{-1} f(x)\,dx \ ,$$

that is, the focus would have been on integrating the function $f(-2u+1)$, using the substitution $x = -2u+1$. Note particularly the inversion in the limits of integration that typically appears using the old approach from single-variable calculus when the map T has a negative derivative.

3. Let us compute

$$\int_{\mathcal{B}} \cos(x+y)\cos(x-y)\,dx\,dy \ ,$$

where $\mathcal{B}$ is the square region with vertices at $(2,0)$, $(0,2)$, $(-2,0)$, and $(0,-2)$. Let T be rotation counterclockwise about the origin through an angle of 45 degrees. Then $\mathcal{B} = T(\mathcal{R})$, where $\mathcal{R} = [-\sqrt{2},\sqrt{2}] \times [-\sqrt{2},\sqrt{2}]$. The matrix of T' is

$$\begin{bmatrix} \frac{\sqrt{2}}{2} & -\frac{\sqrt{2}}{2} \\ \frac{\sqrt{2}}{2} & \frac{\sqrt{2}}{2} \end{bmatrix} ,$$

which has determinant 1. In coordinate form,

$$(x,y) = T(u,v) = \frac{\sqrt{2}}{2}(u-v, u+v) \ ,$$

so

$$\cos(x-y) = \cos(-\sqrt{2}v) = \cos(\sqrt{2}v) \text{ and } \cos(x+y) = \cos(\sqrt{2}u) \ .$$

Thus,

$$\int_{\mathcal{B}} \cos(x+y)\cos(x-y)\,dx\,dy$$

$$\begin{aligned}
&= \int_{-\sqrt{2}}^{\sqrt{2}} \int_{-\sqrt{2}}^{\sqrt{2}} \cos(\sqrt{2}u) \cos(\sqrt{2}v) \, dv \, du \\
&= \left(\int_{-\sqrt{2}}^{\sqrt{2}} \cos(\sqrt{2}u) \, du \right)^2 \\
&= 2\sin^2(2) .
\end{aligned}$$

Rotations are often a useful way to simplify integrals.

4. Consider the improper integral

$$\int_{\mathcal{B}} e^{-z^2} \, dV ,$$

where $\mathcal{B} = \{(x, y, z) : x, y \geq 0, z \geq x + y\}$. This is the region in the positive octant of $\mathbb{R}^3$ above the plane $z = x + y$. We will not compute this integral exactly, but we will show that it is finite.

We may think of $\mathcal{B}$ as the parallelepiped region generated by the vectors $(a, 0, a)$, $(0, a, a)$, and $(0, 0, a)$, where a is very large (a approaches ∞). Thus, $\mathcal{B}$ is the image of the positive octant of $\mathbb{R}^3$ under the linear map T whose matrix is

$$[T] = \begin{bmatrix} 1 & 0 & 0 \\ 0 & 1 & 0 \\ 1 & 1 & 1 \end{bmatrix} .$$

The determinant of T is 1. In coordinate form,

$$(x, y, z) = T(u, v, w) = (u, v, u + v + w) .$$

In particular, $z = u + v + w$. According to the change of variables formula,

$$\int_{\mathcal{B}} e^{-z^2} \, dV = \int_0^\infty \int_0^\infty \int_0^\infty e^{-(u+v+w)^2} \, dw \, dv \, du .$$

Now

$$0 \leq e^{-(u+v+w)^2} \leq e^{-(u^2+v^2+w^2)}$$

for nonnegative u, v, and w. Therefore,

$$\begin{aligned}
\int_{\mathcal{B}} e^{-z^2} \, dV &\leq \int_0^\infty \int_0^\infty \int_0^\infty e^{-(u^2+v^2+w^2)} \, dw \, dv \, du \\
&= \left(\int_0^\infty e^{-u^2} \, du \right)^3 .
\end{aligned}$$

According to the calculation done in Example 8 of Section 9.2c, this last expression, involving an integral from 0 to ∞ rather than from $-\infty$ to ∞ as in that example, equals $(\sqrt{\pi}/2)^3$, so our original improper integral is finite.

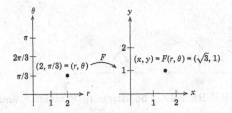

FIGURE 9.40. Two different coordinate planes, related through F

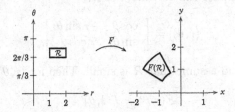

FIGURE 9.41. A rectangular region and its image under F

9.4B   ANOTHER LOOK AT POLAR COORDINATES

Imagine two copies of $\mathbb{R}^2$. We label points in the first copy by the usual (x, y). We label those in the second copy by (r, θ), that is, r is the first coordinate (usually pictured as being along the horizontal axis), and θ is the second coordinate (along the vertical axis). Let F be the map from the second copy of $\mathbb{R}^2$ to the first defined by

$$(x, y) = F(r, \theta) = (r \cos \theta, r \sin \theta) .$$

We recognize the polar coordinates formulas from Section 9.2c (and also Section 5.1e). As we have already seen, there is a simple geometric relationship between the values of r and θ and the position of the point $(x, y) = F(r, \theta)$: the quantity r equals the distance from the origin to (x, y) in $\mathbb{R}^2$, and the quantity θ equals the angle, measured counterclockwise in radians, between the positive x-axis and the vector (x, y). This relationship makes it easy to plot the point $(x, y) = F(r, \theta)$ if you know r and θ. Nevertheless, it is important to think of (x, y) and (r, θ) as existing *separately* in their own copies of $\mathbb{R}^2$. The point (r, θ) is found in the $r\theta$-plane by moving from the origin r units horizontally and θ units vertically. See Figure 9.40.

In Section 9.2c, we used polar coordinates to transform certain double integrals. We now take another look at this transformation in light of the work that we did in Section 9.4a concerning change of variables.

Let $\mathcal{R}$ be a rectangular region in the (r, θ)-copy of $\mathbb{R}^2$. If $\mathcal{R}$ does not contain the origin, then $F(\mathcal{R})$ is a "curved rectangular" region $\mathcal{B}$ in the other copy of $\mathbb{R}^2$, as in Figure 9.41. Let us find an approximation for the area of $\mathcal{B}$ in terms of the area of $\mathcal{R}$, by using the best affine approximation

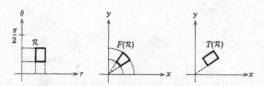

FIGURE 9.42. The images of $\mathcal{R}$ under F and T

to F. First we compute the derivative of F:

$$[F'(r,\theta)] = \begin{bmatrix} \cos\theta & -r\sin\theta \\ \sin\theta & r\cos\theta \end{bmatrix}.$$

Fix $(r_0,\theta_0) \in \mathcal{R}$ and assume that $\mathcal{R}$ is small. Then for $(r,\theta) \in \mathcal{R}$,

$$F(r,\theta) \approx T(r,\theta),$$

where

$$[T(r,\theta)] = \begin{bmatrix} \cos\theta & -r\sin\theta \\ \sin\theta & r\cos\theta \end{bmatrix} \begin{bmatrix} r - r_0 \\ \theta - \theta_0 \end{bmatrix} + [F(r_0,\theta_0)].$$

Thus, when $\mathcal{R}$ is small, the image of $\mathcal{R}$ under F is approximately the same as the image of $\mathcal{R}$ under T. It seems intuitively reasonable that the areas of these two images are also nearly the same. See Figure 9.42. We already know that the area of $T(\mathcal{R})$ is $|\det T'|$ times the area of $\mathcal{R}$. Since $T' = F'(r_0,\theta_0)$, we conclude that

$$\text{area of } F(\mathcal{R}) \approx |\det F'(r_0,\theta_0)| \times (\text{area of } \mathcal{R}).$$

The determinant of $F'(r_0,\theta_0)$ is easily computed from its matrix:

$$\det F'(r_0,\theta_0) = r_0.$$

Therefore,

$$\text{area of } F(\mathcal{R}) \approx |r_0| \times (\text{area of } \mathcal{R}).$$

If we assume that r_0 is positive, as we usually will, and if we let the sides of the rectangular region $\mathcal{R}$ have dimensions Δr and $\Delta\theta$, then our formula becomes

$$\text{area of } F(\mathcal{R}) \approx r_0\,\Delta r\,\Delta\theta,$$

in agreement with the formula in Section 9.2c.

Let us now try to mimic the reasoning used in Section 9.4a to obtain a change of variables formula for polar coordinates. We will let $\mathcal{R}$ denote a region in the $r\theta$-plane, and let $\mathcal{B} = F(\mathcal{R})$. In order to ensure that each point in $\mathcal{B}$ corresponds to only one point in $\mathcal{R}$, we assume that $\mathcal{R}$ is contained in the strip

$$\{(r,\theta) : 0 \le r \text{ and } 0 \le \theta < 2\pi\}.$$

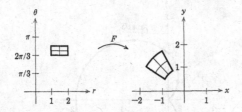

FIGURE 9.43. Partitioning $\mathcal{R}$ produces a partition of $\mathcal{B}$

Note that the image under F of this strip is all of $\mathbb{R}^2$. We do not assume that $\mathcal{R}$ is a rectangular region in the $r\theta$-plane, but we do assume that $\mathcal{R}$ can be approximated with arbitrary precision as the disjoint union of many small rectangular regions $\mathcal{R}_i$, such as can be done if $\mathcal{R}$ is a regular region. For each i, let $\mathcal{B}_i = F(\mathcal{R}_i)$. Also let (r_i, θ_i) be a point in $\mathcal{R}_i$, and let $(x_i, y_i) = F(r_i, \theta_i)$ be the corresponding point in $\mathcal{B}_i$. Since each set $\mathcal{B}_i$ has area approximately equal to r_i times the area of $\mathcal{R}_i$, we have

$$\sum_i f(x_i, y_i)(\text{area of } \mathcal{B}_i) \approx \sum_i (f \circ F)(r_i, \theta_i) r_i (\text{area of } \mathcal{R}_i) \,,$$

where f is a function defined on $\mathcal{B}_i$. The sum on the right-hand side is the Riemann sum approximation of the double integral

$$\int_{\mathcal{R}} (f \circ F)(r, \theta) r \, dA \,.$$

The sum on the left-hand side is an approximation of the double integral

$$\int_{\mathcal{B}} f(x, y) \, dA \,,$$

where $\mathcal{B} = F(\mathcal{R})$. Note that $\mathcal{B}$ is approximately the union of the sets $\mathcal{B}_i$, just as $\mathcal{R}$ is approximately the union of the sets $\mathcal{R}_i$. See Figure 9.43.

For suitably nice functions f (for example, bounded and continuous on $\mathcal{B}$), all these approximations become exact as $\mathcal{R}$ is partitioned into smaller and smaller rectangular regions $\mathcal{R}_i$. We will not prove this fact. Nevertheless, you should try to develop confidence in the following formula, through careful rereading of the argument just given.

$$\int_{\mathcal{R}} (f \circ F)(r, \theta) r \, dA = \int_{\mathcal{B}} f(x, y) \, dx \, dy \,.$$

Since $(f \circ F)(r, \theta) = f(r \cos \theta, r \sin \theta)$, this formula is similar to the one given in Section 9.2c. The difference is that we are not assuming that $\mathcal{R}$ is necessarily a rectangular region in the $r\theta$-plane. We merely assumed that $\mathcal{R}$ is contained in the strip $[0, \infty) \times [0, 2\pi)$, and that $\mathcal{R}$ can be approximated with arbitrary precision as the union of small rectangular regions.

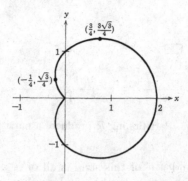

FIGURE 9.44. The cardioid

EXAMPLE:

5. Let $\mathcal{R}$ be the region in the first quadrant of the $r\theta$-plane that lies to the left of the curve $r = 1 + \cos(\theta)$, with $0 \le \theta < 2\pi$. See Figure 9.44, where the image of $\mathcal{R}$ under F in the xy-plane is shown. The boundary of $F(\mathcal{R})$ is called a **cardioid**, due to its heart-like shape. It is good practice to try plotting the image of $\mathcal{R}$ under F.

Let us find the area of $F(\mathcal{R})$. According to our change of variables formula, this area is given by

$$\int_{F(\mathcal{R})} 1 \, dA = \int_{\mathcal{R}} (1 \circ F)(r, \theta) r \, dA \, .$$

Since $(1 \circ F) = 1$, the integral on the right-hand side is just

$$\int_{\mathcal{R}} r \, dA \, .$$

We may evaluate this last integral using the methods of Section 9.2. That is, we set up an iterated integral, using limits of integration appropriate to the region $\mathcal{R}$. We obtain

$$\int_0^{2\pi} \left(\int_0^{1+\cos\theta} r \, dr \right) d\theta \, .$$

After carrying out the inside integration, we are left with the single integral

$$\int_0^{2\pi} \frac{(1 + \cos\theta)^2}{2} \, d\theta \, .$$

This last integral can be evaluated with methods of single-variable calculus. The answer is $3\pi/2$.

9.4C THE GENERAL CHANGE OF VARIABLES FORMULA

As in the preceding section, we start with two copies of $\mathbb{R}^2$. This time we use (x, y) to label the points in one copy, and (u, v) to label the points in the other copy. Let $\mathcal{R}$ be a regular region in the uv-plane, and let F be a differentiable function from $\mathcal{R}$ to $\mathbb{R}^2$. Let us assume that F is one-to-one, so that for each point (x, y) in $F(\mathcal{R})$, there is a unique point $(u, v) \in \mathcal{R}$ such that $(x, y) = F(u, v)$. The function F is a change of variables transformation from the uv-plane to the xy-plane.

Now imagine that $\mathcal{R}$ is partitioned into small rectangular regions $\mathcal{R}_i$. (If $\mathcal{R}$ has a curved boundary, these rectangular regions may not quite fill out all of $\mathcal{R}$.) Reasoning as we did for the polar coordinates transformation, if $\mathcal{R}_i$ is small, we can approximate the area of $F(\mathcal{R}_i)$ by

$$\text{area of } F(\mathcal{R}_i) \approx |\det F'(u_i, v_i)|(\text{ area of } \mathcal{R}_i) \,,$$

where (u_i, v_i) is a point in $\mathcal{R}_i$. If we write $(x_i, y_i) = F(u_i, v_i)$ and let f be a function defined on $F(\mathcal{R})$, we have

$$\sum_i f(x_i, y_i)(\text{area of } F(\mathcal{R}_i)) \approx \sum_i (f \circ F)(u_i, v_i)|\det F'(u_i, v_i)|\Delta A_i \,,$$

where ΔA_i is the area of $\mathcal{R}_i$. If f is a suitable real-valued function defined on $F(\mathcal{R})$, then as the rectangular regions in the partition of $\mathcal{R}$ become smaller and smaller, the expression on the right-hand side converges to the double integral

$$\int_{\mathcal{R}} (f \circ F)(u, v) \, |\det F'(u, v)| \, dA \,,$$

the expression on the left-hand side converges to the double integral

$$\int_{F(\mathcal{R})} f(x, y) \, dA \,,$$

and the difference between the two sides goes to zero. This fact is summarized in the following formulas, which contain as special cases all previous change of variables formulas that we have developed so far. We assume throughout that the region $\mathcal{R}$ and the functions F and f are sufficiently nice.

1. For $F : [a, b] \to \mathbb{R}$,

$$\int_{F(\mathcal{R})} f(x) \, dx = \int_a^b f(F(u)) \, |F'(u)| \, du \,.$$

(Note that $F'(u) = \det F'(u)$ in this one-dimensional case.)

2. For $\mathcal{R} \subseteq \mathbb{R}^2$ and $F : \mathcal{R} \to \mathbb{R}^2$,

$$\int_{F(\mathcal{R})} f(x, y) \, dA = \int_{\mathcal{R}} f(F(u, v)) \, |\det F'(u, v)| \, dA \,.$$

3. For $\mathcal{R} \subseteq \mathbb{R}^3$ and $F : \mathcal{R} \to \mathbb{R}^3$,

$$\int_{F(\mathcal{R})} f(x, y, z) \, dV = \int_{\mathcal{R}} f(F(u, v, w)) \, |\det F'(u, v, w)| \, dV \; .$$

EXAMPLES:

6. In dimensions higher than one, these formulas usually work best from left to right. That is, one starts with the integral on the left-hand side, and then uses the right-hand side of the formula to obtain an integral that is easier to evaluate. In 1 dimension, however, as the reader will recall from single-variable calculus, the formula also works well from right to left. For example, integration of the function $\sin(u) \cos(u)$ is often carried by using the substitution $x = \sin u$ as follows:

$$\int_0^{\pi/2} \sin(u) \cos(u) \, du = \int_0^1 x \, dx = 1/2 \; .$$

In this example, $f(x) = x$, $F(u) = \sin(u)$, and $|\det F'(u)| = \cos(u)$. The reader should ponder what happens if the region of integration in this example is enlarged, so that F is no longer one-to-one, and $F'(u)$ is no longer nonnegative.

7. Let $\mathcal{B}$ be the region inside the ellipse $(x/a)^2 + (y/b)^2 = 1$. Then $\mathcal{B} = F(\mathcal{R})$, where

$$\mathcal{R} = \{(u, v) : 0 \le u \le 1 \text{ and } 0 \le v < 2\pi\} \; ,$$

and

$$(x, y) = F(u, v) = (au \cos v, bu \sin v) \; .$$

Note that as v varies in the interval $[0, 2\pi)$, $F(1, v)$ gives a parametrization of the ellipse. Note also the similarity to polar coordinates. The determinant of $F'(u, v)$ is abu. We can quickly compute the area of the elliptical region using the change of variables formula as follows:

$$\int_{\mathcal{B}} 1 \, dA = \int_0^{2\pi} \int_0^1 abu \, du \, dv = \pi ab \; .$$

The reader is asked in the exercises to integrate other functions over this region.

9.4D   CYLINDRICAL COORDINATES

In this section, we introduce one of the three most important coordinate systems in $\mathbb{R}^3$ (counting the rectangular coordinate system) and the corresponding change of variables formula. Both of the new coordinate systems are extensions of the polar coordinate system.

Let $(x, y, z) = F(u, v, w) = (u \cos v, u \sin v, w)$. Because of the similarity to polar coordinates, it is customary to use the letter r instead of u, and the letter θ instead of v. Since $z = w$ in this transformation, it is also customary to write z for w. This last notational convention does not properly keep the two coordinates systems separate, but it usually does not cause any confusion. Thus, F is a map from $r\theta z$-space to xyz-space, and

$$(x, y, z) = F(r, \theta, z) = (r \cos \theta, r \sin \theta, z) . \qquad (9.2)$$

The matrix of $F'(r, \theta, z)$ is

$$\begin{bmatrix} \cos \theta & -r \sin \theta & 0 \\ \sin \theta & r \cos \theta & 0 \\ 0 & 0 & 1 \end{bmatrix},$$

which is easily seen to have determinant r, just as with polar coordinates.

The change of variables formula is

$$\int_{F(\mathcal{R})} f(x, y, z) \, dV = \int_{\mathcal{R}} f(r \cos \theta, r \sin \theta, z) r \, dV .$$

Of course, we must restrict $\mathcal{R}$ so that F is one-to-one on $\mathcal{R}$. We usually accomplish this by only considering nonnegative values of r, and by restricting θ to some interval of length less than or equal to 2π, such as $[0, 2\pi)$, or $(-\pi, \pi]$. (Note that F is not one-to-one when $r = 0$, but this is all right, since the line $r = 0$ in the $r\theta$-plane has area 0.) There is no need to restrict the values of z. When a point in xyz-space is specified by giving values of r, θ, and z for which (9.2) holds, r, θ, and z are called the **cylindrical coordinates** of the point. Of course, these same numbers r, θ, and z are the rectangular coordinates of a corresponding point in $r\theta z$-space.

The cylindrical coordinate change of variables formula is most often useful when the region $F(\mathcal{R})$ in $\mathbb{R}^3$ has a base that looks like the type of region that would be amenable to the polar coordinate change of variables.

EXAMPLE:

8. Let $\mathcal{B}$ be the region in the positive octant of $\mathbb{R}^3$ that lies below the graph of the cone $z = \sqrt{x^2 + y^2}$, inside the cylinder $x^2 + y^2 = 4$, and is bounded by the planes $x = y$ and $y = 0$. To visualize this region, imagine that a conical hole has been bored into the top of a cylinder, and that the resulting figure is then divided by vertical cuts into 8 identical wedge-shaped pieces. The region $\mathcal{B}$ is one of those pieces.

The region $\mathcal{B}$ is the image, under the cylindrical coordinate transformation, of a certain region in $r\theta z$-space. How do we find that region? We first try to represent each of the boundary pieces of $\mathcal{B}$ as images of regions in $r\theta z$-space. This is accomplished by substituting $r \cos \theta$ and $r \sin \theta$ for x and y in the equations for these boundary pieces. For example, the surface $z = \sqrt{x^2 + y^2}$ is the image of the plane $z = r$ in $r\theta z$-space, since

$$\sqrt{x^2 + y^2} = \sqrt{r^2(\cos^2 \theta + \sin^2 \theta)} = r \ .$$

Similarly, the cylinder $x^2 + y^2 = 4$ is the image of the surface $r^2 = 4$, which is the same as the plane $r = 2$, since we are restricting r to be nonnegative. The plane $x = y$ is the image of $r \cos \theta = r \sin \theta$. This last equation is satisfied when $r = 0$ and also when $\theta = \pi/4$ and $\theta = 5\pi/4$. Since we are concerned only with the positive octant in xyz-space, both x and y must be nonnegative. Thus θ must be between 0 and $\pi/2$, ruling out $\theta = 5\pi/4$. So, the equation $r \cos \theta = r \sin \theta$ is equivalent to the pair of equations $r = 0$ and $\theta = \pi/4$. The plane $y = 0$ is the image of the pair of planes $r = 0, \theta = 0$. Of course the plane $z = 0$ in xyz-space is the image of the plane $z = 0$ in $r\theta z$-space.

We summarize: the region $\mathcal{B}$ is the image of the region in $r\theta z$-space bounded by the planes $r = 0$, $r = 2$, $\theta = 0$, $\theta = \pi/4$, $z = 0$, and $z = r$. Thus, under the usual restrictions on the function f,

$$\int_{\mathcal{B}} f(x, y, z) \, dV = \int_0^{\pi/4} \int_0^2 \int_0^r f(r \cos \theta, r \sin \theta, z) r \, dz \, dr \, d\theta \ .$$

For functions f that have some symmetry about the z-axis, the integral on the right-hand side is often quite easy to calculate. For example, suppose $f(x, y, z) = z \sin((x^2 + y^2)^2)$. Then

$$f(r \cos \theta, r \sin \theta, z) = z \sin(r^4) \ ,$$

so the integral on the right-hand side is evaluated as follows:

$$
\begin{aligned}
\int_0^{\pi/4} \int_0^2 \int_0^r z \sin(r^4) r \, dz \, dr \, d\theta &= \int_0^{\pi/4} \int_0^2 \frac{r^3}{2} \sin(r^4) \, dr \, d\theta \\
&= \int_0^{\pi/4} \frac{-\cos(16) + 1}{8} \, d\theta \\
&= \frac{\pi(1 - \cos(16))}{32} \ .
\end{aligned}
$$

9.4E SPHERICAL COORDINATES

For this transformation, it is customary to use (ρ, θ, φ) instead of (u, v, w). Define F by

$$(x, y, z) = F(\rho, \theta, \varphi) = (\rho \cos \theta \sin \varphi, \rho \sin \theta \sin \varphi, \rho \cos \varphi) \ .$$

Note that

$$\sqrt{x^2 + y^2 + z^2} = \sqrt{\rho^2[\sin^2(\varphi)(\cos^2(\theta) + \sin^2(\theta)) + \cos^2(\varphi)]}$$

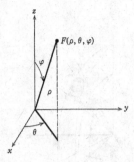

FIGURE 9.45. The geometric relationship between (x, y, z) and (ρ, θ, φ)

$$= \rho\sqrt{\cos^2(\varphi) + \sin^2(\varphi)}$$
$$= \rho .$$

Thus, ρ corresponds to the distance from the origin to the point (x, y, z). As with polar coordinates, θ is the angle between the vector $(x, y, 0)$ and the positive x-axis. A little trigonometry shows that φ is the angle between the vector (x, y, z) and the positive z-axis. The numbers ρ, θ, and φ are called the **spherical coordinates** of the corresponding point (x, y, z) in xyz-space. See Figure 9.45. We remind the reader that although such a figure aids in the visualization of the relationship $(x, y, z) = F(\rho, \theta, \varphi)$, it is important to keep separate mental pictures of xyz-space and $\rho\theta\varphi$-space, and that ρ, θ, and φ are the rectangular coordinates of points in $\rho\theta\varphi$-space.

The matrix of $F'(\rho, \theta, \varphi)$ is

$$\begin{bmatrix} \cos\theta\sin\varphi & -\rho\sin\theta\sin\varphi & \rho\cos\theta\cos\varphi \\ \sin\theta\sin\varphi & \rho\cos\theta\sin\varphi & \rho\sin\theta\cos\varphi \\ \cos\varphi & 0 & -\rho\sin\varphi \end{bmatrix} .$$

The determinant of this matrix is

$$-\rho^2 \sin\varphi .$$

The reader should verify this formula.

The change of variables formula for spherical coordinates is

$$\int_{F(\mathcal{R})} f(x, y, z)\, dV = \int_{\mathcal{R}} f(\rho\cos\theta\sin\varphi, \rho\sin\theta\sin\varphi, \rho\cos\varphi)\rho^2 \sin\varphi\, dV .$$

In order to keep F one-to-one on $\mathcal{R}$, we usually restrict ρ to be nonnegative, θ to lie in some interval of length less than or equal to 2π, and φ to lie in the interval $[0, \pi]$ (as with polar coordinates, there is a lack of one-to-oneness for certain values of ρ and φ which we choose not to worry about). This last restriction on φ ensures that $\rho^2 \sin\varphi$ is nonnegative, eliminating the need for absolute value signs in the change of variables formula.

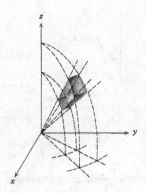

FIGURE 9.46. A spherical rectangular region

The image in xyz-space of a typical rectangular region in $\rho\theta\varphi$-space is a wedge-shaped portion cut from a ball centered at the origin. See Figure 9.46. The spherical coordinates transformation works best for integrals over such regions.

EXAMPLE:

9. Let $\mathcal{B}$ be the region in the positive octant in $\mathbb{R}^3$ which is inside the sphere of radius 2 centered at the origin, below the surface $z = \sqrt{x^2 + y^2}$, and between the planes $y = 0$ and $x = y$. This region is similar to the region in Example 8, except that it is bounded on the sides by part of a sphere rather than a cylinder.

We express each of the boundary equations in spherical coordinates. The sphere of radius 2 centered at the origin in xyz-space is the image of the plane $\rho = 2$ in $\rho\theta\varphi$-space. This fact may be derived either from the obvious appeal to geometric intuition, or by the more mechanical method of converting the equation $x^2 + y^2 + z^2 = 4$ into spherical coordinates. The surface $z = \sqrt{x^2 + y^2}$ is the image of the surface

$$\rho \cos\varphi = \sqrt{\rho^2 (\sin^2\varphi)(\cos^2\theta + \sin^2\theta)} = \rho \sin\varphi \, .$$

Thus, either $\rho = 0$, or

$$\cos\varphi = \sin\varphi \, .$$

Since φ is restricted between 0 and π, this last equation is equivalent to $\varphi = \pi/4$, a plane in $\rho\theta\varphi$-space. (It is instructive to compare this result with the cylindrical coordinates version found in Example 8.) The bottom of the region $\mathcal{B}$ is the plane $z = 0$, which is the image of the surface $\rho \cos\varphi = 0$. Simplifying, we obtain $\rho = 0$ and $\varphi = \pi/2$. The plane $y = 0$ is the image of the plane $\theta = 0$, as in cylindrical coordinates. Finally, the plane $x = y$ is the image of the surface

$$\rho \cos \theta \sin \varphi = \rho \sin \theta \sin \varphi \,,$$

which simplies to the three planes $\rho = 0$, $\theta = \pi/4$, and $\varphi = 0$. We can rule out the plane $\varphi = 0$ because we already have the restriction that $\varphi \geq \pi/4$, which comes from the cone-shaped surface on top of B.

To summarize, the region B is the image of a region R in $\rho\theta\varphi$-space whose boundary is made up of the planes $\rho = 0, \rho = 2, \theta = 0, \theta = \pi/4, \varphi = \pi/4, \varphi = \pi/2$. Thus, R is a rectangular region in $\rho\theta\varphi$-space. For a suitable function f defined on B the change of variables formula for spherical coordinates tells us that

$$\int_B f(x, y, z) \, dV =$$

$$\int_0^{\pi/4} \int_{\pi/4}^{\pi/2} \int_0^2 f(\rho \cos \theta \sin \varphi, \rho \sin \theta \sin \varphi, \rho \cos \varphi) \rho^2 \sin \varphi \, d\rho \, d\varphi \, d\theta.$$

If f is a function that has some spherical symmetry, then the integral on the right-hand side is often easy to compute. As a simple example, consider $f(x, y, z) = x^2 + y^2 + z^2$. In this case, the integrand of the integral on the right-hand side (including the determinant factor $\rho^2 \sin \varphi$) becomes $\rho^4 \sin \varphi$. The resulting triple iterated integral is easily calculated. The answer is $4\pi\sqrt{2}/5$.

Exercises

1. Calculate the volume of the solid in $\mathbb{R}^3$ that lies above the parallelogram region in the xy-plane with vertices at $(1, 3, 0)$, $(-1, 8, 0)$, $(5, 9, 0)$, $(7, 4, 0)$ and below the plane $z = x + 2y + 3$.

2. Use the methods of this section to evaluate the last integral in Exercise 2 in Section 9.2.

3. Let R be a region in $\mathbb{R}^2$ that is symmetric about the origin (that is, $(x, y) \in R \Leftrightarrow (-x, -y) \in R$). Use a linear change of variables to show that for any bounded continuous function defined on R,

$$\int_R f(x, y) \, dA = \int_R f(-x, -y) \, dA \,.$$

4. The equation

$$\frac{5}{8}x^2 + \frac{3\sqrt{3}}{8}xy + y^2 = 1$$

represents an ellipse in $\mathbb{R}^2$ whose axes are rotated $\pi/3$ radians counterclockwise with respect to the coordinate axes. Let B be the region inside this ellipse.

(a) Find a map F from the uv-plane to the xy-plane such that B is the image under F of an ellipse $\mathcal{R}$ in the uv-plane, and such that the axes of $\mathcal{R}$ are parallel to the u- and v-axes.

(b) Calculate the integral of the function $f(x, y) = x^2 + y^2$ over the region B.

5. Find *one* double iterated integral that represents the area of the region in $\mathbb{R}^2$ which contains the origin and is bounded by the hyperbola $xy = 1$ and the lines $y = x - 1$ and $y = x + 1$. You need not evaluate the iterated integral. *Hint:* Use a rotation so that the area can be represented by only one double iterated integral.

6. Find the area of the part of the cardioidial region in Example 5 of Section 9.4b that lies outside the circle $x^2 + y^2 = 1$.

7. Let $\mathcal{R}$ be the region in the $r\theta$-plane that is bounded by the lines $r = 0, r = \theta$ and $\theta = 2\pi$.

(a) Draw $\mathcal{R}$.

(b) Draw the image $F(\mathcal{R})$ in the xy-plane, where F is the polar coordinates transformation.

(c) Find the area of $F(\mathcal{R})$.

(d) Find the area of the part of $F(\mathcal{R})$ that lies outside the circle $x^2 + y^2 = 1$.

(e) Calculate the integral of $f(x, y) = x^2 + y^2$ over the region $F(\mathcal{R})$.

(f) Replace the line $\theta = 2\pi$ by the line $\theta = 4\pi$. Again draw $\mathcal{R}$ and the image $F(\mathcal{R})$.

(g) How would you find the area of the region $F(\mathcal{R})$ in part (f)? Note that F is not one-to-one on $\mathcal{R}$ in part (f).

8. Let $\mathcal{R}$ be the region in the $r\theta$-plane bounded by the lines $r = 0, \theta = 0, \theta = \pi/2$ and the curve $r = \sin 2\theta$.

(a) Draw $\mathcal{R}$.

(b) Draw the image $F(\mathcal{R})$ in the xy-plane, where F is the polar coordinates transformation.

(c) Find the area of $F(\mathcal{R})$.

(d) Calculate the integral over $F(\mathcal{R})$ of the function $f(x, y) = x$.

(e) Replace the line $\theta = \pi/2$ by the line $\theta = 2\pi$. Again draw $\mathcal{R}$ and the image $F(\mathcal{R})$. Note that r is negative for certain values of θ.

(f) How would you find the area of the region $F(\mathcal{R})$ in part (e)? Is F one-to-one on $\mathcal{R}$?

9. Find the volume of the solid in $\mathbb{R}^3$ bounded below by the plane $z = 0$, on the sides by the cylinder $x^2 + y^2 = 1$, and on top by the sphere $x^2 + y^2 + z^2 = 3$, using

(a) polar coordinates

(b) cylindrial coordinates

(c) spherical coordinates.

10. Use the "elliptical coordinates" from Example 7 of Section 9.4c, to set up a double integral for the volume of the ellipsoidal region $(x/a)^2 + (y/b)^2 + (z/c)^2 \leq 1$. Evaluate the integral.

11. A thin plate has the shape of the elliptical region $(x/a)^2 + (y/b)^2 \leq 1$. The mass per area density on this plate at the point (x, y) equals $(x/a)^2 + (y/b)^2$. Find the mass of the plate.

12. Let
$$(x, y) = F(u, v) = (au \cosh v, bu \sinh v) ,$$
where a and b are positive scalars.

   (a) Show that the image under F of the line $u = 1$ is the hyperbola $(x/a)^2 - (y/b)^2 = 1$.

   (b) Draw in the xy-plane the images under F of the lines $u = c$ for $c = 1$, $c = 1/2$, and $c = 2$. Note that all these images are hyperbolas, and that they all have the same asymptotes. Determine these asymptotes.

   (c) Show that the images in the xy-plane under F of the lines $v = c$ are also lines. Draw these lines for the cases $v = 0$, $v = 1$, and $v = -1$.

   (d) Determine what happens to the image under F in the xy-plane of the line $v = c$ as c approaches ∞ and also as c approaches $-\infty$.

   (e) Draw the image in the xy-plane under F of a "typical" rectangular region $\mathcal{R}$ in the uv-plane.

   (f) What is the approximate ratio between the area of a small rectangular region $\mathcal{R}$ in the uv-plane and the area of its image $F(\mathcal{R})$ in the xy-plane?

   (g) Let $\mathcal{R}$ be the region in the uv-plane that is bounded by the lines $u = 0$, $u = 1$, $v = -1$, and $v = 1$. Draw $F(\mathcal{R})$ in the xy-plane and find its area.

   (h) Let $\mathcal{B}$ be the region in the first quadrant of the xy-plane that is bounded between the hyperbola $(x/a)^2 - (y/b)^2 = 1$ and the asymptote of that hyperbola that has positive slope. Show that $\mathcal{B}$ has infinite area.

   (i) Find the integral over $\mathcal{B}$ of the function $f(x, y) = \sqrt{(x/a)^2 - (y/b)^2}$.

13. Prove the fact alluded to in the discussion following Theorem 3.7.6 that if $T(X) = L(X) + Y$ is an affine map from $\mathbb{R}^3$ to $\mathbb{R}^3$ and if $\mathcal{R}$ is a region in $\mathbb{R}^3$ with volume V, then $T(\mathcal{R})$ has volume $V| \det L|$.

14. Let F be the cylindrical coordinates transformation from $r\theta z$-space to xyz-space. For each of the following points (x, y, z), find a corresponding point (r, θ, z) such that $(x, y, z) = F(r, \theta, z)$.

   (a) $(1, -1, 5)$

   (b) $(0, 2, -1)$

   (c) $(0, 0, 0)$

   (d) $(1, -2, 3)$

15. Let F be the cylindrical coordinates transformation from $r\theta z$-space to xyz-space. For each of the following points (r, θ, z), find a corresponding point (x, y, z) such that $(x, y, z) = F(r, \theta, z)$.

    (a) $(1, \pi, -2)$

    (b) $(0, 5, 10)$

    (c) $(6, 10\pi/3, 1)$

    (d) $(1, -2, 3)$

16. Let F be the cylindrical coordinates transformation from $r\theta z$-space to xyz-space. Draw the image under F in xyz-space of a "typical" rectangular box $\mathcal{R}$ in $r\theta z$-space.

17. Let F be the cylindrical coordinates transformation from $r\theta z$-space to xyz-space. For each of the following surfaces $\mathcal{B}$ in xyz-space, find a corresponding surface $\mathcal{R}$ in $r\theta z$-space such that $\mathcal{B} = F(\mathcal{R})$ (that is, find an equation for such a surface). Wherever possible, draw or give a description of the surface $\mathcal{R}$.

    (a) the sphere $x^2 + y^2 + z^2 = 9$

    (b) the plane $z = x + 2$

    (c) the plane $y = ax$, where a is a scalar

    (d) the plane $y = ax + b$, where a and b are scalars

    (e) the surface $z^3 + x + y = 5$.

18. Let F be the cylindrical coordinate transformation from $r\theta z$-space to xyz-space. For each of the following surfaces $\mathcal{R}$ in $r\theta z$-space, find a corresponding surface $\mathcal{B}$ in xyz-space such that $\mathcal{B} = F(\mathcal{R})$ (that is, find an equation for such a surface). Wherever possible, draw or describe the surface $\mathcal{B}$.

    (a) the plane $\theta = c$, where c is a scalar between 0 and 2π

    (b) the plane $z = c$, where c is any scalar

    (c) the plane $r = c$, where c is any nonnegative scalar (careful with the case $c = 0$)

    (d) the surface $z = 3r \cos \theta - 1$

    (e) the surface $z = \cos 2\theta$

    (f) the plane $z = \theta$.

19. Let $\mathcal{R}$ be the region in $r\theta z$-space bounded by the planes $r = 1$, $\theta = 0$, $\theta = \pi$, and $z = 0$ and by the surfaces $z = r^2$ and $r = 1 + \sin \theta$. Let F be the cylindrical coordinate transformation from $r\theta z$-space to xyz-space.

    (a) Draw a picture of $\mathcal{R}$ in $r\theta z$-space.

    (b) Draw a picture of (or describe in words) the image $F(\mathcal{R})$ in xyz-space.

    (c) Find the volume of $F(\mathcal{R})$.

    (d) Find the integral of the function $f(x, y, z) = x + z$ over the region $F(\mathcal{R})$.

20. The mass density per unit volume at any point (x, y, z) of a certain solid in $\mathbb{R}^3$ is equal to the distance from that point to the z-axis. The solid is bounded above by the surface $z = 4 - x^2 - y^2$, below by the surface $z = y$, and on the sides by the cylinder $x^2 + (y + 1)^2 = 1$. Set up triple iterated integrals for the mass of the solid, with the appropriate limits of integration, in both xyz- and $r\theta z$-coordinates. Unless you want lots of practice in the techniques of finding anti-derivatives, do not bother to evaluate either integral.

21. Find the volume of one of the two congruent solids in $\mathbb{R}^3$ bounded by the surfaces $z^2 - x^2 - y^2 = 1$, $z^2 = x^2 + y^2$ and $x^2 + y^2 = 1$.

22. Two circular cylindrical regions of radius 1 are located so that their axes intersect at right angles. Find the volume of the intersection of the two cylinders. *Hint:* Let the axes of the two cylinders be the z- and x-axes in $\mathbb{R}^3$.

23. Let F be the spherical coordinates transformation from $\rho\theta\varphi$-space to xyz-space. For each of the following points (x, y, z), find a corresponding point (ρ, θ, φ) such that $(x, y, z) = F(\rho, \theta, \varphi)$.

   (a) $(1, -1, 1)$

   (b) $(0, 2, -2\sqrt{3})$

   (c) $(0, 0, 0)$

   (d) $(1, -2, 3)$ (this one is somewhat messy).

24. Let F be the spherical coordinates transformation from $\rho\theta\varphi$-space to xyz-space. For each of the following points (ρ, θ, φ), find a corresponding point (x, y, z) such that $(x, y, z) = F(\rho, \theta, \varphi)$.

   (a) $(1, \pi, -2)$

   (b) $(0, 5, 10)$

   (c) $(6, 10\pi/3, \pi)$

   (d) $(1, -2, 3)$.

25. Let F be the spherical coordinates transformation from $\rho\theta\varphi$-space to xyz-space. Draw the images under F in xyz-space of a several "typical" rectangular boxes $\mathcal{R}$ in $\rho\theta\varphi$-space. In particular, consider cases in which the dimension in the φ direction of the box $\mathcal{R}$ is larger than $\pi/2$, and the dimension in the θ direction of the box $\mathcal{R}$ is larger than π.

26. Let F be the spherical coordinates transformation from $\rho\theta\varphi$-space to xyz-space. For each of the following surfaces $\mathcal{B}$ in xyz-space, find a corresponding surface $\mathcal{R}$ in $\rho\theta\varphi$-space such that $\mathcal{B} = F(\mathcal{R})$ (that is, find an equation for such a surface). Wherever possible, draw or give a description of the surface $\mathcal{R}$.

   (a) the sphere $x^2 + y^2 + z^2 = 9$

   (b) the plane $z = x + 2$

   (c) the plane $y = ax$, where a is a positive scalar

   (d) the cone $z = a\sqrt{x^2 + y^2}$, where a is a scalar

(e) the paraboloid $z = x^2 + y^2$.

27. Let F be the spherical coordinates transformation from $\rho\theta\varphi$-space to xyz-space. For each of the following surfaces $\mathcal{R}$ in $\rho\theta\varphi$-space, find a corresponding surface $\mathcal{B}$ in xyz-space such that $\mathcal{B} = F(\mathcal{R})$ (that is, find an equation for such a surface). Wherever possible, draw or describe the surface $\mathcal{B}$.

   (a) the plane $\theta = c$, where c is a scalar between 0 and 2π

   (b) the plane $\varphi = c$, where c is any scalar between 0 and π

   (c) the plane $\rho = c$, where c is any nonnegative scalar (careful with the case $c = 0$)

   (d) the plane $\varphi = \theta$.

28. Use spherical coordinates to describe the sections of an orange. Imagine that the orange is spherical, and that it has six identically shaped sections.

29. Find the volume of the intersection of the ball $\bar{B}((0,0,0);3)$ and the set of those (x, y, z) for which $z^2 \leq x^2 + y^2$. *Hint:* For drawing pictures it is useful to notice that this last set is the complement of the interior of double conical region.

30. Evaluate the integral of the function

$$f(x, y, z) = \frac{1}{\sqrt{x^2 + y^2 + z^2}}$$

over the unit ball centered at the origin in $\mathbb{R}^3$. (Note that this integral is improper, because the function f is not defined at the origin. This impropriety "disappears" when the integral is transformed to spherical coordinates.)

31. Let α be a real number. Evaluate, in terms of α, the following integrals, whether finite or equal to $+\infty$:

   (a) $\int_{B(0;1)} |x|^\alpha dx$

   (b) $\iint_{B((0,0);1)} \|X\|^\alpha dA$

   (c) $\iiint_{B((0,0,0);1)} \|X\|^\alpha dV$.

32. Do the problem obtained by replacing the balls in the preceding exercise by their complements.

33. Do the problem obtained by replacing the balls in Exercise 31 by $\mathbb{R}$, $\mathbb{R}^2$, and $\mathbb{R}^3$, respectively.

34. The determinant of the derivative of the spherical coordinates transformation F is negative for φ between 0 and π. Discuss the significance of this fact, using appropriate illustrations. *Hint:* Draw three line segments of unit length in $\rho\theta\varphi$-space, emanating from the point $(1, \pi/2, 0)$, oriented in the same directions as the positive ρ, θ and φ coordinate axes. Draw the images of these line segments in xyz-space. Why do you think we suggested that you use the point $(1, \pi/2, 0)$ rather than the origin?

9.5 Surface Area

Recall from Section 8.3 that a surface $\mathcal{B}$ in $\mathbb{R}^n$ is the image of a function $G : \mathcal{R} \to \mathbb{R}^n$, where $\mathcal{R}$ is a region in $\mathbb{R}^2$. Thus $\mathcal{B} = G(\mathcal{R})$. In this section, we will develop a formula for the area of $\mathcal{B}$ in terms of the integral of a certain function over the region $\mathcal{R}$.

We first consider the case in which the target space of G is $\mathbb{R}^3$ and $\mathcal{R}$ is a rectangular region in $\mathbb{R}^2$ with sides parallel to the coordinate axes and centered at a point (x_0, y_0). Let Δx and Δy be the lengths of the sides of this rectangular region. We think of Δx and Δy as being small positive numbers. Assume that G is differentiable on $\mathcal{R}$, and let T be the best affine linear approximation to G near (x_0, y_0):

$$T(x_0, y_0) = G'(x_0, y_0)(x - x_0, y - y_0) + G(x_0, y_0) \, .$$

It seems reasonable that if $\mathcal{R}$ is small enough, then however we define surface area, the surface area of $G(\mathcal{R})$ should be approximately the same as the surface area of $T(\mathcal{R})$, which should be the same as the surface area of $L(\mathcal{R})$, where L is the linear map from $\mathbb{R}^2$ to $\mathbb{R}^3$ defined by

$$L(x_0, y_0) = G'(x_0, y_0)(x, y) \, .$$

Since L is linear, $L(\mathcal{R})$ is the interior of the parallelogram region in $\mathbb{R}^3$ generated by the vectors $L(\Delta x, 0)$ and $L(0, \Delta y)$. Applying the formula in Section 2.4 for the area of a parallelogram region in $\mathbb{R}^3$, we obtain the following formula:

$$\text{surface area of } G(\mathcal{R}) \approx \|L(\Delta x, 0) \times L(0, \Delta y)\| \, .$$

We wish to rewrite the right-hand side of this last formula in terms of G. The vector $L(\Delta x, 0)$ equals Δx times the vector represented by the first column of the 3-by-2 matrix $[G'(x_0, y_0)]$. That is,

$$L(\Delta x, 0) = \Delta x \, (\frac{\partial g_1}{\partial x}(x_0, y_0), \ \frac{\partial g_2}{\partial x}(x_0, y_0), \ \frac{\partial g_3}{\partial x}(x_0, y_0)) \, ,$$

where g_1, g_2, g_3 are the coordinate functions of G. Similarly,

$$L(0, \Delta y) = \Delta y \, (\frac{\partial g_1}{\partial y}(x_0, y_0), \ \frac{\partial g_2}{\partial y}(x_0, y_0), \ \frac{\partial g_3}{\partial y}(x_0, y_0)) \, .$$

As in Section 8.3 we write

$$\frac{\partial G}{\partial x}(x_0, y_0) = (\frac{\partial g_1}{\partial x}(x_0, y_0), \ \frac{\partial g_2}{\partial x}(x_0, y_0), \ \frac{\partial g_3}{\partial x}(x_0, y_0))$$

and

$$\frac{\partial G}{\partial y}(x_0, y_0) = (\frac{\partial g_1}{\partial y}(x_0, y_0), \ \frac{\partial g_2}{\partial y}(x_0, y_0), \ \frac{\partial g_3}{\partial y}(x_0, y_0)) \, .$$

Then we have

$$\text{surface area of } G(\mathcal{R}) \approx \|\frac{\partial G}{\partial x}(x_0, y_0) \times \frac{\partial G}{\partial y}(x_0, y_0)\|\Delta x \Delta y .$$

Now we suppose that the region $\mathcal{R}$ is approximated by the union of a collection of small disjoint rectangular regions $\mathcal{R}_i$ centered at (x_i, y_i) and having sides of length Δx_i and Δy_i parallel to the coordinate axes. If the rectangular regions $\mathcal{R}_i$ are sufficiently small, our previous formula leads us to expect that

$$\text{surface area of } G(\mathcal{R}) \approx \sum_i \|\frac{\partial G}{\partial x}(x_i, y_i) \times \frac{\partial G}{\partial y}(x_i, y_i)\|\Delta x_i \Delta y_i .$$

The right-hand side of this expression is a Riemann sum for the integral

$$\int_{\mathcal{R}} \|D_1 G \times D_2 G\| \, dA ,$$

where we have written $D_1 G$ for $\partial G/\partial x$ and $D_2 G$ for $\partial G/\partial y$. As we partition $\mathcal{R}$ into smaller and smaller rectangular regions, the corresponding Riemann sums converge to the integral. As might be expected, they also converge to the surface area of $G(\mathcal{R})$. This fact is expressed in the following formula, which is valid for regular regions $\mathcal{R} \subseteq \mathbb{R}^2$ and functions $G : \mathbb{R}^2 \to \mathbb{R}^3$ which are one-to-one and differentiable on $\mathcal{R}$:

$$\text{surface area of } G(\mathcal{R}) = \int_{\mathcal{R}} \|D_1 G \times D_2 G\| \, dA .$$

The precise definition of surface area, along with the proof of this formula, may be found in a book on advanced calculus.

If G is $\mathbb{R}^n$-valued for $n \neq 3$, then the formula just given does not make sense, since the cross product is only defined in $\mathbb{R}^3$. We now consider the general case in which $n \geq 2$. A careful examination of the heuristics leading to the formula for the case $n = 3$ shows that we only used the special properties of $\mathbb{R}^3$ when we obtained the formula for the area of the parallelogram region generated by the vectors $L(\Delta x, 0)$ and $L(0, \Delta y)$. In general, we may compute this area without using the cross product. It is

$$\|L(\Delta x, 0)\| \, \|L(0, \Delta y)\| \, |\sin \theta| ,$$

where θ is the angle between the two vectors (see Section 2.4). The quantity $|\sin \theta|$ may be computed in terms of the inner product of the vectors $\partial G/\partial x$ and $\partial G/\partial y$ as follows:

$$
\begin{aligned}
|\sin \theta| &= \sqrt{1 - \cos^2 \theta} \\
&= \sqrt{1 - \left(\frac{\langle \frac{\partial G}{\partial x}(x_0, y_0), \frac{\partial G}{\partial y}(x_0, y_0) \rangle}{\|\frac{\partial G}{\partial x}(x_0, y_0)\| \|\frac{\partial G}{\partial y}(x_0, y_0)\|} \right)^2} .
\end{aligned}
$$

After combining these formulas and simplifying, we obtain the following expression for the area of the parallelogram region:

$$\sqrt{\|\frac{\partial G}{\partial x}(x_0, y_0)\|^2 \|\frac{\partial G}{\partial y}(x_0, y_0)\|^2 - \langle\frac{\partial G}{\partial x}(x_0, y_0), \frac{\partial G}{\partial y}(x_0, y_0)\rangle^2}\, \Delta x \Delta y \,.$$

This is the expression that replaces

$$\|\frac{\partial G}{\partial x}(x_0, y_0) \times \frac{\partial G}{\partial y}(x_0, y_0)\|\Delta x \Delta y$$

in our previous discussion. The rest of the heuristic justification is unchanged. Thus, we have the following formula, which is valid for regular regions $\mathcal{R} \subseteq \mathbb{R}^2$ and one-to-one differentiable functions $G : \mathcal{R} \to \mathbb{R}^n$, $n \geq 2$:

$$\text{surface area of } G(\mathcal{R}) = \int_{\mathcal{R}} \sqrt{\|D_1 G\|^2 \|D_2 G\|^2 - \langle D_1 G, D_2 G\rangle^2}\, dA \,.$$

EXAMPLES:

1. Let $0 < b < a$. Let us find the area of the torus defined parametrically by

    $$G(\theta, \varphi) = ((a + b\cos\varphi)\cos\theta \,, (a + b\cos\varphi)\sin\theta \,, b\sin\varphi) \,,$$

    for $0 \leq \theta < 2\pi$ and $0 \leq \varphi < 2\pi$. (By using the condition "$< 2\pi$" rather than "$\leq 2\pi$" we have arranged for G to be one-to-one as needed in order to apply the methods of this section. Of course, when we integrate over the square region $[0, 2\pi) \times [0, 2\pi)$ we will get the same answer as that obtained by integrating over the square region $[0, 2\pi] \times [0, 2\pi]$.)

    The derivatives of G, first viewed as a parametrized curve with respect to the variable θ (with φ fixed) and then as a parametrized curve with respect to the variable φ are

    $$\begin{aligned} D_1 G(\theta, \varphi) &= (-(a + b\cos\varphi)\sin\theta \,, (a + b\cos\varphi)\cos\theta \,, 0) \,, \\ D_2 G(\theta, \varphi) &= ((-b\sin\varphi)\cos\theta \,, (-b\sin\varphi)\sin\theta \,, b\cos\varphi) \,. \end{aligned}$$

    The cross product of these two vectors equals

    $$(b(a + b\cos\varphi)\cos\theta\cos\varphi, b(a + b\cos\varphi)\sin\theta\cos\varphi, b(a + b\cos\varphi)\sin\varphi),$$

    the norm of which equals $b(a + b\cos\varphi)$. Thus the surface area of the torus equals

    $$\int_0^{2\pi}\int_0^{2\pi} b(a + b\cos\varphi)\, d\theta\, d\varphi = 2\pi b \int_0^{2\pi} (a + b\cos\varphi)\, d\varphi = 4\pi^2 ab \,.$$

    The reader might compare the preceding calculation with the standard method for treating solids of revolution described in many single-variable calculus texts.

2. Let us calculate the area of the surface given by

$$z = \sqrt{4x^2 + y^2} \text{ and } z \leq 2 .$$

(The graph of $z = \sqrt{4x^2 + y^2}$ is called an "elliptical cone".) We can view the surface of interest as the image of the function F into $\mathbb{R}^3$ given by

$$F(x, y) = (x, y, \sqrt{4x^2 + y^2})$$

with

$$\mathcal{D}(F) = \{(x, y) : 4x^2 + y^2 \leq 4\} .$$

The two derivatives of F with respect to each of its two variables (with the other variable being held constant) are

$$\frac{\partial F}{\partial x} = \left(1, 0, \frac{4x}{\sqrt{4x^2 + y^2}}\right) ,$$

$$\frac{\partial F}{\partial y} = \left(0, 1, \frac{y}{\sqrt{4x^2 + y^2}}\right) .$$

The cross product of these two vectors is

$$\left(-\frac{4x}{\sqrt{4x^2 + y^2}}, -\frac{y}{\sqrt{4x^2 + y^2}}, 1\right) ,$$

the norm of which equals

$$\sqrt{\frac{20x^2 + 2y^2}{4x^2 + y^2}} . \tag{9.3}$$

To integrate this function over $\mathcal{D}(F)$ we use an elliptical change of variables

$$(x, y) = (\frac{r}{2} \cos \theta, r \sin \theta) .$$

The function (9.3) becomes $\sqrt{5 \cos^2 \theta + 2 \sin^2 \theta}$, the absolute value of the determinant of the Jacobian equals $r/2$, and the region of integration in the $r\theta$-plane is the square region defined by $0 \leq r \leq 2$ and $0 \leq \theta < 2\pi$. Thus the surface area of the elliptical cone truncated at the height 2 equals

$$\int_0^{2\pi} \int_0^2 \frac{r}{2} \sqrt{5 \cos^2 \theta + 2 \sin^2 \theta} \, dr \, d\theta = \int_0^{2\pi} \sqrt{5 \cos^2 \theta + 2 \sin^2 \theta} \, d\theta$$

$$= 4\sqrt{5} \int_0^{\pi/2} \sqrt{1 - \frac{3}{5} \sin^2 \theta} \, d\theta .$$

This last single-variable integral, which is called a "complete elliptic integral of the second kind", cannot be simplified. However, this type of integral arises so often that extensive tables of numerical approximations have been published. Such a table gives an approximation 1.298428 for the integral and thus an approximation 11.6135 for the surface area of the truncated conical surface that we are treating.

3. Let us calculate the area of the surface S in $\mathbb{R}^4$ described parametrically by
$$G(x, y) = (x^2 + y^2, \, x^2 - y^2, \, x + y, \, x - y)$$
with
$$\mathcal{D}(G) = \{(x, y) : 0 \le x \le 1 \text{ and } 0 \le y \le 1\} \, .$$

Focusing on only the last two coordinates of G and noting that there is a one-to-one linear map described there, we conclude that F is one-to-one. Therefore the methods of this section apply for calculating the area of the surface S. The partial derivatives of G equal

$$\begin{aligned} D_1 G(x, y) &= (2x, 2x, 1, 1) \, , \\ D_2 G(x, y) &= (2y, -2y, 1, -1) \, . \end{aligned}$$

The inner product of these two vectors is 0, so the sine of the angle between them is 1. Thus the surface area of S is the integral over $\mathcal{D}(F)$ of the product of their norms. Hence the area of S equals

$$\begin{aligned} \int_0^1 \int_0^1 & \sqrt{2 + 8x^2} \sqrt{2 + 8y^2} \, dx \, dy \\ &= \left(\int_0^1 \sqrt{2 + 8x^2} \, dx \right) \left(\int_0^1 \sqrt{2 + 8y^2} \, dy \right) \\ &= 2 \left(\int_0^1 \sqrt{1 + 4x^2} \, dx \right)^2 \, . \end{aligned}$$

The last single-variable integral can be treated by the substitution $2x = \tan \theta$ giving rise to an integral of $\sec^3 \theta$. We omit the details.

Exercises
Some calculations of surface area.

1. Calculate the area of the circular cone $z = \sqrt{x^2 + y^2}$ truncated by the plane

   (a) $z = 3$

   (b) $z = c$, where c is a fixed positive constant

   (c) $2z = x + 2$, in which case a single-variable integral is a satisfactory answer (but carry it through to an exact answer if you want some practice in antidifferentiation).

2. Calculate the area of the circular parabola $z = 3x^2 + 3y^2$ truncated by the plane

   (a) $z = 4$

   (b) $z = c$, where c is a fixed positive constant.

3. For each of part (b) of Exercise 1 and part (b) of Exercise 2 find positive numbers γ and l such that $f(c)/c^\gamma \to l$ as $c \to \infty$, where $f(c)$ denotes the answer to the appropriate part (b). But before carrying out your calculations make an astute guess of the value of γ.

4. Let $G(x, y) = x(1, 3, -4) + y(-2, 3, 0) + (1, 0, 3)$ with $\mathcal{D}(G)$ equal to a set having area α. Use a formula from this section to calculate the area of $\mathcal{I}(G)$ in terms of α.

5. Let $G(x, y) = xX + yY + Z$, where X, Y, and Z are members of $\mathbb{R}^q$, with $\mathcal{D}(G)$ equal to a set having area α. Calculate the area of $\mathcal{I}(G)$ in terms of α, X, Y, and Z.

6. Calculate the area of that part of the torus of Example 1 that one can view from a point in the plane $z = 0$ "infinitely far away" from the torus.

7. Calculate the area of that protion of the sphere $S((0, 0, 0); a)$ that is inside a cone with vertex $(0, 0, 0)$ and angle ψ. *Hint:* Parametrize the sphere as $(a \cos \theta \sin \varphi, a \sin \theta \sin \varphi, a \cos \varphi)$ and orient the cone so that its axis lies on the z-axis.

8. Find a single-variable integral which equals the area of the *graph* in $\mathbb{R}^5$ of the function G of Example 1.

9. Find a single-variable integral which equals the area of the elliptical cone $z = \sqrt{9x^2 + y^2}$ truncated by the plane $z = 6$. Then use a table of elliptical integrals to find an approximation of the area.

10. Calculate the area of the surface in $\mathbb{R}^4$ which is the image of the function

$$(\cos(x + y),\ \sin(x + y),\ \cos x,\ \sin x)$$

for $0 \le x \le \frac{\pi}{2}$ and $0 \le y \le \frac{\pi}{2}$.

10

Vector Integrals and Stokes' Theorem

10.0   Introduction

This chapter is about "vector integration". For this new kind of integration, both the regions of integration and the integrands are of a different type from those considered in Chapter 9. We will learn to calculate integrals for which the region of integration is a curve in $\mathbb{R}^q$ (so-called "line integrals") and integrals for which the region of integration is a surface in $\mathbb{R}^3$ ("surface integrals"). The integrands will be of the form $\langle F(X), V(X)\rangle$, where F and V are vector-valued functions.

In this introductory section, we will describe briefly the kinds of curves and surfaces that we will use for regions of integration in our vector integrals. For line integrals, the region of integration will be a "smooth oriented curve" $\mathcal{C}$. See Figure 10.1. The adjective "smooth" means that there is a tangent line at each point on the curve, except of course at its endpoints. The word "oriented" means that we will choose a direction along the curve, as indicated by the arrow in the figure. In surface integration, we will integrate over "smooth oriented surfaces" in $\mathbb{R}^3$, which are surfaces that have tangent planes at each point (except at the "surface boundary" or "edges"), and which are oriented by choosing one side as the "positive" side. For example, a sphere is a smooth surface which can be oriented by choosing either the inside or the outside as the positive side. It has no surface boundary. As a second example, consider a parallelogram region in $\mathbb{R}^3$, which is a smooth surface whose surface boundary consists of four line segments. If you think of this region as a thin wafer that is blue on one side and red on the other, you can orient it by choosing either the blue or the red side as its positive side.

As mentioned earlier, our integrands will be inner products of the form $\langle F(X), V(X)\rangle$, where F and V are vector fields defined on the region of integration. The vector field F may be thought of as a force field that surrounds the region of integration. See Figure 10.2, where F is indicated by the curved arrows ("streamlines"), and the region of integration is the surface $\mathcal{B}$, with surface boundary $\partial\mathcal{B}$. For each point X on the surface, the vector $V(X)$ is a special unit vector that indicates the orientation of the region of integration at X.

FIGURE 10.1. A curve in the plane

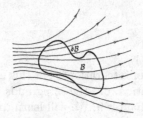

FIGURE 10.2. A vector field defined on $\mathcal{B}$

Line and surface integrals arise often in physics and engineering as well as in mathematics. In Section 10.1, we will discuss smooth oriented curves in more detail, and define the line integral. In Section 10.2, an important theorem (Stokes' Theorem) concerning line integrals is proved. It is a beautiful and useful generalization of the Fundamental Theorem of Calculus. In Section 10.3, smooth surfaces and surface integrals are treated, along with generalizations of Stokes' Theorem. We will see that both line and surface integrals can be calculated in terms of ordinary integrals (including double and triple integrals). The final section, Section 10.4, introduces an important type of vector field, called a "conservative force field", or a "potential field". Calculations of line integrals in conservative force fields are quite easy, since the answer only depends on the locations of the endpoints of the curve of integration. Throughout the chapter, there are discussions of applications of vector integration.

10.1   Line Integrals

10.1A   SMOOTH ORIENTED CURVES

Let $\mathcal{C}$ be a curve in $\mathbb{R}^q$. For example, see Figure 10.3, where a half circle in $\mathbb{R}^2$ is shown. The curve $\mathcal{C}$ may have two endpoints, as shown in the figure, or it may have one or no endpoints, for example, it may be a ray or a full circle. Let $\mathcal{C}_o$ be that part of the curve $\mathcal{C}$ that doesn't include the endpoints. Suppose for each point X of $\mathcal{C}_o$, there exists a *unit* vector $V = V(X)$ such that the line parametrized by $tV + X$ is the tangent line of $\mathcal{C}$ at X. We imagine that the unit vector $V(X)$ is attached to the curve at the point

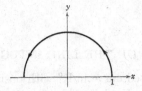

FIGURE 10.3. An oriented half circle

X. Note that $V : \mathcal{C}_o \to \mathbb{R}^q$ is a vector-valued function. If V is a continuous function, we say that $\mathcal{C}$ is a **smooth oriented curve** with **(positive) unit tangent vector** $V(X)$ at X. The direction of $V(X)$ is the orientation of the curve $\mathcal{C}$ at X. In Figure 10.3, the curve has been oriented so that it travels counterclockwise around the origin. You might imagine the positive tangent vectors attached to various points on the half circle in the picture. These tangent vectors point upward near the point $(1, 0)$, downward near $(-1, 0)$, and toward the left near $(0, 1)$.

Most often, we will work with curves in terms of parametrizations. A parametrization of a curve provides a natural orientation for that curve. If the curve $\mathcal{C}$ is parametrized by the function $\alpha : [a, b] \to \mathbb{R}^q$, and if $\alpha'(t)$ exists and is not $\mathbf{0}$ for $t \in (a, b)$, then a unit tangent vector at $X = \alpha(t)$ is given by

$$V(X) = \frac{\alpha'(t)}{\|\alpha'(t)\|}.$$

If $\alpha'(t)$ is continuous and not $\mathbf{0}$ at all $t \in (a, b)$, then $V(X)$ will also be continuous, and thus will provide an orientation for the curve. In this case, we will say that $\mathcal{C}$ is **oriented by** the parametrization α.

EXAMPLES:

1. Let $\alpha(t) = (\cos t, \sin t)$, $0 \le t \le \pi$. Then α parametrizes the curve in Figure 10.3. Since $\alpha'(t) = (-\sin t, \cos t)$ is a unit vector, we take $V(X) = \alpha'(t)$ for $X = \alpha(t)$. As you may easily verify, this parametrization provides the same orientation as that shown in the figure.

2. The half circle is also parametrized by $\alpha(t) = (\cos 2t, \sin 2t)$, $0 \le t \le \pi/2$. You should check that it also provides the same orientation.

3. The half circle in the figure is parametrized with the *opposite* orientation by $\alpha(t) = (\cos(\pi - t), \sin(\pi - t))$, $0 \le t \le \pi$.

Some curves, such as circles, have no endpoints. We call such curves **closed** curves. We will always insist in this chapter that if $\alpha : [a, b] \to \mathbb{R}^q$ is a parametrization of a closed curve $\mathcal{C}$ in $\mathbb{R}^q$, then $\alpha(a) = \alpha(b)$. The usual parametrization of the circle

$$\alpha(t) = (\cos t, \sin t), \quad 0 \le t \le 2\pi$$

satisfies this condition.

10.1B DEFINITION OF THE LINE INTEGRAL

Let C be a smooth oriented curve in $\mathbb{R}^q$ with positive unit tangent vectors $V(X)$ for $X \in C_o$. Also let $F : C \to \mathbb{R}^q$ be a continuous vector field. Imagine that C has been partitioned into small pieces C_i, and let X_i be a point in the piece C_i. Let Δs_i be the arc length of the piece C_i. Our approximation for the line integral of the vector field F along the oriented curve C is

$$\sum_i \langle F(X_i), V(X_i) \rangle \Delta s_i .$$

This formula has a somewhat more familiar appearance if we use the alternate notation

$$X \cdot Y = \langle X, Y \rangle$$

for the inner product of two vectors X and Y, in which case the synonym "dot product" is often used for "inner product". Then our approximating sum becomes

$$\sum_i F(X_i) \cdot V(X_i) \Delta s_i ,$$

which looks a lot like an approximating Riemann sum. Under appropriate conditions, the approximating sums converge as the pieces C_i become smaller and more numerous to a quantity which we denote by

$$\int_C F(X) \cdot V(X) \, ds = \int_C F \cdot V \, ds .$$

This quantity is called the **line integral** of F along the oriented curve C.

REMARK:

- There are several varations on the notation for the line integral. For example, we may write dX for $V \, ds$, leading to the notation

$$\int_C F \cdot V \, ds = \int_C F(X) \cdot dX .$$

Of course, dX is not really a vector, so $F \cdot dX$ is not defined. Nevertheless, this notation is well-established, and reminds us of the notation for ordinary single-variable integrals. It is also common to write the line integral in terms of components. For example, if $F = (p, q)$ is a vector field in $\mathbb{R}^2$, then we write

$$\int_C F(X) \cdot dX = \int_C p(x, y) \, dx + q(x, y) \, dy ,$$

where we have expressed the "vector" dX as $dX = (dx, dy)$. We will switch among these various ways of writing the line integral, depending on which seems most convenient.

We need a way to calculate the line integral. This will be done in terms of a parametrization of the curve $\mathcal{C}$. Suppose that $\mathcal{C}$ is oriented by the parametrization α. Let the domain of α (an interval) be denoted by $\mathcal{R}$, and let $\mathcal{R}$ be partitioned into small intervals $\mathcal{R}_i$ centered at t_i and having arc length Δt_i. We know from our work with arc length (Section 4.3) that the arc length of the image of $\mathcal{R}_i$ under α is approximated by $\|\alpha'(t_i)\|\Delta t_i$. Let $X_i = \alpha(t_i)$. Since $V(X_i) = \alpha'(t_i)/\|\alpha'(t_i)\|$, the approximating sum for the line integral is itself approximated by

$$\sum_i F(X_i) \cdot \frac{\alpha'(t_i)}{\|\alpha'(t_i)\|} \|\alpha'(t_i)\|\Delta t_i = \sum_i F(\alpha(t_i)) \cdot \alpha'(t_i)\Delta t_i .$$

This last sum is an approximation for the (ordinary) integral

$$\int_{\mathcal{R}} F(\alpha(t)) \cdot \alpha'(t)\, dt .$$

Note that the integrand in this last integral is a *scalar*, since it is the inner product of two vector-valued functions. This reasoning justifies the following theorem:

Theorem 10.1.1 Let $\mathcal{C}$ be a smooth curve in $\mathbb{R}^q$ which is oriented by a parametrization α whose domain is an interval $[a, b]$, and let F be a continuous vector field whose domain includes $\mathcal{C}$. Then

$$\int_{\mathcal{C}} F \cdot V\, ds = \int_a^b F(\alpha(t)) \cdot \alpha'(t)\, dt .$$

As in our previous work with integration, we have not been completely rigorous in our definitions of line integrals, so it is not appropriate to try to give a proof of this theorem. There is however one aspect of the theorem that we wish to justify here. The right-hand side of the formula given in the theorem appears to depend on the parametrization α. We would like to know that if we use another parametrization $\beta : [c, d] \to \mathbb{R}^q$ that orients $\mathcal{C}$ with the same orientation, then

$$\int_a^b F(\alpha(t)) \cdot \alpha'(t)\, dt = \int_c^d F(\beta(\tau)) \cdot \beta'(\tau)\, d\tau.$$

(A similar situation occurred when we found a formula for arc length in Section 4.3c.) For simplicity, let us assume that $\beta = \alpha \circ \gamma$, where γ is an increasing function from $[c, d]$ to $[a, b]$. (It turns out that this assumption always holds if α' and β' are both never equal to $\mathbf{0}$.) Using the single-variable chain rule and the substitution $t = \gamma(\tau)$, we obtain

$$\int_c^d F(\beta(\tau)) \cdot \beta'(\tau)\, d\tau = \int_c^d F(\alpha(\gamma(\tau))) \cdot \alpha'(\gamma(\tau))\gamma'(\tau)\, d\tau$$

$$= \int_a^b F(\alpha(t)) \cdot \alpha'(t)\, dt .$$

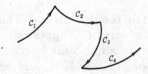

FIGURE 10.4. A piecewise smooth curve

EXAMPLE:

4. Let C be the oriented curve $\alpha(t) = (\cos 2t, \sin 2t)$ for $0 \le t \le \pi/2$. See Figure 10.3. (The image of α is the upper half-unit circle traversed counterclockwise.) Let us evalute

$$\int_C (x_2, x_1) \cdot dX ,$$

which we can also write as

$$\int_C x_2 \, dx_1 - x_1 \, dx_2 .$$

Since $\alpha'(t) = (-2\sin 2t, 2\cos 2t)$,

$$\int_C x_2 \, dx_1 - x_1 \, dx_2 = \int_0^{\pi/2} (-2\sin^2 2t - 2\cos^2 2t) \, dt$$

$$= \int_0^{\pi/2} -2 \, dt = -\pi.$$

There is an obvious way to extend the definition of the line integral to finite unions of smooth oriented curves. If we have two curves C_1 and C_2, then we write $C_1 + C_2$ for their union, and define

$$\int_C F(X) \cdot dX = \int_{C_1} F(X) \cdot dX + \int_{C_2} F(X) \cdot dX.$$

An important case in which it is appropriate to think of C as a union of smooth oriented curves is when C is a **piecewise smooth**. See Figure 10.4. Such curves have tangent lines except at a finite number of points along the curve (shown as "corners" in the illustration). We think of a piecewise curve C as a union of finitely many smooth curves, in the obvious way. A line integral along a piecewise smooth curve is calculated by integrating separately over each smooth piece, with each successive integration picking up where the previous one left off, and then adding the results. Incidently, the reason for our increased generality is not obscure. We merely want to include curves like triangles and squares.

EXAMPLE:

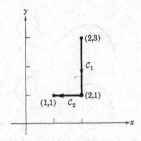

FIGURE 10.5. A curve composed of two line segments

5. Let $C = C_1 + C_2$, where C_1 is defined by $\alpha_1(t) = (2, 3-t)$ for $0 \le t \le 2$, and C_2 by $\alpha_2(t) = (4-t, 1)$ for $2 \le t \le 3$, and so C is composed of two straight-line segments. See Figure 10.5. Then, because

$$\int_C = \int_{C_1} + \int_{C_2},$$

For $F(x, y) = (x^2, -xy)$ we evaluate

$$\int_{C_1} x^2 \, dx - xy \, dy = \int_0^2 2(3-t) \, dt = 8,$$

and

$$\int_{C_2} x^2 \, dx - xy \, dy = \int_2^3 (4-t)^2(-dt) = -\frac{7}{3}.$$

Consequently

$$\int_C x^2 \, dx - xy \, dy = 8 - \frac{7}{3} = \frac{17}{3}.$$

10.1C PROPERTIES OF LINE INTEGRALS

The first and most basic property of a line integral is its linearity described in the following theorem.

Theorem 10.1.2 Let F and G be continuous vector fields, k a constant, and C a piecewise smooth oriented curve. Then the following properties hold:

1. $\int_C [F(X) + G(X)] \cdot dX = \int_C F(X) \cdot dX + \int_C G(X) \cdot dX$,

2. $\int_C kF \cdot dX = k \int_C F \cdot dX$, .

The next property is an inequality.

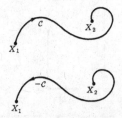

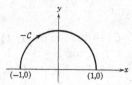

FIGURE 10.6. Two orientations of the same curve

FIGURE 10.7. Reversing the orientation of the half circle

Theorem 10.1.3 Let F be a continuous vector field and let C be a piecewise smooth oriented curve in $\mathcal{D}(F)$. Assume that $\|F\| \leq M$ for all X on the curve C. Then

$$\left| \int_C F(X) \cdot dX \right| \leq ML\,,$$

where L is the arc length of C.

PROOF: Let α be a differentiable parametrization of the curve C and let $a < b$ denote the endpoints of $\mathcal{D}(\alpha)$. Then

$$\left| \int_C F(X) \cdot dX \right| = \left| \int_a^b F(\alpha(t)) \cdot \alpha'(t)\, dt \right| \leq \int_a^b |F(\alpha(t)) \cdot \alpha'(t)|\, dt\,.$$

By the Cauchy-Schwarz inequality, we have

$$|F(\alpha(t)) \cdot \alpha'(t)| \leq \|F(\alpha(t))\|\, \|\alpha'(t)\| \leq M\|\alpha'(t)\|\,.$$

Therefore

$$\left| \int_C F(X) \cdot dX \right| \leq M \int_a^b \|\alpha'(t)\|\, dt = ML\,.$$

Done. $<<$

If C is an oriented curve, then we denote by $-C$ the oriented curve that is obtained by reversing the orientation of C. See Example 3, illustrated by Figure 10.7, and also see Figure 10.6.

Theorem 10.1.4 If C is any piecewise smooth curve, then

$$\int_{-\mathcal{C}} F(X) \cdot dX = -\int_{\mathcal{C}} F(X) \cdot dX.$$

PROOF: This is essentially identical to the argument that the formula for the line integral does not depend on the parametrization, except that now $t = h(\tau) = a + b - \tau$, $a \leq \tau \leq b$, and $-\mathcal{C}$ is parametrized as $\beta = \alpha \circ h$. The key change is in the limits of integration. Here $a = h(d)$ and $b = h(c)$. Thus

$$\begin{aligned}
\int_{\mathcal{C}} F(X) \cdot dX &= \int_{d}^{c} F(\beta(\tau)) \cdot \beta'(\tau) \, d\tau \\
&= -\int_{c}^{d} F(\beta(\tau)) \cdot \beta'(\tau) \, d\tau \\
&= -\int_{-\mathcal{C}} F(X) \cdot dX.
\end{aligned}$$

Done. $\ll$

EXAMPLES:

6. One parametrization of the oriented line segment $\mathcal{C}$ from $(2, 3)$ to $(1, 1)$ shown in Figure 10.8 is

$$\alpha(t) = (2 - t, 3 - 2t), \ 0 \leq t \leq 1.$$

(Note that the words "from $(2, 3)$ to $(1, 1)$" indicate the orientation.) Then

$$\int_{\mathcal{C}} x^2 \, dx - xy \, dy = \int_{0}^{1} [-(2 - t)^2 + 2(2 - t)(3 - 2t)] \, dt = 4.$$

7. Let us evaluate

$$\int_{\mathcal{C}} (2y - 3) \, dx + x^2 \, dy,$$

where $\mathcal{C}$ is the curve from $(-1, 1)$ to $(1, 1)$ along the parabola $y = x^2$. We will use the parametrization

$$\beta(r) = (r, r^2), \ -1 \leq r \leq 1.$$

We obtain

$$\int_{\mathcal{C}} (2y - 3) \, dx + x^2 \, dy = \int_{-1}^{1} [(2r^2 - 3) + 2r^3] \, dr = -\frac{14}{3}.$$

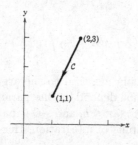

FIGURE 10.8. The line segment from $(2,3)$ to $(1,1)$

8. Let Γ denote the boundary of the triangular region with vertices at $(1,1)$, $(2,1)$, and $(2,3)$ traversed once counterclockwise. See Figure 10.9. We will compute

$$\int_\Gamma x^2 \, dx - xy \, dy \, .$$

Again, this makes sense with any parametrization for Γ. Now $\Gamma = \Gamma_1 + \Gamma_2 + \Gamma_3$, where Γ_1 is the line segment from $(1,1)$ to $(2,1)$, Γ_2 is the line segment from $(2,1)$ to $(2,3)$, and Γ_3 is the line segment from $(2,3)$ to $(1,1)$. Thus $\Gamma_1 = -C_2$ and $\Gamma_2 = -C_1$ where C_1 and C_2 are the oriented curves in Example 5, and $\Gamma_3 = C$ the oriented curve in Example 6. Thus

$$\int_\Gamma = \int_{\Gamma_1} + \int_{\Gamma_2} + \int_{\Gamma_3}$$

$$= \int_{-C_2} + \int_{-C_1} + \int_C = -\int_{C_2} - \int_{C_1} + \int_C$$

$$= \frac{7}{3} - 8 + 4 = -\frac{5}{3}.$$

Our calculations show that the line integrals along two different curves $\Gamma_1 + \Gamma_2$ and $-\Gamma_3$ from $(1,1)$ to $(2,3)$ are *different even though* the oriented curves begin and end at the same places; the values along the curves are $\frac{7}{3} - 8 = -\frac{17}{3}$ and -4, respectively. (This does *not* mean that different parametrizations of the same oriented curve would give different results.)

9. Let C denote the line segment from $(2,1)$ to $(1,1)$. One convenient parametrization is $\alpha(s) = (-s,1)$, $-2 \le s \le -1$. Thus

$$\int_C x^2 \, dx - xy \, dy = \int_{-2}^{-1} -s^2 \, ds = -\frac{7}{3}.$$

Of course, this should have been anticipated, because α is just a reparametrization of the curve C_2 in Example 5. Another method

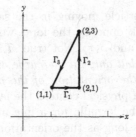

FIGURE 10.9. Traversing the boundary of a triangular region

for the same problem is to observe that $-\mathcal{C}$ is somewhat simpler to parametrize: $\beta(t) = (t, 1)$, $1 \leq t \leq 2$. Then we find that

$$\int_{\mathcal{C}} = -\int_{-\mathcal{C}} = -\int_{1}^{2} t^2 \, dt = -\frac{7}{3}.$$

10.1D   WORK

If the vector field F is a force field so that $F(X)$ is the force "felt" by a particle at the point X, then the line integral

$$\int_{\mathcal{C}} F \cdot V \, ds$$

is defined in physics to be the **work** done by the force in moving the particle along the curve $\mathcal{C}$. Since V is a unit vector, the dot product $F \cdot V$ is the length of the component of F in the direction of V, so only the component of the vector field F *tangent* to the curve influences the value of the line integral, and hence the work done by the force. In particular, if a force field is perpendicular to the curve $\mathcal{C}$ traveled by a moving particle, then this force does zero work.

For example, if the force field, like gravity, is vertical, then this force does no work in any horizontal motion of the object. Work done by this force is nonzero only when moving an object up or down. It is friction, not gravity directly, that causes us to expend work in sliding objects horizontally. On smooth ice, where sliding friction is quite small, it is very easy to move objects horizontally.

If the curve $\mathcal{C}$ is parametrized by a function $\alpha : [a, b] \to \mathbb{R}^q$, then the formula for the work can be written as

$$\text{work} = \int_{a}^{b} F(\alpha(t)) \cdot \alpha'(t) \, dt.$$

Since the velocity $\alpha'(t)$ appears in the formula for work, one might be led to believe that work depends on the velocity, and so if $\beta(\tau)$ describes the motion of another particle moving in the same direction along the same curve, then the work done by the force would be different for the two particle motions $\alpha(t)$ and $\beta(\tau)$. Not true. *The work does not depend on the particular motions but only on the curve and the direction traveled along the curve (that is, the orientation of the motion).* More precisely, we are just reasserting in a physical context the fact discussed earlier that the value obtained from the formula for a line integral does not depend on the parametrization, as long as the orientation remains the same. We can interpret Theorem 10.1.4 as stating that if a force F does work W in moving a particle along a curve C, then in running the particle backward, that is, along $-C$, the force does work $-W$.

Exercises

1. Compute $\int_C 2x\, dx + 6(x-y)\, dy$ where C is the oriented curve parametrized by

   (a) $\alpha(t) = (t, t),\ 0 \le t \le 1$

   (b) $\alpha(t) = (t, t^2),\ 0 \le t \le 1$

   (c) $\beta(s) = (\sin s, s),\ 0 \le s \le 2\pi$

   (d) $\Phi(t) = (e^t, 1),\ 0 \le t \le 1;$

   (e) $\alpha(\theta) = (1 - \cos\theta, \sin\theta),\ 0 \le \theta \le \pi/2$

   (f) $\Psi(r) = (e^r, e^{-r}),\ -1 \le r \le 1$.

2. Compute $\int_C (x_2, x_3, x_1) \cdot dX$, where C is the oriented curve parametrized by

   (a) $\alpha(t) = (t, t, t),\ 0 \le t \le 1$

   (b) $\alpha(t) = tZ,\ 0 \le t \le 1$, where Z is a fixed, but arbitrary member of $\mathbb{R}^3$

   (c) $(\cos t, \sin t, t),\ 0 \le t \le 2\pi$

   (d) $(\cos 2t \sin t,\ \sin 2t \sin t,\ \cos t)$

3. For the oriented curve C parametrized by
   $$\alpha(t) = (\cos^2 t,\ \cos t \sin t,\ \cos t \sin t,\ \sin^2 t),\ 0 \le t \le \pi,$$
   compute $\int_C F(X) \cdot dX$, where

   (a) $F(X) = X$

   (b) $F(X) = X + (1, 1, 0, 0)$

   (c) $F(X) = (x_1, -x_2, -x_3, x_4)$

   (d) $F(X) = (x_1, x_2, -x_3, -x_4)$.

4. Compute $\int_C 2x\, dx + 6(x-y)\, dy$ where

   (a) C is the graph of $y = 2x$ for $1 \le x \le 2$ and oriented in the direction of increasing x

   (b) C is the line segment beginning at $(-1, 1)$ and ending at $(2, -2)$

(c) C is the shortest curve from $(0,0)$ to $(1,1)$ that passes through the point $(1,0)$

(d) C is the shortest curve from $(0,0)$ to $(1,1)$ that passes through the point $(0,1)$

(e) C follows the parabola $x = y^2$, from $(1,-1)$ to $(1,1)$

(f) C is the larger arc of the circle $(x-1)^2 + y^2 = 1$ from $(0,0)$ to $(1,1)$.

5. Compute $\int_C (6x - y^2)\, dx - 2xy\, dy$ for the curves C in parts (a) and (b) of Exercise 1 and parts (c) and (d) of Exercise 4. Compare your four answers.

6. Compute $\int_C X \cdot dX$, where C is a circle of radius $r \le 1$ on $S((0,0,0);1)$.

7. Evaluate $\int_C y^2\, dx + 3x^2\, dy$ for each of the following curves:

   (a) C is the line segment from $(-1,1)$ to $(2,-2)$,

   (b) C is the line segment from $(2,-2)$ to $(-1,1)$.

8. Use answers to parts (a) of Exercise 1 and (d) of Exercise 4 to calculate

$$\int_C 2x\, dx + 6(x-y)\, dy\,,$$

where C is the counterclockwise oriented boundary of the triangular region with vertices at $(0,0)$, $(0,1)$, and $(1,1)$.

9. Use parts (e) of Exercise 1 and (f) of Exercise 4 to calculate

$$\int_C 2x\, dx + 6(x-y)\, dy\,,$$

where C is the entire circle $(x-1)^2 + y^2 = 1$ oriented counterclockwise.

10. Evaluate

$$\int_C e^x \sin y\, dx + e^x \cos y\, dy\,,$$

where C is the portion of the circle $x^2 + y^2 = 1$ with $x \ge 0$, oriented so that it begins at $(0,1)$ and ends at $(0,-1)$. *Hint:* In calculating the eventual single-variable integral avoid the temptation to represent the integral of a sum as the sum of the integrals.

11. Evaluate $\int_C (x_1^2 + x_2^2)\, dx_1 + 2x_1 x_2\, dx_2$, where C is the oriented curve represented parametrically as $(1 - t^2,\, 3 - 2t)$ for $1 \le t \le 2$.

12. Let a force field be $F(x,y) = (xy, x^2 - 3y^2)$. Compute the work done by this force to move a particle of unit mass from $(0,0)$ to $(1,1)$ along each of the oriented curves in Exercise 1, parts (a) and (b).

13. Let a force field be $F(x,y) = (xy, x^2 - 3y^2)$. Compute the work done by this force to move a particle of unit mass from $(0,0)$ to $(1,1)$ along each of the oriented curves in Exercise 4, parts (c) and (d).

14. If a force field is $F(x,y) = (ye^{xy}, xe^{xy})$, find the work done by this force to move a particle of mass 1 counterclockwise around the boundary of the square region $|x| \le 1$, $|y| \le 1$.

15. Here is a review problem. Let α be a differentiable parametrized curve whose image does not include 0, and let h be a scalar-valued differentiable function on the interval $(0, \infty)$. Show that

$$\frac{d}{dt} h(\|\alpha(t)\|) = \frac{h'(\|\alpha(t)\|)}{\|\alpha(t)\|} \alpha(t) \cdot \alpha'(t).$$

16. Let C be a smooth curve in $\mathbb{R}^n$ with initial point U_0 and terminal point U_1. Assume that $0 \notin C$. Use the preceding exercise to show that

$$\int_C \frac{X}{\|X\|^3} \cdot dX = \frac{1}{\|U_0\|} - \frac{1}{\|U_1\|}.$$

17. Generalize the preceding exercise by replacing $\|X\|^3$ on the left-hand side by an arbitrary power of $\|X\|$ and changing the right-hand side appropriately.

18. For C, U_0, and U_1 as in Exercise 16 and ψ an arbitrary continuous scalar-valued function on the interval $(0, \infty)$, find a formula for

$$\int_C (\psi(\|X\|)X) \cdot dX.$$

    Hint: To express your answer you will need to introduce a function not explicitly mentioned in the preceding sentence.

19. For each of the following statements decide whether it is true or false. If it is false, give a counterexample. If it is true, give a proof. Here F is a continuous vector field whose domain contains the smooth curve C.

    (a) If C is a vertical line segment in $\mathbb{R}^2$ and f_2 is the zero function, then $\int_C F(X) \cdot dX = 0$.

    (b) If C is a circle in the $x_2 x_3$-plane in $\mathbb{R}^3$ and f_2 and f_3 are the zero function, then $\int_C F(X) \cdot dX = 0$.

    (c) If C is a circle in the $x_2 x_3$-plane in $\mathbb{R}^3$ and f_2 and f_3 are negatives of each other, then $\int_C F(X) \cdot dX = 0$.

    (d) If each f_j is nonnegative, then $\int_C F(X) \cdot dX \geq 0$.

20. Let C denote the quarter-circle $x^2 + y^2 = 1$ in the first quadrant oriented counterclockwise.

    (a) Show that

    $$\int_C -e^{(xy)^2} dx + e^{-(xy)^2} dy \leq \frac{\pi}{2} \sqrt{e^{1/16} + 1}.$$

    (b) Show that

    $$\int_C -e^{(xy)^2} dx + e^{-(xy)^2} dy \leq \frac{\pi}{2} \sqrt{e^{1/16} + e^{-1/16}}.$$

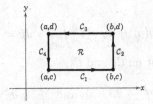

FIGURE 10.10. Stokes' Theorem on a rectangular region

21. Let $F(X)$ be a continuous force field in the plane. Suppose that a particle of mass m moves so that its position $\alpha(t)$ is determined by Newton's second law:

$$m\alpha'' = F \circ \alpha .$$

Show that the work done by the force F during the time interval $t_1 \leq t \leq t_2$ is equal to the change in kinetic energy of the particle:

$$\int_{t_1}^{t_2} ((F \circ \alpha)\alpha')(t)\, dt = \frac{m}{2}\|\alpha'(t_2)\|^2 - \frac{m}{2}\|\alpha'(t_1)\|^2.$$

Hint: First show that $\frac{1}{2}m(d/dt)\|\alpha'(t)\|^2 = ((F \circ \alpha)\alpha')(t)$.

10.2   Stokes' Theorem in the Plane

10.2A   STOKES' THEOREM FOR A RECTANGULAR REGION

The time has come to generalize the Fundamental Theorem of Calculus to multiple integrals. Throughout this section, we will use the following notation: if $\mathcal{R}$ is a closed set in R^2 whose interior is pathwise connected, then we will denote the boundary of $\mathcal{R}$ by $\partial\mathcal{R}$. Here is the simplest case.

Theorem 10.2.1 Let $\mathcal{R}$ denote the rectangle $a \leq x \leq b$, $c \leq y \leq d$ and $F = (p, q)$ a vector field with a continuous derivative. Then

$$\iint_{\mathcal{R}} \left(\frac{\partial q}{\partial x} - \frac{\partial p}{\partial y}\right) dA = \int_{\partial\mathcal{R}} p\, dx + q\, dy,$$

where the boundary $\partial\mathcal{R}$ is oriented counterclockwise.

PROOF: We treat the p and q terms separately. Now

$$\iint_{\mathcal{R}} \frac{\partial q}{\partial x}\, dA = \int_c^d \left(\int_a^b \frac{\partial q}{\partial x}(x, y)\, dx\right) dy = \int_c^d [q(b, y) - q(a, y)]\, dy.$$

In the notation of Figure 10.10, however, we have

$$\int_c^d q(b, y)\, dy = \int_{C_2} q\, dy$$

and

$$-\int_c^d q(a,y)\, dy = -\int_{-C_4} q\, dy = \int_{C_4} q\, dy,$$

and since $y \equiv$ constant on C_1 and C_3,

$$\int_{C_1} q\, dy = 0 \quad \text{and} \quad \int_{C_3} q\, dy = 0.$$

Therefore

$$\iint_\mathcal{R} \frac{\partial q}{\partial x}\, dA = \int_{\partial\mathcal{R}} q\, dy.$$

Similar considerations give

$$\iint_\mathcal{R} \frac{\partial p}{\partial y}\, dA = -\int_{\partial\mathcal{R}} p\, dx.$$

Subtracting these two equations, we arrive at the result. $\ll$

10.2B MORE GENERAL REGIONS

Precisely the same result holds for more general regions than rectangles. We prove this in two steps. First is the case where a region $\mathcal{B} \subseteq \mathbb{R}^2$ can be written both as

$$\varphi(x) \le y \le \psi(x), \quad a \le x \le b,$$

so that φ and ψ are the lower and upper boundary curves, respectively, or as

$$\Phi(y) \le x \le \Psi(y), \quad c \le y \le d,$$

so that Φ and Ψ are the left-hand and right-hand boundary curves, respectively. See Figure 10.11. We also suppose that the boundary is a piecewise smooth curve; that is, the boundary $\partial\mathcal{B}$ consists of a finite number of smooth curves. Such regions are called **simple regions**.

Theorem 10.2.2 Let $\mathcal{B} \subseteq \mathbb{R}^2$ be a simple region and $F = (p, q)$ a vector field with a continuous derivative. Then

$$\iint_\mathcal{B} \left(\frac{\partial q}{\partial x} - \frac{\partial p}{\partial y} \right) dA = \int_{\partial\mathcal{B}} p\, dx + q\, dy,$$

where the boundary $\partial\mathcal{B}$ is oriented counterclockwise.

PROOF: We discuss only the p term. The q term is treated similarly. Now

$$\iint_\mathcal{B} \frac{\partial p}{\partial y}\, dA = \int_a^b \left(\int_{\varphi(x)}^{\psi(x)} \frac{\partial p}{\partial y}(x,y)\, dy \right) dx$$

$$= \int_a^b [p(x,\psi(x)) - p(x,\varphi(x))]\, dx.$$

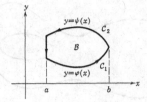

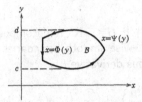

FIGURE 10.11. Stokes' Theorem on a region with curved sides

Let C_1 denote the bottom curve of $\mathcal{B}$, and C_2 the top curve, oriented so that $\partial\mathcal{B}$ is oriented counterclockwise. Then

$$\int_a^b p(x, \psi(x))\, dx = \int_{-C_2} p\, dx, \qquad \int_a^b p(x, \varphi(x))\, dx = \int_{C_1} p\, dx.$$

Thus, as claimed,

$$\iint_{\mathcal{B}} \frac{\partial p}{\partial y}\, dA = -\int_{C_1 + C_2} p\, dx = -\int_{\partial\mathcal{B}} p\, dx$$

(observe that $\int p\, dx = 0$ over any vertical line segment). $<<$

Finally, we extend this to more general regions, possibly containing holes. The idea is to dissect the region into simple regions and apply Theorem 10.2.2 to these simple regions separately. An example makes this clear. Let $\mathcal{B} \subseteq \mathbb{R}^2$ be the region indicated in Figure 10.12. We have dissected $\mathcal{B}$ into a number of simple regions. If one applies Stokes' Theorem to each of these simple regions separately, then line integrals along the interior dotted lines appear. The integrals along the dotted lines cancel, however, since, for example, the line shared by $\mathcal{B}_1$ and $\mathcal{B}_2$ is traversed once in each direction. By Theorem 10.1.4, the net result is zero. Thus, only integration along $\partial\mathcal{B}$ remains, that along the "outer" portion of $\partial\mathcal{B}$ is traversed counterclockwise and that along the "inner" portion is traversed clockwise. There is a mnemonic device that enables us to remember the orientation of the boundary: if you walk along $\partial\mathcal{B}$ so that your left hand is in $\mathcal{B}$, then you are walking in the positive direction of $\partial\mathcal{B}$. We call this the **positive orientation** of $\partial\mathcal{B}$.

The discussion above has established

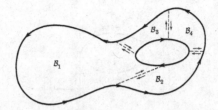

FIGURE 10.12. Dissecting $\mathcal{B}$ into simpler regions

Theorem 10.2.3 (Stokes') If $\mathcal{B} \subseteq \mathbb{R}^2$ is decomposable into a finite number of simple regions with piecewise smooth boundaries, then for any vector field $F = (p, q)$ with a continuous derivative,

$$\iint_{\mathcal{B}} \left(\frac{\partial q}{\partial x} - \frac{\partial p}{\partial y} \right) dA = \int_{\partial \mathcal{B}} p(x, y) \, dx + q(x, y) \, dy,$$

where $\partial \mathcal{B}$ has positive orientation.

EXAMPLES:

1. Let $\mathcal{B}$ denote the triangular region with vertices at $(0, 0)$, $(1, 0)$, and $(0, 1)$, with positively oriented boundary. We evaluate

$$\int_{\partial \mathcal{B}} (e^x + y - 2x) \, dx + (7x - \sin y) \, dy.$$

Let $p = e^x + y - 2x$ and $q = 7x - \sin y$. Then, by Stokes' Theorem,

$$\int_{\partial \mathcal{B}} (e^x + y - 2x) \, dx + (7x - \sin y) \, dy = \iint_{\mathcal{B}} 6 \, dA.$$

But

$$\iint_{\mathcal{B}} dA = \text{area} \, (\mathcal{B}) = \frac{1}{2}.$$

Thus

$$\int_{\partial b} (e^y + y - 2x) \, dx + (7x - \sin y) \, dy = 3.$$

The theorem greatly simplified the amount of computation.

2. Let $\mathcal{B}$ denote the disk of radius a centered at the origin and let $p(r) = p(x, y)$, $q(r) = q(x, y)$ be functions that depend only on the distance r from the origin. Assume that p and q have continuous derivatives. Then by Stokes' Theorem

$$\iint_{\mathcal{B}} \left(\frac{\partial q}{\partial x} - \frac{\partial p}{\partial y} \right) dA = \int_{\partial \mathcal{B}} p \, dx + q \, dy \, .$$

But on ∂B = circle of radius a, $p = p(a)$, $q = q(a)$ are constant. Therefore

$$\iint_B (q_x - p_y)\, dA = \int_{\partial B} p(a)\, dx + q(a)\, dy.$$

To evaluate the line integrals, we could introduce a parametrization and go through the easy computation. Instead, we are sneaky and again apply Stokes' Theorem, using the fact that $p(a)$ and $q(a)$ are constants. A moment's thought (do not write anything) reveals that the value of the integral is zero; that is, if p and q are constant on the circle,

$$\iint_B (q_x - p_y)\, dA = 0.$$

10.2c OTHER VERSIONS

In practice, people often use several different variants of Stokes' Theorem. For the first variant we need the idea of the **unit outer normal** vector to the boundary of a region B. If $V(X)$ is the positive unit tangent vector at a point X on ∂B, then the unit outer normal vector $N(X)$ at X is the unit vector obtained by rotating $V(X)$ 90 degrees *clockwise* in $\mathbb{R}^2$. In other words, $N(X)$ is perpendicular to $V(X)$, and hence perpendicular to the curve ∂B. Furthermore, if you are facing in the direction of $V(X)$, then $N(X)$ will point toward your right, which is "outward" or away from the region B. See Figure 10.13. If $X = \alpha(t) = (\alpha_1(t), \alpha_2(t))$ is a parametrization of ∂B, oriented positively, then

$$\frac{\alpha'(t)}{\|\alpha'(t)\|} = \frac{(\alpha_1'(t), \alpha_2'(t))}{\|(\alpha_1'(t), \alpha_2'(t))\|}$$

equals $V(X)$. If we apply the linear transformation with matrix

$$\begin{bmatrix} 0 & 1 \\ -1 & 0 \end{bmatrix}$$

to this vector (thus rotating it 90 degrees clockwise), we obtain

$$N(X) = \frac{(\alpha_2'(t), -\alpha_1'(t))}{\|(\alpha_2'(t), \alpha_1'(t))\|}.$$

EXAMPLE:

3. For a simple example of a unit outer normal vector, let B be the disk of radius 2 centered at the origin, so that ∂B is the circle of radius 2. See Figure 10.14. At a point (x, y) on ∂B, the unit outer normal is given by $N = \frac{1}{2}(x, y) = (\cos t, \sin t)$, where ∂B is parametrized by $\alpha(t) = (2\cos t, 2\sin t)$. Note here that the unit tangent is $(-\sin t, \cos t) = \frac{1}{2}(-y, x)$.

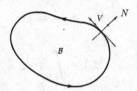

FIGURE 10.13. The outer normal vector N

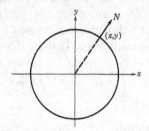

FIGURE 10.14. Stokes' Theorem on a circle

Let us now state the **divergence form** of Stokes' Theorem.

Theorem 10.2.4 (Divergence Theorem in $\mathbb{R}^2$) If $F = (f_1, f_2)$ is a vector field with a continuous derivative, and if $\mathcal{B} \subseteq \mathbb{R}^2$ satisfies the assumptions of Theorem 10.2.2, then

$$\iint_{\mathcal{B}} \operatorname{div} F \, dA = \iint_{\mathcal{B}} \left(\frac{\partial f_1}{\partial x} + \frac{\partial f_2}{\partial y} \right) dA = \int_{\partial \mathcal{B}} F \cdot N \, ds,$$

where N is the unit outer normal to $\partial \mathcal{B}$ and s is arc length on $\partial \mathcal{B}$.

PROOF: The first equality is the definition of the divergence:

$$\operatorname{div} F = \frac{\partial f_1}{\partial x} + \frac{\partial f_2}{\partial y}.$$

To prove the second equality, we let $q = f_1$, $p = -f_2$ in Theorem 10.2.2. Then

$$\iint_{\mathcal{B}} \left(\frac{\partial f_1}{\partial x} + \frac{\partial f_2}{\partial y} \right) dA = \int_{\partial \mathcal{B}} -f_2 \, dx + f_1 \, dy.$$

Thus, if $\alpha(t) = (\alpha_1(t), \alpha_2(t))$, $a \leq t \leq b$, is a parametrization of $\partial \mathcal{B}$, we have

$$\iint_{\mathcal{B}} \left(\frac{\partial f_1}{\partial x} + \frac{\partial f_2}{\partial y} \right) dA = \int_a^b (-f_2 \alpha_1' + f_1 \alpha_2') \, dt = \int_a^b F \cdot N \, ds.$$

Done. $<<$

This theorem has a nice physical interpretation. Think of F as the velocity vector of a fluid—a gas, say—moving in the plane, and assume unit

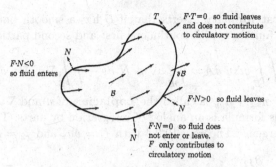

FIGURE 10.15. Fluid flow in a region

FIGURE 10.16. Stokes' Theorem in one dimension

density. Then div $F(X)$ measures how the gas is moving from the point X. Integrating this over the whole region $\mathcal{B}$, we obtain the net change in the amount of fluid in $\mathcal{B}$. But there is another way to measure the same quantity: merely stand by the walls of $\mathcal{B}$ and see how much leaves. See Figure 10.15. In computing this, we of course need only the component of the velocity F perpendicular to the boundary (the component of the velocity tangent to the boundary affects only the circulatory, that is, whirlpool-like, motion of the fluid in $\mathcal{B}$). This explains why we use the unit outer normal vector in the line integral. Theorem 10.2.4 is just a statement of the equality of these two ways of measuring the change in the amount of fluid in $\mathcal{B}$. If div $F = 0$, then the net fluid flow across any closed curve is zero. Thus, we refer to a vector field satisfying div $F = 0$ as having no sources or sinks in $\mathcal{B}$.

Perhaps a more familiar statement of the same idea is to let $\mathcal{B}$ denote a room in which a party is taking place and $F(X)$ the velocity of the person standing at X. Then one can measure the change in the number of people at the party either by (1) counting the people inside the room $\mathcal{B}$ or (2) standing at the doors ($= \partial \mathcal{B}$) and counting how many people enter and leave. The number from (1) is the left-hand side of Theorem 10.2.4, and (2) the right-hand side.

One can also see how the single-variable theorem

$$\int_a^b f'(x)\, dx = f(b) - f(a)$$

fits into this pattern. See Figure 10.16. The minus sign in front of $f(a)$ arises because the "outer normal" to the interval $a \leq x \leq b$ at a is in the negative direction.

We conclude this section with two important identities due to Green.

Green's first identity asserts that if $\mathcal{B}$ has a smooth boundary and if u and v are functions with continuous first and second partial derivatives, then

$$\iint_{\mathcal{B}} v \Delta u \, dA = \int_{\partial \mathcal{B}} v \nabla u \cdot N \, ds - \iint_{\mathcal{B}} \nabla v \cdot \nabla u \, dA,$$

where $\Delta u = u_{xx} + u_{yy}$ is called the **Laplacian** of u and $\nabla v = \operatorname{grad} v = (v_x, v_y)$. This formula is an analog of integration by parts. To prove it, we use the Divergence Theorem 10.2.4 with $f_1 = vu_x$ and $f_2 = vu_y$. Then

$$\frac{\partial f_1}{\partial x} + \frac{\partial f_2}{\partial y} = vu_{xx} + v_x u_x + vu_{yy} + v_y u_y = v \Delta u + \nabla v \cdot \nabla u,$$

and the result follows. It is often useful to note that the term $\nabla u \cdot N$ in the boundary integral is the directional derivative of u in the direction of the outer normal to $\partial \mathcal{B}$. This is sometimes written $\partial u / \partial N$, and so Green's first identity reads

$$\iint_{\mathcal{B}} v \Delta u \, dA = \int_{\partial \mathcal{B}} v \frac{\partial u}{\partial N} \, ds - \iint_{\mathcal{B}} \nabla v \cdot \nabla u \, dA.$$

Green's second identity asserts that

$$\iint_{\mathcal{B}} (v \Delta u - u \Delta v) \, dA = \int_{\partial \mathcal{B}} \left(v \frac{\partial u}{\partial N} - u \frac{\partial v}{\partial N} \right) ds.$$

This is proved by writing Green's first identity with the roles of u and v reversed and then substracting the two equations.

10.2D APPLICATIONS

Now let us see how Stokes' Theorem may be applied to some issues in mathematical physics. Let $\mathcal{B} \subseteq \mathbb{R}^2$ be a given region with smooth boundary, and let $u(x, y, t)$ denote the temperature at the point $(x, y) \in \mathcal{B}$ at time t. In physics, it is shown that a simple model for heat flow requires that u satisfy the differential equation $u_t = \Delta u$; that is,

$$\frac{\partial u}{\partial t} = \frac{\partial^2 u}{\partial x^2} + \frac{\partial^2 u}{\partial y^2}.$$

To determine the temperature $u(x, y, t)$ for $t \geq 0$, it is physically plausible to require prior knowledge of both the initial temperature (at $t = 0$)

$$u(x, y, 0) = f(x, y), \quad (x, y) \in \mathcal{B},$$

and the temperature on the boundary for all time $t \geq 0$

$$u(x, y, t) = \varphi(x, y, t), \quad (x, y) \in \partial \mathcal{B}, \quad t \geq 0.$$

Although we do not prove that a solution does exist, it is easy to show that there is at most one solution—a uniqueness theorem. Thus, once one solution satisfying the initial and boundary conditions has been found, by whatever means, you are guaranteed that there is no other one.

Suppose that v and w are two solutions, and let $u = v - w$. We show that $u(x, y, t) = 0$ for all $(x, y) \in \mathcal{B}$. This, of course, proves that $v = w$. Now, since $v_t = \Delta v$ and $w_t = \Delta w$, we see that

$$u_t - \Delta u = v_t - w_t - (\Delta v - \Delta w) = 0,$$

and for $(x, y) \in \mathcal{B}$,

$$u(x, y, 0) = v(x, y, 0) - w(x, y, 0) = f(x, y) - f(x, y) = 0,$$

and similarly for $(x, y) \in \partial \mathcal{B}$ and $t \geq 0$,

$$u(x, y, t) = v - w = \varphi - \varphi = 0.$$

Now—and here is the trick—we define a kind of "energy" function $E(t)$ by

$$E(t) = \frac{1}{2} \int \int_{\mathcal{B}} u^2(x, y, t) \, dx \, dy.$$

Then, differentiating with respect to t under the integral sign (see the theorem in Section 9.3b), we find that

$$\frac{dE}{dt} = \frac{1}{2} \int \int_{\mathcal{B}} \frac{\partial}{\partial t} u^2(x, y, t) dx \, dy = \int \int_{\mathcal{B}} u u_t \, dx \, dy.$$

But $u_t = \Delta u$, and so by Green's first identity with $v = u$ there,

$$\frac{dE}{dt} = \int \int_{\mathcal{B}} u \Delta u \, dx \, dy = \int_{\partial \mathcal{B}} u \nabla u \cdot N \, ds - \int \int_{\mathcal{B}} \|\nabla u\|^2 \, dx \, dy.$$

Now we recall that $u = 0$ on $\partial \mathcal{B}$, and so the boundary integral is zero. Thus

$$\frac{dE}{dt} = - \int \int_{\mathcal{B}} \|\nabla u\|^2 \, dx \, dy \leq 0.$$

Consequently, $E(t)$ is a decreasing function, $E(t) \leq E(0)$—physically, energy is dissipated. Since $u(x, y, 0) = 0$, however, we see that $E(0) = 0$. Moreover, it is evident from the definition that $E(t) \geq 0$. Hence for all $t \geq 0$

$$0 \leq E(t) \leq E(0) = 0;$$

that is, $E(t) = 0$ for all $t \geq 0$. This implies that $u(x, y, t) = 0$ for all $t \geq 0$, since if not, then $E(t) > 0$ for some $t \geq 0$.

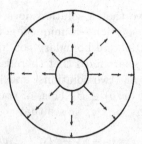

FIGURE 10.17. A region Ω with no sources or sinks

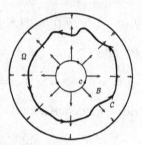

FIGURE 10.18. A closed curve in Ω

As our second application, we consider a vector field F satisfying div $F = 0$ in a region $\Omega \subseteq \mathbb{R}^2$; that is F has no sources or sinks in Ω. See Figure 10.17. For instance, we might have

$$F(x, y) = \left(\frac{x}{x^2 + y^2}, \ \frac{y}{x^2 + y^2} \right)$$

with Ω the region $1 \le x^2 + y^2 \le 9$ between two circles of radius 1 and 3, respectively. It is easy to check that, for $(x, y) \ne \mathbf{0}$,

$$\text{div } F = \frac{\partial}{\partial x} \left(\frac{x}{x^2 + y^2} \right) + \frac{\partial}{\partial y} \left(\frac{y}{x^2 + y^2} \right) = 0.$$

Let $\mathcal{C}$ be the curve in Figure 10.18. We claim that

$$\int_{\mathcal{C}} F \cdot N \, ds = \int_c F \cdot N \, ds = 2\pi,$$

where c is the unit circle $\|X\| = 1$ oriented counterclockwise. If F represents the velocity vector of a fluid in Ω, this asserts that the net flow across the curve $\mathcal{C}$ equals that across the unit circle c. This is plausible if one observes that the vector field F is radial outward of length 1 at every point on c. It also appears that there is a source at the origin.

The computation is easy. Let $\mathcal{B}$ denote the region between $\mathcal{C}$ and the unit circle $\|X\| = 1$. Then by the Divergence Theorem 10.2.4,

$$0 = \iint_{\mathcal{B}} \text{div } F \, dA = \int_{\partial \mathcal{B}} F \cdot N \, ds.$$

But $\partial \mathcal{B} = \mathcal{C} - c$, where we have $-c$, not c, since its orientation is reversed when considered part of the boundary of $\mathcal{B}$. Therefore

$$0 = \int_{\partial \mathcal{B}} F \cdot N \, ds = \int_{\mathcal{C}} F \cdot N \, ds - \int_{c} F \cdot N \, ds.$$

We now evaluate the integral around c. There are two methods. The easiest is to observe that on c, the radius vector $X = (x, y)$ is the unit outer normal. Thus $F = N$ is also the unit outer normal, so that $F \cdot N = 1$ on c. Consequently

$$\int_{c} F \cdot N \, ds = \int_{c} ds = 2\pi.$$

The second method is to parametrize c as $x = \cos\theta$, $y = \sin\theta$, $0 \le \theta \le 2\pi$. Then $F = (\cos\theta, \sin\theta)$ on c and, as we saw just before Theorem 10.2.4, $N = (\cos\theta, \sin\theta)$. Therefore $F \cdot N = 1$, and

$$\int_{c} F \cdot N \, ds = \int_{0}^{2\pi} 1 \, d\theta = 2\pi.$$

This number 2π represents the magnitude of the source at the origin.

Other applications of the various forms of Stokes' Theorem appear in later sections.

Exercises

1. As a check of Stokes' Theorem 10.2.2 or 10.2.3, compute both sides separately and showing that they are equal, for each of the following:

   (a) $F(x, y) = (1 - y, \, x)$, $\mathcal{B}$ is the disk $x^2 + y^2 \le 4$

   (b) $F(x, y) = (x + y, \, x - 6y)$, $\mathcal{B}$ is the triangular region with vertices at $(-4, 1)$, $(2, 1)$, and $(2, 5)$

   (c) $F(x, y) = (xy, -xy)$, $\mathcal{B}$ is the disk $x^2 + y^2 \le 9$

   (d) $F(x, y) = (y \sin \pi x, \, y \cos \pi x)$, $\mathcal{B}$ is the rectangular region given by $1 \le x \le 2$ and $-1 \le y \le 1$

   (e) $F(x, y) = (x^2 + y, \, e^y - x)$, $\mathcal{B}$ is the region above the parabola $y = x^2$ and below the line $y = 1$

   (f) $F(x, y) = (x^2 + y^2, \, -2xy)$, $\mathcal{B}$ is the half-disk $x^2 + y^2 \le 16$, $y \ge 0$

   (g) $F(x, y) = (x, \, 2x + y)$, $\mathcal{B}$ is the ring region $1 \le x^2 + y^2 \le 4$

   (h) $F(x, y) = (y^2, \, x^2)$, $\mathcal{B}$ is the region outside the circle $x^2 + y^2 = 1$ but inside the rectangle $-2 \le x \le 3$, $-4 \le y \le 5$.

2. (a) Given a bounded region $B \subseteq \mathbb{R}^2$, show that

$$\text{Area } (B) = \frac{1}{2} \int_{\partial B} x \, dy - y \, dx \,.$$

(b) Use part (a) to find the area inside the ellipse

$$(x, y) = (a \cos \varphi, b \sin \varphi), \ 0 \le \varphi \le 2\pi \,.$$

(c) Use part (a) to find the area of the region $1 \le x^2 + y^2 \le 25$.

3. If $u : \mathbb{R}^2 \to \mathbb{R}$ is a function with continuous first and second partial derivatives, prove that

$$\iint_B \Delta u \, dA = \int_{\partial B} \frac{\partial u}{\partial N} \, ds.$$

4. Let a force field $F = (2 - 5x + y, x)$. Find the work expended in moving a particle of unit mass counterclockwise around the boundary of the triangle with vertices at $(1, 1)$, $(3, 1)$, and $(3, 5)$.

5. Let C_1 denote the circle $x^2 + y^2 = 1$ and C_2 the circle $(x-2)^2 + (y-1)^2 = 25$, both oriented counterclockwise, and let B denote the ring region between these curves. If a vector field F satisfies div $F = 0$ in B, show that

$$\int_{C_1} F \cdot N \, ds = \int_{C_2} F \cdot N \, ds.$$

(Note that the proof of this extends immediately to the case where C_1 and C_2 are replaced by more general curves and B is the region between them.)

6. For certain kinds of heat flow, the temperature $u(x, y, t)$ in a region $B \subset \mathbb{R}^2$ satisfies

$$u_t = \Delta u - u.$$

Prove that there is at most one solution that has a given initial temperature $u(x, y, 0) = f(x, y)$, $(x, y) \in B$, and a given boundary temperature $u(x, y, t) = \varphi(x, y, t)$ for $(x, y) \in \partial B$ and $t \ge 0$. (The same function $E(t)$ given in Section 10.2d works here.)

7. Let $F = (p, q)$ be a vector field with a continuous derivative in $\mathbb{R}^2$, and let $B \subseteq \mathbb{R}^2$ be a region with a smooth boundary ∂B. If F is a constant vector on ∂B, show that

$$\iint_B (q_x - p_y) \, dA = 0 \,.$$

10.3   Surface integrals

10.3A   SMOOTH ORIENTED SURFACES IN $\mathbb{R}^3$

Stokes' Theorem generalizes to three and higher dimensions. We content ourselves with an intuitive description of the $\mathbb{R}^3$ case without proofs. For this case we require the notion of integration on an oriented smooth surface Σ in $\mathbb{R}^3$.

Let Σ be a surface in $\mathbb{R}^3$. We picture Σ as consisting of two disjoint pieces, the **surface interior**, which we will denote by Σ_o, and the **surface boundary**, which we denote by $\partial\Sigma$. Suppose that the surface boundary $\partial\Sigma$ is the union of a finite number of closed curves, and further suppose that at each point X on the surface interior Σ_o, there is a tangent plane. Then we call Σ a **smooth surface**. If the surface boundary of Σ is empty, then we call Σ a **closed surface**.

EXAMPLES:

1. The torus and the sphere in $\mathbb{R}^3$ are examples of smooth surfaces Σ such that $\partial\Sigma = \emptyset$, since both of these surfaces have tangent planes at every point. Thus they are both closed surfaces.

2. The upper hemisphere

$$\Sigma = \{(x, y, z) : x^2 + y^2 + z^2 = 1 \text{ and } z \geq 0\}$$

has the set

$$\Sigma_o = \{(x, y, z) : x^2 + y^2 + z^2 = 1 \text{ and } z > 0\}$$

as its surface interior, and the circle

$$\partial\Sigma = \{(x, y, z) : x^2 + y^2 + z^2 = 1 \text{ and } z = 0\}$$

as its surface boundary.

3. Consider the set

$$\Sigma = \{(x, y, z) : z^2 = x^2 + y^2 \text{ and } a \geq z \geq b\},$$

where $a > b > 0$. This is the circular cone $z = \|(x, y)\|$ truncated by the planes $z = a$ and $z = b$. It looks somewhat like an upside-down lampshade. Its surface interior is

$$\Sigma_o = \{(x, y, z) : z^2 = x^2 + y^2 \text{ and } a > z > b\},$$

and its surface boundary consists of two horizontal circles centered on the z-axis. One of these circles has radius a and lies in the plane $z = a$, and the other has radius b and lies in the plane $z = b$.

4. If we let $b = 0$ in the preceding example, we obtain a surface that looks like an ice cream cone (or a dunce cap). One piece of the surface boundary is the same as before, namely a circle of radius a in the plane $z = a$. The other piece has shrunk to a single point, namely the origin. You may not find it natural to think of this point as a closed curve, or even as a surface boundary point. However, it is the image of the function $F(t) = (0, 0, 0)$ (with $0 \leq t \leq 1$, say), so it is a curve. And for the purposes of this discussion, we want to consider it a surface boundary point, since there is no tangent plane to the cone at the origin.

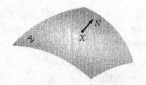

FIGURE 10.19. A surface in $\mathbb{R}^3$

To orient a smooth surface Σ, we must designate one side of the surface as the positive side. We do this by supposing that at each point X on the surface interior Σ_o there is attached a unit vector $N = N(X)$ that is perpendicular to the tangent plane at X. See Figure 10.19. The direction of N is called the **orientation** of Σ at X, and N is called the **positive unit normal vector** of Σ at X. The adjective "positive" signifies that N points in the direction of the orientation of Σ. The vector $-N$ is the **negative** unit normal of Σ at X. Note that N is a vector-valued function from Σ_o to $\mathbb{R}^3$. If N is a *continuous* function, we say that Σ is **oriented** by N.

EXAMPLE:

5. Not all surfaces can be oriented. The "Moebius strip" is a famous example of this. The Moebius strip may be parametrized as follows:

$$G(\theta,t) = ((1 + t \cos \theta) \cos 2\theta, (1 + t \cos \theta) \sin 2\theta, t \sin \theta),$$

where $0 \le \theta \le 2\pi$ and $-\frac{1}{2} \le t \le \frac{1}{2}$. Ask someone to show you a few of the strange and fascinating properties of this surface. We guarantee that he or she will be delighted to do so.

Once a surface Σ is oriented, we are automatically led to an orientation of the surface boundary $\partial\Sigma$ by thinking of Σ as a piece of a "curved plane". See Figure 10.20, where Σ looks like part of a curved plane with a hole cut out of it. Imagine that you are standing on the positive side of an oriented surface, with your right foot on one of the curves that makes up its surface boundary, and your left side toward the surface interior. Then you are facing in the direction of the orientation of the curve. Try to visualize yourself doing this near each of the two boundary curves in Figure 10.20. Note that if you stand on the opposite (negative) side of Σ near one of the curves and face in the direction of orientation of the curve, then the surface interior will be on your *right*, and your left foot will be on the boundary curve.

We now describe a natural way in which a surface can be oriented by using a parametrization. Suppose $G = (g_1, g_2, g_3)$ is a parametrization of the surface Σ in $\mathbb{R}^3$. Let (s_0, t_0) be a point in the do-

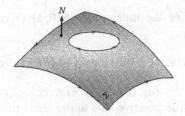

FIGURE 10.20. An oriented surface

main of G where G is differentiable. We recall the notation

$$D_1 G(s,t) = \frac{\partial G}{\partial s}(s,t) = \left(\frac{\partial g_1}{\partial s}(s,t), \ \frac{\partial g_2}{\partial s}(s,t), \ \frac{\partial g_3}{\partial s}(s,t) \right)$$

and

$$D_2 G(s,t) = \frac{\partial G}{\partial t}(s,t) = \left(\frac{\partial g_1}{\partial t}(s,t), \ \frac{\partial g_2}{\partial t}(s,t), \ \frac{\partial g_3}{\partial t}(s,t) \right)$$

from Sections 8.3 and 9.5. If the linear transformation $G'(s_0, t_0)$ has rank 2, or in other words, if the vectors $D_1 G(s_0, t_0)$ and $D_2 G(s_0, t_0)$ are linearly independent, Theorem 8.3.2 tells us that the plane $\mathcal{P}$ parametrized by

$$s D_1 G(s_0, t_0) + t D_2 G(s_0, t_0) + G(s_0, t_0)$$

is tangent to the image of G at the point (s_0, t_0). Let N be the unit vector that has the same direction as the cross product $D_1 G(s_0, t_0) \times D_2 G(s_0, t_0)$. Then N is orthogonal to the two vectors $D_1 G(s_0, t_0)$ and $D_2 G(s_0, t_0)$, and since those two vectors have directions parallel to the tangent plane $\mathcal{P}$, it follows that N points in a direction that is perpendicular to the tangent plane of Σ at $X = G(s_0, t_0)$. Thus the parametrization G leads to an orientation of the surface at the point $G(s_0, t_0)$.

Now suppose for each point (s,t) of the domain of G such that $G(s,t)$ is on the surface interior Σ_o, the derivative $G'(s,t)$ is continuous and has rank 2. For $X = G(s,t)$, let

$$N(X) = \frac{D_1 G(s,t) \times D_2 G(s,t)}{\| D_1 G(s,t) \times D_2 G(s,t) \|} \ .$$

Since $G'(s,t)$ is continuous, the vector-valued function N is continuous, *provided* it is well-defined. A problem with the definition may arise if G maps more than one point in its domain to the same point X on the surface, for then it would be possible to get two different normal vectors $N(X)$. If we obtain the same vector $N(X)$ from the above formula for all points in the domain of G that are mapped to X, then $N(X)$ is well-defined on Σ_o, and we say that Σ is oriented **by the parametrization** G.

EXAMPLES:

6. The upper half of the unit sphere $S(\mathbf{0}; 1)$ in $\mathbb{R}^3$ is oriented by the parametrization

$$G(\theta, \varphi) = (\cos\theta \sin\varphi, \sin\theta \sin\varphi, \cos\varphi) \,,$$

where $0 \leq \theta \leq 2\pi$ and $0 \leq \varphi \leq \pi/2$. Let us determine which side of the sphere is the positive side under this orientation. We compute the positive normal vector at the point $(\sqrt{2}/2, 0, \sqrt{2}/2) = G(0, \pi/4)$. The vectors $D_1 G(0, \pi/4)$ and $D_2 G(0, \pi/4)$ are computed by taking partial derivatives with respect to θ and φ. We get

$$D_1 G(0, \pi/4) = (0, \sqrt{2}/2, 0) \text{ and } D_2 G(0, \pi/4) = (\sqrt{2}/2, 0, -\sqrt{2}/2) \,,$$

so $N(\sqrt{2}/2, 0, \sqrt{2}/2)$ is the unit vector in the direction of the vector

$$(0, \sqrt{2}/2, 0) \times (\sqrt{2}/2, 0, -\sqrt{2}/2) = (-1/2, 0, -1/2) \,.$$

The third coordinate of this vector is negative, so it points toward the "inside" of the hemisphere. Similar computations can be made at other points, and in each case, the positive normal vector has a negative third coordinate. Thus, the inside of the hemisphere is its positive side under this particular orientation. The orientation of the surface boundary, the "equator", is clockwise around the vertical axis (when viewed from above). Note that if we parametrize the equator by setting $\varphi = \pi/2$,

$$F(\theta) = G(\theta, \pi/2) \,,$$

then F gives us the reverse of this orientation. (This is a coincidence, not a general rule.)

7. You will be asked to show in the exercises that the parametrization given earlier for the Moebius band *does not* produce a well-defined unit normal vector at every point. It is in fact true that if a smooth surface Σ has a parametrization G that has a continuous derivative of rank 2 at every point in its domain and produces two different normal vectors $N(X)$ at some point X on Σ_o, then Σ cannot be oriented.

REMARKS:

- If a surface Σ is oriented by a parametrization $G(s, t)$, then it is oriented in the opposite direction by the parametrization $H(s, t) = G(t, s)$, since the positive normal vector for the parametrization H points in the direction of

$$\begin{aligned} D_1 H(s, t) \times D_2 H(s, t) &= D_2 G(t, s) \times D_1 G(t, s) \\ &= -D_1 G(s, t) \times D_2 G(s, t) \,. \end{aligned}$$

In other words, simply by switching the order of the variables, we change the orientation to the opposite direction. Try this with the parametrization of the hemisphere given above.

- In the example of the hemisphere, we glossed over a minor technical point, namely that the vectors D_1G and D_2G are not linearly independent when $\varphi = 0$, since $D_1G(\theta, 0) = \mathbf{0}$ for all θ. When $\varphi = 0$, $G(\theta, \varphi)$ is the "north pole" $(0, 0, 1)$, so G fails to provide us with a unit normal vector at $(0, 0, 1)$. There are several ways to take care of this problem, but perhaps it is best to either not worry about it, or to think of such a point as a surface boundary point, such as we did for the vertex of the cone in Example 4. Note that there is in fact a tangent plane at the north pole, and that the orientation provided by G can be extended continuously to give the unit normal vector $N(0, 0, 1) = (0, 0, -1)$ at that point. A similar technicality occurs at the south pole when we extend this parametrization to the whole sphere by letting φ range between 0 and π.

10.3B DEFINITION OF A SURFACE INTEGRAL

We are now ready to discuss the integral of a vector field F in $\mathbb{R}^3$ over an oriented surface Σ in $\mathbb{R}^3$. We assume that Σ is a smooth oriented surface, with positive unit normal vector $N(X)$ at each point X in Σ_o. Imagine that Σ is partitioned into tiny pieces Σ_i that have surface area ΔA_i. Let X_i be a point on Σ_i. Then the sum

$$\sum_i F(X_i) \cdot N(X_i) \Delta A_i$$

is to be our approximation of the **surface integral**

$$\int_\Sigma F(X) \cdot N(X) \, dA .$$

You should think about the analogy between this description of the surface integral and our development of the line integral. A completely rigorous definition of the surface integral is beyond the scope of this book.

REMARK:

- One physical interpretation of the quantity $F(X) \cdot N(X) \, dA$ is in terms of wind pressure on a surface, such as a sail. The vector field F indicates the direction and strength of the wind at each point. The inner product $F(X) \cdot N(X)$ gives the force per unit area exerted on a surface perpendicular to $N(X)$, and the quantity $F(X) \cdot N(X) \, dA$ is the force exerted on the small piece of the sail at the point X with surface area dA. Note that if the force $F(X)$ is perpendicular to $N(X)$, then the inner product is 0, and no force is exerted on the surface at X. This is as expected.

Just as with line integrals, in order to compute a surface integral, we need a parametrization. Thus we assume that Σ is oriented by a parametrization G. To derive a formula for surface integrals on Σ, we think of each of the small pieces Σ_i as images under G of rectangles in the domain of G. Let $\mathcal{R}$ be the domain of G, and partition $\mathcal{R}$ (at least approximately) into small rectangles $\mathcal{R}_i$ centered at (s_i, t_i), with sides of length Δs_i and Δt_i parallel to the coordinate axes in $\mathbb{R}^2$. Let $\Sigma_i = G(\mathcal{R}_i)$. Then Σ_i has approximately the same area as the parallelogram $G'(s_i, t_i)(\mathcal{R}_i)$ in $\mathbb{R}^3$. According to Theorem 3.7.5, the area of the parallelogram is

$$\|D_1 G(s_i, t_i) \times D_2 G(s_i, t_i)\| \cdot (\text{ area of } \mathcal{R}_i) .$$

Letting $X_i = (s_i, t_i)$, it follows that

$$\sum_i F(X_i) \cdot N(X_i) \Delta A_i \approx$$

$$\sum_i F(G(s_i, t_i)) \cdot N(G(s_i, t_i)) \|D_1 G(s_i, t_i) \times D_2 G(s_i, t_i)\| \Delta s_i \Delta t_i .$$

Using our formula for N in terms of the parametrization G, we have

$$\sum_i F(X_i) \cdot N(X_i) \Delta A_i \approx \sum_i F(G(s_i, t_i)) \cdot (D_1 G(s_i, t_i) \times D_2 G(s_i, t_i)) .$$

This reasoning justifies the following theorem:

Theorem 10.3.1 Let F be a continuous vector field defined on $\mathbb{R}^3$, and let Σ be a smooth surface in $\mathbb{R}^3$ that is oriented by a parametrization G with domain $\mathcal{R}$. Then

$$\int_\Sigma F(X) \cdot N(X) \, dA \;\; = \;\; \int\int_\mathcal{R} F(G(s,t)) \cdot (D_1 G(s,t) \times D_2 G(s,t)) \, dA$$

$$= \;\; \int\int_\mathcal{R} [F \circ G, \, D_1 G, \, D_2 G] \, dA .$$

(Recall that $[X, Y, Z]$ is the notation for the triple product of three vectors in $\mathbb{R}^3$.)

EXAMPLE:

8. We compute the surface integral of $F(x, y, z) = (y, x, z)$ over the portion of the cone $z = \|(x, y)\|$ where $z \leq 3$. We orient the cone so that the positive side is the "outside". Thus all the positive unit normal vectors will have *negative* third coordinate. A parametrization for this orientation is

$$G(\theta, t) = (t \cos \theta, t \sin \theta, t) ,$$

where $0 \leq \theta \leq 2\pi$ and $0 \leq t \leq 3$. At the point $G(\theta, t)$, we have

$$
\begin{aligned}
D_1 G(\theta, t) \times D_2 G(\theta, t) &= (-t \sin \theta, t \cos \theta, 0) \times (\cos \theta, \sin \theta, 1) \\
&= (t \cos \theta, t \sin \theta, -t) .
\end{aligned}
$$

Note that this vector has negative third coordinate. Now we compute the integrand

$$
F(G(\theta, t)) \cdot D_1 G(\theta, t) \times D_2 G(\theta, t) = t^2 (\sin 2\theta - 1) .
$$

This integrand is easily integrated over the rectangle determined by $0 \leq \theta \leq 2\pi$, $0 \leq t \leq 3$ in the θt-plane. The answer is -18π.

10.3c STOKES' THEOREM IN $\mathbb{R}^3$

In this section we give two versions of Stokes' Theorem for surface integrals. For the first version, we need a definition. If $F = (p, q, r)$ is a differentiable vector field in $\mathbb{R}^3$, we define curl F to be the vector field

$$
\text{curl } F = (r_y - q_z, p_z - r_x, q_x - p_y) .
$$

EXAMPLE:

9. Since the curl F is a new object, we compute it for $F(X) = (xyz, y - 3z, 2y)$. Then $p = xyz$, $q = y - 3z$, and $r = 2y$, so that curl $F = (2 + 3, xy - 0, 0 - xz) = (5, xy, -xz)$.

We remark that the complicated formula for curl F is usually remembered by thinking of it as the symbolic cross product of the operator $\nabla = (\partial/\partial x)\mathbf{i} + (\partial/\partial y)\mathbf{j} + (\partial/\partial z)\mathbf{k}$ with the vector F and is written

$$
\text{curl } F = \nabla \times F = \begin{vmatrix} \mathbf{i} & \mathbf{j} & \mathbf{k} \\ \frac{\partial}{\partial x} & \frac{\partial}{\partial y} & \frac{\partial}{\partial z} \\ p & q & r \end{vmatrix} .
$$

Theorem 10.3.2 (Stokes' or Ostrogradsky's Theorem) Let

$$
F(x, y, z) = (p(x, y, z), q(x, y, z), r(x, y, z))
$$

be a differentiable vector field in $\mathbb{R}^3$. Then for any smooth oriented surface Σ

$$
\iint_\Sigma (\text{curl } F) \cdot N \, dA = \int_{\partial \Sigma} F(X) \cdot dX .
$$

REMARK:

- In the special case when Σ is a region in the xy-plane, so that $N = (0, 0, 1)$, $r = 0$, and $F = (p, q, 0)$, then

$$(\text{curl } F) \cdot N = q_x - p_y,$$

and the formula simplifies to

$$\iint_\Sigma (q_x - p_y) \, dA = \int_{\partial \Sigma} F(X) \cdot dX,$$

which is just our Theorem 10.2.2.

The second generalization of Stokes' Theorem is the three-dimensional version of the Divergence Theorem. Let $F = (p, q, r)$ be a differentiable vector field in $\mathbb{R}^3$. Then, by definition,

$$\text{div } F = \frac{\partial p}{\partial x} + \frac{\partial q}{\partial y} + \frac{\partial r}{\partial z}$$

Let $\mathcal{B} \subset \mathbb{R}^3$ be a solid region whose "exterior" $\partial \mathcal{B}$ is an oriented surface with positive unit normal N pointing "away from" $\mathcal{B}$, then the Divergence Theorem states that

$$\iiint_\mathcal{B} \text{div } F \, dV = \iint_{\partial \mathcal{B}} F \cdot N \, dA,$$

where $dV = dx \, dy \, dz$ is the element of volume in $\mathbb{R}^3$. The physical interpretation made before in terms of fluid flow carries over to this three-dimensional case without change.

It is straightforward to prove the three-dimensional versions of Green's identities by imitating the two-dimensional proofs. For example, Green's first identity becomes

$$\iiint_\mathcal{B} v \Delta u \, dV = \iint_{\partial \mathcal{B}} v \nabla u \cdot N \, dA - \iiint_\mathcal{B} \nabla v \cdot \nabla u \, dV,$$

where $\Delta u = u_{xx} + u_{yy} + u_{zz}$ and $\nabla v = (v_x, v_y, v_z)$.

Although it might be difficult to believe, these theorems are quite easy to use in most standard applications. Here are a few.

EXAMPLES:

10. Let $F = (x, 3y, -z)$, and let Σ be the sphere $S((0, 0, 0); a)$. Then, if $\mathcal{B} = B((0, 0, 0); a)$ is the solid ball of radius a, we have $\Sigma = \partial \mathcal{B}$, and

$$\iint_\Sigma F \cdot N \, dA = \iiint_\mathcal{B} \text{div } F \, dV = 4\pi a^3.$$

To see this, we compute $\text{div } F = 1 + 3 - 1 = 3$, and so the right side is three times the volume of the ball; that is, $3(4\pi a^3/3) = 4\pi a^3$ as claimed.

11. Let $u(x, y, z)$ be a function that satisfies the **Laplace equation**

$$\frac{\partial^2 u}{\partial x^2} + \frac{\partial^2 u}{\partial y^2} + \frac{\partial^2 u}{\partial z^2} = 0;$$

that is, $\Delta u = 0$, in a domain $\mathcal{B}$ in $\mathbb{R}^3$. Such functions are called **harmonic functions**. They arise often in applications. For example, u might be the temperature distribution in a body at thermal equilibrium (this means that the temperature is independent of time). If the temperature is zero on the boundary of the body, $u = 0$ on $\partial\mathcal{B}$, we claim that it must be zero throughout the body. This physically plausible assertion is easy to prove. We apply Green's first identity, stated above, in the special case when $u = v$. Since $\Delta u = 0$, the integral on the left is zero. Moreover, $u = 0$ on $\partial\mathcal{B}$. Thus the double integral over $\partial\mathcal{B}$ is also zero. This leaves

$$\iiint_{\mathcal{B}} \|\nabla u\|^2 \, dV = 0,$$

which implies that $\nabla u = 0$. That is, $u' = O$ throughout $\mathcal{B}$. Therefore $u \equiv$ constant in $\mathcal{B}$. But $u = 0$ on $\partial\mathcal{B}$, and so the value of the constant is zero; that is, $u \equiv 0$ throughout $\mathcal{B}$. The proof is complete.

10.3D GRAVITATIONAL FORCE

The purpose of this section is to briefly investigate the gravitational force due to a spherically symmetric solid. We will apply some of the results obtained so far to prove that in this case the gravitational force is the same as that due to a point mass. Although the problem caused Newton considerable difficulty, we find it quite easy. To use Newton's phrase, however, "We are standing on the shoulders of giants".

Recall (Section 5.3) Newton's law of gravitation for the force F on a mass m at X due to a mass M at Y:

$$F = -\gamma \frac{mM}{\|X - Y\|^3}(X - Y).$$

It is convenient to introduce the force at X per unit mass, $G = F/m$,

$$G(X) = -\gamma \frac{M}{\|X - Y\|^3}(X - Y).$$

$G(X)$ is called the **gravitational field** at X. By a straightforward computation, one sees that

$$\operatorname{div} G(X) = 0, \quad \text{for } X \neq Y.$$

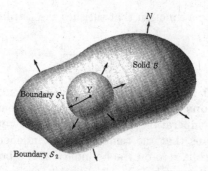

FIGURE 10.21. A point mass contained in a hole in $\mathcal{B}$

Thus if $\mathcal{B} \subseteq \mathbb{R}^3$ is any region whose boundary is a finite union of closed smooth surfaces, and if $\mathcal{B}$ does not contain Y (see Figure 10.21, where Y is in a hole in $\mathcal{B}$), then the Divergence Theorem yields

$$0 = \iiint_{\mathcal{B}} \operatorname{div} G \, dV = \iint_{\partial \mathcal{B}} G \cdot N \, dA = \iint_{\mathcal{S}_2} G \cdot N \, dA - \iint_{\mathcal{S}_1} G \cdot N \, dA.$$

Here we have used $-\mathcal{S}_1$, since the orientation given to $\mathcal{S}_1$ in Figure 10.21 (its positive normal points into $\mathcal{B}$) is the negative normal when regarded as part of the boundary of $\mathcal{B}$ (denoted $\partial \mathcal{B}$ above). Consequently, for the smooth surfaces $\mathcal{S}_1, \mathcal{S}_2$, with $Y \notin \mathcal{B}$,

$$\iint_{\mathcal{S}_1} G \cdot N \, dA = \iint_{\mathcal{S}_2} G \cdot N \, dA.$$

Now we pick a convenient surface for $\mathcal{S}_1$: a sphere of radius r with center at Y. Then for $X \in \mathcal{S}_1$, the positive unit normal to $\mathcal{S}_1$ is $N = (X - Y)/(\|X - Y\|)$ and $r = \|X - Y\|$. Therefore, if $X \in \mathcal{S}_1$,

$$G \cdot N = -\frac{-\gamma M}{\|X - Y\|^2} = -\frac{\gamma M}{r^2}.$$

Now, since the area of $\mathcal{S}_1 = 4\pi r^2$, we have

$$\iint_{\mathcal{S}_2} G \cdot N \, dA = \iint_{\mathcal{S}_1} G \cdot N \, dA = -\frac{\gamma M}{r^2} \iint_{\mathcal{S}_1} dA = -4\pi \gamma M.$$

All this has assumed that the mass M (at Y) lies inside the surface $\mathcal{S}_1$. On the other hand, if M lies outside $\mathcal{S}_2$, then $\operatorname{div} G = 0$ throughout $\mathcal{B}$. See Figure 10.22. Hence

$$\iiint_{\mathcal{B}} \operatorname{div} G \, dV = \iint_{\mathcal{S}_2} G \cdot N \, dA = 0.$$

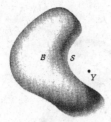

FIGURE 10.22. A point mass at Y outside the surface $\mathcal{S}$

Combining these two cases, we conclude that if $\mathcal{S}$ is any closed surface, then

$$\iint_{\mathcal{S}} G \cdot N \, dA = \left\{ \begin{array}{ll} 0, & M \text{ outside } \mathcal{S}, \\ -4\pi\gamma M, & M \text{ inside } \mathcal{S}. \end{array} \right.$$

It is important to observe that this integral, called the total "flux" of the gravitational field through the surface $\mathcal{S}$, depends only on the mass M and whether M is inside or outside $\mathcal{S}$, but not on the particular location Y of the mass nor on the shape of the surface.

Now suppose that there are lots of masses $M_1, M_2, \ldots, M_n$ at the points $Y_1, Y_2, \ldots, Y_n$. Let $G_j(X)$ be the gravitational field at X due to the mass M_j at Y_j. Since force is a vector, the gravitational field G due to all these masses is the sum

$$G = G_1 + \ldots + G_n.$$

Consequently, if $\mathcal{S}$ is any closed surface (see Figure 10.23), then the result above applied to each mass separately yields

$$\iint_{\mathcal{S}} G \cdot N \, dA = \sum_{j=1}^{n} \iint G_j \cdot N \, dA$$

$$= -4\pi\gamma \cdot (\text{total mass inside } \mathcal{S});$$

that is,

$$\iint_{\mathcal{S}} G \cdot N \, dA = -4\pi\gamma \cdot (\text{total mass inside } \mathcal{S}).$$

This is called **Gauss's Law**.

More generally yet, if the mass distribution in a region $\mathcal{B}$ is given in terms of a density ρ, then

$$M = \text{ total mass in } \mathcal{B} = \iiint_{\mathcal{B}} \rho \, dV.$$

Gauss's Law clearly extends to this case, as one can see by thinking of M as the sum $\sum \Delta M$, where $\Delta M = \rho \, \Delta V$ is the mass in a small element of volume ΔV.

We now apply this to the case of a spherical body $\mathcal{B}$, like the earth, and see how to compute the gravitational field. We assume that the mass

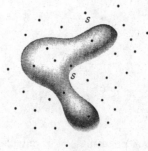

FIGURE 10.23. Many point masses outside the surface S

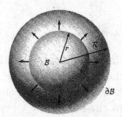

FIGURE 10.24. Spherically symmetric mass distribution

distribution is symmetric and place the origin at the center of $\mathcal{B}$. See Figure 10.24. This implies, by symmetry, that the gravitational force $F(X)$ and hence the gravitational field $G(X)$ are in the radial direction. Consequently, if $\mathcal{S}_r$ is any sphere of radius r concentric with $\partial\mathcal{B}$, possibly inside $\mathcal{B}$, *we have* $G \cdot N = -\|G\|$, where N is the limit outer normal, in the radial direction, to $\mathcal{S}_r$. Moreover, again by symmetry, we see that $\|G(X)\|$ is constant for all X on $\mathcal{S}_r$. Therefore

$$\pm \iint_{\mathcal{S}_r} G \cdot N \, dA = \|G\| \iint_{\mathcal{S}_r} dA = 4\pi r^2 \|G\|.$$

Combined with Gauss's Law, we find that

$$G(X) \cdot N(X) = \|G(X)\| = -\frac{\gamma M_r}{r_2} = -\frac{\gamma M_r}{\|X\|_2},$$

for X on $\mathcal{S}_r$. Here M_r is the total mass inside the sphere $\mathcal{S}_r$. Since the direction of $G(X)$ is radial, that is, in the direction of the unit vector $-X/\|X\|$, we finally conclude that

$$G(X) = -\frac{\gamma M_r}{\|X\|^3} X,$$

and so

$$F(X) = -\frac{\gamma m M_r}{\|X\|^3} X.$$

Of course, if $r \geq R$, where R is the radius of $\mathcal{B}$, then $M_r = M$, the total mass of $\mathcal{B}$. Since, given any point X, we can construct the sphere S_r with $r = \|X\|$, the last formula proves that the gravitational force at X due to a spherically symmetric mass distribution is the same as that from a point mass M_r at the center of the body, where M_r is the mass of the body at distance $\leq r = \|X\|$ from the center of the body.

Exercises

1. Let $F(x, y, z) = (2xy, yz, z)$ be a vector field in $\mathbb{R}^3$. Use the divergence form of Stokes' Theorem to evaluate

$$\iint_\Sigma F \cdot N \, dA,$$

   where Σ is the boundary of the cubical region $\|x\| \leq 1$, $\|y\| \leq 1$, $\|z\| \leq 1$.

2. Let $F(x, y, z) = (x - y, y^2 - z, x^2 + 5z)$ be a vector field in $\mathbb{R}^3$. Evaluate

$$\iint_\Sigma F \cdot N \, dA,$$

   where Σ is the sphere $x^2 + y^2 + z^2 = 9$.

3. Compute curl F for the following vector fields F:

   (a) $F(x, y, z) = (y, \ z, \ 2 - z)$

   (b) $F(x, y, z) = (xy + z, \ yz + x, \ zx + y)$

   (c) $F(x, y, z) = (x^2 + y^2, \ 0, \ 0)$

   (d) $F(x, y, z) = (y - z, \ z + x, \ x + y)$.

4. Let $F = (p, q, r)$ be a twice differentiable vector field in $\mathbb{R}^3$ and u a twice differentiable real-valued function. Prove the following:

   (a) div (curl F) = 0

   (b) curl (grad u) = 0

   (c) curl $(uF) = u$ curl $F + $ (grad u) $\times F$

   (d) if U is a constant vector, then curl $(U \times (x, y, z)) = 2U$.

5. Let F and G be differentiable vector fields in $\mathbb{R}^3$. Show that:

   (a) div $(F \times G) = ($curl $F) \cdot G - F \cdot$ curl G

   (b) curl $(F + G) = $ curl $F + $ curl G

   (c) if curl $F = G$, then div $G = 0$.

6. Is there a vector field $F = (p, q, r)$ such that

   (a) curl $F = (x, y, z)$? Why?

   (b) curl $F = 2(z - y, \ x - z, \ y - x)$? Why?

7. Indicate on a sketch the orientation given to the torus by the parametrization in Example 2 of Section 8.0.

8. Find a parametrization of $S((0,0,0);1)$ that gives it an outward orientation.

9. Let

$$\Sigma = \{(x,y,z) : x^2 + y^2 = 1 \text{ and } -1 \le z \le 1\}.$$

(a) Draw Σ.

(b) Find a parametrization of Σ that gives it an "outward" orientation. Indicate this orientation on your sketch of Σ, along with the corresponding orientation of the surface boundary curves.

(c) Calculate the surface integral of $F(x,y,z) = (x,y,z)$ over Σ. Give an interpretation of this result. *Hint:* What is the surface area of Σ?

(d) Calculate the surface integral of $F(x,y,z) = (x^2 + y^2, 0, xyz)$ over Σ.

(e) Show that the third coordinate function of F in the preceding part is irrelevant. Can you explain why?

(f) Use your answers to parts (d) and (e) to determine the value of the surface integral of the constant function $F(x, y, z) = (1, 0, 0)$ over Σ without further integration.

(g) Compute the surface integral of

$$F(x,y,z) = (5, xy, -xz) = \text{curl } (xyz, y - 3z, 2y)$$

over Σ, first by using the general formula for surface integrals, and second by using Stokes' Theorem.

10. Calculate a unit normal vector at each point of the surface interior of the Moebius strip of Example 5. Verify that the parametrization given there produces two different normal vectors at some points on the surface interior.

11. Let g be a scalar-valued function defined on a simple region $\mathcal{R}$ in $\mathbb{R}^2$, and let Σ be the graph of g. Assume that ∇g exists, is not equal to $\mathbf{0}$, and is continuous at all points in the interior of the domain of g.

(a) Show that Σ is oriented by the unit normal vector

$$N(x,y,z) = \frac{(-D_1 g(x,y), -D_2 g(x,y), 1)}{\|(-D_1 g(x,y), -D_2 g(x,y), 1)\|}.$$

This orients Σ so that its "top" is the positive side.

(b) Use the answer to the previous part to derive a formula for the surface integral of a vector field F over the surface Σ.

12. Calculate the surface integrals of the vector field F over the oriented surface Σ.

(a) $F(x,y,z) = (e^{x+y+z}, xy, z + 1)$, Σ oriented by the parametrization $G(s,t) = s(0, 1, -2) + t(1, 1, 3) + (2, -1, 0)$, where (s,t) lies in the triangle bounded by the lines $s = 0, t = 0, s + t = 1$

(b) $F(x,y,z) = (yz, x + z, 2)$, Σ is the graph of $g(x,y) = x^2 - y^2$, $0 \le x, y \le 1$, oriented upward

(c) $F(X) = X$, Σ is the torus of Example 2 in Section 8.0, oriented outward.

13. Let Σ be the hemisphere $x^2 + y^2 + z^2 = 1$, and $z \geq 0$, oriented outward, and let $F = (x - y, 3z^2, -x)$. Evaluate

$$\iint_\Sigma (\text{curl } F) \cdot N \, dA.$$

14. Let Σ be the part of the plane $x + y + z = 1$ in the first octant of $\mathbb{R}^3$, oriented so that the positive unit normal is $N = (1, 1, 1)/\sqrt{3}$. As a check on Stokes' Theorem, evaluate both sides of

$$\iint_\Sigma (\text{curl } F) \cdot N \, dA = \int_{\partial\Sigma} F(X) \cdot dX,$$

where $F = (z - y, x - z, y - x)$.

15. Let $\mathcal{B} \subseteq \mathbb{R}^3$ be a solid region, say a ball, and F a vector field with a continuous derivative. Show that

$$\iint_{\partial\mathcal{B}} (\text{curl } F) \cdot N \, dA = 0.$$

Hint: Let $\Sigma = \partial\mathcal{B}$, and observe that $\partial\Sigma$ has no points.

16. Let $\mathcal{B} \subseteq \mathbb{R}^3$ denote the solid pyramid bounded by the planes $x + y + 2z = 6$, $x = 0$, $y = 0$, and $z = 0$. Use the Divergence Theorem to evaluate

$$\iint_{\partial\mathcal{B}} F \cdot N \, dA,$$

where $F = (2x, 2y, 4z)$.

17. Let F and G be two vector fields with continuous derivatives in a ball $\mathcal{B} \subseteq \mathbb{R}^3$ such that (1) $F = \nabla\varphi$, $G = \nabla\psi$, and (2) $\text{div } F = \text{div } G$ in $\mathcal{B}$, while (3) $F \cdot N = G \cdot N$ on the boundary of $\mathcal{B}$, where N is the positive unit normal to $\partial\mathcal{B}$. Prove that $F \equiv G$ in $\mathcal{B}$.

18. (a) Let $\Sigma \subseteq \mathbb{R}^3$ be the sphere $S((0, 0, 0); 1)$ and Σ' be some smooth closed surface outside the sphere Σ, so that for instance Σ' might some larger sphere. If $\text{div } F = 0$ in the region between Σ and Σ', show that

$$\iint_\Sigma F \cdot N \, dA = \iint_{\Sigma'} F \cdot N \, dA.$$

(b) Let $\mathcal{B} \subseteq \mathbb{R}^3$ be the region $1 \leq \|X\| \leq 3$. If $F(X) = X/\|X\|^3$ in $\mathcal{B}$, show by direct computation of both sides that

$$\iiint_\mathcal{B} \text{div } F \, dV = \iint_{\partial\mathcal{B}} F \cdot N \, dA.$$

19. Prove the Divergence Theorem in $\mathbb{R}^3$ if $\mathcal{B}$ is a box: $a_1 \leq x \leq b_1$, $a_2 \leq y \leq b_2$, $a_3 \leq z \leq b_3$.

20. Let $\mathcal{B}$ be a spherical shell with uniform density. Explain why the gravitational field inside $\mathcal{B}$ due to the mass of $\mathcal{B}$ is zero.

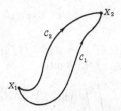

FIGURE 10.25. Two different paths from X_1 to X_2

10.4 Independence of Path; Potential Functions

10.4A THE ISSUE

Say that X_1 and X_2 are points in $\mathbb{R}^q$ and F is a force field on $\mathbb{R}^q$. Then the work done by the force F in moving a unit mass along a curve C from one endpoint X_1 to the other endpoint X_2, is

$$\text{Work} = \int_C F(X) \cdot V(X)\, ds.$$

The question is, under what conditions on F is the work independent of the particular oriented curve C from X_1 to X_2? See Figure 10.25. In physical problems, it is customary to refer to an oriented curve as a **path**. We then rephrase the question: under what conditions of F is the work independent of the path? In Example 8 in Section 10.1c, the line integral *does* depend on the path, and so it would be wrong to believe that one has independence of the path unless further restrictions are imposed on the force F. If the line integral is independent of the path C, we say that F is a **conservative force field**. The reason for the name is easy to explain. If F is conservative, then the line integrals along any two paths C_1 and C_2 from X_1 to X_2 are equal. Therefore, if we go from X_1 to X_2 along C_1 and return from X_2 to X_1 along $-C_2$, the net result is zero; that is, no work is needed to traverse the closed curve $C_1 - C_2$. Consequently, the force F could not have any dissipative aspects, like friction, since dissipation causes an irreversible loss of energy.

The question of a line integral being independent of the path can be raised about any line integral. It does not need the physical conception of work.

10.4B THE MAIN THEOREMS

We would like some criterion to determine when a line integral is independent of the path. First we show that some line integrals are indeed independent of the path.

Theorem 10.4.1 Let $F = \nabla\varphi$, where φ is a scalar-valued function defined on $\mathcal{D}(F) \in \mathbb{R}^n$, and let X_1 and X_2 be any two points in $\mathcal{D}(F)$. Then

$$\int_{X_1}^{X_2} F \cdot V \, ds = \int_{X_1}^{X_2} \nabla\varphi \cdot V \, ds = \varphi(X_2) - \varphi(X_1) \, ,$$

where the integration on the left-hand side is over any piecewise smooth path C from X_1 to X_2 that lies in $\mathcal{D}(F)$. In particular the integral on the left-hand side depends on the path only through its endpoints, provided that F is a gradient.

PROOF: First assume that C is smooth, and let $X = \alpha(t)$, $a \leq t \leq b$, be a parametrization of C. Note that $\alpha(a) = X_1$, $\alpha(b) = X_2$. Then

$$\int_C F \cdot V \, ds = \int_a^b \nabla\varphi(\alpha(t)) \cdot \alpha'(t) \, dt \, .$$

But by the Chain Rule

$$\frac{d}{dt}\varphi(\alpha(t)) = \nabla\varphi(\alpha(t)) \cdot \alpha'(t)$$

Therefore, by the Fundamental Theorem of Calculus,

$$\int_C F \cdot V \, ds = \int_a^b \frac{d}{dt}\varphi(\alpha(t)) \, dt = \varphi(X_2) - \varphi(X_1) \, .$$

Since the right-hand side is independent of the path C from X_1 to X_2, we conclude that the line integral is independent C..

If C is only piecewise smooth, then let its smooth segments be

$$X_1Z_1, \ Z_1Z_2, \ldots, Z_nX_2 \, .$$

See Figure 10.26. We apply the result above to each smooth segment.

$$\int_{X_1}^{X_2} \nabla\varphi \cdot V \, ds = \int_{X_1}^{Z_1} + \int_{Z_1}^{Z_2} + \cdots + \int_{Z_n}^{X_2}$$
$$= [\varphi(Z_1) - \varphi(X_1)] + [\varphi(Z_2) - \varphi(Z_1)] + \cdots + [\varphi(X_2) - \varphi(Z_n)] \, .$$

After cancellation, we find that

$$\int_{X_1}^{X_2} \nabla\varphi \cdot V \, ds = \varphi(X_2) - \varphi(X_1) \, ,$$

just as desired. This completes the proof. $<<$

We remark that this theorem is itself a generalization of the Fundamental Theorem of Calculus, to which it reduces if C is an interval on the x_1-axis

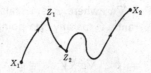

FIGURE 10.26.

and $f(X)$ depends on x_1 only. The next result is our main theorem. In it we collect a variety of conditions for the line integral of a given vector field F to be independent of the path of integration. It asserts that essentially the only way this can happen is if there is some function φ such that $F = \nabla\varphi$, in which case Theorem 10.4.1 gives the independence of path. Recall that a curve $\mathcal{C}$, parametrized by $X = \alpha(t)$, $a \leq t \leq b$, is closed if the initial and final points coincide: $\alpha(a) = \alpha(b)$.

Theorem 10.4.2 Let $F = (f_1, \ldots, f_n)$ be a vector field in a pathwise connected open subset $\mathcal{B}$ in $\mathbb{R}^n$. Assume that each of the coordinate functions of F has continuous first partial derivatives in $\mathcal{B}$. The following conditions are equivalent:

1. $\int_{\mathcal{C}} F \cdot V \, ds = 0$ for any piecewise smooth closed curve in $\mathcal{B}$.

2. $\int_{\mathcal{C}} F \cdot V \, ds$ is independent of the path for any piecewise smooth curve $\mathcal{C}$ joining two fixed points in $\mathcal{B}$.

3. There is a function φ such that $\nabla\varphi = F$;

Any of these implies the following condition on F:

4. $D_j f_i = D_i f_j$ for $1 \leq i, j \leq n$.

Moreover, if $\mathcal{B}$ is an open ball, then condition (4) implies conditions (1), (2), and (3). Thus, for an open ball $\mathcal{B}$, conditions (1) through (4) are equivalent.

PROOF: We prove the chain of implications (1) $\Rightarrow$ (2) $\Rightarrow$ (3) $\Rightarrow$ (1). This will prove the equivalence of (1), (2), and (3).

(1) $\Rightarrow$ (2). Let $\mathcal{C}_1$ and $\mathcal{C}_2$ be any two piecewise smooth curves from X_1 to X_2. Then $\mathcal{C} = \mathcal{C}_1 - \mathcal{C}_2$ is a piecewise smooth closed curve in $\mathcal{B}$, which we have pictured as having a "hole." See Figure 10.27. Therefore, by (1),

$$0 = \int_{\mathcal{C}} F \cdot V \, ds = \int_{\mathcal{C}_1} + \int_{-\mathcal{C}_2} = \int_{\mathcal{C}_1} - \int_{\mathcal{C}_2},$$

So that

$$\int_{\mathcal{C}_1} F \cdot V \, ds = \int_{\mathcal{C}_2} F \cdot V \, ds.$$

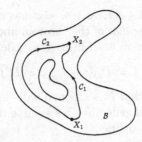

FIGURE 10.27. A piecewise smooth closed curve in $\mathcal{B}$

$(2) \Rightarrow (3)$. We must exhibit the function $\varphi(X)$. Pick a point X_0 in $\mathcal{B}$ and fix it throughout the discussion. If X is in $\mathcal{B}$, let

$$\varphi(X) = \int_{X_0}^{X} F(Y) \cdot dY ,$$

where we need not specify the path of integration since, by part (2), the result does not depend on the particular path chosen. Note that there is at least one path from X_0 to X since $\mathcal{B}$ is pathwise connected. It can be shown that since $\mathcal{B}$ is open, such a path can be chosen to be smooth.

Fix a point $X_1 \in \mathcal{B}$. We must show that φ is differentiable at X_1 and that $\nabla\varphi(X_1) = F(X_1)$, that is, $T(X) = \varphi(X_1) + F(X_1) \cdot (X - X_1)$ is the best affine approximation to φ at X_1. Now

$$\varphi(X) - T(X) = \int_{X_0}^{X} F(Y) \cdot dY - \int_{X_0}^{X_1} F(Y) \cdot dY - F(X_1) \cdot (X - X_1)$$

so

$$\frac{\varphi(X) - T(X)}{\|X - X_1\|} = \frac{1}{\|X - X_1\|} \int_{X_1}^{X} [F(Y) - F(X_1)] \cdot dY .$$

Since F is differentiable, and hence continuous, one can pick a ball $B(X_1; \delta)$ so small that $\|F(Y) - F(X_1)\| < \varepsilon$ for $Y \in B(X_1; \delta)$. By Theorem 10.1.3,

$$\left| \int_{X}^{X_1} [F(Y) - F(X_1)] \cdot dY \right| \le \|X - X_1\|\varepsilon ,$$

so

$$\frac{|\varphi(X) - T(X)|}{\|X - X_1\|} \le \varepsilon$$

for all $X \in \mathcal{B}$.

$(3) \Rightarrow (1)$. This follows from Theorem 10.4.1, since for a closed curve $\mathcal{C}$ in that theorem, we have $X_2 = X_1$.

To prove part (4), assume that (3) holds. That is, assume that $F = \nabla\varphi$. We observe that, by part (3), $D_i\varphi = f_i$, and $D_j\varphi = f_j$. By assumption, F

has continuous first partial derivatives, so φ has continuous second partial derivatives. Therefore, because of the theorem about the equality of mixed partial derivatives (Theorem 6.3.1 in Section 6.3c), we have

$$D_j f_i = D_{ij}\varphi = D_{ji}\varphi = D_i f_j .$$

Finally, let us assume that $\mathcal{B}$ is an open ball, and prove in this case that $(4) \Rightarrow (3)$. Without loss of generality, we assume that the ball $\mathcal{B}$ is centered at the origin. For $X \in \mathcal{B}$, define

$$\begin{aligned}
\varphi(X) &= \int_0^1 F(tX) \cdot X \, dt \\
&= \int_0^1 [x_1 f_1(tX) + \ldots + x_n f_n(tX)] \, dt .
\end{aligned}$$

We are using the fact that $\mathcal{B}$ is a ball to ensure that the point tX is in $\mathcal{B}$ for all $t \in [0,1]$.

We wish to compute the partial derivatives of φ. To accomplish this, we use the theorem in Section 9.3 about differentiating under the integral sign. We have

$$\frac{\partial \varphi}{\partial x_i}(X) = \int_0^1 \frac{\partial}{\partial x_i} [x_1 f_1(tX) + \cdots + x_n f_n(tX)] \, dt .$$

Using the chain rule to differentiate term by term in the integrand on the right-hand side, we obtain

$$\frac{\partial \varphi}{\partial x_i}(X) = \int_0^1 [f_i(tX) + t(x_1 D_i f_1(tX) + \ldots + x_n D_i f_n(tX))] \, dt .$$

Now we use the hypothesis that $D_i f_j = D_j f_i$. The integrand becomes

$$f_i(tX) + t(x_1 D_1 f_i(tX) + \ldots + x_n D_n f_i(tX)) ,$$

and according to Example 4 in Section 8.2, this equals

$$\psi(X,t) + t\frac{\partial \psi(X,t)}{\partial t} = \frac{d[t\psi(X,t)]}{dt} ,$$

where $\psi(X,t) = f_i(tX)$. Substitute this expression into the integral and use the Fundamental Theorem of Calculus to obtain

$$\frac{\partial \varphi}{\partial x_i}(X) = t f_i(tX)\Big|_{t=0}^{t=1} = f_i(X) .$$

It follows that $\nabla \varphi(X) = F(X)$ as desired. Done. $<<$

REMARK:

- If the line integral of a force field F is independent of the path, we say that the force is conservative and call the function φ such that $\nabla \varphi = F$ the **potential function** of F.

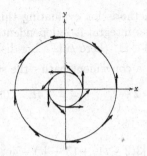

FIGURE 10.28. A ring in $\mathbb{R}^2$

10.4c EXAMPLES

One wonders if the proof that (4) $\Rightarrow$ (3) can be extended to more general regions than balls. The following example shows that it cannot always be generalized.

EXAMPLE:

1. Let $\mathcal{B} \subseteq \mathbb{R}^2$ be the ring domain $1 \leq x^2 + y^2 \leq 9$, and let $F(x, y) = (p(x, y), q(x, y))$, where

$$p(x, y) = -\frac{y}{x^2 + y^2}, \quad q(x, y) = \frac{x}{x^2 + y^2}.$$

Then a computation shows that $p_y = q_x$. See Figure 10.28. If we integrate $F = (p, q)$ around the circle $x = 2\cos\theta$, $y = 2\sin\theta$, $0 \leq \theta \leq 2\pi$, then

$$\int_C F \cdot V \, ds = \int_0^{2\pi} (\sin^2\theta + \cos^2\theta) \, d\theta = 2\pi .$$

In other words, the line integral of F around closed curve C is not zero. Thus (4) does not imply (1) in this case.

It turns out that (4) implies (1) only if the region $\mathcal{B}$ has no "holes" in it. In the example just done, $\mathcal{B}$ does have a hole, namely the disk $x^2 + y^2 < 1$, and at the center of this hole, the given vector field F "blows up". This is typical of the problems that arise if the region has one or more holes.

EXAMPLES:

2. Let C be a smooth curve from $X_1 = (2, 3)$ to $X_2 = (3, -1)$. We evaluate

$$\int_C (3x^2 + y) \, dx + (e^y + x) \, dy .$$

There are three methods for evaluating this. All three rely on the observation that the integral is independent of the path, since $p_y = 1 = q_x$ for all $(x, y) \in \mathbb{R}^2$ (here $\mathcal{B}$ is a big disk, or even all of $\mathbb{R}^2$).

Method 1. We pick a convenient path—the straight line $\mathcal{C}$:

$$x = 2 + t, \quad y = 3 - 4t, \quad 0 \le t \le 1.$$

Then

$$\int_{\mathcal{C}} = \int_0^1 [3(2+t)^2 + (3-4t) - 4e^{3-4t} - 4(2+t)]\, dt$$
$$= 10 + e^{-1} - e^3.$$

Method 2. Since $p_y = q_x$ and $\mathcal{B}$ is a disk containing X_1 and X_2, by Theorem 10.4.2, there is a function φ such that $\nabla\varphi = F$,

$$\nabla\varphi = (\varphi_x, \varphi_y) = (3x^2 + y,\ e^y + x).$$

We find φ by the method used in the proof of Theorem 10.4.2. Thus

$$\varphi(x, y) = \int_0^1 (3t^2x^2 + ty,\ e^{ty} + tx) \cdot (x, y)\, dt$$
$$= (x^3 + (xy/2)) + (e^y - 1 + (xy/2))$$
$$= x^3 + xy - 1 + e^y.$$

The line integral we are calculating equals

$$\varphi(3, -1) - \varphi(2, 3)$$
$$= (27 - 3 - 1 + e^{-1}) - (8 + 6 - 1 + e^3) = 10 + e^{-1} - e^3.$$

Method 3. The beginning and end of this method is similar to the beginning and end of the preceding method. The difference is in the way that the potential function φ is found. Since $\varphi_x = 3x^2 + y$, by integrating with respect to x and holding y constant, we find that

$$\varphi(x, y) = \int (3x^2 + y)\, dx = x^3 + xy + K(y),$$

where $K(y)$ is a function (the "constant" of integration) that depends only on y. Now we compute φ_y from this expression and compare it with what φ_y is known to be, that is, $\varphi_y = e^y + x$:

$$e^y + x = \varphi_y = x + \frac{dK}{dy}.$$

This yields $K'(y) = e^y$, and so $K(y) = e^y$, where we have ignored the integration constant since we are only looking for *one* potential function, that is,

$$\varphi(x, y) = x^3 + xy + e^y.$$

Even though the potential function obtained here differs by a constant from the one found by the preceding method, they give identical answers for the line integral.

3. Let $F(x_1, x_2, x_3) = (2x_2 + 2x_3,\ 2x_3 + 2x_1,\ 2x_1 + 2x_2)$. We show that F is a conservative vector field and find its potential function. Since $D_i f_j = D_j f_i$ for each i and j, we conclude from Theorem 10.4.2 that F is conservative. We use the method given in the proof of Theorem 10.4.2 to calculate a potential function φ. Thus

$$\varphi(x_1, x_2, x_3)$$
$$= \int_0^1 (2tx_2 + 2tx_3,\ 2tx_3 + 2tx_1,\ 2tx_1 + 2tx_2) \cdot (x_1, x_2, x_3)\, dt$$
$$= (x_2 + x_3)x_1 + (x_3 + x_1)x_2 + (x_1 + x_2)x_3$$
$$= 2(x_1x_2 + x_2x_3 + x_3x_1).$$

You might wonder why one prefers to work with potential functions instead of vector fields. One answer is that potential functions are scalars and it is easier to work with scalar-valued functions, like $2(x_1x_2 + x_2x_3 + x_3x_1)$, than with vector-valued functions.

Exercises

1. Evaluate the line integrals $\int_C F \cdot V\, ds$ by finding and using a potential function:

   (a) $F(x, y) = (x, y)$, C is the image of $(\cos t, 2\sin t)$ for $0 \le t \le \pi/2$

   (b) $F(X) = (x_3,\ x_2 + x_3,\ x_1 + x_2 + 3)$, C is the image of $(t^2, t, t+1)$ for $-1 \le t \le 2$

   (c) $F(x, y) = (e^y, xe^y)$, C is the boundary of the rectangular region given by $-1 \le x \le 2$ and $5 \le y \le 9$, beginning at $(2, 5)$ and oriented counterclockwise

   (d) $F(x, y) = (e^y, xe^y)$, C is the boundary of the rectangular region given by $-1 \le x \le 2$ and $5 \le y \le 9$, beginning at $(-1, 9)$ and oriented counterclockwise

   (e) $F(x_1, x_2) = (x_2^2,\ 2x_1x_2 + 2x_2)$, C is the semicircle defined by $x_1^2 + x_2^2 = 9$ and $x_1 \ge 0$, oriented clockwise

   (f) $F(x, y) = (x + y, x - 7y)$, C is a pentagon with successive vertices at $(2, 0)$, $(1, 1)$, $(0, 1)$, $(-1, 0)$, and $(0, -1)$, oriented clockwise.

2. (a) Show that the force field $F(X) = (x + e^x \sin y,\ e^x \cos y)$ is conservative and find a potential function for it.

   (b) Let C be a smooth curve from $(1, \pi)$ to $(0, \pi/2)$. Evaluate the work performed by the force to move a particle of unit mass along C. Recall that

$$\text{Work} = \int_C F \cdot V\, ds.$$

3. Let C be a smooth path in the disk $x^2 + (y-3)^2 \leq 6$ joining the two points $(-2, 2)$ and $(2, 2)$. For which of the following vector fields does the line integral depend on C only through its endpoints?

(a) $F(x, y) = (x - 1, \ e^y)$

(b) $F(x, y) = (3x - y, \ 2y + x)$

(c) $F(x, y) = \left(-\frac{y}{x^2 + y^2}, \ \frac{x}{x^2 + y^2} \right)$

(d) $F(x, y) = (e^y \cos x, \ e^y \sin x)$

(e) $F(x, y) = (e^y, e^x)$

(f) $F(x, y) = \left(\frac{x}{1 + x^2 + y^2}, \ \frac{y}{1 + x^2 + y^2} \right)$.

4. Let $F(x, y) = \left(-\frac{y}{x^2 + y^2}, \ \frac{x}{x^2 + y^2} \right)$. Evaluate $\int_C F \cdot V \, ds$, where:

(a) C is the shorter arc of the circle $x^2 + y^2 = 8$ from $(-2, 2)$ to $(2, 2)$.

(b) C is the longer arc of the circle $x^2 + y^2 = 8$ from $(-2, 2)$ to $(2, 2)$.

(c) Is the line integral independent of the path C in the disk $x^2 + y^2 \leq 100$ from $(-2, 2)$ to $(2, 2)$. Does this agree with the "moreover" part of Theorem 10.4.2?

5. The gravitational force between a mass M at the origin in $\mathbb{R}^3$ and a mass m at X in $\mathbb{R}^3$ is

$$F(X) = \frac{-\gamma m M X}{\|X\|^3},$$

where γ is a constant. Show that F is conservative by finding a potential function φ such that $F = \text{grad } \varphi$ at all points other than $\mathbf{0}$. See Exercise 16 in Section 10.1.

6. Let C be a smooth closed curve, and let p and q be functions having continuous first derivatives on $\mathbb{R}$. Show that

$$\int_C p(x) \, dx + q(y) \, dy = 0.$$

7. Show that $\int_C (x - y) \, dx + 2y \, dy$ is not independent of the path C by showing that condition (4) in Theorem 10.4.2 does not hold.

8. Decide whether the following statement is true or false and explain your answer, including a counterexample in case your answer is "false". If F is a conservative vector field and G is any continuous vector field, then for any smooth closed curve C

$$\int_C (F(X) + G(X)) \cdot V(X) \, ds = \int_C G(X) \cdot V(X) \, ds.$$

9. Decide whether the following statement is true or false and explain your answer, including a counterexample in case your answer is "false". Let F and G be continuous vector fields in $\mathbb{R}^2$. If F is conservative and if the vectors $F(X)$ and $G(X)$ have the same direction at every point $X \in \mathbb{R}^2$, then G is conservative too.

10. Let ψ be a scalar-valued function with a continuous derivative on the interval $(0, \infty)$. Let $F(X) = \psi(\|X\|)X$ for $X \in \mathbb{R}^n$ different from $\mathbf{0}$. Prove that F is conservative by verifying that $F(X) = \nabla \varphi(\|X\|)$, where

$$\varphi(r) = \int_1^r \rho \psi(\rho) \, d\rho.$$

In particular, since any central force field F is of the form above, this shows that *central force fields are conservative*. Compare with Exercise 18 of Section 10.1.

Answers and Partial Answers to Selected Exercises

Chapter 1

1.1

1. (a) $(0,5)$, (c) $(0,9)$, (e) $(3,-9)$, (g) $(-1,4)$, (h) $(0,0) = \mathbf{0}$
2. (a) $(-1,-\frac{10}{3})$, (c) $(-\frac{9}{2},-\frac{19}{2})$
3. (c) $X = \frac{1}{2}Z - \frac{5}{2}Y$
5. (a) length $= \sqrt{5}$
8. (b) No; the second coordinate of $\alpha Y + \beta Z$ is twice the first coordinate.
9. $\alpha = -5$, $\beta = -4$

1.2

1. (a) $(1,4,-1)$, (c) $(2,0,0)$, (e) $(2,12,0)$, (g) $(7,0,-21)$
2. (a) $(1,4,-2,1)$, (c) $(1,-2,1,1)$, (e) $(-1,-4,2,-1)$
4. (b) $\alpha = 1$, $\beta = -2$, $\gamma = 3$
6. (b) yes   (c) no   (d) a, b, c arbitrary, $d = 0$

1.3

1. (a) subspace,   (c) not subspace,   (e) subspace,   (f) not subspace,
   (g) not subspace

2. (a) not subspace,   (c) not subspace,   (e) subspace,   (g) not subspace   (h) subspace,   (j) subspace

3. yes, no.

4. (a) false,   (c) true,   (e) false,   (g) true,   (h) false,   (k) true, (l) false

5. many answers, one of which is $\gamma = 13\alpha_1 - 13\alpha_2$, $\delta = 13\beta_1 - 13\beta_2$

7. (a) all lines through $\mathbf{0} \in \mathbb{R}^2$

8. There are two: $\mathbb{R}^2$ itself and the line passing through $\mathbf{0}$ and $(2, -1)$.

1.4

1. (a) not these,   (c) these; $Y = 0X$,   (g) these; $X = (4/3)Y$

2. (a) in, $2X_1$ (c) not in,   (e) not in,   (f) in, $0X_1 + 0X_2$ (g) in, $-X_2$ (j) in, $X_1 - 2X_2$

3. (a) $(2, 0, 8) = 2(1, 0, 4)$ and $(-4, 0, -16) = -4(1, 0, 4)$ constitute one of three possible answers.   (b) $(0, 0, 0) = 0(1, -1, 2)$ and $(-2, 2, -4) = -2(1, -1, 2)$ constitute one of two possible answers.

4. (a) $(3, 1, 2) = (1, 1, 0) + 2(1, 0, 1)$ is one of three possible answers. (c) $(3, 3, 1) = 2(1, 1, 1) + (1, 1, -1)$ is one of three possible answers.

5. (a) $\mathbb{R}^3$,   (b) point—namely $(0, 0, 0)$, (c) plane that is the linear span of any two of the given three vectors, (d) line that is the linear span of any one of the three given vectors, (f) $\mathbb{R}^3$,   (g) plane that is the linear span of third of the given three vectors together with either of the first two vectors

6. (many correct answers)   (a) $X_1 = (1, 0, 0)$, $X_2 = (0, 1, 0)$; (c) $X_1 = (1, 1, 0)$, $X_2 = (0, 1, 1)$

1.5

1. (a) line, $X_1 = (0, 1, 0)$, $(0, \alpha, 0)$; (d) plane, $X_1 = (1, 1, 0)$, $X_2 = (0, 0, 1)$, $(\alpha_1, \alpha_1, \alpha_2)$; (e) line, $X_1 = (2, 1, 0)$, $\alpha(2, 1, 0)$; (g) plane, $X_1 = (-1, 0, 3\sqrt{2} + 2)$, $X_2 = (\sqrt{2} + 1, 3\sqrt{2} + 2, 0)$, $\alpha_1(-1, 0, 3\sqrt{2} + 2) + \alpha_2(\sqrt{2} + 1, 3\sqrt{2} + 2, 0)$

2. (a) $a_1 = -\frac{2}{3}a_3$, $a_2 = \frac{1}{3}a_3$, a_3 arbitrary

3. (c) $(0, -\frac{5}{2}, 0, 0) + \alpha(0, 0, 1, 0) + \beta(\frac{1}{2}, \frac{3}{4}, 0, 1)$, α, β arbitrary;
 (e) $(3, 5, 0) + \alpha(1, -3, 1)$, α arbitrary;
 (g) $(\frac{1}{3}, 0, 0, 0, 0) + \alpha(\frac{2}{3}, 1, 0, 0, 0) + \beta(\frac{-4}{3}, 0, 1, 0, 0)$
  $+ \gamma(\frac{1}{3}, 0, 0, 1, 0) + \delta(\frac{-2}{3}, 0, 0, 0, 1)$, $\alpha, \beta, \gamma, \delta$ arbitrary

4. (a) $-2x_1 + x_2 + 3x_3 = 0$

6. (many correct answers)   (a) $(2, -1, 0)$, $(1, 0, -1)$

7. (a) $2x_1 - x_2 = 0$, $x_1 - x_3 = 0$

8. (d) $0x_1 + 0x_2 + 0x_3 = 32$ is one correct answer.

1.6

1. (a) $x_1 + 2x_2 = 0$,   (c) yes

2. (a) $x_1 - x_2 + x_3 = 0$

3. (a) $x_2 = 0$

4. (a) $2x_1 + 2x_2 + x_3 = 0$,   (b) $2x_1 + 2x_2 + x_3 = 2$

5. (b) $x_1 + x_2 + x_3 = 0$

6. (c) One correct answer is given by the planes corresponding to $-x_1 + 5x_2 + 3x_3 = 0$ and $x_1 - 2x_2 = 3$.

1.7

1. (a) 3; $(\frac{-1}{5}, \frac{-4}{5}, 1, 0, 0)$, $(\frac{2}{5}, \frac{3}{5}, 0, 1, 0)$, $(\frac{-2}{5}, \frac{7}{5}, 0, 0, 1)$;
 (d) 2; $(0, \frac{1}{2}, 1, 0)$, $(-1, \frac{1}{2}, 0, 1)$

2. In $\mathbb{R}^3$ only (b), (c), and (d) are possible.

3. (a) is,   (d) is not,   (f) is not

Chapter 2

2.1

2. (a) 4,   (c) $\sqrt{2}$

3. (a) 4,   (d) 1

4. (a) 4,   (b) 2

5. (b) 1,

   (c) $\sqrt{\frac{1-4^{-n}}{3}}$, which is very close to $\sqrt{1/3}$ when n is large.

7. (a) 4,   (c) $\sqrt{2}$

10. $(0,0,-1)$, $(\frac{1}{2},\frac{1}{2},\frac{1}{2},\frac{1}{2})$

2.2

1. (a) 0,   (c) 3

3. $\cos\theta = 1/\sqrt{2}$

4. (b)(b) $(\frac{7}{11},\frac{21}{11},\frac{7}{11})$, (b)(d) $\frac{3}{11}\sqrt{330}$, (d)(a) 0,
   (d)(c) $P = \mathbf{0}$, $Q = (1,-1,1,0,4,1)$

5. (a) $Y = (5,-1)$ is one such vector; the other correct answers are the
   nonzero multiples of $(5,-1)$,   (b) two, $Y = \pm(5,-1)/\sqrt{26}$

6. $V = (-\frac{3}{5},-\frac{4}{5})$

8. (b) projection: $(\frac{5}{7},\frac{13}{7},\frac{17}{7})$, distance: $2\sqrt{2/7}$ (d) $(\frac{4}{9},\frac{8}{9},\frac{8}{9},\frac{12}{9})$, 2/3

11. (b) $\pi/4$, $\pi/2$, $\pi/4$
   (d) $\arccos\frac{3}{\sqrt{209}}$, $\arccos\frac{4}{\sqrt{66}}$, $\arccos\frac{8}{\sqrt{114}}$;
   (e) $\arccos\frac{2}{3}$, $\arccos\frac{1}{\sqrt{6}}$, $\arccos\frac{1}{\sqrt{6}}$

15. (b) $(\frac{1}{2},\frac{1}{2})$, $\frac{1}{\sqrt{2}}$ (d) $(\frac{14}{11},\frac{9}{11},\frac{19}{11})$, $\frac{3}{11}\sqrt{11}$ (f) $(0,1,1,0)$, 2

2.3

1. (b) A correct answer consists of $X_1 = (1,2,0)$ and $X_2 = (0,3,1)$.

2. (a) S' is the line through the origin containing Z and thus a linear
   subspace. Each of the two sets S and S' is the set of vectors which
   are perpendicular to every vector in the other of the two sets.

3. (b) e_2

5. (a) $Z = (-4,1,2)$ is one correct answer. It is not the only correct answer, but all other correct answers are nonzeromultiples of $(-4,1,2)$.

6. (a) false,   (b) false,   (c) true,   (e) true,   (g) false

8. first projection: $(\frac{1}{2}, \frac{1}{2}, \frac{3}{2}, \frac{1}{2})$, first distance: 1

2.4

1. (a) $-3i - 3j - 3k$,   (c) $-4i + 2j$,   (e) $\mathbf{0}$,   (g) $-9i + 6j + k$,
   (i) $9i - 6j - k$,   (k) $-10i - 4j - 6k$,   (m) $6i + 6j + 6k$

6. $\pm(5i - 2j + 3k)/\sqrt{38}$

8. $\pm(4i - 3j)/5$

13. $\sqrt{77}$

15. (a) $X = 0$

16. $c(2j - k)$ for any c

18. (b) 36

Chapter 3

3.1

1. (a) linear,   (c) not linear

2. (b) linear

3. (a) not linear,   (c) not linear,   (e) not linear

4. (b) linear

6. (a) $L(X) = (3x_1 + x_2, 4x_1 - x_2)$, $L((13, -2)) = (37, 54)$

12. (a) $\bar{e}_1 + 2\bar{e}_2$,   (c) $3\bar{e}_1 + 20\bar{e}_2$

14. $[\ 4\ \ 3\ ]$

16. (a) 2,   (c) 0

18. (a) $\bar{e}_2$,   (c) $\bar{e}_1 + \bar{e}_2$

19. (b) $x_1 + x_3$

20. (a) false,   (c) true,   (e) false,   (g) true,   (i) true,   (k) false,
    (m) true,   (o) true

3.2

1. (b) basis for $\mathcal{N}(L)$: $(-1, 0, 2)$ and $(1, 2.0)$,
   basis for $\mathcal{I}(L)$: $(1, -1)$, $\dim(\mathcal{N}(L)) = 2$, $\dim(\mathcal{I}(L)) = 1$

4. (a) no (b) yes

6. whenever $y_1 = 2y_2$

7. (a) yes (c) no (e) no (g) no

8. (a) yes (b) no

10. basis for graph: $(1, 2, 3)$

11. basis for graph: $(1, 0, 1)$ and $(0, 1, 1)$

12. (b) $(1, 1, 1, \ldots, 1) \in \mathbb{R}^{q+1}$
    (c) $(1,0,0,0)$, $(0,1,0,1)$, and $(0,0,1,0)$
    (f) $(1,0,0,\ldots,0,0,1)$, $(0,1,0,\ldots,0,1,0)$, $(0,0,1,\ldots,1,0,0)$,\ldots,   and
    $(0,0,0,\ldots, 0,1,1,0,\ldots,0,0,0,)$, where the last two ones are in the
    center positions.

15. (a) false—if L is the zero map and $B = \mathbf{0}$, then every member of the
    domain is a solution,   (f)false—same counterexample as for part (a)

3.3

1. (a) linear,   (c) linear,   (e) not linear,   (g) not linear

2. $\mathbb{R}$

3. (a) $(2, -5)$,   (c) $[\, 2 \quad -5 \,]$

4. (b) $\mathcal{N}(L)$ is the set of vectors perpendicular to Z.

8. $L(X) = -\frac{1}{3}x_1 - 2x_2 + \frac{4}{3}x_3 + \frac{5}{3}x_4 - \frac{5}{3}x_5$

9. One of infinitely many correct answers is $L(X) = x_1 - x_2 - x_3$.

3.4

1. (a) $3\mathbf{e}_1 + \mathbf{e}_2$,   (c) $8\mathbf{e}_1 + 4\mathbf{e}_2$,   (e) $2\mathbf{e}_1 + 4\mathbf{e}_2$,   (g) $\mathbf{e}_1 - 3\mathbf{e}_2$,   (i)
   $2\mathbf{e}_1 + 2\mathbf{e}_2$,   (k) $2\mathbf{e}_1$

3. (a) $[L] = \begin{bmatrix} 1 & 1 \\ 1 & -1 \end{bmatrix}$, $[M] = \begin{bmatrix} 2 & 1 \\ 0 & 3 \end{bmatrix}$

5. (b) Each of L and M commute with P, but not with each other.

6. (a) true,   (c) true,   (e) true,   (g) true

7. (a) nonsense,   (c) sense,   (e) nonsense,   (g) nonsense

10. (b) (one part) $(L + M)^2 = L + M$

11. (a) false; $L(x_1, x_2) = (x_2, 0)$ gives $L^2 = 0$.   (c) false; $L(x_1, x_2) = (-x_1, x_2)$ gives $L^2 = I$.

3.5

1. $A = \begin{bmatrix} 2 & 0 & 4 \\ -3 & -1 & 144 \end{bmatrix}$

2. (a) $\begin{bmatrix} 3 & 1 & 6 \\ -3 & 0 & 146 \end{bmatrix}$,   (c) $\begin{bmatrix} 1 \\ 0 \\ 0 \\ 1 \end{bmatrix}$

3. (a) $\begin{bmatrix} 1 & -2 \\ 6 & 5 \end{bmatrix}$,   (c) $\begin{bmatrix} 21 & 28 \\ 7 & -7 \\ 7 & 0 \end{bmatrix}$

4. (a) $\begin{bmatrix} -3 & 5 \\ 3 & 9 \end{bmatrix}$,   (c) $\begin{bmatrix} 2 \\ 0 \end{bmatrix}$,   (e) $\begin{bmatrix} 1 \\ 3 \end{bmatrix}$ (g) $\begin{bmatrix} 1 & -1 & 3 \\ 0 & 0 & 2 \\ 3 & -3 & -3 \end{bmatrix}$,

(i) $\begin{bmatrix} 5 & 10 & 0 & 3 \\ 15 & 5 & 10 & 6 \end{bmatrix}$

5. (a) $a = \frac{1}{2}$, $b = -\frac{1}{2}$, $c = 0$, $d = 1$

6. (a) $\begin{bmatrix} -2 & 1 \\ 1 & 3 \end{bmatrix}$,   (e) $\begin{bmatrix} 4 \\ 5 \end{bmatrix}$

7. (b) $\begin{bmatrix} -3 & 5 \\ 3 & 9 \end{bmatrix}$

8. (a) $L(e_1) = -2e_1 + e_2$, $L(e_2) = e_1 + 3e_2$,   (b) $(-2x_1 + x_2, x_1 + 3x_2)$

10. $A^2 = A$, $B^2 = O$, $AB = O$, $BA = B$, $ABA = BAB = (AB)^2 = (BA)^2 = O$, $A^5 = A$, $(A + 2B)^2 = A + 2B$, $(3A - B)^2 = 9A - 3B$

13. (c) plane spanned by $(4, 0, 1)$ and $(-3, 1, 0)$,   (f) line spanned by $(-18, 157, 84, 137)$.

14. (c) nullity $= 2$ and rank $= 1$, affine subspace passing through $(-2, 0, 0)$ and parallel to the linear subspace spanned by $(4, 0, 1)$ and $(-3, 1, 0)$,   (f) The solution set is the empty set.

15. $A = \begin{bmatrix} 0 & -1 \\ 1 & 0 \end{bmatrix}$, which is a counterclockwise rotation by $\pi/2$.

3.6

1. (a) not affine,   (c) affine

2. (b) affine

3. (a) affine,   (c) affine

4. (a) not affine,   (c) affine,   (e) affine

5. (b) not affine,   (c) affine

6. (b) affine

7. (a) 6,   (c) 7,   (e) $L(x_1, x_2) = (-x_1 + 4x_2)$

8. $T(X) = (7 - 3x_1 + x_2)/6$

10. (a) $T(X) = 7 - 8x_1 - 3x_2$ (b) yes

11. $Y_0 = (7)$, $[L] = [\,1 \quad 4\,]$

12. (a) false,   (c) false,   (e) false,   (g) true,   (i) true,   (k) false

14. $T(X) = (1 - x_1, -2 + 3x_1 + x_2, 3 - 2x_1 - 3x_2)$,
$T(1, 2) = (0, 3, -5)$

3.7

1.

$$\text{(a) } A^{-1} = \begin{bmatrix} \frac{1}{5} & 0 \\ 0 & \frac{1}{6} \end{bmatrix} \quad \text{(b) } X = (\frac{12}{5}, -\frac{13}{6})$$

2.

$$\text{(b) } B = \begin{bmatrix} -\frac{1}{3} & \frac{1}{3} \\ \frac{1}{3} & \frac{2}{3} \end{bmatrix} \quad \text{(e) no inverse}$$

3. (b) $x_1 = -\frac{1}{3}y_1 + \frac{1}{3}y_2$, $x_2 = \frac{1}{3}y_1 + \frac{2}{3}y_2$

6.

(a) has an inverse equal to $\begin{bmatrix} \frac{3}{4} & -\frac{5}{4} \\ -\frac{1}{4} & \frac{3}{4} \end{bmatrix}$

7. (b) $L^{-1}(e_1) = e_1$, $L^{-1}(e_2) = -e_1 + e_2$

8. (b) $3,712e_1 - 1,984e_2$

11. (a) true,   (c) true,   (e) true,   (g) false

12. (a) $A^{-1} = \begin{bmatrix} 1 & -2 & -2 \\ 0 & 1 & 1 \\ 0 & 0 & 1 \end{bmatrix}$,   (d) $A^{-1} = \begin{bmatrix} 1 & -1 & 1 & -1 \\ 0 & 1 & -1 & 1 \\ 0 & 0 & 1 & -1 \\ 0 & 0 & 0 & 1 \end{bmatrix}$

13. (b) The determinant equals 0, so A is singular.

14.

(a) $\begin{bmatrix} \frac{1}{2} & \frac{\sqrt{3}}{2} \\ -\frac{\sqrt{3}}{2} & \frac{1}{2} \end{bmatrix}$ (c) $\begin{bmatrix} 2^{99} & 0 \\ 0 & 6^{99} \end{bmatrix}$

(f) $\begin{bmatrix} 2^{97} + 27 \times 6^{97} & \sqrt{3}\left(-2^{97} + 9 \times 6^{97}\right) \\ \sqrt{3}\left(-2^{97} + 9 \times 6^{97}\right) & 3 \times 2^{97} + 9 \times 6^{97} \end{bmatrix}$

15. (c) not necessarily true   (e) true   (f) not necessarily true

16. (a) $C = A^{-1}B = \begin{bmatrix} -1 & -2 \\ 1 & 1 \end{bmatrix}$,   (c) $C = O$,   (e) $C = A^{-2} = \begin{bmatrix} 7 & -12 \\ -4 & 7 \end{bmatrix}$

18. $A - 2I$

20. (a) true,   (c) true,   (e) true,   (g) true,   (h) false, for example $-I$ or

$$\begin{bmatrix} 0 & 1 \\ 1 & 0 \end{bmatrix}$$

23. (b) no inverse   (c) $X_0 = (\frac{2}{7}, \frac{1}{7}, -\frac{2}{7})$ and

$$[M] = \begin{bmatrix} -\frac{2}{7} & \frac{3}{7} & 0 \\ -\frac{15}{7} & \frac{5}{7} & -1 \\ -\frac{12}{7} & \frac{4}{7} & -1 \end{bmatrix}$$

(e) $[M] = [I]$ and $X_0 = -e_1$ (f) $[M] = [L]$ and $X_0 = (0,1,0,1)$

24. (a) $(-\frac{1}{5}, -\frac{1}{5})$ (d) e_1 (e) no X

26. no fixed point: $y = x + b$ with $b \neq 0$
one fixed point: $y = mx + b$ with $m \neq 1$
every point fixed: $y = x$

27. (a) area $= 2$ and direction of traverse is counterclockwise   (d) area $= 1$ and direction of traverse is counterclockwise   (g) area $= c(a-b)$ and direction of traverse is clockwise

28. (a) 28   (c) 56   (f) $14c(b - a)$

29. $|c|^3$

30. (a) 6   (c) 12

3.8

1.

$$\text{(b)} \begin{bmatrix} 14 \\ 14 \\ -18 \\ 4 \end{bmatrix}, \quad \text{(e) not defined,} \quad \text{(f)} \begin{bmatrix} -6 & 2 \\ 0 & 0 \\ 30 & -10 \\ 12 & -4 \end{bmatrix}$$

(i) $\begin{bmatrix} 56 & -84 \end{bmatrix}$, (k) not defined, (l) not defined

2. (a) does,   (c) does not,   (e) does,   (f) does,   (h) does not

3. (a) both it and inverse represent rigid motions—inverse equals

$$\begin{bmatrix} \frac{3}{5} & -\frac{4}{5} \\ \frac{4}{5} & \frac{3}{5} \end{bmatrix}, \quad \text{the given isometry} = L_{\tan^{-1}(-4/3)},$$

(e) neither it nor its inverse represent rigid motions—inverse equals

$$\begin{bmatrix} -1 & 0 \\ 0 & 1 \end{bmatrix}, \quad \text{the given isometry} = M_{\pi/2},$$

(f) both it and inverse represent rigid motions—inverse equals

$$\begin{bmatrix} \sin\gamma & \cos\gamma \\ -\cos\gamma & \sin\gamma \end{bmatrix}, \quad \text{the given isometry} = L_{(\pi/2)-\gamma}.$$

4.

$$I = \begin{bmatrix} 1 & 0 \\ 0 & 1 \end{bmatrix} \quad \text{and} \quad M_0 = \begin{bmatrix} 1 & 0 \\ 0 & -1 \end{bmatrix}$$

6. (c)

$$\begin{bmatrix} -\frac{1}{2} & \frac{\sqrt{3}}{2} \\ -\frac{\sqrt{3}}{2} & -\frac{1}{2} \end{bmatrix} = L_{-2\pi/3}$$

7. (b) The four symmetries are I, $-I$, M_0, and $M_{\pi/2}$. Multiplication is commutative in this case. The product of any matrix with I is itself. The square of any of the four symmetries is the identity. The product of two distinct symmetries neither of which is the identity is the remaining symmetry that is not the identity.

9. (b) does not,   (d) does

10. (c) neither it nor its inverse represent rigid motions—inverse equals

$$\begin{bmatrix} \frac{2}{3} & -\frac{2}{3} & -\frac{1}{3} \\ -\frac{1}{3} & -\frac{2}{3} & \frac{2}{3} \\ \frac{2}{3} & \frac{1}{3} & \frac{2}{3} \end{bmatrix}$$

13. (a) does,   (b) does

Chapter 4

4.0

1. (a) neither,   (c) neither,   (e) affine,   (g) linear (and hence affine too),   (i) neither,   (l) neither,   (o) linear (affine too)

2. (d) $f_1(X) = x_1 x_2$, $f_2(X) = 0$ (f) $f_1(x_1, x_2) = 0$, $f_2(x_1, x_2) = x_1$, $f_3(x_1, x_2) = x_1 x_2$

3. a,d,e,j, and o in case $n = 1$

4. In each case f is its own coordinate function.

4.1

1. (a) a circle, radius 1, traversed twice counterclockwise beginning at $(1, 0)$.

3. $F(t) = (\cos t, -\sin t, t)$ is one correct answer.

4. an ellipse, the one which is the solution set of the equation

$$\frac{x_1^2}{9} + \frac{x_2^2}{4} = 1.$$

F is one-to-one.

5. more than one correct answer for each part
(a) $F(t) = (2t, 5t, -3t)$, $t \geq 0$,   (c) $F(t) = (t, 1 - t, t, 1 - t)$, $t \geq 0$

6. more than one correct answer for each part (a) $F(t) = (1 + 2t, 3 - 2t)$, $0 \leq t \leq 1$;   (c) $F(t) = (1 + 4t, 3 + 3t)$, $0 \leq t \leq 2$. Another correct answer is $F(t) = (1 + 8t, 3 + 6t)$, $0 \leq t \leq 1$.   (f) $F(t) = (t, 1, 1, \ldots, 1)$, $0 \leq t \leq 1$.

7. $F(t) = (2t^{3/2}, t^3)$; $F(4) = (16, 64)$

8. $F(t) = (\sqrt[3]{t^3 + 7}, t^3)$; $F(4) = (\sqrt[3]{71}, 64)$

12. The first limit equals $(0, 0)$ and the second fails to exist.

13. answer to the question: 1

14. (a) true,   (d) false,   (i) true,

17.

$$\text{(b) } F(t) = \frac{(t_2 - t)P + (t - t_1)Q}{t_2 - t_1} \; ; \; t \in \mathbb{R}$$

21. center: Y;   radius: $\|Z_1\|$

22. radius of sphere: $\sqrt{\|Y\|^2 + \|Z_1\|^2}$

23.

$$F(t) = (\tfrac{1}{2}, 0, 0, \tfrac{1}{2}) + (\tfrac{1}{2}, 0, 0, -\tfrac{1}{2})\cos t + (0, \tfrac{1}{2}, \tfrac{1}{2}, 0)\sin t \, , \; t \in [0, 2\pi)$$

Exercise 22 is relevant. The radius of the sphere is 1.

25. four points of intersection:

$$\frac{1}{\sqrt{35}}(\pm 4\sqrt{2}, \pm 3\sqrt{3})$$

4.2

1. (a) $T(x) = 1$,   (c) $T(x) = -3 + 4(x + 2)$

2. (a) $F'(t) = (-7,9)$,   (c) $H'(\theta) = (e^\theta, 3 - 14\theta)$,   (e) $\Phi'(t) =$
$(-1,4,-9)$,   (g) $Q'(r) = (2r\sin r + r^2\cos r, 2re^{3r} + 3r^2 e^{3r})$,   (h)
$(-3t,9)/\|(3,t)\|^3$

3. (a) $T(t) = (1-7t, 5+9t)$ for $t_0 = 0, 1, -2$,   (c) $T(\theta) = \theta(1,3)+(3,0)$,
(g) $T(r) = 0$

4. (b) at $t_0 = 1$: $G(t)-T(t) = (0, t^2-2t+1)$ so $\frac{G(t)-T(t)}{t-1} = (0, t-1) \to 0$
as $t \to 1$.

5. (a) at $t = 0$: position is $(1,5)$, velocity is $(-7,9)$, and speed is $\sqrt{130}$.
at $t = 1$: position is $(-6,14)$, velocity is $(-7,9)$, and speed is $\sqrt{130}$.
(h) at $t = 0$: position is $(1,0)$, velocity is $(0,\frac{1}{3})$, and speed is $1/3$. at
$t = 1$: position is $(\frac{3}{\sqrt{10}}, \frac{1}{\sqrt{10}})$, velocity is $(-\frac{3}{10\sqrt{10}}, \frac{9}{10\sqrt{10}})$, and speed
is $3/10$.

6. (a) $F''(t) = (0,0)$,   (c) $H''(\theta) = (e^\theta, -14)$,   (e) $\Phi''(t) = (0,0,0)$,
(g) $Q''(r) = ((2 - r^2)\sin r + 4r\cos r, (2 + 12r + 9r^2)e^{3r})$,
(h) $(6t^2 - 27, -27t)/\|(3,t)\|^5$

7. (a) $F(t) = (1+t+t^2, 3+3t)$,   (c) $F(t) = (2+t-2t^2, 3t+t^2, -2+t+t^3)$
(e) $F(t) = (1+t, 1+\frac{t^3}{3}, 1+\frac{t^4}{4})$

8. (a) $(3,8,3)$

9. (a) $F(t) = (6t + 3t^2 + 2t^3, 18t + 9t^2)/6$
(c) $F(t) =$
$(-\frac{11}{6} + 2t + \frac{1}{2}t^2 - \frac{2}{3}t^3, -\frac{11}{6} + \frac{3}{2}t^2 + \frac{1}{3}t^3, \frac{9}{4} - 2t + \frac{1}{2}t^2 + \frac{1}{4}t^4)$

10. (a) $(-16, 27, 21)$

11. (a) $27t^2 - 32t - 60$.

15. (b) at $t = 1$: $T(t) = (2,2)t$ so $x = 2t, y = 2t$ and $x = y$; at
$t = -1 : T(t) = (2,-2)t$ so $x = 2t, y = -2t$ and $x = -y$

16. interval (possibly unbounded).

20. in first and third quadrants: $\arccos \dfrac{33}{\sqrt{97 \cdot 29 \cdot 3}}$
in second and fourth quadrants: $\arccos -\dfrac{33}{\sqrt{97 \cdot 29 \cdot 3}}$

25. $(5, 12, -8)$

27. $\langle X, Z \rangle = \langle F(0), Z \rangle$

29. time $= \frac{1}{2} + \frac{7\pi}{4}$.

4.3

1. (a) 15,   (c) 12,   (e) $\frac{14}{3}$,   (g) 36

2. (a) 72

3. $\|P - Q\|$

4. (c) $\frac{56}{3}$

7. $(0, 0)$

14. $\left(\cos \frac{\tau}{\sqrt{2}}, \sin \frac{\tau}{\sqrt{2}}, \frac{\tau}{\sqrt{2}}\right)$;   radius of curvature equals 2

16.

$$\frac{|f''|}{[1 + f'^2]^{3/2}},$$

where f denotes the function in question. The answer has been written as a function without the independent variable being mentioned. Plug in a particular value of the independent variable x to obtain the radius of curvature at the point $(x, f(x))$.

17. $y = c(x^2 + y^2)$

18.

$$y = \frac{2 + 3(2cx)^{2/3}}{4c}$$

19. affirmative

20.

$$\mathcal{L}(F; 0, t) = \frac{t}{2}\sqrt{10 + t^2} + 5\ln(t + \sqrt{10 + t^2}) - \frac{5}{2}\ln 10$$

The calculation showing that

$$\lim_{t \to \infty} \frac{2\mathcal{L}(F; 0, t)}{t^2} = 1$$

indicates that $\mathcal{L}(F; 0, t)$ behaves like a constant multiple of t^2 when t is large.

21. ice cream cone;   ice cream

24. $F(0) = F(1) = (0, 0)$ and $\mathcal{L}(F; 0, 1) = 1/\sqrt{2}$

25. (b) 48

26. A low-side approximation to the sum of the lengths of the ten line segments is 2.190 which, thus, is a low-side estimate of the length of the curve. The trapezoidal estimate of the integral

$$\int_0^1 \sqrt{1 + 4t^2 + 9t^4 + 16t^6}\, dt$$

is approximately 2.204.

Chapter 5

5.1

1. (b) same line,   (c) do not meet,   (f) meet in exactly one point at an angle of arccos $\dfrac{43}{\sqrt{38\cdot509}}$,   (g) meet in exactly one point at an angle of arccos $\dfrac{43}{\sqrt{38\cdot509}}$

2. (b) arccos $\dfrac{4\sqrt{10}}{15}$,   (e) meet in a line at an angle of arccos $\dfrac{4}{\sqrt{13\cdot29}}$,   (h) same plane

4. (c) $\dfrac{10}{\sqrt{17}}$,   (g) 0

5. (d) $\dfrac{\sqrt{3}}{6}$

6. (b) intersect in one point at a right angle,   (d) lies in the plane,   (e) intersects in one point at an angle of arcsin $\dfrac{13}{9\sqrt{19}}$

7. (b) arccos $\dfrac{1}{\sqrt{3}}$

8. (c) $\dfrac{3}{\sqrt{2}}$,   (d) 1 (for each θ),   (f) 0

11. (d) $1/\sqrt{n}$,   (g) $\dfrac{6}{\sqrt{14}}$

12. (b) one edge is given by $(0, -1 + t, t)$, $0 \le t \le 1$,   (c) one plane is given by $-x_1 + x_2 - x_3 = 1$,   (g) arcsin $\sqrt{2/3}$

15. 3

17. $\dfrac{|1-\kappa|}{2b}\sqrt{1 + 2\kappa}$, where $\kappa = 2^{1/3}b^{2/3}k^{2/3}$

19. 0

21. (a) 1/2,   (d) 1

22. (d) -4,   (e) $\dfrac{23}{65} + \dfrac{41}{65}i$

24. $-\frac{1}{2} - \frac{\sqrt{3}}{2}i$

26. (c) $(5,0)$,   (d) $(0,1)$

29. (b) $(-2+i)z$,   (d) $(2-i)z + (5+5i)$,   (e) $(2-i)z + (3-i)$

31. (c) $25e^{i\arcsin(-7/25)}$

32. (c) $(25, \arcsin(-7/25))$

33. (b) $L(z) = e^{i2\arctan(1/2)}z$

34. $T(z) = -z + (1+i)$

5.2

3. angle size: $\arccos\frac{1}{3}$

4.

$$
\begin{array}{lll}
(1,-1,0) + (t/\sqrt{2})(0,1,-1) & \text{if} & 0 \le t < \sqrt{2} \\
(2,-1,-1) + (t/\sqrt{2})(-1,1,0) & \text{if} & \sqrt{2} \le t < 2\sqrt{2} \\
(2,1,-3) + (t/\sqrt{2})(-1,0,1) & \text{if} & 2\sqrt{2} \le t < 3\sqrt{2} \\
(-1,4,-3) + (t/\sqrt{2})(0,-1,1) & \text{if} & 3\sqrt{2} \le t < 4\sqrt{2} \\
(-5,4,1) + (t/\sqrt{2})(1,-1,0) & \text{if} & 4\sqrt{2} \le t < 5\sqrt{2} \\
(-5,-1,6) + (t/\sqrt{2})(1,0,-1) & \text{if} & 5\sqrt{2} \le t < 6\sqrt{2}
\end{array}
$$

curve is regular hexagon

5. $2r\omega$ attained when the second coordinate of the blue dot is $2r$; that is, when the blue dot is at the top of the wheel

6. $t_0 = 2\pi/\omega$, $x_0 = 2\pi r$, $y_0 = 0$

7. $8r$

8. $2r(0, 1 - \cos t)$, although an alternative interpretation of the problem is conceivable, yielding the answer $2r(0, 1 - \cos\frac{t}{2})$.

5.3

1.

$$
\text{maximal height: } \frac{s^2 \sin^2\alpha}{2g} \quad \text{horizontal distance: } \frac{s^2 \sin 2\alpha}{g}
$$

2. for maximum maximal height: $\frac{\pi}{2}$,
for maximal horizontal distance: $\frac{\pi}{4}$

3.

$$\frac{1}{2}\cos^{-1}\left(\frac{\delta}{\delta+2}\right) \text{ where } \delta = \frac{2gh}{s^2}$$

5. 2.05×10^4 miles per hour

6. 1.59×10^4 miles per hour

8. 2.64×10^4 miles, 6.91×10^3 miles per hour

9. 4.01 hours

12. constant, ; constant multiple of t plus a constant

5.4

2. (a) $(0,-1)$ and $(0,3)$ (b) $(0,0)$
(e) $(0,0,\ldots,-1)$ and $(0,0,\ldots,3)$ (g) none

4. one version of the answer: the solution set of the system

$$\begin{aligned} y_1 + y_2 &= 1 \\ y_3^2 &= 2y_1 y_2 \end{aligned}$$

10. One of the two equations is $1 = x_3 + \|(x_1, x_2)\|$, where the norm is the norm for $\mathbb{R}^2$.

12. one correct answer is $(2\cos t, 2\sin t)$, $-\sin^{-1}\frac{\sqrt{15}}{8} < t < \sin^{-1}\frac{\sqrt{15}}{8}$

13. one correct answer is $(2 + 2\cos t, 2\sin t)$, $\cos^{-1}(-\frac{7}{8}) \le t \le 2\pi - \cos^{-1}(-\frac{7}{8})$

14. more than one correct answer

$$(1 - 2t, 1 - t, -t), \frac{3 - \sqrt{3}}{6} < t < \frac{3 + \sqrt{3}}{6}$$

Chapter 6

6.0

6. complement is a three-dimensional linear subspace of $\mathbb{R}^6$

6.1

1. Theorem 6.1.1 shows that x_1 and x_2 are continuous, two applications of product rule in Theorem 6.1.2 show that $x_1^2 x_2$ is continuous, two further such applications shows that x_2^3 is continuous, the constant function -1 is continuous although the theorem asserting the continuity of constant functions is not in the text, an application of product rule in Theorem 6.1.2 shows that $-x_2^3$ is continuous, and an application of the sum rule in Theorem 6.1.2 shows that $x_1^2 x_2 - x_2^3$ is continuous.

4. limit equals 5

8. all

9. itself if its dimension is positive, the empty set if its dimension is 0

12. $\mathbb{R}$

17. domain is complement of $\{0\}$, f is continuous, $\mathbb{R}^2$ is set of limit points, limit equals $f(U)$ if $U \neq 0$ and equals 0 if $U = 0$

24. a single point

25. $\mathbb{R}^n$, $\mathbb{R}^n$, empty

6.2

1. (a) $\frac{6}{5}$,   (c) $-\frac{1}{18}$,   (e) -1,   (g) $\frac{6}{7}$

2. (b) 2, $2\sqrt{2}$, 2, 0, -2, $-2\sqrt{2}$, -2, 0,   (c) increase most rapidly in direction $(1,1)$,   decrease most rapidly in direction $-(1,1)$,   not change in directions $(-1,1)$ and $-(-1,1)$

6. (a) $\nabla_{-e} f(X) = -\nabla_e f(X)$

6.3

1. (a) $f_1 = x_2^2 + 2$, $f_2 = 2x_1 x_2$,
   (c) $h_x = (1+x)e^{x+2y}$, $h_y = 2xe^{x+2y} + 2y$,
   (e) $g_\theta = -\sin(\theta - 3\varphi)$, $g_\varphi = 3\sin(\theta - 3\varphi)$,
   (g) $g_1 = x_2 x_3$, $g_2 = x_1 x_3$, $g_3 = x_1 x_2$,
   (i) $v_x = 2x/(x^2 + y^2 + z^2 + 3)$, $v_y = 2y/(\cdots)$, $v_z = 2z/(\cdots)$

2. (a) $f_{11} = 0$, $f_{21} = f_{12} = 2x_2$, $f_{22} = 2x_1$,
   (c) $h_{xx} = (2+x)e^{x+2y}$, $h_{xy} = h_{yx} = 2(1+x)e^{x+2y}$,
   $h_{yy} = 4xe^{x+2y} + 2$,
   (e) $g_{\theta\theta} = -\cos(\theta - 3\varphi)$, $g_{\theta\varphi} = g_{\varphi\theta} = 3\cos(\theta - 3\varphi)$,
   $g_{\varphi\varphi} = -9\cos(\theta - 3\varphi)$,
   (g) $g_{11} = g_{22} = g_{33} = 0$, $g_{12} = g_{21} = x_3$,
   $g_{13} = g_{31} = x_2$, $g_{23} = g_{32} = x_1$,
   (i) $v_{xx} = (-4x^2 + 2g)/g^2$, $v_{yy} = (-4y^2 + 2g)/g^2$,
   $v_{zz} = (-4z^2 + 2g)/g^2$, $v_{xy} = v_{yx} = -4xy/g^2$,
   $v_{xz} = v_{zx} = -4xz/g^2$, $v_{yz} = v_{zy} = -4yz/g^2$, where $g = x^2 + y^2 + z^2 + 3$

3. (a) All are zero except $f_{122} = f_{212} = f_{221} = 2$.   (g) All are zero
   except $D_{123}g$ and the other five involving exactly one derivative with
   repsect to each variable which all equal 1.

4. $f(x,y) = -x + 3y - 3$

5. same answer as preceding problem

14. (b) $0, 0$ (c) limit equals 1

18. (a) $u_t(x,t)$ is the velocity, $u_{tt}(x,t)$ the acceleration, and $u_x(x,t)$ the
    slope at the point x at time t.   (b) $u(0,t) = u(\pi,t) = 0$, so the
    end points are fixed. The initial position is $u(x,0) = 3\sin 2x$, ini-
    tial velocity $u_t(x,0) = 0$, while $u_t(\pi/4, 3) = 0$. The slope there is
    $u_x(\pi/4, 3) = 0$.

6.4

1. (a) $T(x) = 1 + 2(x - 1)$,   (c) $T(x) = 1 + (x + 1)$

2. (a) $[f'(X_0)] = [\ 0\quad 0\ ]$;
   $f(X) - T(X) = 2(x_1 - 1)^2 + 2x_2^2 = 2\|X - X_0\|^2$ so $\lim_{X \to X_0} \frac{f(X) - T(X)}{\|X - X_0\|} =$
   $\lim_{X \to X_0} \frac{2\|X - X_0\|^2}{\|X - X_0\|} = \lim_{X \to X_0} 2\|X - X_0\| = 0$
   (c) $[f'(X_0)] = [\ -1\quad 0\ ]$; $f(X) - T(X) = (x - 1)^2$ so $\lim_{X \to X_0} \frac{f(X) - T(X)}{\|X - X_0\|} =$
   $\lim_{X \to X_0} \frac{(x-1)^2}{\|X - X_0\|} = 0$ (note that $|x - 1| \le \|X - X_0\|$)   (e) $[f'(X_0)] =$
   $[\ 2\quad -1\ ]$

6. (a) $3 - 2x_1 - 2x_2$,   (c) $12 + 3\sqrt{13} - (3 + \sqrt{13})x_1 + (3 + \sqrt{13})x_2$ and
   $12 - 3\sqrt{13} - (3 - \sqrt{13})x_1 + (3 - \sqrt{13})x_2$,   (f) $2 - 2\langle X, e \rangle$ for any
   unit vector e

7. (a) 12.979,   (d) 6.981,   (e) 11.021

6.5

1. (a) $[f'(2,-1)] = [\ -2 \quad 3\ ]$,   (c) $[h'(1,2)] = [\ -\frac{1}{18} \quad -\frac{1}{9}\ ]$,
   (e) $[k'(1,\pi)] = [\ 0 \quad -1\ ]$,   (g) $[u'(1,0,3)] = [\ 0 \quad 3 \quad 0\ ]$

3. (a) $T(X) = -6 - 2(x_1 - 2) + 3(x_2 + 1)$,   (c) $T(X) = (3 - (x-1) - 2(y-2))/18$,   (e) $T(X) = \pi - y$,   (g) $T(X) = 3y$

4. (b) $T(X) = 2 - 2(x-1) - 2(y-1)$

6. (b) $T(X) = -2(x-1) + 2(y-1)$

10. (a) $2[X]^T$,   (c) $-\frac{1}{\|X\|}e^{-\|X\|}[X]^T$

11. The condition that $g'(0) = 0$ is necessary and sufficient in which case $f'(0) = O$.

12. (c) h is not differentiable at $\mathbf{0}$,   (d) $k'(0) = O$

13. e^t

15. (c) $f(x_1, x_2) = \sqrt{(x_2/x_1)}$ near $(4,4)$, 0.975,   (f) $f(x,y) = x^y$ near $(4, \frac{3}{2})$, $7.94 - 0.08\ln 4$, which using tables or a calculator to approximate $\ln 4$ gives 7.829

16. (a) increase: $e = (-2,5)/\sqrt{29}$; decrease: $e = -(-2,5)/\sqrt{29}$, maximal rate of increase: $\sqrt{29}$
    (c) increase: $e = (4,7,-10)/\sqrt{165}$; decrease: $e = -(4,7,-10)/\sqrt{165}$, maximal rate of increase: $\sqrt{165}$ (e) standard methods fail—but observe $p(x,y) = (x+1)(y-1) - 1$ so by inspection, increase $e = \pm(1,1)/\sqrt{2}$; decrease: $e = \pm(1,-1)/\sqrt{2}$
    maximal rate of increase: 0, but, the maximal rate is positive at points near $(-1,1)$ even though the rate in all directions is 0 at $(-1,1)$.

Chapter 7

7.0

1. true

3. true

5. true

7. false, $f(x) = x^4$ is a counterexample

9. true

11. true

13. true

15. true

17. false

19. false

21. true

23. true

25. true

27. true

28. false

29. true

7.1

1. (a) $(3,0)$,   (c) $(0, n\pi)$, $n = 0, \pm 1, \pm 2, \ldots$,   (e) $(0,0)$, $\pm(1,1)$, $\pm(1,-1)$
(g) all points on the ξ and η axes,   (i) $(-1,0,2)$,   (l) no critical points

2. (a) $\sqrt{\frac{1369}{374}}$,   (c) $\sqrt{\frac{4}{11}}$,   (e) 1

3. 2

4. (a) $\varphi(h,k) = h^2 + 2k^2 \geq 0$, local minimum,   (e) At $(0,0)$: for $|h|, |k| < 1$, $\varphi(h,k) = -h^2 - k^2 + h^2 k^2 \leq -(h^2 + k^2)/2 \leq 0$ so p has a local maximum at $(0,0)$; at $(1,1)$: $\varphi(h,k) = hk(h+2)(k+2)$ which can be either positive or negative for small $|h|$ and $|k|$ so p has a saddle at $(1,1)$. Similarly, p has a saddle at $(-1,-1)$ and $\pm(1,-1)$.   (g) For a critical point (a,b), $\varphi(h,k) = (a+h)^2(b+k)^2 \geq 0$ so u has local minima at all critical points, that is, on the lines $\xi = 0$ and $\eta = 0$.   (i) $\varphi(h,k,l) = h^2 + 3k^2 + l^2 \geq 0$ so f has a local minimum at $(-1,0,2)$.

5. $(0, 0, \frac{1}{4})$

8. The orgin is a critical point at which the global (and local) minimum value of 0 is attained. Each point on $S(0; 2)$ is a critical point at which the global (and local) maximum value of $4e^{-2}$ is attained.

7.2

1. (a) $\begin{bmatrix} 2 & -4 \\ -4 & 2 \end{bmatrix}$,   (c) $\begin{bmatrix} 9e^{3x-2y} & -6e^{3x-2y} \\ -6e^{3x-2y} & 4e^{3x-2y} \end{bmatrix}$,

   (e) $\begin{bmatrix} 0 & x_3 & x_2 \\ x_3 & 0 & x_1 \\ x_2 & x_1 & 0 \end{bmatrix}$

2. (c)(a) the function $9x_1^2 - 12x_1x_2 + 4x_2^2 = (3x_1 - 2x_2)^2$,
   (e)(b) the function $2c_3x_1x_2 + 2c_2x_1x_3 + 2c_1x_2x_3$, where $(c_1, c_2, c_3) = X_0$

3. (a) By the Mean Value Theorem, $f(X) = f(X_0)$ for any X, X_0.
   (b) Let $h = f - g$, observe $h' = 0$ and apply (a).

4. $f(X) = 3x - 2y - 1$, unique by part (b) of preceding exercise

6. $\frac{1}{2}(x^2 + 4xy - 3y^2 - 6x + 6y + 19)$

7. Let $T(x,y)$ be the best affine approximation to f at (x_0, y_0). Then $z = T(x,y)$ is the tangent plane at $(x_0, y_0, f(x_0,y_0))$. We must show $f(x,y) - T(x,y) \geq 0$ for all (x,y). But, by the Taylor's Theorem, $f(x,y) - T(x,y) = \frac{1}{2}\langle f''(x^*,y^*)(x - x_0, y - y_0), (x - x_0, y - y_0)\rangle = 2(x - x_0)^2 + 3(y - y_0)^2 \geq 0$.

7.3

1. (a) positive definite,   (c) indefinite,   (e) semidefinite,   (g) semidefinite,   (i) indefinite,   (k) positive definite

2. (a) $(-2, 0)$ minimum,   (c) $(-1, 1)$ saddle point,   (e) $(0, \frac{1}{3})$ maximum, $(0, 1)$ saddle,   (g) $(3, 3)$ minimum, $(-1, -1)$ saddle,   (i) $(0, 0)$ saddle,   (k) $(0, 1)$ minimum, $(0, -1)$ maximum   (m) $(0, 0, -1)$ and $(-1, -1, -1)$ saddles, $(1, -1, -1)$ minimum,   (o) $(1, 0, 0)$ minimum, $(-1, 0, 0)$ saddle

3. $\varphi(x) \equiv f(x, b)$ has a local minimum at $x = a$. Hence $f_{xx}(a, b) = \varphi''(a) \geq 0$. Similarly, $f_{yy}(a, b) \geq 0$.

13. (a) counterexample: $\begin{bmatrix} 2 & -1 \\ -1 & 2 \end{bmatrix}$

7.4

1. (a) unbounded, not open, closed, pathwise connected   (e) unbounded, not open, closed, not pathwise connected   (f) unbounded, not open, not closed, not pathwise connected   (j) unbounded, not open, closed, pathwise connected

2. (a) bounded, not open, not closed, pathwise connected   (b) unbounded, open, not closed, pathwise connected

3. (d) global maximum equal to 1 at $(1, \frac{1}{2})$, global minimum equal to $\frac{\sqrt{2}}{4}$ at $(\frac{1}{2}, \frac{1}{2})$

4 no, no

7.5

1. (a) global maximum at $(3, 4)/5$,   global minimum at $-(3, 4)/5$,   (c) global maximum at $\pm(1, 1)$ and $\pm(1, -1)$,   global minimum at $(0, 0)$,   (e) global maximum at $(2, 3)$,   global minimum at $(-1, 5)$

2. (a) no global maximum, global minimum at $(3, -3)/2$,   (c) no global maximum, global minimum at $(1, 1)$ (e) global maximum at $(2, \pm 2, \pm 2)$ and $(-2, \mp 2, \pm 2)$, global minimum at $(2, \mp 2, \pm 2)$ and $(-2, \pm 2, \pm 2)$

5. 2 feet $\times$1 foot $\times$1 foot

7. 2 feet $\times$2 feet $\times$2 feet

9. $x = 12,\ y = 7$

Chapter 8

8.0

4. Let F be an $\mathbb{R}^q$-valued function with domain $\mathcal{D}(F) \subseteq \mathbb{R}^n$ for some n. Let X_0 be a limit point of $\mathcal{D}(F)$. Then the **limit** of $F(X)$ as X approaches X_0 is equal to L, written as

$$\lim_{X \to X_0} f(X) = L ,$$

if and only if for every positive number ε, there is a positive number δ such that $|F(X) - L| < \varepsilon$ whenever $0 < \|X - X_0\| < \delta$ and $X \in \mathcal{D}(f)$.

Here is an alternative correct way to follow the first two sentences of the definition just given.·

Then the **limit** of $F(X)$ as X approaches X_0 is equal to L if and only if for every open ball $B(L;\varepsilon)$ centered at L there is an open ball $B(X_0;\delta)$ centered at X_0 such that $F(X) \in B(L;\varepsilon)$ whenever $X \in B(X_0;\delta) \cap \mathcal{D}(F)$ and $X \neq X_0$.

6. a theorem about compositions: Let F be a continuous function from a subset $\mathcal{D}(F)$ of $\mathbb{R}^n$ into $\mathbb{R}^q$ and suppose that $\mathcal{I}(F)$ is a subset of the domain $\mathcal{D}(G)$ of a continuous function G into $\mathbb{R}^r$. Then $G \circ F$ is a continuous function from $\mathcal{D}(F)$ into $\mathbb{R}^r$.

8.1

1. (a) $\begin{bmatrix} 1 & 1 \\ 1 & -1 \end{bmatrix}$, (c) $\begin{bmatrix} 1 & 0 \\ 0 & 2\pi \end{bmatrix}$, (e) $\begin{bmatrix} -\frac{1}{2} & \frac{1}{2} \\ \frac{1}{2} & \frac{1}{2} \\ 0 & -1/\sqrt{2} \end{bmatrix}$,

2. (a) $T(x_1, x_2) = (1 + x_1 + x_2, -2 + x_1 - x_2)$

   (c) $T(y_1, y_2) = (y_1, 2\pi y_2)$

   (e) $T(\theta, \varphi) = \frac{1}{2}(1 - \theta + \varphi, 1 - \pi/2 + \theta + \varphi, \sqrt{2}(1 + \pi/4 - \varphi))$

8.2

2. yes, yes, the zero matrix

3. (b) the linear map whose matrix has the entry $2x_i x_j$ at position i, j if $i \neq j$ and the entry $2x_i^2 + \|X\|^2$ at position i, i

4. (c) yes, no

5. The following make sense: (a) $G \circ F$, (c) $F \circ G$, (e) both

6. (a) $[F'(X)] = \begin{bmatrix} e^{x+y^2} & 2ye^{x+y^2} \\ 2xe^{x^2+y} & e^{x^2+y} \end{bmatrix}$ $[G'(r,s,t)] = \begin{bmatrix} 1 & 2s & 2t \\ 3r^2 & 1 & 2t \end{bmatrix}$

   (b) $F \circ G$ makes sense $G(-1,0,0) = (-1,-1)$,

   $$[(F \circ G)'(-1,0,0)] = [F'(-1,-1)][G'(-1,0,0)]$$
   $$= \begin{bmatrix} 1 & -2 \\ -2 & 1 \end{bmatrix} \begin{bmatrix} 1 & 0 & 0 \\ 3 & 1 & 0 \end{bmatrix} = \begin{bmatrix} -5 & -2 & 0 \\ 1 & 1 & 0 \end{bmatrix}$$

7. (a) At $(\pi, 1)$, $u = 0$, $v = -1$. If $f(x,y) = g(u(x,y), v(xy))$,

$$[f'(\pi,1)] = [g_u(0,-1)\ \ g_v(0,-1)]\begin{bmatrix} u_x(\pi,1) & u_y(\pi,1) \\ v_x(\pi,1) & v_y(\pi,1) \end{bmatrix}$$

$$= [1\ 0]\begin{bmatrix} -1 & 0 \\ 0 & -1 \end{bmatrix} = [-1\ 0]$$

so $f_x(\pi,1) = -1$, $f_y(\pi,1) = 0$.

9. (a) $\varphi'(t) = [\ f_x(1-t^2, 2t-3)\ \ f_y(1-t^2, 2t-3)\ ]\begin{bmatrix} -2t \\ 2 \end{bmatrix} = -6t^2 + 6t + 2 + 2te^{(1-t^2)}$

10. $h'(1) = -1$

12. (b) $u_x = a\varphi'(ax+by)$, $u_y = b\varphi'(ax+by)$ so $bu_x - au_y = 0$.

14. At $t = 1$, where $X(1) = (0,-1,1)$ and temperature $f(X(1)) = 2$.

16. $g_s(1,1) = -2$, $g_t(1,1) = -2$ so $[g'(1,1)] = [-2,-2]$.

21. (a) $r^2 = x^2 + y^2$ so $r_x = x/r$, $u_x = \varphi'(r)r_x = \varphi'(r)x/r$, $u_{xx} = (r^2 - x^2)\varphi'(r)/r^3 + x^2\varphi''(r)/r^2$. u_{yy} is similar. Thus, $0 = u_{xx} + u_{yy} = \varphi''(r) + \varphi'(r)/r$. (b) Let $\psi = \varphi'$. Then $r\psi'(r) + \psi(r) = 0$ so $\psi(r) = a/r$, that is, $\varphi'(r) = a/r$. Thus $\varphi(r) = a\log r + b$.

8.3

1. (b) does not exist,
   (e) $\frac{1}{2}(\sqrt{2}\,a + b,\ \sqrt{2}\,a + b,\ \sqrt{2}\,b) + \frac{t}{\sqrt{2}}(-a - \sqrt{2}\,b,\ a,\ b)$

3. (a) $z = \sqrt{6} + \frac{2(x-2)}{\sqrt{6}} + \frac{y-1}{\sqrt{6}}$,   (d) $\mathcal{I}(G)$,   (g) does not exist

7. (b) A function that almost works is $y/\|X\|$. The problem is that this function is not defined at the origin, which is a point on the cone. The remainder of the cone is its level set at level $\cot\psi$. A function that does work is $\frac{(y+1)}{(\|X\|\cot\psi+1)}$. Its level set at level one is the given cone

8. (c) $\frac{x}{a} + \frac{y}{b} + \frac{z}{c} = \sqrt{3}$

10. (d) $T(\theta,\varphi) = (1, \theta, \pi/2 - \varphi)$, the plane $x = 1$,
    (e) $T(\theta,\varphi) = \frac{1}{\sqrt{2}}(\pi/2 - \theta, 1 - \pi/4 + \varphi, 1 + \pi/4 - \varphi)$, the plane $y + z = \sqrt{2}$

8.4

2. (a) 0,   (c) 2,   (e) 0,   (g) 0,   (i) $2x$.

3. (a) $(2x, 2y) = \text{grad } (x^2 + y^2)$ (b) If $F = \text{grad } \varphi = (\varphi_x, \varphi_y)$, then $p = \varphi_x$, $q = \varphi_y$ so $p_y = \varphi_{xy} = q_x$. (c) 1–a, d, e are not irrotational since $p_y \neq q_x$ (1–b) $(1, 2) = \text{grad } (x + 2y)$ (1–c) $(x, y) = \text{grad } (x^2 + y^2)/2$ (1–f) $(y, -x)/(x^2 + y^2) = \text{grad } (\arctan(x/y))$ (1–g) $(x, -y) = \text{grad } (x^2 - y^2)/2$ (1–h) $(2x, y) = \text{grad } (x^2 + y^2/2)$ (1–i) $(x^2, 0) = \text{grad } x^3/3$

Chapter 9

9.1

13. (c) 4

17. (a) 704,   (c) $\frac{104}{3}$

19. (a) 64,   (c) -120

9.2

1. (a) 9,   (c) $\frac{64}{5}$

2. (a) $\int_1^3 \left(\int_1^5 xy \, dy \right) dx = \int_1^5 \left(\int_1^3 xy \, dx \right) dy = 48$

(c) $\int_1^2 \left(\int_{x^3}^8 xy \, dy \right) dx = \int_1^8 \left(\int_1^{y^{1/3}} xy \, dx \right) dy = \frac{513}{16}$

(e) $\int_{-1}^0 \left(\int_{-x}^1 xy \, dy \right) dx + \int_0^2 \left(\int_{x/2}^1 xy \, dy \right) dx = \int_0^1 \left(\int_{-y}^{2y} xy \, dx \right) dy = \frac{3}{8}$

(g) $\int_0^1 \left(\int_{x/2}^{2x} xy \, dy \right) dx + \int_1^2 \left(\int_{x/2}^{(x+3)/2} xy \, dy \right) dx + \int_2^3 \left(\int_{2x-3}^{(x+3)/2} xy \, dy \right) dx$
$= \int_0^1 \left(\int_{y/2}^{2y} xy \, dx \right) dy + \int_1^2 \left(\int_{y/2}^{(y+3)/2} xy \, dx \right) dy + \int_2^3 \left(\int_{2y-3}^{(y+3)/2} xy \, dx \right) dy$

3. 0

5. (a) 6,   (c) $\frac{4}{15}$

6. (a) 36,   (c) 0,   (e) 6π (the $dx \, dy$ integral is most easily evaluated using polar coordinates)

7. (a) π,   (c) 0,   (e) $\pi(e^{50} - e^8)/16$

8. $\ln\left(\frac{3}{2}\right)$, evaluate the x integral first

10. $125\pi/3$, use polar coordinates

11. (a) $(e-1)/6$

13. $4\pi abc/3$

15. (b) use $xy = \frac{\partial^2}{\partial x\,\partial y}(x^2 y^2/4)$

16. volume $= \pi^2 a^4/2$

18. (a) 6

9.3

1. $J = 3\iint_B dA = 3(\text{area } B) = 36$

4. $f'(t) = \int_1^4 x^2 e^{tx^2}\,dx$

6. $u'(x) = 2\int_0^x (x-t)\sin t\,dt,\ \ u''(x) = 2 - 2\cos x$

9. $v_x(x,t) = 2g(2x+t) - 2g(2x-t),\ \ v_t(x,t) = g(2x+t) + g(2x-t),$
   $v_{xx}(x,t) = 4g'(2x+t) - 4g'(2x-t) = 4v_{tt}(x,t)$

9.4

1. 576

5. After using a rigid motion change of variables and doing one integral in the resulting iterated integral, one can obtain

$$4\int_0^{1/\sqrt{2}} \sqrt{2+v^2}\,dv\ .$$

With a different change of variables one can end up with something different, for example, a formula that is similar to the formula given here, but with different constants at various places. The single-variable integral can be evaluated using a trigonometric substitution.

6. $2 + \frac{\pi}{4}$

7. (c) $\frac{4}{3}\pi^3$,   (d) $\frac{1}{3}(4\pi^3 - 3\pi + 1)$, (g) integrate θ between 2π and 4π

8. (c) $\frac{\pi}{8}$,   (d) $\frac{16}{105}$

(f) F is one-to-one on the interior of $\mathcal{R}$, so from a conceptual stand-point the area of $F(\mathcal{R})$ can be obtained in the standard way. From a calcuational standpoint it is best to write the area as the sum of four integrals, since for θ between $\frac{\pi}{2}$ and π and also between $\frac{3\pi}{2}$ and 2π the lower endpoint for r is $\sin 2\theta$ and the upper endpoint is 0. It is easy to see, either geometrically or from the forms of the integrals that the values of all four of the integrals are identical and, thus, that the area of $F(\mathcal{R})$ is four times the answer to part (c), namely $\frac{\pi}{2}$.

9. $\frac{2\pi}{3}(3^{3/2} - 2^{3/2})$

10. $\frac{4\pi abc}{3}$

12. (i) $+\infty$

14. (a) $(\sqrt{2}, -\pi/4, 5)$ (c) $(0, 134, 0)$ is one correct answer

15. (a) $(-1, 0, -2)$ (d) $(\cos 2, -\sin 2, 3)$

17. (b) $z = r\cos\theta + 2$ (c) $\tan\theta = a$, which is the union of parallel planes in $r\theta z$)-space

18. (c) $x^2 + y^2 = c^2$, which is a cylinder if $c > 0$ and the z-axis if $c = 0$,
(e) $z(x^2 + y^2) = x^2 - y^2$

19. (c) $\frac{10}{3} + \frac{27\pi}{32}$

21. $\frac{4\pi}{3}(\sqrt{2} - 1)$

23. (a) $(\sqrt{3}, \frac{7\pi}{4}, \arccos\frac{1}{\sqrt{3}})$

24. (b) $(0, 0, 0)$,   (c) $(0, 0, -6)$

26. (d) $\cot\varphi = a$. Thus, by comparaison with Section 5.4c, φ is the angle of the cone.

27. (b) a cone (compare with part (d) of the preceding exercise)

29. $18\pi\sqrt{2}$

30. 2π

9.5

1. (a) $9\pi\sqrt{2}$

2. (b) $\frac{\pi}{54}((12c + 1)^{3/2} - 1)$

3. (2-b) $\gamma = \frac{3}{2}, l = \frac{4\pi\sqrt{3}}{9}$

5. $\alpha\sqrt{||X||^2||Y||^2 - (X \cdot Y)^2}$

6. $\pi^2 ab + 2\pi b^2$

8. $2\pi\sqrt{1 + b^2} \int_0^{2\pi} \sqrt{1 + a^2 + 2ab\cos\varphi + b^2\cos^2\varphi}\, d\varphi$

10. $\frac{\pi^2}{4}$

Chapter 10

10.1

1. (a) 1,   (c) $-12\pi^2$,   (e) $4 - 3\pi/2$

2. (a) $\frac{3}{2}$ (c) $-\pi$

3. (a) 0   (c) 0

4. (a) -15,   (c) 4,   (e) 4

5. All equal 2. Observe that $(6x - y^2, -2xy) = \nabla(3x^2 - xy^2)$ and see Theorem 10.4.2 in Section 10.4.

6. 0

8. 3

10. $-2\sin 1$

12. (a) $-\frac{1}{3}$

13. (d) $-\frac{1}{2}$

14. work $= 0$

19. (b) true,   (d) false

10.2

1. (a) 8π,   (c) 0,   (e) $-\frac{8}{3}$,   (g) 6π

2. (b) πab,   (c) 24π

4. 0

5. Apply the Divergence Theorem to $\mathcal{B}$, observing $\partial\mathcal{B} = \mathcal{C}_2 - \mathcal{C}_1$.

10.3

2. -216π or 216π depending on orientation

3. (a) $(-1, 0, -1)$,   (c) $(0, 0, -2y)$

9. (d) 0,   (g) 0

11. (b) $\iint_{\mathcal{R}} F(x, y, g(x, y)) \cdot (-D_1 g(x, y), -D_2 g(x, y), 1)\, dA$

12. (b) $7/3$

13. π. Use Stokes' Theorem and integrate over $\partial\Sigma = \{(x, y, z) : x^2 + y^2 = 1,\ z = 0\}$, counterclockwise as viewed from the positive z-axis.

16. 144

10.4

1. (a) $(x^2 + y^2)/2$, $\frac{3}{2}$,   (c) xe^y, 0,   (f) $(x^2 - 7y^2)/2 + xy$, 0

2. (a) $F(X) = \nabla(x^2/2 + e^x \sin y)$,   (b) $\frac{1}{2}$

3. (a) independent,   (c) independent,   (e) dependent

5. $F(X) = \operatorname{grad}(\gamma mM/\|X\|)$

Index

Undergraduate Texts in Mathematics

(continued)

Undergraduate Texts in Mathematics

Toth: Glimpses of Algebra and Geometry.
  Readings in Mathematics
Troutman: Variational Calculus and
  Optimal Control. Second edition.

Valenza: Linear Algebra: An Introduction
  to Abstract Mathematics.
Whyburn/Duda: Dynamic Topology.
Wilson: Much Ado About Calculus.